CELL BIOLOGY
A Short Course

SECOND EDITION

Stephen R. Bolsover
Department of Physiology
University College London

Jeremy S. Hyams
Department of Biology
University College London

Elizabeth A. Shephard
Department of Biochemistry and Molecular Biology
University College London

Hugh A. White
Department of Biochemistry and Molecular Biology
University College London

Claudia G. Wiedemann
Department of Physiology
University College London

WILEY-LISS

A JOHN WILEY & SONS, INC., PUBLICATION

Library of Congress Cataloging-in-Publication Data:

Cell biology : a short course / Stephen R. Bolsover . . . [et al.].—2nd ed.
   p.   cm.
Includes bibliographical references and index.
ISBN 0-471-26393-1 (Paper)
   1. Cytology.   I. Bolsover, Stephen R., 1954–
QH581.2.C425   2003
571.6—dc21   2003000577

Printed in the United States of America

10   9   8   7   6   5   4   3   2   1

CONTENTS IN BRIEF

CONTENTS

PREFACE

Cell Biology, A Short Course aims to cover a wide area of cell biology in a form especially suitable for first year undergraduates. We have deliberately kept the book to a manageable size so that neither the cost, the content, nor the weight is too daunting for the student.

The overall theme for the book is the cell as the unit of life. We begin (Chapters 1–3) by describing the components of the cell as seen under the microscope. We then (Chapters 4–8) turn to the central dogma of molecular biology and describe how DNA is used to make RNA which in turn is used to make protein. The next section (Chapters 9–11) describes how proteins are delivered to the appropriate location inside or outside the cell, and how proteins perform their many functions. We then (Chapters 12–14) turn to cell energetics and metabolism. Signaling within and between cells is covered in Chapters 15 through 17. To conclude the book, Chapter 18 describes the composition and function of the cytoskeleton, Chapter 19 covers cell birth and cell death, while Chapter 20 uses the example of the common and severe genetic disease cystic fibrosis to illustrate many of the themes discussed earlier in the book.

Boxed material throughout the book is divided into *examples* to illustrate the topics covered in the main text, explanations of the *medical relevance* of the material, and *in depth* sections that extend the coverage beyond the content of the main text. *Questions* are provided at the end of each chapter to help the reader assess how well they have assimilated and understood the material.

As well as giving references to printed material, we reference *material available on the internet* in many places in the book. Rather than give detailed addresses, we provide links to all these sites and many others from the book's homepage at http://www.physiol.ucl.ac.uk/sbolsover/teaching/cbasc/cbasc.html.

ACKNOWLEDGMENTS

We thank all the students, colleagues and family members who read the initial versions of the book and whose suggestions and constructive criticism helped enormously.

INSTRUCTOR NOTES

Molecular cell biology courses now form a foundation for many subsequent specializations in areas outside cell biology. We therefore cover molecular genetics, metabolic pathways and electrophysiology in sufficient detail to make *Cell Biology* a suitable course book for first year students who will later specialize in genetics, biochemistry, pharmacology or physiology.

Each chapter comprises:

- The *main text,* with figures and tables.
- A numbered *summary*.
- *Review questions* with *answers* for student self-assessment. These questions concern the main text only; no knowledge of the boxed material is required.
- *Example* boxes that illustrate the points made in the main text.
- *Medical relevance* boxes to show how basic cell biological knowledge illuminates medical problems or has provided solutions.
- *In Depth* boxes that extend the content.

Self-assessment questions can form the basics for tutorials, with students asked to defend the correct answer. They are also easily modified to generate new questions for student assessment. Instructors are encouraged to submit new questions for inclusion on the CBASC website.

Instructors may wish to specify parts of *Cell Biology* as core material for courses targeted to particular specialties. The parts chosen can be customized to the particular specialty in two ways:

1. By selecting from the complete set of twenty chapters. The following sections could be used to support particular teaching modules:

Chapters 4 through 7	DNA, RNA and genetic engineering.
Chapters 8 through 10	Protein synthesis, structure and trafficking.
Chapters 11 through 13	Metabolism and cellular energetics.
Chapters 14 through 17	Electrophysiology and cell signaling.
Chapter 18	The cytoskeleton and cell motility.
Chapter 19	Cell division and apoptosis.

    Chapters 16, 17 and 19 might in contrast be selected in a module concentrating on the control of development, since these describe how growth factors and other extracellular chemicals regulate cell division and cell death.

2. By including In Depth boxes. The following boxes are especially to be noted:

In Depth 1.2:	Fluorescence Microscopy
In Depth 8.1:	How We Study Proteins in One Dimension
	describes SDS-PAGE
In Depth 9.1:	Chirality and Amino Acids
In Depth 9.2:	Hydropathy Plotting—The PDGF Receptor
In Depth 9.3:	Curing Mad Mice with Smelly Fish
	introduces the concept of osmolarity and osmosis and extends the coverage of chaotropic and structure stabilizing agents
In Depth 11.1:	What to Measure in an Enzyme Assay
In Depth 11.2:	Determination of V_m and K_M
	the Lineweaver-Burk plot
In Depth 12.2:	ATP Synthase, Rotary Motor, and Synthetic Machine
In Depth 12.3:	Can It Happen? The Concept of Free Energy
In Depth 13.1:	The Urea Cycle—The First Metabolic Cycle Discovered
In Depth 13.2:	The Glyoxylate Shunt
In Depth 14.1:	The Nernst Equation
In Depth 14.2:	Measuring the Transmembrane Voltage
In Depth 15.1:	Frequency Coding in the Nervous System
In Depth 16.1:	Ryanodine Receptors
In Depth 19.1:	A Worm's Eye View of Cell Death
In Depth 20.1:	Lipid Bilayer Voltage Clamp

For example, a course emphasizing protein structure would include In Depth 8.1, 9.1, 9.2 and 9.3, while a course concentrating on metabolic pathways would include In Depth 13.1 and 13.2.

The CBASC website is maintained by the authors. As well as providing over one hundred links to sites with information that extends or illustrates the material in the book, we will use the site to post typological or other errors, comments and test questions sent to us by readers. The full address is http://www.physiol.ucl.ac.uk/sbolsover/teaching/cbasc/cbasc.html or simply type 'CBASC' into a search engine.

1

CELLS AND TISSUES

The **cell** is the basic unit of life. Microorganisms such as bacteria, yeast, and amoebae exist as single cells. By contrast, the adult human is made up of about 30 trillion cells (1 trillion $= 10^{12}$) which are mostly organized into collectives called **tissues.** Cells are, with a few notable exceptions, small (Fig. 1.1) with lengths measured in micrometers (μm, where 1000 μm $= 1$ mm) and their discovery stemmed from the conviction of a small group of seventeenth-century microscope makers that a new and undiscovered world lay beyond the limits of the human eye. These pioneers set in motion a science and an industry that continues to the present day.

The first person to observe and record cells was Robert Hooke (1635–1703) who described the *cella* (open spaces) of plant tissues. But the colossus of this era of discovery was a Dutchman, Anton van Leeuwenhoek (1632–1723), a man with no university education but with unrivaled talents as both a microscope maker and as an observer and recorder of the microscopic living world. van Leeuwenhoek was a contemporary and friend of the Delft artist Johannes Vermeer (1632–1675) who pioneered the use of light and shade in art at the same time that van Leeuwenhoek was exploring the use of light to discover the microscopic world. Sadly, none of van Leeuwenhoek's microscopes have survived to the present day. Despite van Leeuwenhoek's Herculean efforts, it was to be another 150 years before, in 1838, the botanist Matthias Schleiden and the zoologist Theodor Schwann formally proposed that all living organisms are composed of cells. Their "cell theory," which nowadays seems so obvious, was a milestone in the development of modern biology. Nevertheless general acceptance took many years, in large part because the **plasma membrane,** the membrane

Cell Biology: A Short Course, Second Edition, by Stephen R. Bolsover, Jeremy S. Hyams, Elizabeth A. Shephard, Hugh A. White, Claudia G. Wiedemann
ISBN 0-471-26393-1 Copyright © 2004 by John Wiley & Sons, Inc.

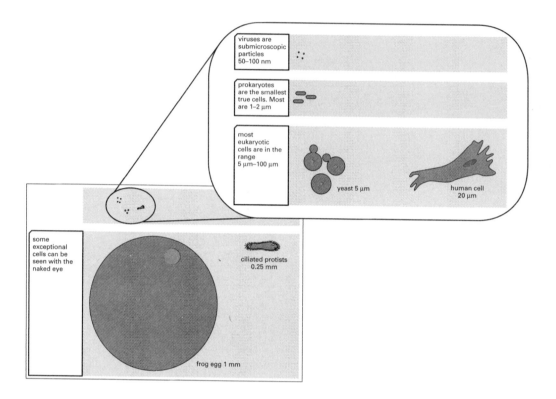

<u>Figure 1.1.</u> Dimensions of some example cells. 1 mm $= 10^{-3}$ m; 1 μm $= 10^{-6}$ m; 1 nm $= 10^{-9}$ m.

surrounding the cell that divides the living inside from the nonliving **extracellular medium** (Fig. 1.2) is too thin to be seen using a light microscope.

✻ PRINCIPLES OF MICROSCOPY

Microscopes make small objects appear bigger. A light microscope will magnify an image up to 1500 times its original size. Electron microscopes can achieve magnifications up to 1 million times. However, bigger is only better when more details are revealed. The fineness of detail that a microscope can reveal is its resolving power. This is defined as the smallest distance that two objects can approach one another yet still be recognized as being separate. The resolution that a microscope achieves is mainly a function of the wavelength of the illumination source it employs. The smaller the wavelength, the smaller the object that will cause diffraction, and the better the resolving power. The light microscope, because it uses visible light of wavelength around 500 nanometers (nm, where 1000 nm $= 1$ μm), can distinguish objects as small as about half this: 250 nm. It can therefore be used to visualize the smallest cells and the major intracellular structures or organelles. The microscopic study of cell structure organization is known as **cytology.** An electron microscope is required to reveal the **ultrastructure** (the fine detail) of the organelles and other cytoplasmic structures (Fig. 1.2).

The wavelength of an electron beam is about 100,000 times less than that of white light. In theory, this should lead to a corresponding increase in resolution. In practice, the

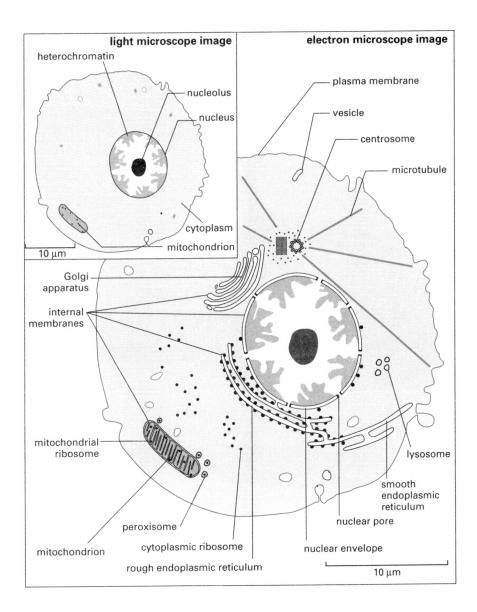

Figure 1.2. Cell structure as seen through the light and transmission electron microscopes.

electron microscope can distinguish structures about 1000 times smaller than is possible in the light microscope, that is, down to about 0.2 nm in size.

The Light Microscope

A light microscope (Figs. 1. 3*a* and 1.4) consists of a light source, which may be the sun or an artificial light, plus three glass lenses: a **condenser lens** to focus light on the specimen, an **objective lens** to form the magnified image, and a **projector lens,** usually called the eyepiece, to convey the magnified image to the eye. Depending on the focal length of the various lenses and their arrangement, a given magnification is achieved. In **bright-field**

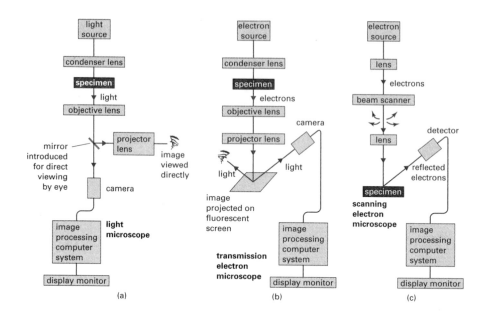

Figure 1.3. Basic design of light and electron microscopes.

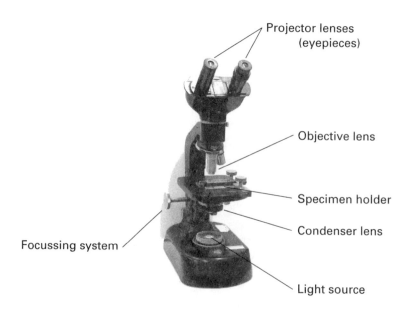

Figure 1.4. Simple upright light microscope.

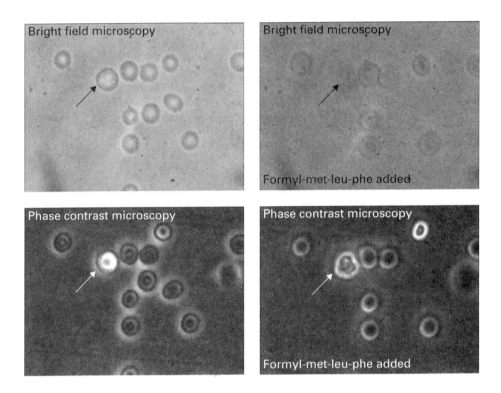

Figure 1.5. Human blood cells viewed by bright-field and phase-contrast light microscopy. Arrow indicates a white blood cell. Formyl-met-leu-phe (page 171) causes the white blood cell to spread out and become very thin. It becomes almost invisible by bright-field microscopy but can still be detected by phase-contrast microscopy.

microscopy, the image that reaches the eye consists of the colors of white light less that absorbed by the cell. Most living cells have little color (plant cells are an obvious exception) and are therefore largely transparent to transmitted light. This problem can be overcome by **cytochemistry,** the use of colored stains to selectively highlight particular structures and organelles. However, many of these compounds are highly toxic and to be effective they often require that the cell or tissue is first subjected to a series of harsh chemical treatments.

A different approach, and one that can be applied to living cells, is the use of **phase-contrast microscopy.** This relies on the fact that light travels at different speeds through regions of the cell that differ in composition. The phase-contrast microscope converts these differences in refractive index into differences in contrast, and considerably more detail is revealed (Fig. 1.5). Light microscopes come in a number of physical orientations (upright, inverted, etc.) but whatever the orientation of the microscope the optical principles are the same.

●●● IN DEPTH 1.1 Fluorescence Microscopy

Fluorescent molecules emit light when they are illuminated with light of a shorter wavelength. Familiar examples are the hidden signature in bank passbooks, which is written in fluorescent ink that glows blue (wavelength about 450 nm) when illuminated with ultraviolet light (UV) (wavelength about 360 nm), and the whitener in fabric detergents that causes your white shirt to glow blue when illuminated by the ultraviolet light in a club. The fluorescent dye Hoechst 33342 has a similar wavelength dependence: It is excited by UV light and emits blue light. However, it differs from the dyes used in ink or detergent in that it binds tightly to the DNA in the nucleus and only fluoresces when so bound. Diagram a shows the optical path through a microscope set up so as to look at a preparation stained with Hoechst. White light from an arc lamp passes through an excitation filter that allows only UV light to pass. This light then strikes the heart of the fluorescent microscope: a special mirror called a dichroic mirror that reflects light of wavelengths shorter than a designed cutoff but transmits light of longer wavelength. To view Hoechst, we use a dichroic mirror of cutoff wavelength 400 nm, which therefore reflects the UV excitation light up through the objective lens and onto the specimen. Any Hoechst bound to DNA in the preparation will emit blue light. Some of this will be

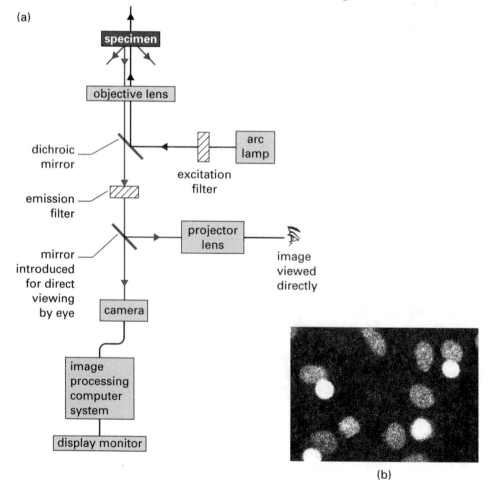

(a)

(b)

captured by the objective lens and, because its wavelength is greater than 400 nm, will not be reflected by the dichroic mirror but will instead pass through. An emission filter, set to pass only blue light, cuts out any scattered UV light. The blue light now passes to the eye or camera in the usual way. Image b shows a field of cells cultured from rat brain (gift of Dr. Charles Krieger, Simon Fraser University) after staining with Hoechst. Only the nuclei are seen, as bright ovals.

Although some of the structures and chemicals found in cells can be selectively stained by specific fluorescent dyes, others are most conveniently revealed by using **antibodies.** In this technique an animal (usually a mouse, rabbit, or goat) is injected with a protein or other chemical of interest. The animal's immune system recognizes the chemical as foreign and generates antibodies that bind to (and therefore help neutralize) the chemical. Some blood is then taken from the animal and the antibodies purified. The antibodies can then be labeled by attaching a fluorescent dye. Images c and d show the same field of brain cells but with the excitation filter, dichroic mirror, and emission filter changed so as to reveal in c a protein called ELAV that is found only in nerve cells; then in d an intermediate filament protein (page 000) found only in glial cells. The antibody that binds to ELAV is labeled with a fluorescent dye that is excited by blue light and emits green light. The antibody that binds to the glial filaments is labeled with a dye that is excited by green light and emits red light. Because these wavelength characteristics are different, the location of the three chemicals—DNA, ELAV, and intermediate filament—can be revealed independently in the same specimen. See the CBASC website for an image of all three signals in color and superimposed.

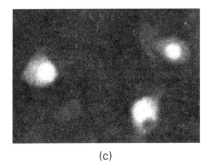
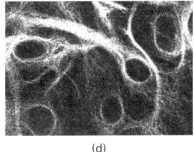

(c)   (d)

The technique just described is called **primary immunofluorescence** and requires that the antibody to the chemical of interest be labeled with a dye. Only antibodies to chemicals that many laboratories study are so labeled. In order to reveal other chemicals, scientists use **secondary immunofluorescence.** In this approach, a commercial company injects an animal (e.g., a goat) with an antibody from another animal (e.g., a rabbit). The goat then makes "goat anti rabbit" antibody. This, called the **secondary antibody,** is purified and labeled with a dye. All the scientist has to do is make or buy a rabbit antibody that binds to the chemical of interest. No further modification of this specialized, **primary antibody** is necessary. Once the primary antibody has bound to the specimen and excess antibody rinsed off, the specimen is then exposed to the secondary antibody that binds selectively to the primary antibody. Viewing the stained preparation in a fluorescence microscope then reveals the location of the chemical of interest. The same dye-labeled secondary antibody can be used in other laboratories or at other times to reveal the location of many different chemicals because the specificity is determined by the unlabeled primary antibody.

The Electron Microscope

The most commonly used type of electron microscope in biology is called the **transmission electron microscope** because electrons are transmitted through the specimen to the observer. The transmission electron microscope has essentially the same design as a light microscope, but the lenses, rather than being glass, are electromagnets that bend beams of electrons (Fig. 1.3b). An electron gun generates a beam of electrons by heating a thin, V-shaped piece of tungsten wire to 3000°C. A large voltage accelerates the beam down the microscope column, which is under vacuum because the electrons would be slowed and scattered if they collided with air molecules. The magnified image can be viewed on a fluorescent screen that emits light when struck by electrons. While the electron microscope offers great improvements in resolution, electron beams are potentially highly destructive, and biological material must be subjected to a complex processing schedule before it can be examined. The preparation of cells for electron microscopy is summarized in Figure 1.6.

A small piece of tissue (~1 mm³) is immersed in glutaraldehyde and osmium tetroxide. These chemicals bind all the component parts of the cells together; the tissue is said to be **fixed**. It is then washed thoroughly.

The tissue is **dehydrated** by soaking in acetone or ethanol.

The tissue is **embedded** in resin which is then baked hard.

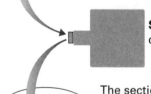
Sections (thin slices less than 100 nm thick) are cut with a machine called an ultramicrotome.

The sections are placed on a small copper grid and **stained** with uranyl acetate and lead citrate. When viewed in the electron microscope, regions that have bound lots of uranium and lead will appear dark because they are a barrier to the electron beam.

Figure 1.6. Preparation of tissue for electron microscopy.

The transmission electron microscope produces a detailed image but one that is static, two-dimensional, and highly processed. Often, only a small region of what was once a dynamic, living, three-dimensional cell is revealed. Moreover, the picture revealed is essentially a snapshot taken at the particular instant that the cell was killed. Clearly, such images must be interpreted with great care. Electron microscopes are large and require a skilled operator. Nevertheless, they are the main source of information on the structure of the cell at the nanometer scale, called the **ultrastructure.**

The Scanning Electron Microscope

Whereas the image in a transmission electron microscope is formed by electrons transmitted through the specimen, in the scanning electron microscope it is formed from electrons that are reflected back from the surface of a specimen as the electron beam scans rapidly back and forth over it (Fig. 1.3*c*). These reflected electrons are processed to generate a picture on a display monitor. The scanning electron microscope operates over a wide magnification range, from 10× to 100,000×. Its greatest advantage, however, is a large depth of focus that gives a three-dimensional image. The scanning electron microscope is particularly useful for providing topographical information on the surfaces of cells or tissues. Modern instruments have a resolution of about 1 nm.

IN DEPTH 1.2 Microscopy Rewarded

Such has been the importance of microscopy to developments in biology that two scientists have been awarded the Nobel prize for their contributions to microscopy. Frits Zernike was awarded the Nobel prize for physics in 1953 for the development of phase-contrast microscopy and Ernst Ruska the same award in 1986 for the invention of the transmission electron microscope. Ruska's prize marks one of the longest gaps between a discovery (in the 1930s in the research labs of the Siemens Corporation in Berlin) and the award of a Nobel prize. Anton van Leeuwenhoek died almost two centuries before the Nobel prizes were introduced in 1901 and the prize is not awarded posthumously.

✳ ONLY TWO TYPES OF CELL

Superficially at least, cells exhibit a staggering diversity. Some lead a solitary existence; others live in communities; some have defined, geometric shapes; others have flexible boundaries; some swim, some crawl, and some are sedentary; many are green (some are even red, blue, or purple); others have no obvious coloration. Given these differences, it is perhaps surprising that there are only two types of cell (Fig. 1.7). Bacterial cells are said to be **prokaryotic** (Greek for "before nucleus") because they have very little visible internal organization so that, for instance, the genetic material is free within the cell. They are also small, the vast majority being 1–2 μm in length.

The cells of all other organisms, from protists to mammals to fungi to plants, are **eukaryotic** (Greek for "with a nucleus"). These are generally larger (5–100 μm, although some eukaryotic cells are large enough to be seen with the naked eye; Fig. 1.1) and structurally more complex. Eukaryotic cells contain a variety of specialized structures known collectively as **organelles,** surrounded by a viscous substance called cytosol. The largest organelle, the **nucleus,** contains the genetic information stored in the molecule deoxyribonucleic

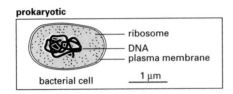

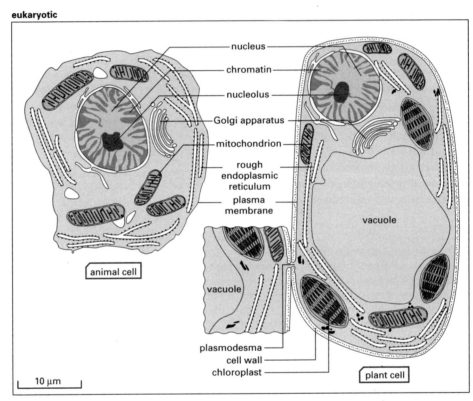

Figure 1.7. Organization of prokaryotic and eukaryotic cells.

acid (DNA). The structure and function of organelles will be described in detail in subsequent chapters. Table 1.1 provides a brief glossary of the major organelles and summarizes the differences between prokaryotic and eukaryotic cells.

Example Box 1.1   Sterilization by Filtration

Because even the smallest cells are larger than 1 μm, harmful bacteria and protists can be removed from drinking water by passing through a filter with 200-nm-diameter holes. Filters can vary in size from huge, such as those used in various commercial processes, to small enough to be easily transportable by backpackers. Filtering drinking water greatly reduces the chances of bringing back an unwanted souvenir from your camping trip!

Table 1.1. Differences Between Prokaryotic and Eukaryotic Cells

	Prokaryotes	Eukaryotes
Size	Usually 1–2 μm	Usually 5–100 μm
Nucleus	Absent	Present, bounded by nuclear envelope
DNA	Usually a single circular molecule (=chromosome)	Multiple molecules (=chromosomes), linear, associated with protein.[a]
Cell division	Simple fission	Mitosis or meiosis
Internal membranes	Rare	Complex (nuclear envelope, Golgi apparatus, endoplasmic reticulum, etc.—Fig. 1.2)
Ribosomes	70S[b]	80S (70S in mitochondria and chloroplasts)
Cytoskeleton	Absent	Microtubules, microfilaments, intermediate filaments
Motility	Rotary motor (drives bacterial flagellum)	Dynein (drives cilia and eukaryote flagellum); kinesin, myosin
First appeared	3.5×10^9 years ago	1.5×10^9 years ago

[a] The tiny chromosomes of mitochondria and chloroplasts are exceptions; like prokaryotic chromosomes they are often circular.

[b] The S value, or Svedberg unit, is a sedimentation rate. It is a measure of how fast a molecule moves in a gravitational field, and therefore in an ultracentrifuge.

Special Properties of Plant Cells

Among eukaryotic cells the most striking difference is between those of animals and plants (Fig. 1.7). Plants have evolved a sedentary lifestyle and a mode of nutrition that means they must support a leaf canopy. Their cells are enclosed within a rigid cell wall that gives shape to the cell and structural rigidity to the organism (page 53). This is in contrast to the flexible boundaries of animal cells. Plant cells frequently contain one or more **vacuoles** that can occupy up to 75% of the cell volume. Vacuoles accumulate a high concentration of sugars and other soluble compounds. Water enters the vacuole to dilute these sugars, generating hydrostatic pressure that is counterbalanced by the rigid wall. In this way the cells of the plant become stiff or turgid, in the same way that when an inner tube is inflated inside a bicycle tire the combination becomes stiff. Vacuoles are often pigmented, and the spectacular colors of petals and fruit reflect the presence of compounds such as the purple anthocyanins in the vacuole. Cells of photosynthetic plant tissues contain a special organelle, the **chloroplast,** that houses the light-harvesting and carbohydrate-generating systems of **photosynthesis** (page 271). Plant cells lack **centrosomes** (page 382) (Fig. 1.2) although these are found in many algae.

VIRUSES

Viruses occupy a unique space between the living and nonliving worlds. On one hand they are made of the same molecules as living cells. On the other hand they are incapable of independent existence, being completely dependent on a host cell to reproduce. Almost all living organisms have viruses that infect them. Human viruses include polio, influenza, herpes, rabies, ebola, smallpox, chickenpox, and the AIDS (acquired immunodeficiency syndrome) virus HIV (human immunodeficiency virus). Viruses are submicroscopic parti-cles consisting of a core of genetic material enclosed within a protein coat called the capsid. Some viruses have an extra membrane layer called the envelope. Viruses are metabolically inert until they enter a host cell, whereupon the viral genetic material directs the host cell machinery to produce viral protein and viral genetic material. Viruses often insert their genome into that of the host, an ability that is widely made use of in molecular genet-ics (Chapter 7). Bacterial viruses, called bacteriophages (page 74), are used by scientists to transfer genes between bacterial strains. Human viruses are used as vehicles for gene therapy. By exploiting the natural infection cycle of a virus such as adenovirus, it is possible to in-troduce a functional copy of a human gene into a patient suffering from a genetic disease such as cystic fibrosis (Chapter 20).

✸ ORIGIN OF EUKARYOTIC CELLS

Prokaryotic cells are simpler and more primitive in their organization than eukaryotic cells. According to the fossil record, prokaryotic organisms antedate, by at least 2 billion years, the first eukaryotes that appeared some 1.5 billion years ago. It seems highly likely that eukaryotes evolved from prokaryotes, and the most likely explanation of this process is the **endosymbiotic theory.** The basis of this hypothesis is that some eukaryotic organelles orig-inated as free-living prokaryotes that were engulfed by larger cells in which they established a mutually beneficial relationship. For example, **mitochondria** would have originated as free-living aerobic bacteria and **chloroplasts** as cyanobacteria, photosynthetic prokaryotes formerly known as blue-green algae.

The endosymbiotic theory provides an attractive explanation for the fact that both mito-chondria and chloroplasts contain DNA and ribosomes of the prokaryotic type (Table 1.1). The case for the origin of other eukaryotic organelles is less persuasive. While it is clearly not perfect, most biologists are now prepared to accept that the endosymbiotic theory pro-vides at least a partial explanation for the evolution of the eukaryotic cell from a prokaryotic ancestor. Unfortunately, living forms having a cellular organization intermediate between prokaryotes and eukaryotes are rare. Some primitive protists possess a nucleus but lack mitochondria and other typical eukaryotic organelles. They also have the prokaryotic type of ribosomes. These organisms are all intracellular parasites and they include *Microspora*, an organism that infects AIDS patients.

✸ CELL SPECIALIZATION

All the body cells that comprise a single organism share the same set of genetic instructions in their nuclei. Nevertheless, the cells are not all identical. Rather, plants and animals are composed of different **tissues,** groups of cells that are specialized to carry out a common function. This specialization occurs because different cell types read out different parts of the DNA blueprint and therefore make different proteins, as we will see in Chapter 6. In animals there are four major tissue types: epithelium, connective tissue, nervous tissue, and muscle.

Epithelia

Epithelia are sheets of cells that cover the surface of the body and line its internal cavities such as the lungs and intestine. The cells may be **columnar,** taller than they are broad, or

squamous, meaning flat. In the intestine, the single layer of columnar cells lining the inside, or **lumen,** has an absorptive function that is increased by the folding of the surface into **villi** (Fig. 1.8). The luminal surfaces of these cells have **microvilli** that increase the surface area even further. The basal surface sits on a supporting layer of extracellular fibers called the **basement membrane.** Many of the epithelial cells of the airways, for instance, those lining the trachea and bronchioles, have **cilia** on their surfaces. These are hairlike appendages that actively beat back and forth, moving a layer of mucus away from the lungs (Chapter 18). Particles and bacteria are trapped in the mucus layer, preventing them from reaching the delicate air exchange membranes in the lung. In the case of the skin, the epithelium is said to be **stratified** because it is composed of several layers.

Medical Relevance 1.1

Smoking Is Bad for Your Cilia

The cilia on the epithelial cells that line our airways move a belt of mucus that carries trapped particles and microorganisms away from the lungs. These cilia are paralyzed by cigarette smoke. As a consequence, smokers have severely reduced mucociliary clearance with the result that they have to cough to expel the mucus that continuously accumulates in their lungs. Because of this impaired ciliary function, smokers are much more susceptible to respiratory diseases such as emphysema and asthma. The cilia of passive smokers, people who breathe smoky air, are also affected.

Connective Tissue

Connective tissues provide essential support for the other tissues of the body. They include bone, cartilage, and adipose (fat) tissue. Unlike other tissues, connective tissue contains relatively few cells within a large volume of **extracellular matrix** that consists of different types of fiber embedded in **amorphous** ground substance (Fig. 1.8). The most abundant of the fibers is **collagen,** a protein with the tensile properties of steel that accounts for about a third of the protein of the human body. Other fibers have elastic properties that permit the supported tissues to be displaced and then to return to their original position. The amorphous ground substance absorbs large quantities of water, facilitating the diffusion of metabolites, oxygen, and carbon dioxide to and from the cells in other tissues and organs. Of the many cell types found in connective tissue, two of the most important are **fibroblasts,** which secrete the ground substance and fibers, and **macrophages,** which remove foreign, dead, and defective material from it. A number of inherited diseases are associated with defects in connective tissue. Marfan's syndrome, for example, is characterized by long arms, legs, and torso and by a weakness of the cardiovascular system and eyes. These characteristics result from a defect in the organization of the collagen fibers.

Nervous Tissue

Nervous tissue is a highly modified epithelium that is composed of several cell types. Principal among these are the **nerve cells,** also called **neurons,** along with a variety of supporting cells that help maintain them. Neurons extend processes called **axons,** which can be over a meter in length. Neurons constantly monitor what is occurring inside and outside the body. They integrate and summarize this information and mount appropriate responses to it (Chapters 15–17). Another type of cell called **glia** has other roles in nervous tissue including forming the electrical insulation around axons.

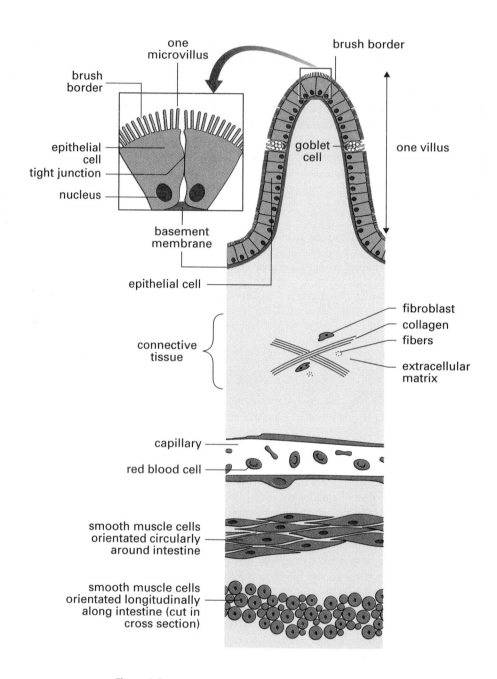

Figure 1.8. Tissues and structures of the intestine wall.

Muscle

Muscle tissue can be of two types, **smooth** or **striated.** Smooth muscle cells are long and slender and are usually found in the walls of tubular organs such as the intestine and many blood vessels. In general, smooth muscle cells contract slowly and can maintain the contracted state for a long period of time. There are two classes of striated muscle: **cardiac** and **skeletal.** Cardiac muscle cells make up the walls of the heart chambers. These

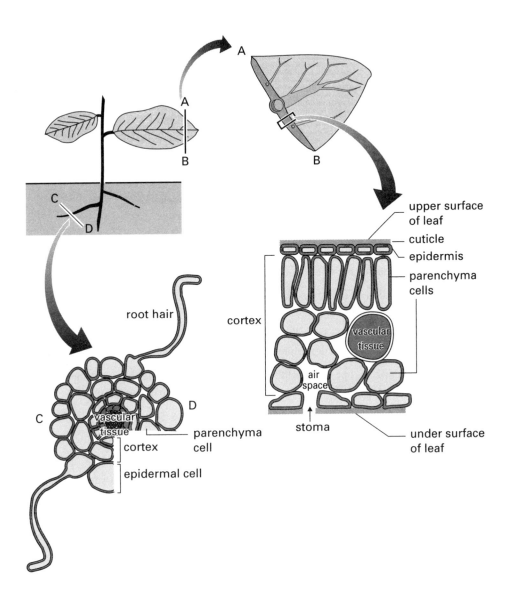

Figure 1.9. Plants are composed of several tissues.

are branched cells that are connected electrically by gap junctions (page 55), and their automatic rhythmical contraction powers the beating of the heart. Each skeletal muscle is a bundle of hundreds to thousands of fibers, each fiber being a giant single cell with many nuclei. This rather unusual situation is the result of an event that occurs in the embryo when the cells that give rise to the fibers fuse together, pooling their nuclei in a common cytoplasm. The mechanism of contraction of skeletal muscle will be described in Chapter 18.

Plants

Plant cells are also organized into tissues (Fig. 1.9). The basic organization of a shoot or root is into an outer protective layer, or **epidermis,** a **vascular tissue** that provides support and transport, and a **cortex** that fills the space between the two. The epidermis consists of one or more layers of closely packed cells. Above the ground these cells secrete a waxy layer, the cuticle, which helps the plant retain water. The cuticle is perforated by pores called stomata that allow gas exchange between the air and the photosynthetic cells and also constitute the major route for water loss by a process called **transpiration.** Below ground, the epidermal cells give rise to root hairs that are important in the absorption of water and minerals. The vascular tissue is composed of **xylem,** which transports water and its dissolved solutes from the roots, and **phloem,** which conveys the products of photosynthesis, predominantly sugars, to their site of use or storage. The cortex consists primarily of **parenchyma cells,** unspecialized cells whose cell walls are usually thin and bendable. They are the major site of metabolic activity and photosynthesis in leaves and green shoots.

SUMMARY

1. All living organisms are made of cells.

2. Our understanding of cell structure and function has gone hand in hand with developments in microscopy and its associated techniques.

3. Light microscopy revealed the diversity of cell types and the existence of the major organelles: nucleus, mitochondrion and, in plants, the vacuole and chloroplast.

4. The electron microscope revealed the detailed structure of the larger organelles and resolved the cell ultrastructure, the fine detail at the nanometer scale.

5. There are two types of cells, prokaryotes and eukaryotes.

6. Prokaryotic cells have very little visible internal organization. They usually measure 1–2 μm across.

7. Eukaryotic cells usually measure 5–100 μm across. They contain a variety of specialized internal organelles, the largest of which, the nucleus, contains the genetic material.

8. The endosymbiotic theory proposes that some eukaryotic organelles, such as mitochondria and chloroplasts, originated as free-living prokaryotes.

9. The cells of plants and animals are organized into tissues. In animals there are four tissue types: epithelium, connective tissue, nervous tissue, and muscle. Plants are formed of epidermis, cortex, and vascular tissues.

✳ REVIEW QUESTIONS

For each question, choose the ONE BEST answer or completion.

1. Eukaryotic cells usually contain
   A. a nucleus.
   B. mitochondria.
   C. ribosomes.

    D. microtubules.

    E. all of the above.

2. The transmission electron microscope

    A. is ideal for looking at living cells.

    B. has a resolution of 0.2 nm.

    C. uses lenses made of glass.

    D. was invented by Anton van Leeuwenhoek.

    E. is easily transportable.

3. Which of the following statements about tissues is incorrect?

    A. Microvilli are typically associated with nervous tissue.

    B. Columnar and squamous are types of epithelial tissue.

    C. Collagen is a component of the extracellular matrix.

    D. Glial cells form the insulation around nerve cells.

    E. Epidermis, vascular tissue, and cortex are types of plant tissue.

4. 1,000,000 nm is equal to

    A. 1 μm.

    B. 10 μm.

    C. 1 mm.

    D. 10 mm.

    E. 1 m.

5. Unlike eukaryotes, prokaryotes lack

    A. a plasma membrane.

    B. DNA.

    C. ribosomes.

    D. nuclei.

    E. molecular motors.

6. Light passes through a light microscope in the following order:

    A. condenser, specimen, objective, projector.

    B. condenser, objective, specimen, projector.

    C. condenser, projector, specimen, objective.

    D. objective, specimen, projector, condenser.

    E. objective, projector, specimen, condenser.

7. Which of the following is not a type of muscle cell?

    A. Smooth.

    B. Ciliated.

    C. Striated.

    D. Cardiac

    E. Skeletal.

ANSWERS TO REVIEW QUESTIONS

1. *E.* These are all typical features of eukaryotic cells.

2. *B.* None of the others are true. In particular the fact that an electron beam is used means that the lenses are electromagnets, not glass, and means that the interior of an electron microscope is held at a vacuum, which means in turn that it is not a suitable environment for living tissue and tends to mean that electron microscopes are bulky, heavy pieces of equipment.

3. *A.* Microvilli are found on epithelia, for example, the epithelium lining the intestine.

4. *C.* The international system prefixes used in this book are m for milli, meaning 10^{-3}, μ for micro, meaning 10^{-6}, and n for nano meaning 10^{-9}. Thus 1,000,000 times 10^{-9} is 10^{-3} so 1,000,000 nm is equal to 1 mm.

5. *D.* Eukaryotes have all of these, but prokaryotes do not have nuclei. Rather, the DNA is free in the cytoplasm.

6. *A*

7. *B.* No muscle cells are ciliated; as we will see in Chapter 18, the molecular motor systems used in cilia and in muscles are different. Muscle cells are divided into smooth and striated; the striated type is then further subdivided into cardiac and skeletal.

2

FROM WATER TO DNA: THE CHEMISTRY OF LIFE

Organisms are made up of a lot of different chemicals. These vary in size, from small molecules like water to large molecules like DNA, and interact and associate in many different ways to generate the processes of life. In this chapter we will introduce some basic concepts of how these chemicals are made and interact. We will then describe the most important of these chemicals: water, carbohydrates, nucleotides, amino acids, and lipids.

✱ THE CHEMICAL BOND: SHARING ELECTRONS

Water is the most abundant substance in organisms. Cells are rich in water. Cytoplasm consists of organelles floating in a watery medium called **cytosol** that also contains proteins. The situation is not so different outside our cells. Although we are land animals living in air, most of our cells are bathed in a watery fluid called **extracellular medium**. We will therefore start by considering water itself.

Figure 2.1*a* shows a molecule of water, consisting of one atom of oxygen and two hydrogen atoms, joined to form an open V shape. The lines represent covalent bonds formed when atoms share electrons, each seeking the most stable structure. Oxygen has a greater affinity for electrons than does hydrogen so the electrons are not distributed equally. The oxygen grabs a greater share of the available negative charge than do the hydrogen atoms. The molecule of water is polarized, with partial negative charge on the oxygen and partial positive charges on the two hydrogens. We write the charge on each hydrogen as $\delta+$

Cell Biology: A Short Course, Second Edition, by Stephen R. Bolsover, Jeremy S. Hyams,
Elizabeth A. Shephard, Hugh A. White, Claudia G. Wiedemann
ISBN 0-471-26393-1 Copyright © 2004 by John Wiley & Sons, Inc.

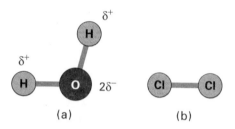

<figure>Figure 2.1. Water is a polar molecule while the chlorine molecule is nonpolar.</figure>

to indicate that it is smaller than the charge on a single hydrogen nucleus. The oxygen atom has the small net negative charge $2\delta-$. Molecules that, like water, have positive regions sticking out one side and negative regions sticking out the other are called **polar**.

Figure 2.1*b* shows a molecule of chlorine gas. It consists of two chlorine atoms, each of which consists of a positively charged nucleus surrounded by negatively charged electrons. Like oxygen, chlorine atoms tend to accept electrons when they become available, but the battle is equal in the chlorine molecule: The two atoms share their electrons equally and the molecule is **nonpolar**.

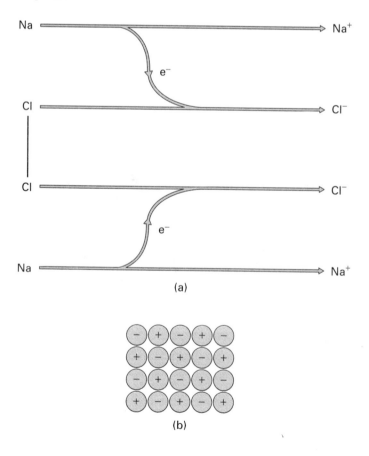

Figure 2.2. Formation of sodium chloride, an ionic compound.

Figure 2.2a shows what happens when a chlorine molecule is allowed to react with the metal sodium. Each atom of chlorine takes over one electron from a sodium atom. This leaves the sodium atoms with a single positive charge because there is now one more positive charge on the sodium nucleus than negatively charged surrounding electrons. Similarly, each chlorine atom now has a single negative charge because it now has one more electron than there are positive charges in its nucleus. Chemical species that have either gained or lost electrons, and that therefore bear an overall charge, are called **ions**. The reaction of chlorine and sodium has produced sodium ions and chloride ions. Positively charged ions like sodium are called **cations** while negatively charged ones like chloride are called **anions**. The positively charged sodium ions and the negatively charged chloride ions now attract each other strongly. If there are no other chemicals around, the ions will arrange themselves to minimize the distance between sodium and chloride, and the resulting well-packed array of ions is a crystal of sodium chloride, shown in Figure 2.2b.

INTERACTIONS WITH WATER: SOLUTIONS

Ionic Compounds Will Dissolve Only in Polar Solvents

Figure 2.3a shows one molecule of octane, the main constituent of gasoline. Octane is an example of a nonpolar solvent. Electrons are shared equally between carbon and hydrogen, and the component atoms do not bear a net charge.

Figure 2.3b shows a small crystal of sodium chloride immersed in octane. At the edge of the crystal, positively charged sodium ions are being pulled in toward the center of the crystal by the negative charge on chloride ions, and negatively charged chloride ions are being pulled in toward the center of the crystal by the positive charge on sodium ions. The sodium and chloride ions will not leave the crystal. Sodium chloride is insoluble in octane. However, sodium chloride will dissolve in water, and Figure 2.4a shows why. The chloride ion at the top left is being pulled into the crystal by the positive charge on its sodium ion neighbors, but at the same time it is being pulled out of the crystal by the positive charge on the hydrogen atoms of nearby water molecules. Similarly, the sodium ion at the bottom left is being pulled into the crystal by the negative charge on its chloride ion neighbors, but at the same time it is being pulled out of the crystal by the negative charge on the oxygen atoms of nearby water molecules. The ions are not held in the crystal so tightly and can leave. Once the ions have left the crystal, they become surrounded by a **hydration shell** of water molecules, all oriented in the appropriate direction (Fig. 2.4b)—oxygen inward for a positive ion like sodium, hydrogen inward for a negative ion like chloride. A chemical species in solution, whether in water or in any other solvent, is called a **solute**. Liquids whose main constituent is water are called **aqueous**.

Acids Are Molecules That Give H^+ to Water

When we exercise, our muscle cells can become acid, and this is what creates the pain of cramping muscles and the heart pain of angina. Acidity is important in all areas of biology, from the acidity gradient that drives our mitochondria (page 261) to the ecological consequences of acid rain.

Acid solutions contain a high concentration of hydrogen ions. The hydrogen atom is unusual in that it only has one electron while, in its most common isotope, its nucleus

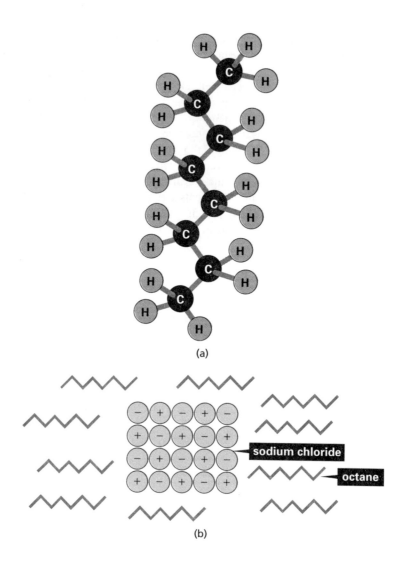

(a)

(b)

<u>Figure 2.3.</u> (a) Structure of the nonpolar compound octane. (b) Ionic compounds are insoluble in nonpolar solvents.

comprises a single proton. In gasses at very low pressure it is possible for bare protons to exist alone and be manipulated, for example, in linear accelerators. However, in water protons never exist alone but always associate with another molecule, for example, with water to create the H_3O^+ ion. Acid solutions are those with an H_3O^+ concentration higher than 100 nmol liter^{-1}.

Sour cream contains lactic acid. Pure lactic acid has the structure shown at the left of Figure 2.5a. The —COOH part in the box is called a **carboxyl group**. Both oxygens have a tendency to pull electrons away from the hydrogen and, in aqueous solution, the hydrogen is donated with a full positive charge to a molecule of water. The electron is left behind on the now negatively charged lactate ion.

$$CH_3CH(OH)COOH + H_2O \rightarrow CH_3CH(OH)COO^- + H_3O^+$$

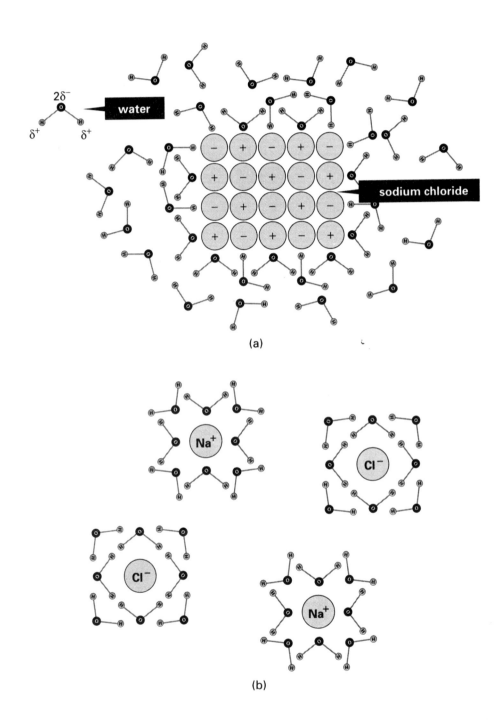

Figure 2.4. Ionic compounds dissolve readily in water.

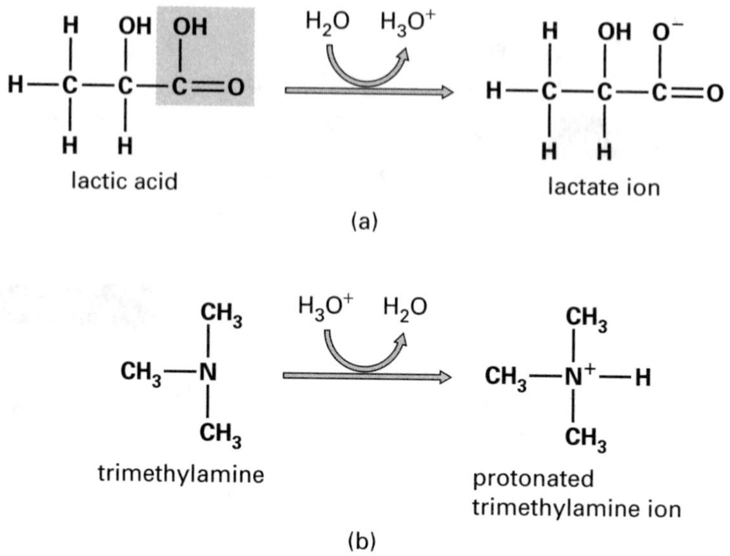

lactic acid

(a)

lactate ion

trimethylamine

protonated
trimethylamine ion

(b)

Figure 2.5. Acids and bases, respectively, give up and accept H⁺ when dissolved in water.

For convenience we often write this as

$$CH_3CH(OH)COOH \rightarrow CH_3CH(OH)COO^- + H^+$$

Here we are using H^+ as a convenient symbol to denote H_3O^+. We do not mean that there are real H^+ ions, that is, bare hydrogen nuclei, in aqueous solutions.

The equilibrium constant, K_a for the dissociation of lactic acid, is defined as

$$K_a = \frac{[CH_3CH(OH)COO^-]_e[H^+]_e}{[CH_3CH(OH)COOH]_e}$$

where the square brackets refer, by convention, to concentrations and the subscripts, e, denote that these are the concentrations of each species at equilibrium.

Dissolving lots of lactic acid in water produces an acid solution, that is, one with a high concentration of H^+ (really H_3O^+) ions. For historical reasons, the acidity of a solution is given as the **pH,** defined thus:

$$pH = -\log_{10}([H^+]) \quad \text{where } [H^+] \text{ is measured in moles per liter}$$

Pure water has a pH of 7, corresponding to $[H^+] = 100$ nmol liter^{-1}—this is said to be neutral as regards pH. If the pH is lower than this, then there is more H^+ about and the solution is acid. Cytosol has a pH that lies very slightly on the alkaline side of neutrality, at about 7.2.

The pH of a solution determines the ratio of lactate to undissociated lactic acid. As the H^+ concentration rises and pH falls, the equilibrium is pushed over from lactate toward lactic acid. At pH 3.9, the concentrations of lactate and lactic acid become equal. Looking at the equation above, we can therefore see that when this happens $K_a = [H^+]$. Just as acidity is given on the logarithmic pH scale so is the scale of strengths of different acids, pK_a being defined as $-\log_{10} K_a$. The pK_a is the pH at which the concentration of the dissociated acid is equal to the concentration of the undissociated acid. For an acid that is weaker than lactic acid, the pK_a is higher than 3.9, meaning that the acid is less readily dissociated and needs a lower concentration of protons before it will give up its H^+. For an acid that is stronger than lactic acid, the pK_a is less than 3.9: Such an acid is more readily dissociated and needs a higher H^+ concentration before it will accept H^+ and form undissociated acid.

Bases Are Molecules That Take H^+ from Water

Trimethylamine is the compound that gives rotting fish its unpleasant smell. Pure trimethylamine has the structure shown at the left of Figure 2.5*b*. When trimethylamine is dissolved in water, it accepts an H^+ to become the positively charged trimethylamine ion shown on the right of the figure. We refer to molecules that have accepted H^+ ions as **protonated**, using "proton" as a short way of saying "hydrogen nucleus." Dissolving lots of trimethylamine in water produces an alkaline solution, that is, one with a low concentration of H^+ ions and hence a pH greater than 7. The solution never runs out of H^+ completely because new H^+ are formed from water:

$$H_2O \rightarrow OH^- + H^+$$

Thus if we keep adding trimethylamine to water and using up H^+, we end up with a low concentration of H^+, but lots of OH^-.

The pH of a solution determines the position of equilibrium between protonated and deprotonated trimethylamine, and as before we define the pK_a as the pH at which the concentration of protonated and deprotonated base are the same. The pK_a of trimethylamine is 9.74, meaning that the concentration of H^+ must fall to the low level of $10^{-9.74}$ mol liter^{-1}, that is, 0.2 nmol liter^{-1}, before half of the trimethylamines will give up their H^+s.

Isoelectric Point

The large molecules called proteins (page 183) have many acidic and basic sites that will give up or accept an H^+ as the pH changes. In alkaline solutions proteins will tend to have an overall negative charge because the acidic sites have lost an H^+ and bear a negative charge. As the pH falls, the acidic sites accept an H^+ and become uncharged, and basic sites also accept an H^+ to gain a positive charge. Thus as the pH falls from an initial high value, the overall charge on the protein becomes less and less negative and then more and more positive. The pH at which the protein has no overall charge is called the **isoelectric point**. The isoelectric points of different proteins are different, and this property is useful in separating them during analysis (page 178). The majority of intracellular proteins have an isoelectric point that is less than 7.2, so that at normal intracellular pH they bear a net negative charge.

A Hydrogen Bond Forms When a Hydrogen Atom Is Shared

We have seen how oxygen tends to grab electrons from hydrogen, forming a polar bond. Nitrogen and sulfur are similarly electron-grabbing. If a hydrogen attached to an oxygen,

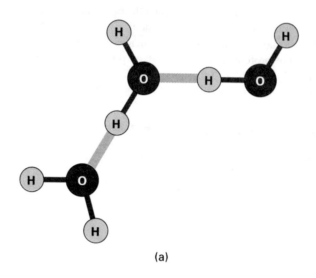

(a)

a hydrogen bond cannot form here because the N H N atoms are not in a straight line

deoxyribose

deoxyribose

thymine (T)

adenine (A)

(b)

Figure 2.6. The hydrogen bond.

nitrogen or sulfur by a covalent bond gets close to a second electron-grabbing atom; then that second atom also grabs a small share of the electrons to form what is known as a **hydrogen bond**. The atom to which the hydrogen is covalently bonded is called the **donor** because it is losing some of its share of electrons; the other electron-grabbing atom is the **acceptor**. For a hydrogen bond to form, the donor and acceptor must be within a fixed distance of one another (typically 0.3 nm) with the hydrogen on a straight line between them.

Liquid water is so stable because the individual molecules can hydrogen bond, as illustrated in Figure 2.6*a*. Hydrogen bonding also plays a critical role in allowing DNA to store and replicate genetic information. Figure 2.6*b* shows how the base pairs (page 69) of DNA form hydrogen bonds in which hydrogen atoms are shared between nitrogen and oxygen and between nitrogen and nitrogen.

✸ BIOLOGICAL MACROMOLECULES

Very large molecules, or **macromolecules,** are central to the working of cells. Large biological molecules are **polymers:** they are assembled by joining together small, simpler molecules, which are therefore called **monomers**. Chemical technology has mimicked nature by producing many important polymers—polyethylene is a polymer of ethylene monomers. Cells make a number of macromolecules that we will introduce, together with their monomer building blocks, in this chapter.

✸ CARBOHYDRATES: CANDY AND CANES

Carbohydrates—sugars and the macromolecules built from them—have many different roles in cells and organisms.

An Assortment of Sweets

All carbohydrates are formed from the simple sugars called **monosaccharides**. Figure 2.7 shows the monosaccharide called glucose. The form shown at the top has five carbons joined each to the other with an oxygen atom completing the ring. As well as the oxygen in the ring, glucose has five other oxygens, each in an −OH (**hydroxyl**) group. Glucose easily switches between the three forms, or **isomers,** illustrated in Figure 2.7. The two ring structures are **stereo isomers:** although they comprise the same atoms connected by the same bonds, they represent two different ways of arranging the atoms in space. The two stereo isomers, named α and β, continually interconvert in solution via the open-chain form.

Figure 2.8 shows five other monosaccharides that we will meet again in this book. Like glucose, each of these monosaccharides can adopt an open-chain form and a number of ring structures. In Figure 2.8 we show each sugar in a form that it adopts quite often. These sugars share with glucose the two characteristics of monosaccharides: they can adopt a form in which an oxygen atom completes a ring of carbons, and they have many hydroxyl groups. The generic names for monosaccharides are derived from the Greek for the number of their carbon atoms, so glucose, galactose, mannose, and fructose are hexoses (6 carbons) while ribose and ribulose are pentoses (5 carbons). Classically, a monosaccharide has the general formula $C_n(H_2O)_n$, hence the name carbohydrate. All the monosaccharides shown in Figures 2.7 and 2.8 fit this rule—the four hexoses can be written as $C_6(H_2O)_6$ and the two pentoses as $C_5(H_2O)_5$.

It is worth noting that although both sugars and organic acids contain an —OH group, the behavior of an —OH group that forms part of a carboxyl group is very different from one that is not next to a double-bonded oxygen. In general the —OH group in a carboxyl group will readily give up an H^+ to water or other acceptors; this is not true of —OH groups

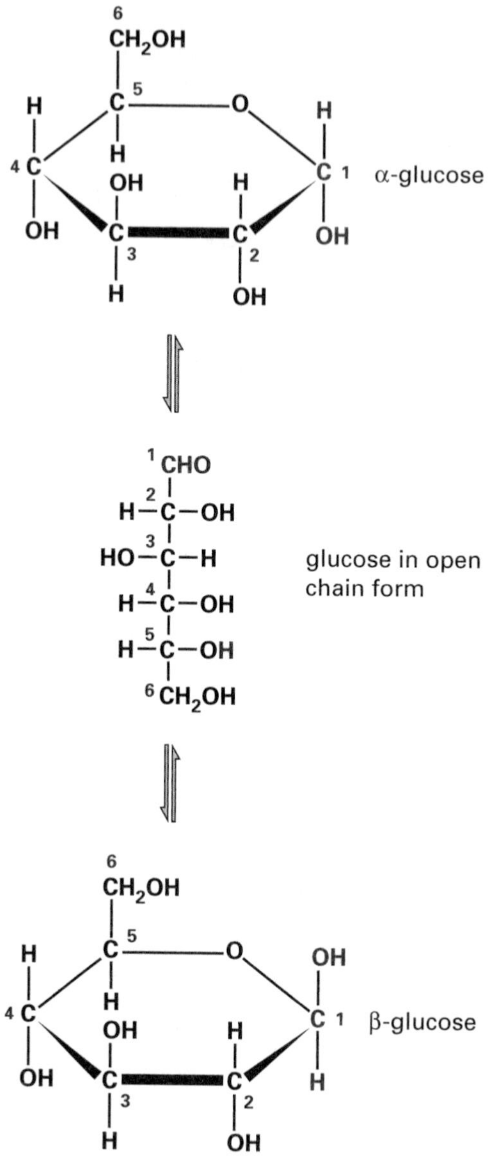

Figure 2.7. Glucose, a monosaccharide, easily switches between three isomers.

in general. We therefore reserve the term *hydroxyl group* for —OH groups on carbon atoms that are not also double-bonded to oxygen.

Disaccharides

Monosaccharides can be easily joined by **glycosidic bonds** in which the carbon backbones are linked through oxygen and a water molecule is lost. The bond is identified by the

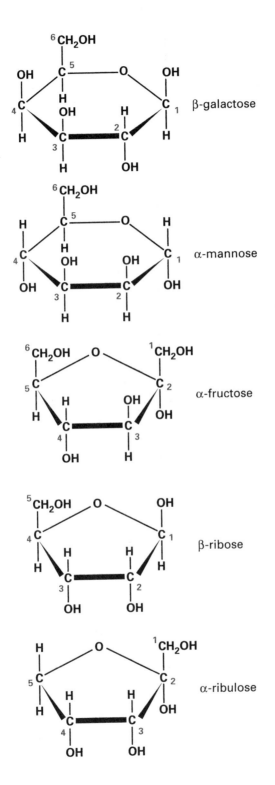

Figure 2.8. Some monosaccharides, each shown as a common isomer.

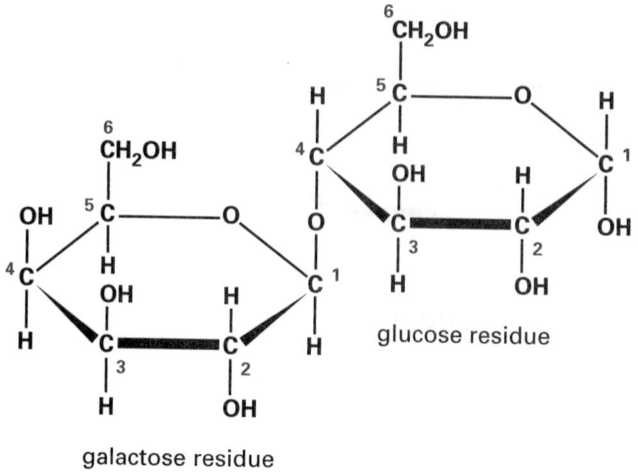

glucose residue

galactose residue

Figure 2.9. The disaccharide lactose.

carbons linked. For example, Figure 2.9 shows lactose, a sugar found in milk that is formed when galactose and glucose are linked by a (1→4) glycosidic bond. Sugars formed from two monosaccharide monomers are called **disaccharides**. Whenever a more complicated molecule is made from simpler building blocks, we call the remnant of the individual building blocks **residues**. Thus lactose is made of one galactose residue and one glucose residue.

A complication arises here. Although free monosaccharides can easily switch between the α and β forms, formation of the glycosidic bond locks the shape. Thus, although galactose in solution spends equal time in the α and β forms, the galactose residue in lactose is locked into the β form, so the full specification of the bond in lactose is $\beta 1 \rightarrow 4$.

Out of the Sweet Comes Forth Strength

Formation of glycosidic bonds can continue almost indefinitely. A chain of up to 100 or so monosaccharides is called an **oligosaccharide,** from the Greek word *oligo* meaning few. Longer polymers are called **polysaccharides** and have characteristics very different from those of their monosaccharide building blocks. Figure 2.10*a* shows part of a molecule of **glycogen,** a polymer made exclusively of glucose monomers organized in long chains with $\alpha 1 \rightarrow 4$ links. The glycogen chain branches at intervals, each branch being an $\alpha(1 \rightarrow 4)$ linked chain of glucose linked to the main chain with an $\alpha(1 \rightarrow 6)$ bond. Solid lumps of glycogen are found in the cytoplasm of muscle, liver, and some other cells. These glycogen granules are 10–40 nm in diameter with up to 120,000 glucose residues. Glycogen is broken down to release glucose when the cell needs energy (Chapter 13).

Cellulose makes up the cell wall of plants and is the world's most abundant macromolecule. Like glycogen, cellulose is a polymer of glucose, but this time the links are $\beta 1 \rightarrow 4$. It is shown in Figure 2.10*b*. Drawn on the flat page, it looks very like glycogen. However, the difference in bond type is critical. Glucoses linked by $\alpha(1 \rightarrow 4)$ links arrange themselves in a floppy helix, while glucoses linked by $\beta(1 \rightarrow 4)$ links form extended chains,

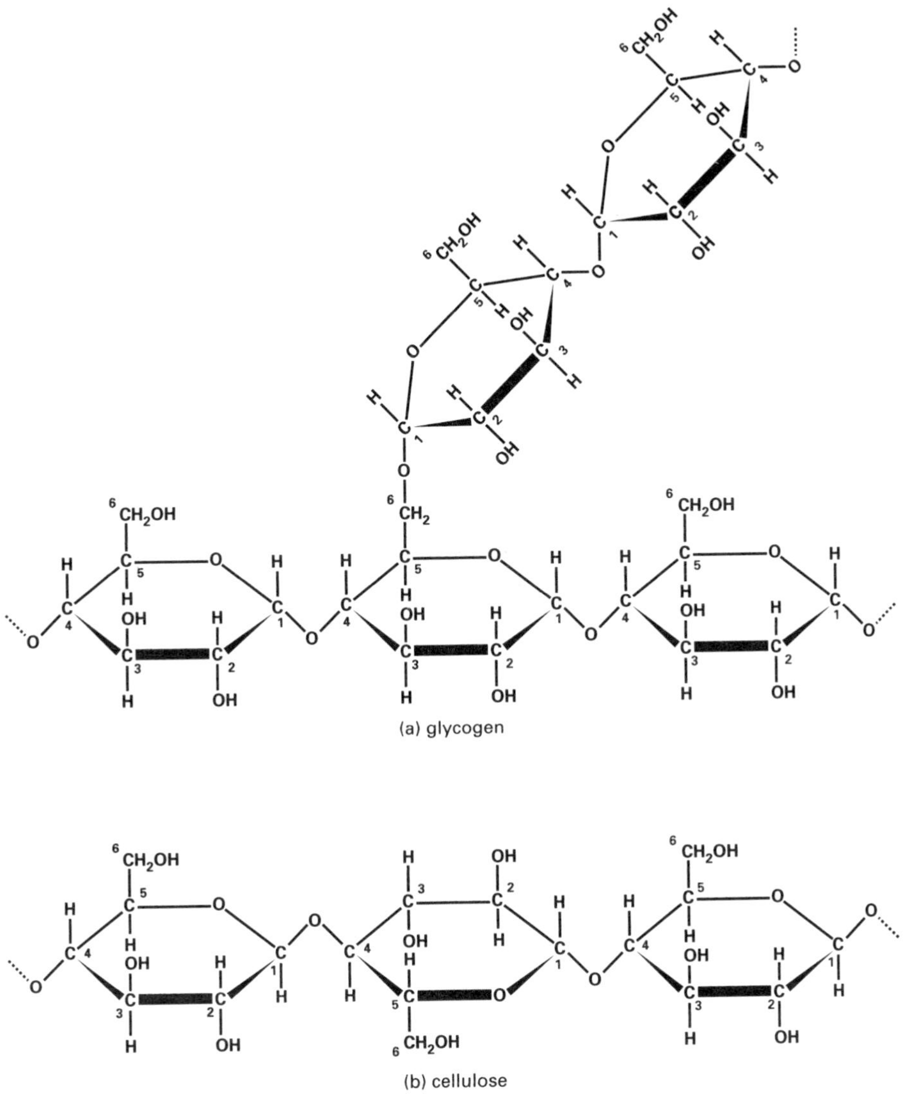

Figure 2.10. The polysaccharides glycogen and cellulose.

ideal for building the rigid plant cell wall. Animals have enzymes (protein catalysts) that can break down the $\alpha(1 \rightarrow 4)$ bond in glycogen, but only certain bacteria and fungi can break the $\beta(1 \rightarrow 4)$ link in cellulose. All animals that eat plants rely on bacteria in their intestines to provide the enzymes to digest cellulose.

Modified Sugars

A number of chemically modified sugars are important in biology. Figure 2.11 shows three. **Deoxyribose** is a ribose that is missing the —OH group on carbon number two. Its

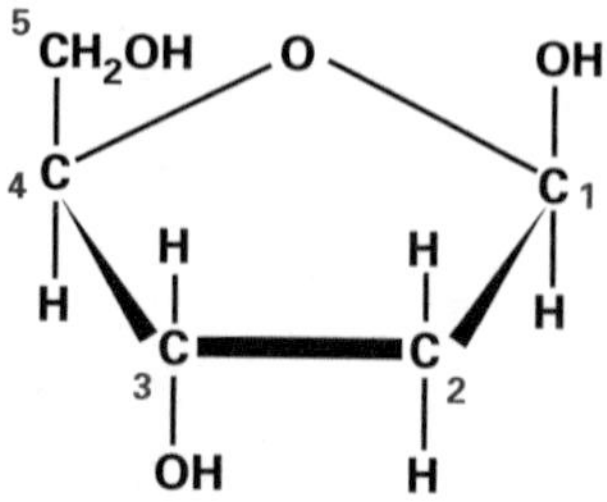

deoxyribose

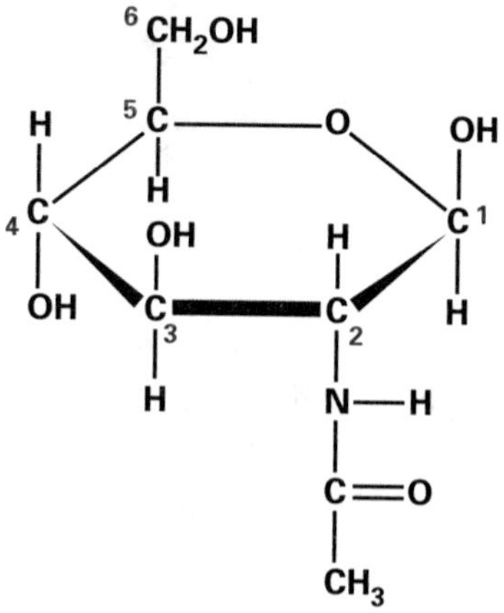

N-acetyl glucosamine

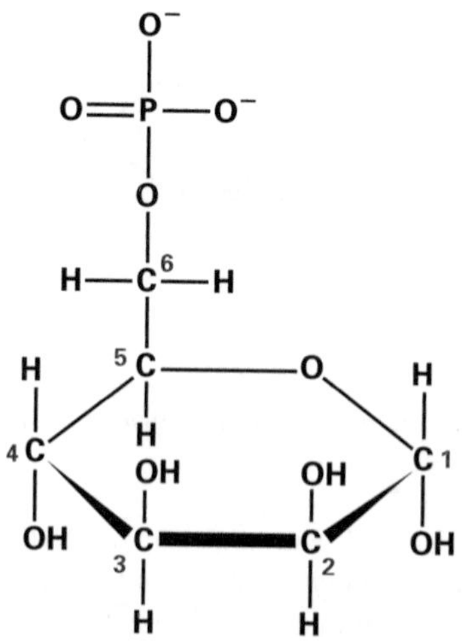

mannose-6-phosphate

Figure 2.11. Deoxyribose, *N*-acetyl glucosamine and mannose-6-phosphate are modified sugars.

main use is in making deoxyribonucleic acid—DNA. *N*-acetyl glucosamine is a glucose in which an —NHCOCH$_3$ group replaces the —OH on carbon two. It is used in a number of oligosaccharides and polysaccharides, for example, in chitin, which forms the hard parts of insects. **Mannose-6-phosphate** is a mannose that has been **phosphorylated,** that is, had a **phosphate** group attached, on carbon number six. We will meet more phosphorylated sugars in the next section.

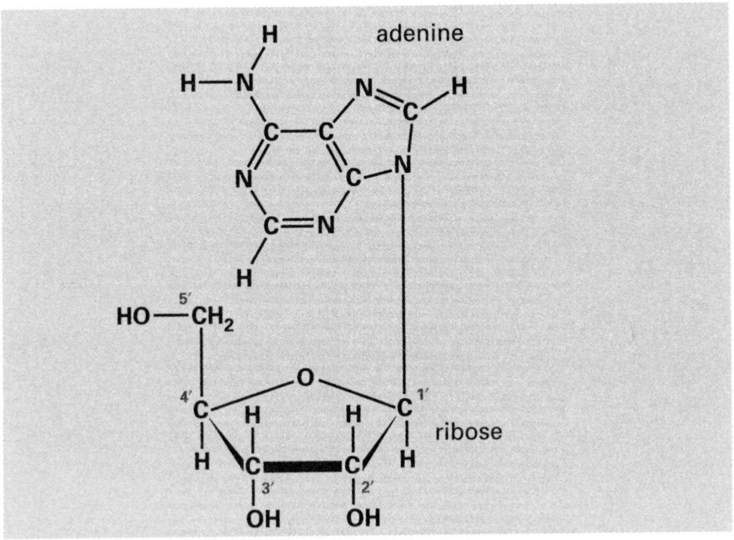

(a) adenosine, a nucleoside

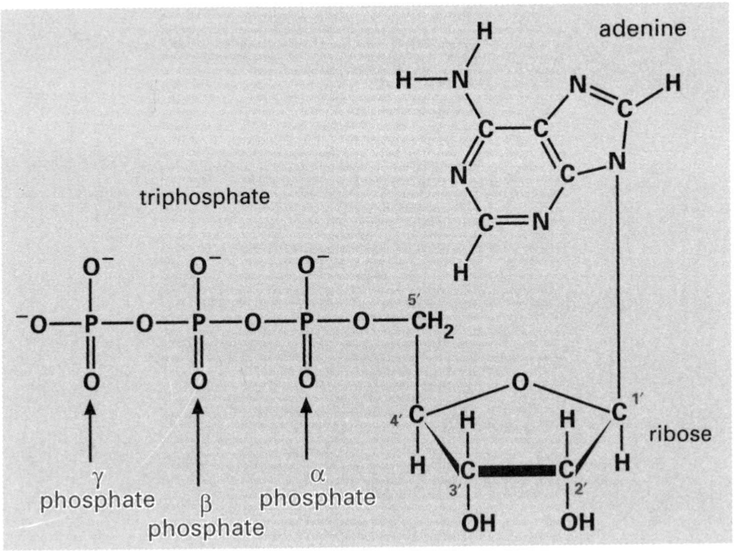

(b) adenosine triphosphate, a nucleotide

Figure 2.12. Adenosine, a nucleoside, and adenosine triphosphate, a nucleotide.

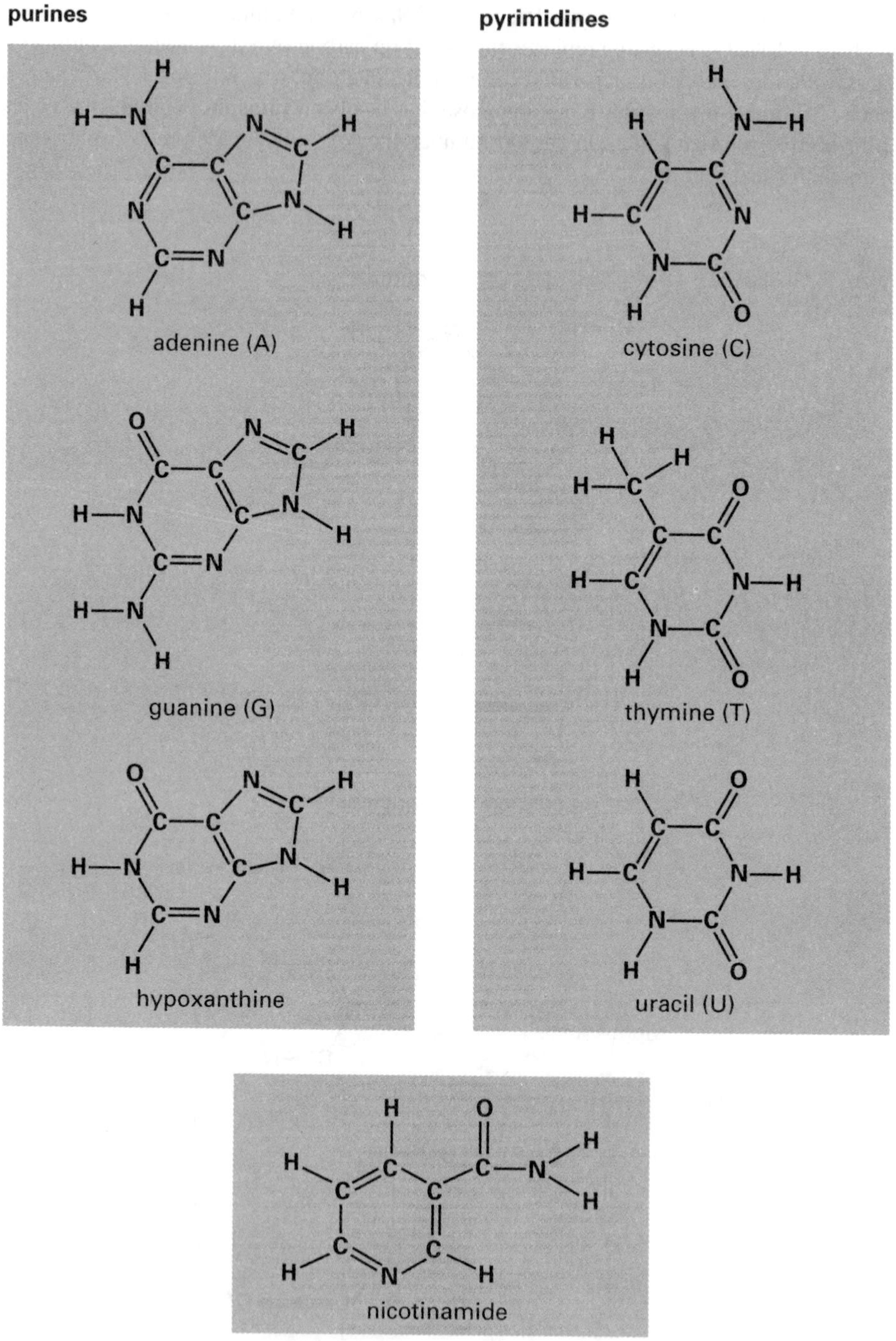

purines

pyrimidines

adenine (A)

guanine (G)

hypoxanthine

cytosine (C)

thymine (T)

uracil (U)

nicotinamide

Figure 2.13. Seven bases found in nucleotides.

✤ NUCLEOSIDES, PHOSPHATE, AND NUCLEOTIDES

Figure 2.12*a* shows a **nucleoside** called **adenosine**. It is composed of ribose coupled to a nitrogen-rich compound called **adenine**. The numbers on the sugar—1′, 2′; and so forth—are the same numbering system we have seen before; the ′symbol is pronounced "prime" and is there to indicate that we are identifying the atoms of the sugar, not the atoms of the adenine. The name nucleoside reflects the fact that phosphorylated nucleosides (see below) are the building blocks of the **nucleic acids** that form the genetic material in the nucleus. However, nucleosides also play important roles in other places inside and outside the cell. Seven different compounds can be used to generate nucleosides (Fig. 2.13). All seven contain many nitrogen atoms and one or more ring structures. They are the three **purines** called adenine, **guanine** and **hypoxanthine,** the three **pyrimidines** called **cytosine, thymine,** and **uracil,** and an odd man out called **nicotinamide**. These ring compounds are called **bases**. Historically, the name arose because the compounds are indeed bases in the sense used earlier in this chapter—they will exchange an H$^+$ with water. The roots of the name are now forgotten by most molecular biologists, who now use the word *base* to mean a purine,

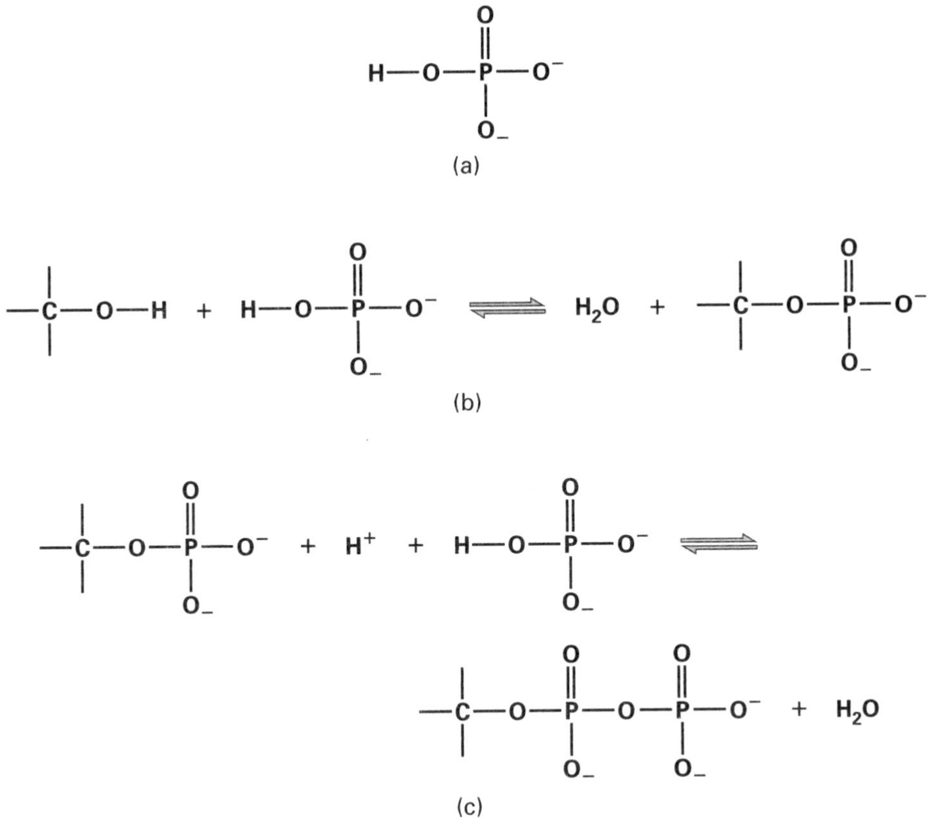

(a)

(b)

(c)

Figure 2.14. Phosphate groups can attach to COH groups or to other phosphate groups.

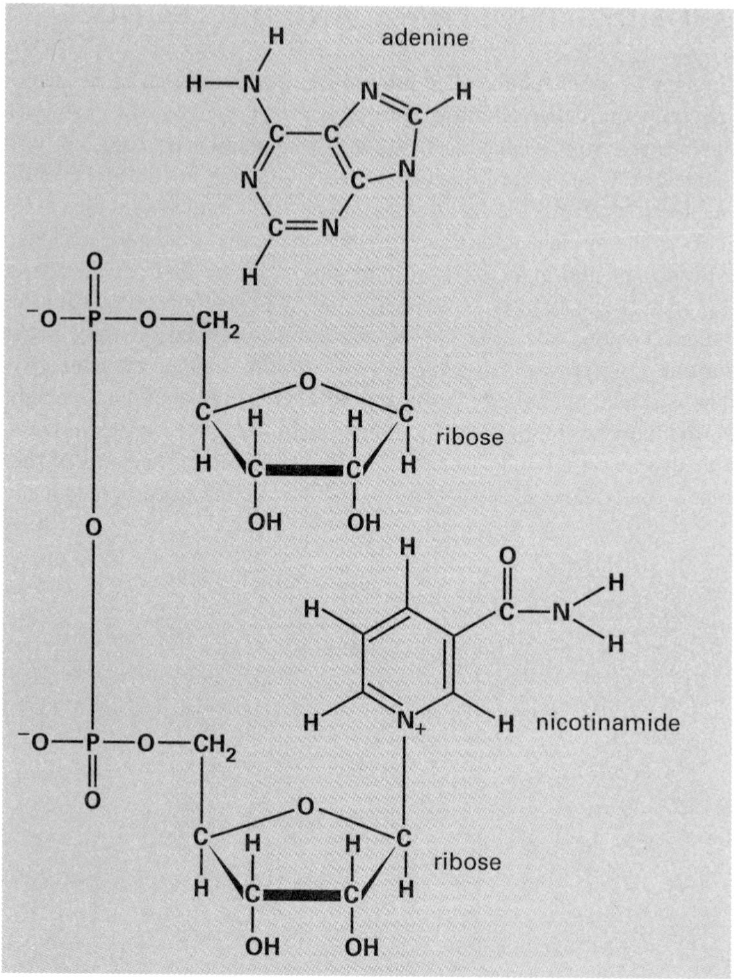

Figure 2.15. Nicotine adenine dinucleotide is formed of two nucleotides joined via their phosphate groups.

a pyrimidine, or nicotinamide. In general, a nucleoside is formed by attaching one of these bases to the 1'-carbon atom of ribose.

Phosphorous, although by weight a relatively minor fraction of the whole cell, plays a number of critical roles. In solution phosphorous is mainly found as a **phosphate** ion with a single hydrogen atom still attached, HPO_4^{2-} (Fig. 2.14a). We often indicate phosphate ions with the symbol Pi, meaning inorganic phosphate. Where phosphate becomes important, however, is when it is attached to an **organic,** that is, carbon-containing, molecule. Phosphate can substitute into any C—OH group with the loss of a water molecule (Fig. 2.14b). The equilibrium in the reaction shown lies far to the left, but cells have other strategies for attaching phosphate groups to organic molecules. Once one phosphate group has been added, more can be added to form a chain (Fig. 2.14c). Once again, the equilibrium in the reaction shown lies far to the left, but cells can achieve this result using other strategies.

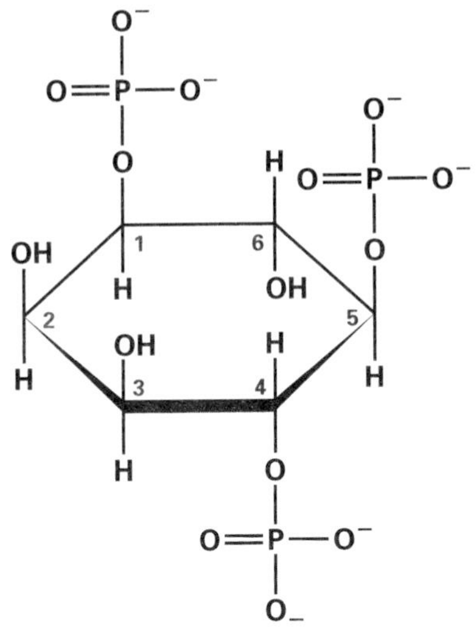

Figure 2.16. Inositol trisphosphate, a multiply phosphorylated polyalcohol.

Figure 2.12*b* shows adenosine with a chain of three phosphate groups attached to the 5′ carbon of the ribose. This important molecule is **adenosine triphosphate** (**ATP**), and we will meet it many times in the course of this book. The three phosphate groups are denoted by the Greek letters α, β, and γ. Phosphorylated nucleosides are called **nucleotides**.

Figure 2.15 shows another important molecule formed from adenosine. Here adenosine and a nicotinamide nucleotide are joined through their phosphate groups. The resulting molecule, called **nicotinamide adenine dinucleotide** (NAD^+), plays a critical role in the cellular energy budget.

Molecules with several OH groups can become multiply phosphorylated. Figure 2.16 shows **inositol trisphosphate** (**IP$_3$**), an important messenger molecule we will meet again in Chapter 16. Both ATP and IP$_3$ have three phosphate groups, but to indicate the fact that in ATP these are arranged in a chain, while in IP$_3$ they are attached to different carbons, we use the prefix *tri* in adenosine *tri*phosphate and the prefix *tris* in inositol *tris*phosphate. Similarly, a compound with two phosphates in a chain is called a *di*phosphate, while one with one phosphate on each of two different carbons is called a *bis*phosphate.

✳ AMINO ACIDS, POLYPEPTIDES, AND PROTEINS

Amino acids contain both a COOH group, which readily gives an H^+ to water and is therefore acidic, and a basic NH_2 group, which readily accepts H^+ to become NH_3^+. Figure 2.17*a* shows two amino acids, leucine and γ-**amino butyric acid** (**GABA**), in the form in which they are found at normal pH: the COOH groups have each lost an H^+ and the NH_2 groups have each gained one, so that the molecules bear both a negative and a positive charge.

We name organic acids by labeling the carbon that bears the carboxyl group α, the next one β, and so on. When we add an amino group, making an amino acid, we state the letter of the carbon to which the amino group is attached. Hence leucine is an α-**amino acid**

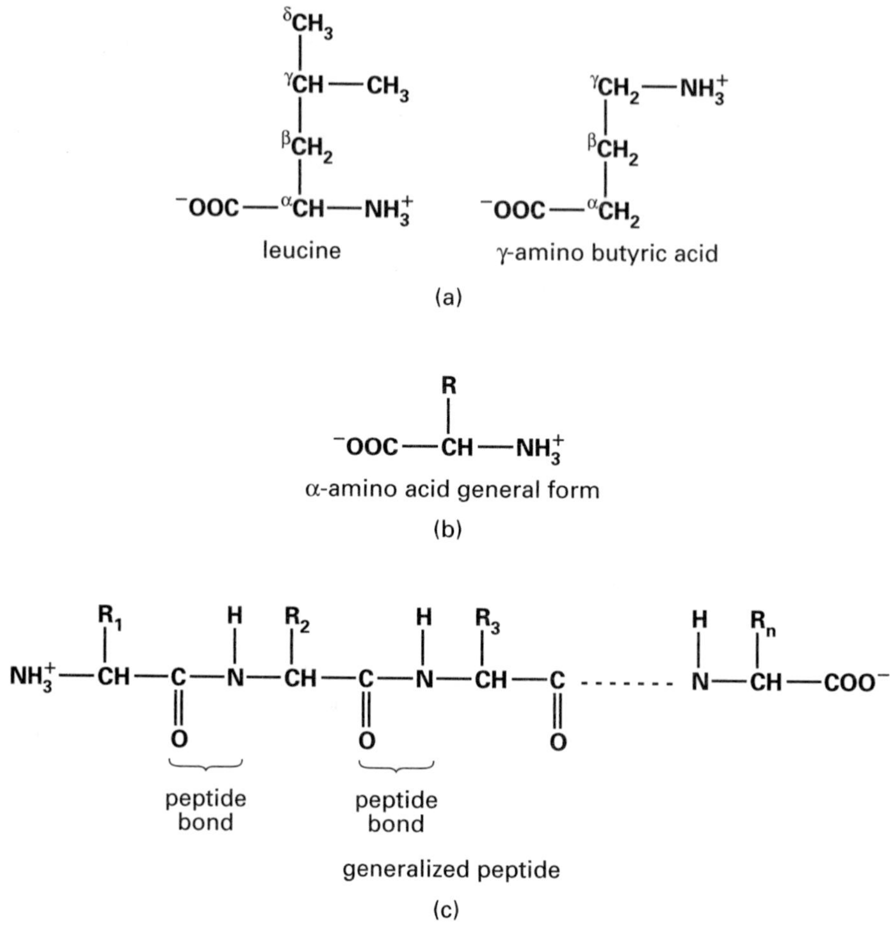

$$^\delta CH_3$$
$$|$$
$$^\gamma CH—CH_3 \qquad\qquad ^\gamma CH_2—NH_3^+$$
$$| \qquad\qquad\qquad\qquad |$$
$$^\beta CH_2 \qquad\qquad\qquad ^\beta CH_2$$
$$| \qquad\qquad\qquad\qquad |$$
$$^-OOC—^\alpha CH—NH_3^+ \qquad ^-OOC—^\alpha CH_2$$

leucine γ-amino butyric acid

(a)

$$R$$
$$|$$
$$^-OOC—CH—NH_3^+$$

α-amino acid general form

(b)

$$R_1 \qquad H \quad R_2 \qquad\quad H \quad R_3 \qquad\qquad\qquad H \quad R_n$$
$$| \qquad | \quad | \qquad\quad | \quad | \qquad\qquad\qquad | \quad |$$
$$NH_3^+—CH—C—N—CH—C—N—CH—C \cdots\cdots N—CH—COO^-$$
$$\| \qquad\qquad \| \qquad\qquad \|$$
$$O \qquad\qquad O \qquad\qquad O$$

peptide   peptide
bond   bond

generalized peptide

(c)

Figure 2.17. Amino acids and the peptide bond.

while GABA stands for *gamma*-amino butyric acid. α-Amino acids are the building blocks of proteins. They have the general structure shown in Figure 2.17*b* where R is called the **side chain**. Leucine has a simple side chain of carbon and hydrogen. Other amino acids have different side chains and so have different properties. It is the diversity of amino side chains that give proteins their characteristic properties (page 185).

  α-Amino acids can link together to form long chains through the formation of a **peptide bond** between the —COO$^-$ group of one amino acid and the —NH$_3^+$ group of the next. Figure 2.17*c* shows the generalized stucture of such a chain of α-amino acids. If there are fewer than about 50 amino acids in a polymer we tend to call it a **peptide**. More and it is a **polypeptide**. Polypeptides that fold into a specific shape are called **proteins**. Most of the components of the machinery of the cell—the chemicals that carry out the processes of life—are proteins. Peptides and polypeptides are formed inside cells by adding specific amino acids to the end of a growing polymer. As we will see in Chapter 4, the particular amino acid to be added at each step is defined by the instructions on a molecule called messenger ribonucleic acid (RNA), which in turn contains a copy of the information recorded on the cell's DNA that is stored in the nucleus. This is the **central dogma** of molecular biology: DNA makes RNA makes protein.

LIPIDS

Figure 2.18*a* shows the **fatty acid** oleic acid. It comprises a carboxyl group plus a long tail of carbons and hydrogens. At neutral pH oleic acid gives up its H^+ to become the oleate ion (Fig. 2.18*b*). The two ends of the oleate ion are very different. The carboxyl group is negatively charged and hence will associate readily with water molecules. The **hydrocarbon tail** is nonpolar and does not readily associate with water. The molecule is said to be **amphipathic,** from the Greek for "hating both," meaning that one half does not like to be in water while the other half does not like to be in a nonpolar environment like octane.

Figure 2.18*c* shows the small molecule **glycerol**. Like sugars, glycerol has many hydroxyl groups and, like sugars, tastes sweet, as its name suggests. However, it is not a sugar because it cannot adopt an oxygen-containing ring structure. Rather, because compounds containing hydroxyl groups are called alcohols, glycerol is a polyalcohol.

As we shall see in Chapter 13, cells can join fatty acids and glycerol to make **glycerides**. The bond is formed by removing the elements of water between the carboxyl group of the fatty acid and a hydroxyl group of glycerol. Any bond of this type, between a carboxyl group and a hydroxyl group, is called an **ester bond**. Figure 2.19 shows two glycerides. In Figure 2.19*a* is trioleoylglycerol, the main component of olive oil. It is a **triacylglycerol** (or **triglyceride**) formed from three molecules of oleic acid and one molecule of glycerol. The fatty acid residues are called **acyl groups**. Formation of the ester bonds has removed the charged carboxyl groups that rendered the oleate amphipathic. Almost all of the triacylglycerol molecule is simple hydrocarbon chain that cannot hydrogen bond with water molecules. For this reason, olive oil and other triacylglycerols do not mix with water. They are therefore said to be **hydrophobic**. Triacylglycerols that are liquid at room temperature are called oils, those that are solid are called fats, but they are all the same sort of molecule.

Example 2.1 Salad Dressing Is a Mixture of Solvents

Vinegar is a dilute solution of acetic acid in water. Acetic acid gives up an H^+ to water to leave the negatively charged acetate ion. The simplest salad dressing is a shaken-up mixture of olive oil and vinegar. Olive oil is hydrophobic and so does not dissolve in the water, the two liquids remain as separate droplets and soon separate again after shaking. However much one shakes, all the acetate will remain in the water because it is an ion. If salt (Na^+Cl^-) is added to the dressing, it also dissolves in the water, with none dissolving in the oil. In contrast if you add chilies to the salad dressing, the active chemical component, capsacin, will dissolve in the oil because it is nonpolar. Only by shaking up the mixture can you get all the tastes together.

Cells contain droplets of triacylglycerol within their cytoplasm. These are usually small, but in fat cells, which are specialized as fat stores, the droplets coalesce into a single large globule so that the cell's cytoplasm is squeezed into a thin layer surrounding the fat

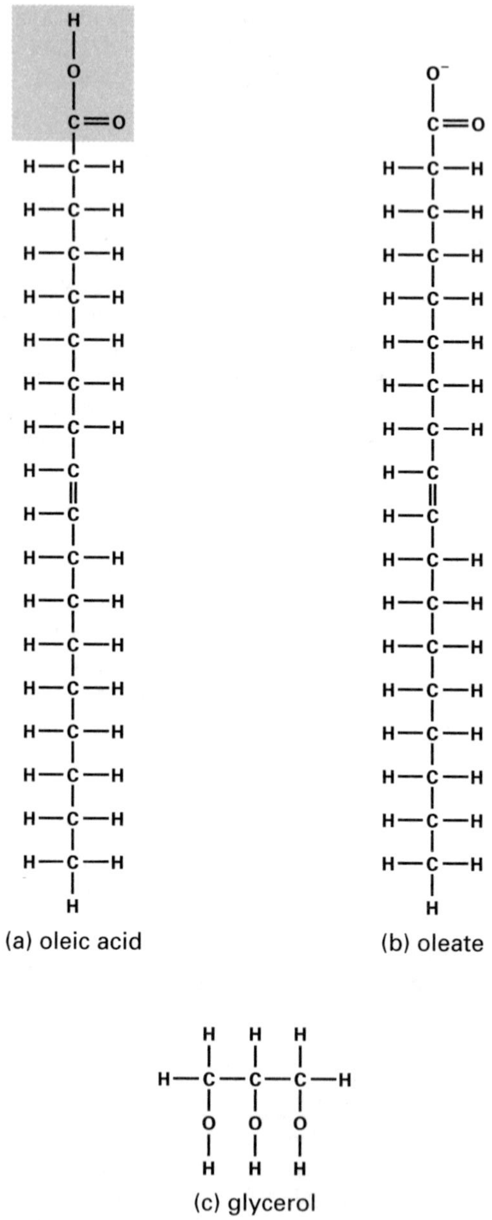

```
        H                                              O⁻
        |                                              |
        O                                              |
        |                                              C=O
        C=O                                            |
        |                                          H—C—H
    H—C—H                                          H—C—H
    H—C—H                                          H—C—H
    H—C—H                                          H—C—H
    H—C—H                                          H—C—H
    H—C—H                                          H—C—H
    H—C—H                                          H—C—H
    H—C—H                                          H—C
     H—C                                             ‖
       ‖                                           H—C
     H—C                                           H—C—H
    H—C—H                                          H—C—H
    H—C—H                                          H—C—H
    H—C—H                                          H—C—H
    H—C—H                                          H—C—H
    H—C—H                                          H—C—H
    H—C—H                                          H—C—H
    H—C—H                                          H—C—H
       H                                              H
```

(a) oleic acid   (b) oleate

```
        H   H   H
        |   |   |
    H—C—C—C—H
        |   |   |
        O   O   O
        |   |   |
        H   H   H
```

(c) glycerol

Figure 2.18. Oleic acid and glycerol.

globule. During fasting triacylglycerols are broken down into free fatty acids and glycerol, a process called lipolysis. The fatty acids and the glycerol then enter the circulation for use by tissues.

Figure 2.19b shows phosphatidylcholine, an example of the phospholipids that make the plasma membrane and other cell membranes. Like triacylglycerols, phospholipids have

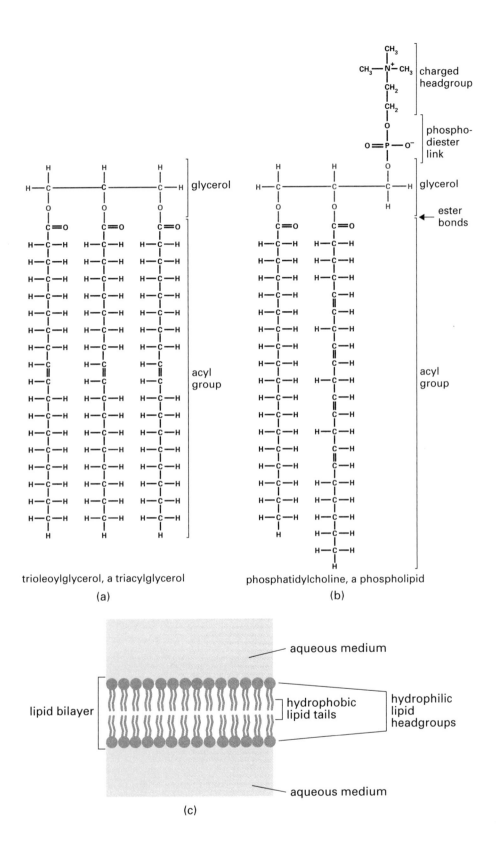

trioleoylglycerol, a triacylglycerol
(a)

phosphatidylcholine, a phospholipid
(b)

(c)

Figure 2.19. Lipids.

fatty acids attached to glycerol, but only two of them. In place of the third fatty acid is a polar, often electrically charged **head group**. The head group is joined to the glycerol through phosphate in a structure called a **phosphodiester link**. The combination of head group and negatively charged phosphate is able to associate strongly with water—it is said to be **hydrophilic**. The two fatty acids, on the other hand, form a tail that, like olive oil, is hydrophobic. Phospholipids can therefore neither dissolve in water (because of their hydrophobic tails) nor remain completely separate, like olive oil (because then the head group could not associate with water). Phospholipid molecules therefore spontaneously form lipid bilayers between 5 and 10 nm thick (Fig. 2.19c) in which each part of the molecule is in its preferred environment. Cell membranes are lipid bilayers plus some added protein.

Medical Relevance 2.1	The Kinks Have It: Double Bonds, Membrane Fluidity, and Evening Primroses

One of the fatty acids commonly found as a component of animal fat is stearic acid, illustrated as (a) opposite. Its 18 carbon atoms are joined by single bonds, making a long straight molecule. In contrast, oleic acid (b) has a double bond between the ninth and tenth carbons. This introduces a kink in the chain. A fatty acid with kinks is less able to solidify because it is less able to pack in a regular fashion. The more double bonds in a fatty acid, the lower its melting point. Thus stearic acid melts at 69.6°C, while oleic acid melts at 13.4°C and is therefore liquid at room temperature. Fatty acids containing double bonds between the carbon atoms are said to be **unsaturated**. **Polyunsaturated** fatty acids have more than one double bond, more than one bend, and therefore even lower melting points. Linoleic acid (c) has two, and melts at −9°C; linolenic acid (d) has three and a melting point of −17°C.

For the triacylglycerols inside our cells, and the phospholipids in our membranes, the same rule applies—the more double bonds in the acyl groups, the lower the melting point. The large masses of triacylglycerols containing stearic acid in animal bodies are liquid at body temperature but form solid lumps of fat at room temperature, while trioleoylglycerol is liquid at room temperature (but should not be kept in the fridge!). Membranes must not solidify—if they did, then they would crack and the cell contents would leak out each time the cell was flexed. Unsaturated fatty acids play an essential role in maintaining membrane liquidity.

Mammals are unable to introduce double bonds beyond carbon 9 in the fatty acid chain. This means that linoleic and linolenic acids must be present in the diet. They are known as **essential fatty acids**. Fortunately, the biochemical abilities of plants are not so restricted, and plant oils form a valuable source of unsaturated fatty acids.

The normal form of linolenic acid, shown in (d), is α-linolenic acid, which has its double bonds between carbons 9 and 10, 12 and 13, and 15 and 16. Some plant seed oils contain an isomer of linolenic acid with double bonds between carbons 6 and 7, 9 and 10, and 12 and 13. This is called γ-linolenic acid (e). The attractive, yellow-flowered garden plant called the evening primrose (*Oenethera perennis*) has seeds that contain an oil with γ-linolenic in its triacyglycerols. The 6 to 7 double bond introduces a kink closer to the glycerol. This is thought to increase fluidity when incorporated into membrane lipids. No one really knows if it really has the marvelous health effects that some claim for it, or if it does, how it works. This ignorance has not stopped people from making lots of money from evening primrose oil. It is included in cosmetics and in alternative medicines for internal and external application. γ-Linolenic acid occurs in other plants—the seeds of borage (*Borago officinalis*) are one of the richest sources. It is even found in some fungi. However, these lack the romance of the evening primrose!

Stearic acid, C18 no double bonds, melts at 69.6°

(a)

Oleic acid, C18 one double bond, melts at 13.4°

(b)

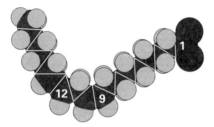

Linoleic acid, C18 two double bonds, melts at –9°

(c)

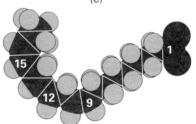

Linolenic acid, C18 three double bonds, melts at –17°

(d)

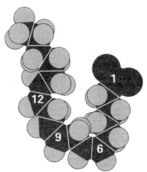

γ-Linolenic acid, C18 three double bonds

(e)

✾ HYDROLYSIS

Many of the macromolecules of which cells are made are generated from their individual building blocks by the removal of the elements of water. Equally, macromolecules can be broken into their individual building blocks by **hydrolysis**—breakage by the addition of water. Figure 2.20 shows four examples. Lactose is hydrolyzed to the monosaccharides galactose and glucose by the addition of one water molecule. Next, we show a dipeptide being broken into individual amino acids by hydrolysis. These two hydrolysis reactions occur all the time in our intestines, though some people lose the ability to hydrolyze lactose as they get older: they are **lactose intolerant**. Third, we show a **dimeric** (formed of two parts) inorganic phosphate ion called **pyrophosphate** being hydrolyzed to regular phosphate ions. The hydrolysis of pyrophosphate is catalyzed by enzymes found throughout the body, both inside cells and out. Later in the book we will meet a number of instances where a reaction creates

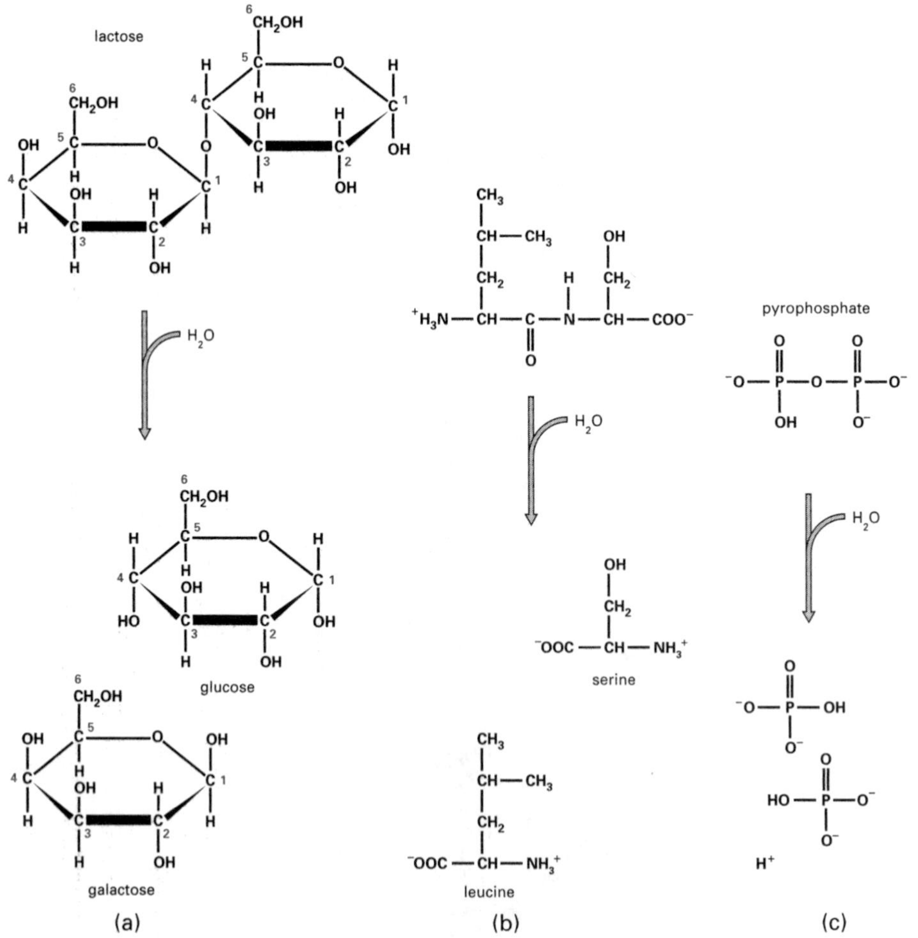

Figure 2.20. Hydrolysis is the breakage of covalent bonds by the addition of the elements of water.

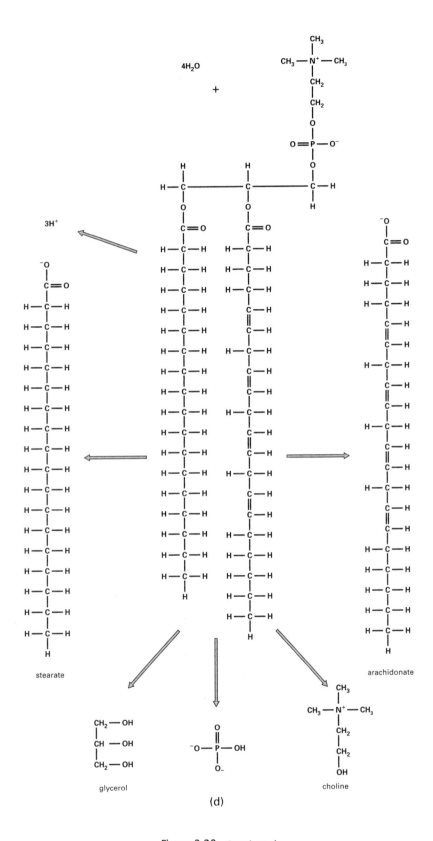

stearate

glycerol

choline

arachidonate

(d)

Figure 2.20. *Continued*

pyrophosphate, but as soon as the pyrophosphate is produced it will be hydrolyzed to regular phosphate. Lastly, we show the complete hydrolysis of the phospholipid phosphatidylcholine. Five molecules result: one phosphate ion together with choline, glycerol, and two fatty acids. This hydrolysis also occurs in our intestines, where it is a multistage process.

Hydrolysis reactions usually proceed rapidly in living tissue when the appropriate catalyst is present, because water is present at high concentration. In contrast the generation of macromolecules from their constituent building blocks, often by the elimination of the elements of water between constituent building blocks, usually requires the expenditure of energy by the cell. We will see how these reactions are performed in later chapters.

SUMMARY

1. When two atoms interact, electrons may be shared between the two to form a covalent bond, or may pass completely from one atom to another, forming ions.

2. In water the electrons are not shared equally but are displaced toward the oxygen atom, so that the molecule is polar.

3. Ionic compounds will only dissolve in polar solvents.

4. Acids are molecules that give an H^+ to water, forming H_3O^+. Dissolving an acid in water produces a solution of pH less than 7.

5. Bases are molecules that accept an H^+ from water, leaving OH^-. Dissolving a base in water produces a solution of pH greater than 7.

6. Pure water has the neutral pH of 7.0.

7. A hydrogen bond can form when a hydrogen atom takes up a position between two electron-grabbing atoms (oxygen, nitrogen, or sulfur), the three forming a straight line.

8. Monosaccharides are compounds with a central skeleton of carbon to which are attached many OH groups. They can switch between two oxygen-containing ring structures via an open-chain configuration.

9. Monosaccharides can join together to form disaccharides such as lactose and to form long chains such as glycogen and cellulose.

10. Adenine, guanine, hypoxanthine, cytosine, thymine, uracil, and nicotinamide are nitrogen-rich ring-shaped molecules called bases. Bases combine with ribose or deoxyribose to form nucleosides. Phosphorylation of a nucleoside on the 5′ carbon produces a nucleotide.

11. Amino acids are compounds with a —COOH acidic group and an —NH_2 basic group. At neutral pH these groups will, respectively, lose and gain an H^+ to become —COO^- and —$NH_3{}^+$.

12. Polypeptides are polymers of α-amino acids linked by peptide bonds. The sequence of amino acids is determined by instructions on the cell's DNA. Proteins are polypeptides that fold into a specific shape.

13. Lipids are formed by the attachment of hydrophobic long-chain fatty acids to a glycerol backbone. Phospholipids also have a hydrophilic head group and spontaneously form lipid bilayers.

14. Hydrolysis is the breakage of covalent bonds by the addition of the elements of water.

FURTHER READING

Voet, D., and Voet, J. 2003. *Biochemistry*. New York: Wiley.

✳ REVIEW QUESTIONS

For each question, choose the ONE BEST answer or completion.

1. Water
   A. can give up an H^+, becoming OH^-.
   B. can accept an H^+, becoming H_3O^+.
   C. can form hydrogen bonds.
   D. is a good solute for ionic substances.
   E. all of the above.

2. The amino acid leucine has an acidic COOH group and a basic NH_2 group. The reaction —COOH $\rightleftarrows$ —COO$^-$ + H^+ has a pK_a of 2.3, while the reaction —NH_3^+ $\rightleftarrows$ —NH_2 + H^+ has a pK_a of 9.7. At pH 1.0
   A. there will be more $^-$OOC—CH(CH_2.CH(CH_3)$_2$)—NH_2 than either
      $^-$OOC—CH(CH_2.CH(CH_3)$_2$)—NH_3^+ or HOOC—CH(CH_2.CH(CH_3)$_2$)—NH_3^+.
   B. there will be equal amounts of $^-$OOC—CH(CH_2.CH(CH_3)$_2$)—NH_2 and
      $^-$OOC—CH(CH_2.CH(CH_3)$_2$)—NH_3^+, but not much HOOC—CH(CH_2.CH(CH_3)$_2$)—NH_3^+.
   C. there will be more $^-$OOC—CH(CH_2.CH(CH_3)$_2$)—NH_3^+ than either
      HOOC—CH(CH_2.CH(CH_3)$_2$)—NH_3^+ or $^-$OOC—CH(CH_2.CH(CH_3)$_2$)—NH_2.
   D. there will be equal amounts of HOOC—CH(CH_2.CH(CH_3)$_2$)—NH_3^+ and
      $^-$OOC—CH(CH_2.CH(CH_3)$_2$)—NH_3^+, but not much $^-$OOC—CH(CH_2.CH(CH_3)$_2$)—NH_2.
   E. there will be more HOOC—CH(CH_2.CH(CH_3)$_2$)—NH_3^+ than either
      $^-$OOC—CH(CH_2.CH(CH_3)$_2$)—NH_3^+ or $^-$OOC—CH(CH_2.CH(CH_3)$_2$)—NH_2.

3. Which of the following statements is incorrect?
   A. Glycogen is formed from units that include glucose.
   B. Cellulose is formed from units that include glucose.
   C. Lactose is formed from units that include ribose.
   D. Adenosine is formed from units that include ribose.
   E. Nicotinamide adenine dinucleotide is formed from units that include ribose.

4. Phospholipids
   A. have hydrophilic fatty acid tails.
   B. have charged, hydrophobic head groups.

C. spontaneously form lipid bilayers in an aqueous environment.
D. form droplets within the cytoplasm.
E. all of the above.

5. When an acid is added to pure water
A. it will dissolve easily.
B. it will give up an H^+ to form a negatively charged ion.
C. it will form a solution with a high concentration of H_3O^+ ions.
D. it will form a solution with a low pH.
E. all of the above.

6. Hydrogen bonds cannot form between
A. water and glucose.
B. water and water.
C. water and phosphate.
D. phosphate and octane.
E. phosphate and glucose.

7. Which of these statements about the polypeptides in cells is wrong?
A. They are a polymer of γ-amino acids linked by peptide bonds.
B. The sequence of amino acids is determined by instructions on the cell's DNA.
C. They have an —$NH_3{}^+$ group at one end of the polymer.
D. They have a —COO^- group at one end of the polymer.
E. They are called proteins if they fold into a specific shape.

ANSWERS TO REVIEW QUESTIONS

1. *E.* All these statements about water are true.

2. *E.* A pH of 1.0 is very acid (an H^+ concentration of 100 mmol liter^{-1}), more acid even than 2.3, the pH at which the COOH group is half-protonated. Both the —COOH and NH_2 groups on leucine are therefore protonated at pH 1.0.

If you are still confused, consider the situation at pH 11. At this pH, the concentration of H^+ is extremely low (10^{-11} mol liter^{-1} or 10 pmol liter^{-1}). Leucine therefore loses all its hydrogen ions and has the form $^-$OOC—CH(CH$_2$.CH(CH$_3$)$_2$)—NH$_2$. Now consider adding H^+ so that the pH falls. The NH_2 group is basic and accepts H^+ readily, even when H^+ is present at quite a low concentration. At pH 9.7 (an H^+ concentration of 200 pmol liter^{-1}) half the NH_2 groups have accepted an H^+ to become $NH_3{}^+$, and at pH values less than 9.7 a majority of the NH_2 groups are protonated. This is the situation at neutral pH: the NH_2 groups are protonated while the COOH groups are deprotonated, so leucine has the form $^-$OOC—CH(CH$_2$.CH(CH$_3$)$_2$)—NH$_3{}^+$. If we keep adding H^+, we eventually approach a pH of 2.3, where the H^+ concentration has the high value of 5 mmol liter^{-1}. Even though the COOH groups like to give up H^+, they are forced to accept H^+ when the concentration of H^+ in solution gets this high. At pH 2.3 half the COOH groups are protonated; half are still COO$^-$. At a pH less than 2.3, that is, at H^+ concentrations higher than 5 mmol liter^{-1}, the majority of the COOH groups are protonated and leucine has the form HOOC—CH(CH$_2$.CH(CH$_3$)$_2$)—NH$_3{}^+$.

3. *C.* Lactose is a disaccharide formed from glucose and galactose.

4. *C.* All the other statements are false. In particular: (A) The fatty acid tails are hydrophobic. (B) The head groups are hydrophilic because they are polar and often are charged. (D) Phospholipids will not form droplets because this would place the hydrophilic head group in a hydrophobic environment.

5. *E.* All these statements are true. In particular, the definition of pH means that a solution with a high concentration of H_3O^+ ions is one with a low pH.

6. *D.* Octane is formed of carbon and hydrogen only. It does not contain any of the electron-grabbing atoms that, together with hydrogen, form a hydrogen bond. Octane therefore cannot form hydrogen bonds with other chemicals.

7. *A.* Cellular polypeptides are polymers of α-amino acids, not γ-amino acids. Changing γ to α in this question generates an accurate summary of the biology of polypeptides.

3

MEMBRANES AND ORGANELLES

The vast majority of the reactions that cells carry out take place in water. Eukaryotic cells are, at any one time, carrying out an enormous range of such chemical manipulations: collectively, these reactions are called metabolism (Chapter 13). In much the same way that our homes are divided into rooms that are adapted for particular activities, so eukaryotic cells contain distinct compartments or **organelles** to house specific functions. The term *organelle* is used rather loosely. At one extreme some scientists use it to mean any distinct cellular structure that has a more or less well-defined job to do; at the other are scientists who would reserve the name organelle for those cellular compartments that contain their own DNA and have some limited genetic autonomy. In this book, we will define organelles as those cellular components whose limits, like those of the cell itself, are defined by membranes. It is first necessary to consider some of the fundamental properties of cell membranes.

✳ BASIC PROPERTIES OF CELL MEMBRANES

It is difficult to overstate the importance of membranes to living cells; without them life could not exist. The plasma membrane, also known as the **cell surface membrane** or **plasmalemma,** defines the boundary of the cell. It regulates the movement of materials into and out of the cell and facilitates electrical signaling between cells. Other membranes define the boundaries of organelles and provide a matrix upon which complex chemical reactions can occur. Some of these themes will be developed in subsequent chapters. In the following section the basic structure of the cell membrane will be outlined.

Cell Biology: A Short Course, Second Edition, by Stephen R. Bolsover, Jeremy S. Hyams,
Elizabeth A. Shephard, Hugh A. White, Claudia G. Wiedemann
ISBN 0-471-26393-1 Copyright © 2004 by John Wiley & Sons, Inc.

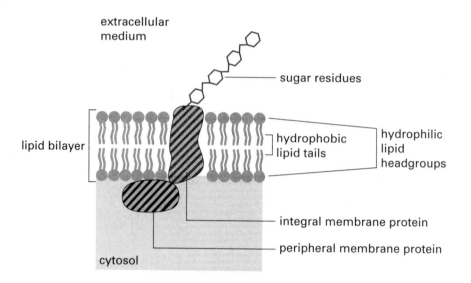

Figure 3.1. Membranes comprise a lipid bilayer plus integral and peripheral proteins.

The basic structure of a biological membrane is shown in Figure 3.1. Approximately half the mass is phospholipid, which spontaneously organizes to form a lipid bilayer (page 42). All the membranes of the cell, including the plasma membrane, also contain proteins. These may be tightly associated with the membrane and extracted from it only with great difficulty, in which case they are called integral proteins (e.g., the gap junction channel, page 55); or they may be separated with relative ease, in which case they are termed peripheral

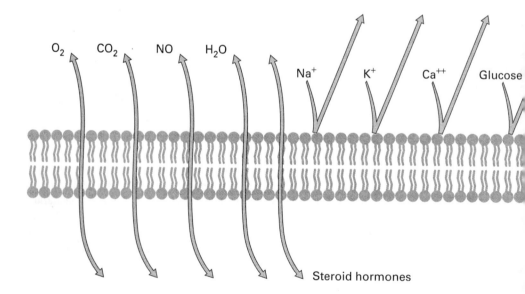

Figure 3.2. Small uncharged molecules can pass through membranes by simple diffusion, but ions cann

proteins (e.g., clathrin adaptor protein, page 229). Membrane proteins are free to move later-ally, within the plane of the membrane. Integral membrane proteins are often **glycosylated**—that is, have sugar residues attached—on the side facing the extracellular medium.

Straight Through the Membrane: Diffusion Through the Bilayer

Molecules of oxygen are uncharged. Although they dissolve readily enough in water, they are also able to dissolve in the hydrophobic interior of lipid bilayers. Oxygen molecules can therefore pass from the extracellular fluid into the interior of the plasma membrane, and from there pass on into the cytoplasm, in a simple diffusion process (Fig. 3.2). Three other small molecules with important roles in biology—carbon dioxide, nitric oxide, and water itself—also pass across the plasma membrane by simple diffusion, as do the uncharged hormones of the steroid family. In contrast, charged ions cannot dissolve in hydrophobic regions (page 21) and therefore cannot cross membranes by simple diffusion.

Example 3.1 **Rapid Diffusion in the Lungs**

Blood is composed of cells, red and white, and plasma, a solution of sodium chloride, other ions, various organic molecules and proteins. Red blood cells are very simple. They have no nucleus and their plasma membrane encloses a cytosol packed with the oxygen-carrying protein hemoglobin, bathed in a salt solution. The sodium concentration in the red blood cytosol is much lower than the sodium concentration in plasma. It is important that the sodium concentration in the cytosol remain low: if it increases, the red blood cells will swell and then burst as water rushes in. As red blood cells pass through the lungs, they quickly gain oxygen because oxygen molecules can pass rapidly across the plasma membrane by simple diffusion. The cells do not, however, gain sodium ions from the plasma as they pass around the body because the lipid bilayer is impermeable to ions. Red blood cells therefore remain the right size to pass though even the tiniest blood vessels in our bodies.

Beyond the Cell Membrane: The Extracellular Matrix

In connective tissue (page 13), the spaces between cells are filled by an extracellular matrix consisting of polysaccharides and proteins such as collagen together with combinations of the two called **proteoglycans.** The proteins provide tensile strength and elasticity while the polysaccharides form a hydrated gel that expands to fill the extracellular space.

The sugar-based extracellular matrix reaches its highest form of expression in the **cell walls** of plants (Fig. 3.3). The plant cell wall consists primarily of cellulose microfibrils linked together by other polysaccharide molecules such as **hemicellulose** and **pectin.** The thickness of the wall is determined largely by its pectin content. In healthy plant cells the plasma membrane is constantly pressing against the cell wall.

In plants the extracellular medium is a much more dilute solution than the intracel-lular solution. If no other forces operated, water would diffuse into plant cells down its concentration gradient, a process called **osmosis.** However, plant cells are prevented from expanding by the inextensible cellulose microfibrils in their cell walls. As water diffuses in, the hydrostatic pressure in the cells rises, making the combination of cell plus cell wall stiff. The same effect can be seen when a bicycle inner tube is blown up inside the tire, and is called turgor. If plants are starved of water, water will leave the cell interior, hydrostatic pressure drops, and the cells lose their turgor pressure, like a tire when the tube inside is deflated. As a result the nonwoody parts of the plant wilt.

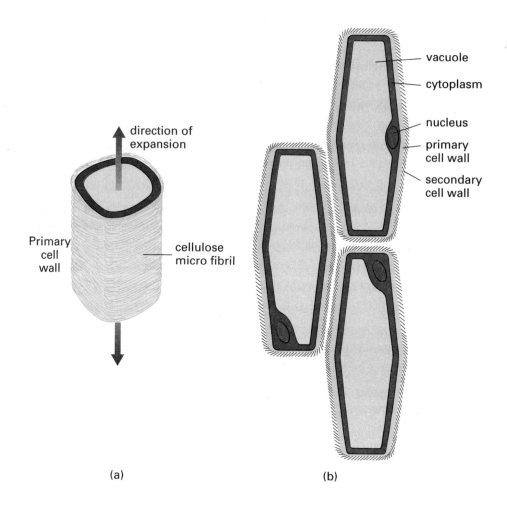

<u>Figure 3.3.</u> The cell wall directs plant cell growth.

Newly generated plant cells first lay down a **primary cell wall** of cellulose microfibrils that are orientated in one direction around the plant cell (Fig. 3.3*a*). Hydrostatic pressure therefore tends to cause a growing plant cell to elongate in one direction, perpendicular to the axis of the microfibrils. Once the growth of the cell is complete, more layers of cellulose and/or other compounds are added, most notably the polyphenolic compound lignin, to form the secondary cell wall (Fig. 3.3*b*).

Cell Junctions

In multicellular organisms, and particularly in epithelia, it is often necessary for neighboring cells within a tissue to be connected together. This function is provided by cell junctions. In animal cells there are three types of junctions. Those that form a tight seal between adjacent cells are known as **tight junctions;** those that allow communication between cells are known as **gap junctions.** A third class of cell junction that anchors cells together, allowing the tissue to be stretched without tearing, are called **anchoring junctions.** Plant cells do not have tight junctions, gap junctions, or anchoring junctions but do contain a unique class of communicating junction known as **plasmodesmata.**

Tight junctions are found wherever flow of extracellular medium is to be restricted and are particularly common in epithelial cells such as those lining the small intestine. The plasma membranes of adjacent cells are pressed together so tightly that no intercellular space exists between them (Fig. 1.8 on page 14). Tight junctions between the epithelial cells of the intestine ensure that the only way that molecules can get from the lumen of the intestine to the blood supply that lies beneath is by passing through the cells, a route that can be selective.

Gap junctions are specialized structures that allow cell-to-cell communication in animals (Fig. 3.4). When two cells form a gap junction, ions and small molecules can pass directly from the cytosol of one cell to the cytosol of the other cell without going into the extracellular fluid. Since ions can move through the junction, changes in electrical voltage

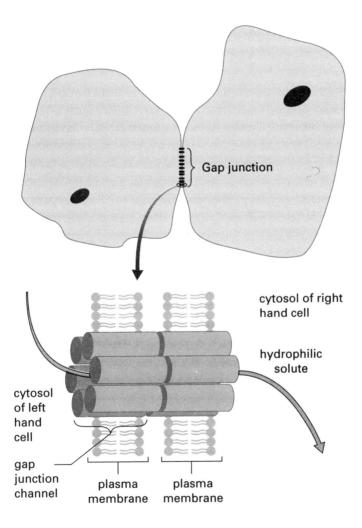

Figure 3.4. Gap junctions allow solute and electrical current to pass from the cytosol of one cell to the cytosol of its neighbor.

are also rapidly transmitted from cell to cell by this route. The structure that makes this possible is the **gap junction channel.** Channels, as we will see in Chapter 14, are water-filled holes through membranes. When two gap junction channels or **connexons** meet, they form a water-filled tube that runs all the way through the plasma membrane of the first cell, across the small gap between the cells, and through the plasma membrane of the second cell. In the middle of the channel is a continuous hole about 1.5 nm in diameter. This hole is large enough to allow small ions through (and therefore to pass electrical current) together with amino acids and nucleotides, but it is too small for proteins or nucleic acids. Gap junctions are especially important in the heart, where they allow an electrical signal to pass rapidly between all the cardiac muscle cells, ensuring that they all contract at the proper time. Each gap junction channel is composed of six protein subunits that can twist against each other to open and close the central channel in a process called gating (page 314) that allows the cell to control the degree to which it shares solute with its neighbor. The plasmodesmata that perforate the cell walls of many plant tissues (Fig. 1.7 on page 10) serve much the same purpose as the gap junctions of animal cells but are much bigger and cannot shut quickly. Some plant viruses use plasmodesmata to spread from cell to cell.

Example 3.2 Gap Junctions Keep Eggs Ready But Waiting

In the days leading up to ovulation, oocytes (cells that undergo meiosis to give rise to eggs) that are to be released from the follicle are kept in a state of suspended development by a chemical in their cytosol called cyclic AMP (page 350). The oocytes themselves do not make cAMP. Rather, the follicle cells that surround them in the ovary make cAMP, which then passes through gap junctions into the oocyte.

Only when the oocyte is released and begins its passage down the Fallopian tube does it prepare for fertilization. In particular, the process of meiosis (page 404) is completed so as to remove half the chromosomes, so that when the sperm adds its complement of chromosomes the fertilized egg will have the correct number.

Anchoring junctions bind cells tightly together and are found in tissues such as the skin and heart that are subjected to mechanical stress. These junctions are described later (page 396).

✸ ORGANELLES BOUNDED BY DOUBLE-MEMBRANE ENVELOPES

Three of the major cell organelles, the nucleus, mitochondrion, and, in plant cells, the chloroplast, share two distinctive features. They are all enclosed within an envelope consisting of two parallel membranes and they all contain the genetic material DNA.

The Nucleus

The nucleus is often the most prominent cell organelle. It contains the **genome,** the cell's database, which is encoded in molecules of the nucleic acid, DNA. The nucleus is bounded by a nuclear envelope composed of two membranes separated by an intermembrane space (Fig. 3.5). The inner membrane of the nuclear envelope is lined by a meshwork of proteins

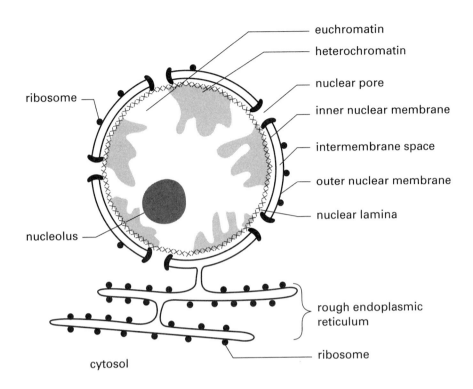

Figure 3.5. The nucleus and the relationship of its membranes to those of the endoplasmic reticulum.

called the nuclear lamina which provides rigidity to the nucleus. A two-way traffic of proteins and nucleic acids between the nucleus and the cytoplasm passes through holes in the nuclear envelope called nuclear pores. The nucleus of a cell that is synthesizing proteins at a low level will have few nuclear pores. In cells that are undergoing active protein synthesis, however, virtually the whole nuclear surface is perforated.

Within the nucleus it is usually possible to recognize discrete areas. Most of it is occupied by chromatin, a complex of DNA and certain DNA-binding proteins such as **histones** (page 73). In most cells it is possible to recognize two types of chromatin; lightly staining **euchromatin** is that portion of the cell's DNA that is being actively transcribed into RNA; **heterochromatin,** on the other hand, represents the inactive portion of the genome where no RNA synthesis is occurring. The DNA in heterochromatin is tightly coiled then supercoiled (see Fig. 4.5 on page 72), leading to its dense appearance in both the light and electron microscopes.

Unlike DNA, RNA is not confined within the nucleus and other organelles but is also found within the cytoplasm associated with particles called ribosomes whose function is to make proteins. Ribosomes are made in the nucleus, in specialized regions called nucleoli that form at specific sites on the DNA called nucleolar organizer regions. These contain blocks of genes that code for the ribosomal RNA.

It should be stressed that the appearance of the nucleus we have described thus far relates to the cell in **interphase,** the period between successive rounds of cell division. As the cell enters mitosis (Chapter 19) the organization of the nucleus changes dramatically. The

DNA becomes more and more tightly packed and is revealed as a number of separate rods called **chromosomes**—46 in human cells. The nucleolus disperses, and the nuclear envelope fragments. Upon completion of mitosis, these structural rearrangements are reversed and the nucleus resumes its typical interphase organization.

Mitochondria and Chloroplasts

Although most of the genetic information of a eukaryotic cell resides in nuclear DNA, some of the information necessary to make both chloroplasts and mitochondria is stored within these organelles themselves. This information resides on small circular DNA molecules that are similar to the chromosomes of bacteria (Table 1.1 on page 11) but very different from the long linear DNA molecules in the nucleus. This is strong evidence for the endosymbiotic theory of the origin of mitochondria and chloroplasts (page 12), which proposes that the small circular DNA molecules found in these organelles is all that is left of the chromosomes of the original symbiotic bacteria. Chloroplasts and mitochondria both contain ribosomes (again, more like those of bacteria than the ribosomes in the cytoplasm of their own cell, Table 1.1 on page 11), and synthesize a small subset of their own proteins, although the great majority of proteins that form chloroplasts and mitochondria are encoded by nuclear genes and synthesized in the cytoplasm. Perhaps the most distinctive feature of these organelles, however, is that the inner of their two membranes is markedly elaborated and folded to increase its surface area.

Mitochondria are among the most easily recognizable organelles due to the extensive folding of their inner membrane to form shelflike projections named cristae (Fig. 3.6). The number of cristae, like the number of mitochondria themselves, depends upon the energy budget of the cell in which they are found. In striated muscle cells, which must contract and relax repeatedly over long periods of time, there are many mitochondria that contain numerous cristae; in fat cells, which generate little energy, there are few mitochondria and their cristae are less well developed. This gives a clue as to the function of mitochondria: they are the cell's power stations. Mitochondria produce the molecule adenosine triphosphate (ATP), one of the cell's energy currencies that provide the energy to drive a host of cellular reactions and mechanisms (Chapter 12). The double-membrane structure provides four distinct domains: the outer membrane, the inner membrane, the intermembrane space, and the matrix. Each domain houses a distinct set of functions, details of which are given in Chapters 12 and 13.

Chloroplasts are found only in photosynthetic protists and plant cells. In addition to the two bounding membranes, they contain internal membranes called thylakoids which, in plants, form stacks called grana (Fig. 3.6). The thylakoids contain the proteins and other molecules responsible for light capture. The dark reactions of photosynthesis (page 302), on the other hand, takes place in the matrix, called the stroma, which also contains the DNA and ribosomes.

✻ ORGANELLES BOUNDED BY SINGLE-MEMBRANE ENVELOPES

Eukaryotic cells contain many sacs and tubes bounded by a single membrane. Although these are often rather similar in appearance, they can be subdivided into different types specialized to carry out distinct functions.

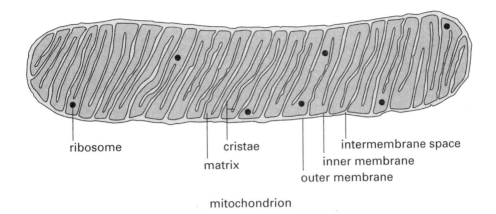

mitochondrion

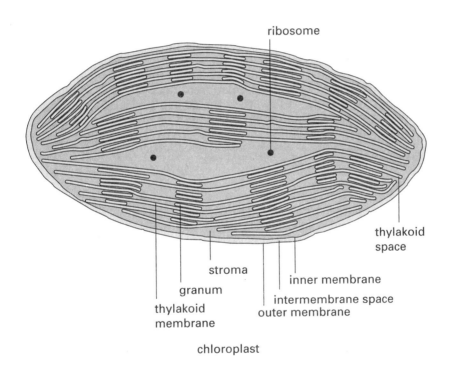

chloroplast

Figure 3.6. Mitochondrion and chloroplast.

Peroxisomes

Mitochondria and chloroplasts are frequently found close to another membrane-bound organelle, the peroxisome. In human cells peroxisomes have a diameter of about 500 nm, and their dense matrix contains a heterogeneous collection of proteins concerned with a variety of metabolic functions, some of which are only now beginning to be understood. Peroxisomes are so named because they are frequently responsible for the conversion of the highly reactive molecule hydrogen peroxide (H_2O_2), which is formed as a by-product of the reactions in the mitochondrion, into water:

$$2H_2O_2 \rightarrow 2H_2O + O_2$$

This reaction is carried out by a protein called catalase, which sometimes forms an obvious crystal within the peroxisome. Catalase is an **enzyme**—a protein catalyst that increases the rate of a chemical reaction (page 241). In fact, it was one of the first enzymes to be discovered. In humans, peroxisomes are primarily associated with lipid metabolism. Understanding peroxisome function is important for a number of inherited human diseases such as X-linked adrenoleukodystrophy where peroxisome malfunction and the consequent inability to metabolise lipid properly typically leads to death in childhood or early adulthood unless dietary lipid is extremely restricted.

●●●● **IN DEPTH 3.1 Nobel Prizes for Organelle Researchers**

In addition to Camillo Golgi's Nobel prize in 1906, modern studies of cell organelles have also been recognized by the Nobel committee. George Palade, Albert Claude, and Christian de Duve shared the 1974 Nobel prize in physiology or medicine for their work on the identification and isolation of cell organelles. Günter Blobel, a former student of Palade's, was awarded the same prize in 1999 for the discovery that proteins carry so-called targeting sequences that determine where in the cell they should reside (page 215).

Endoplasmic Reticulum

The endoplasmic reticulum is a network of membrane enclosed channels that run throughout the cell, forming a continuous network whose lumen (inside) is at all points separated from the cytosol by a single membrane. The membrane of the endoplasmic reticulum is continuous with the outer nuclear membrane (Fig. 3.5). Two regions can be recognized in most cells, known as smooth endoplasmic reticulum and rough endoplasmic reticulum (Fig. 1.2 on page 3). The basic difference is that the rough endoplasmic reticulum is covered in ribosomes, which give it its rough appearance in the electron microscope.

The function of the smooth endoplasmic reticulum varies from tissue to tissue. In the ovaries, testes, and the adrenal gland it is where steroid hormones are made; in the liver it is the site of detoxication of foreign chemicals including drugs. Probably the most universal role of the smooth endoplasmic reticulum is the storage and sudden release of calcium ions. Calcium ions are pumped from the cytosol into the lumen of the smooth endoplasmic reticulum to more than 100 times the concentration found in the cytosol. Many stimuli can cause this calcium to be released back into the cytosol, where it activates many cell processes (Chapter 16).

The rough endoplasmic reticulum is where the cells make the proteins that will end up as integral membrane proteins in the plasma membrane, and proteins that the cell will export to the extracellular medium (such as the proteins of the extracellular matrix, page 13).

Golgi Apparatus

The Golgi apparatus, named after its discoverer, 1906 Nobel prize winner Camillo Golgi, is a distinctive stack of flattened sacks called **cisternae.** The Golgi apparatus is the distribution point of the cell where proteins made within the rough endoplasmic reticulum are further processed and then directed to their final destination.

Appropriately, given this central role, the Golgi apparatus is situated at the so-called **cell center,** a point immediately adjacent to the nucleus that is also occupied by a structure called the centrosome (Fig. 1.2 on page 3). The centrosome helps to organize the cytoskeleton (Chapter 18).

Lysosomes

Lysosomes are sometimes called "cell stomachs" because they contain enzymes that digest cellular components. They are particularly plentiful in cells that digest and destroy other cells, such as macrophages. Lysosomes are roughly spherical and usually 250–500 nm in diameter.

SUMMARY

1. Membranes are made of phospholipids and protein.

2. Cells are bounded by membranes while cell functions are compartmentalized into membrane-bound organelles.

3. Solutes with a significant solubility in hydrophobic solvents can pass across biological membranes by simple diffusion.

4. The extracellular matrix is found on the outside of animal cells. A cellulose-based cell wall is found outside plant cells.

5. Turgor pressure gives nonlignified plant tissue its stiffness.

6. Tight junctions prevent the passage of extracellular water or solute between the cells of an epithelium.

7. Gap junctions allow solute and electrical current to pass from the cytosol of one cell to the cytosol of its neighbor.

8. The nucleus, mitochondrion, and chloroplast are bounded by double-membrane envelopes. In the case of the nucleus, this is perforated by nuclear pores. All three organelles contain DNA.

9. Mitochondria produce the energy currency adenosine triphosphate (ATP).

10. Chloroplasts capture light energy for use by plant cells.

11. Peroxisomes carry out a number of reactions including the destruction of hydrogen peroxide.

12. The endoplasmic reticulum is the site of much protein synthesis. Cell stimulation can often cause calcium ions stored in the endoplasmic reticulum to be released into the cytosol.

13. The Golgi apparatus is concerned with the modification of proteins after they have been synthesized.

14. Lysosomes contain powerful degradative enzymes.

❋ REVIEW QUESTIONS

For each question, choose the ONE BEST answer or completion.

1. Biological membranes
   A. contain phospholipid.
   B. contain protein.
   C. are relatively impermeable to ionic solutes.
   D. are relatively permeable to small uncharged solutes.
   E. all of the above.

2. The extracellular matrix
   A. can contain collagen.
   B. can contain phospholipid.
   C. can contain RNA.
   D. is relatively impermeable to ionic solutes.
   E. all of the above.

3. Which of the following statements about gap junctions is incorrect?
   A. They are formed from six subunits.
   B. They are found in both plants and animals.
   C. They allow small solute molecules to pass between the cytosol of one cell and the cytosol of its neighbor.
   D. They allow electrical current to pass between the cytosol of one cell and the cytosol of its neighbor.
   E. They can switch (gate) to a closed state.

4. A cell with a high density of nuclear pores is likely to be
   A. digesting and destroying other cells.
   B. using lots of the energy currency ATP.
   C. synthesizing lots of protein.
   D. communicating electrically with its neighbors.
   E. undergoing or soon to enter mitosis.

5. In diffusing from the mitochondrial matrix of one animal cell to the interior of the Golgi apparatus of another cell 50 μm away in the same animal a molecule of CO_2 would cross a minimum of
   A. two membranes.
   B. three membranes.
   C. four membranes.
   D. five membranes.
   E. six membranes.

6. One organelle often found at the cell center is the
   A. nucleus.
   B. peroxisome.
   C. endoplasmic reticulum.
   D. Golgi apparatus.
   E. lysosome.

7. A stimulated cell will often release calcium ions from its
   A. nucleus.
   B. endoplasmic reticulum.

    C. mitochondria.
    D. lysosomes.
    E. peroxisomes.

8. In addition to being found free in the cytoplasm, ribosomes are found on the outer surface of the
    A. nucleus.
    B. smooth endoplasmic reticulum.
    C. mitochondrion.
    D. chloroplast.
    E. plasma membrane.

ANSWERS TO REVIEW QUESTIONS

1. *E.* All these statements about biological membranes are true.

2. *A.* Collagen is a major constituent of the extracellular matrix. Considering the other answers: B and C are wrong because phospholipid is found in membranes while RNA is intracellular. Answer D is wrong because the extracellular matrix is an open meshwork through which the aqueous extracellular medium and its dissolved solutes can easily pass.

3. *B.* Plants do not have gap junctions, indeed the presence of the cellulose cell wall means that even if the cells produced the protein subunits neighboring cells never approach each other closely enough to form gap junctions.

4. *C.* Proteins are synthesized using the information on RNA that must be exported from the nucleus to the cytoplasm through nuclear pores. None of the other choices is correct. In particular, the nuclear envelope with all its nuclear pores breaks down in cells undergoing mitosis.

5. *D.* The CO_2 must diffuse across the inner mitochondrial membrane, the outer mitochondrial membrane, the plasma membrane of the first cell, the plasma membrane of the second cell, and the membrane of the Golgi apparatus.

6. *D.* The cell center is occupied by the centrioles (which are not organelles because they are not membrane-bound) and by the Golgi apparatus.

7. *B.* The vast majority of cells are able to release calcium ions from the endoplasmic reticulum when stimulated. The nuclear envelope, mitochondria, and lysosomes may be able to release calcium ions in certain more limited circumstances.

8. *A.* Most ribosomes that are not free in the cytoplasm are found on the rough endoplasmic reticulum, but some are found on the outer nuclear membrane, which is continuous with the membrane of the endoplasmic reticulum (see Fig. 3.5). Mitochondria and chloroplasts have their own ribosomes, but these are found inside these organelles, not on the outer surface.

4

DNA STRUCTURE AND THE GENETIC CODE

✲ INTRODUCTION

Our genes are made of **deoxyribonucleic acid** (**DNA**). This remarkable molecule contains all the information necessary to make a cell, and DNA is able to pass on this information when a cell divides. This chapter describes the structure and properties of DNA molecules, the way in which our DNA is packaged into chromosomes, and how the information stored within DNA is retrieved via the genetic code.

✲ THE STRUCTURE OF DNA

Deoxyribonucleic acid is an extremely long polymer made from units called deoxyribonucleotides, which are often simply called nucleotides. These nucleotides differ from those described in Chapter 2 in one respect: the sugar is deoxyribose, not ribose. Figure 4.1 shows one deoxyribonucleotide, deoxyadenosine triphosphate. Note that deoxyribose, unlike ribose (page 29), has no OH group on its $2'$ carbon. Four bases are found in DNA; they are the two purines adenine (A) and guanine (G) and the two pyrimidines cytosine (C) and thymine (T) (Fig. 4.2). The combined base and sugar is known as a nucleoside to distinguish it from the phosphorylated form, which is called a nucleotide. Four different nucleotides join to make DNA. They are $2'$-deoxyadenosine-$5'$-triphosphate (dATP),

Cell Biology: A Short Course, Second Edition, by Stephen R. Bolsover, Jeremy S. Hyams,
Elizabeth A. Shephard, Hugh A. White, Claudia G. Wiedemann
ISBN 0-471-26393-1 Copyright © 2004 by John Wiley & Sons, Inc.

Example 4.1 **Gift from the Dead**

Deoxyribonucleic acid was known to contain genetic information several years before its structure was finally determined. As early as 1928 Fred Griffith carried out the now famous pneumococcus transformation experiment. There are several strains of the bacterium *Diplococcus pneumoniae*. Some strains cause pneumonia and are said to be virulent; others do not and are nonvirulent. The virulent bacteria possess a polysaccharide coat and, when grown on an agar plate, create colonies with a smooth appearance—these are S bacteria. Nonvirulent bacteria do not have this polysaccharide coat, and their colonies are rough in appearance—these are R bacteria. Mice injected with S bacteria developed pneumonia and died, whereas those injected with R bacteria were unaffected. When S bacteria were killed by heat treatment before injection, the mice remained healthy. However, those mice injected with a mixture of live R bacteria and heat-killed S bacteria died of pneumonia. This observation meant that something in the heat-killed S bacteria carried the information that enabled bacteria to make the polysaccharide coat and therefore to change—transform—the R bacteria into a virulent strain. Later experiments in the mid-1940s by Oswald Avery, Maclyn McCarty, and Colin MacLeod clearly demonstrated that the transforming factor was DNA. They made extracts of S bacteria and treated them with enzymes that destroy either DNA, RNA, or protein. These extracts were then mixed with R bacteria. Only the extract rich in DNA was able to transform R bacteria into S bacteria.

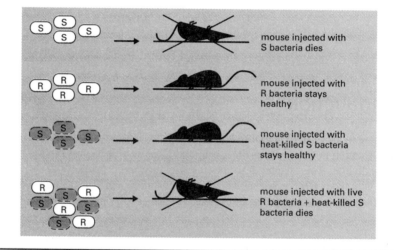

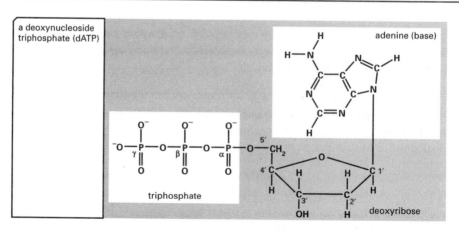

Figure 4.1. 2′-deoxyadenosine-5′-triphosphate.

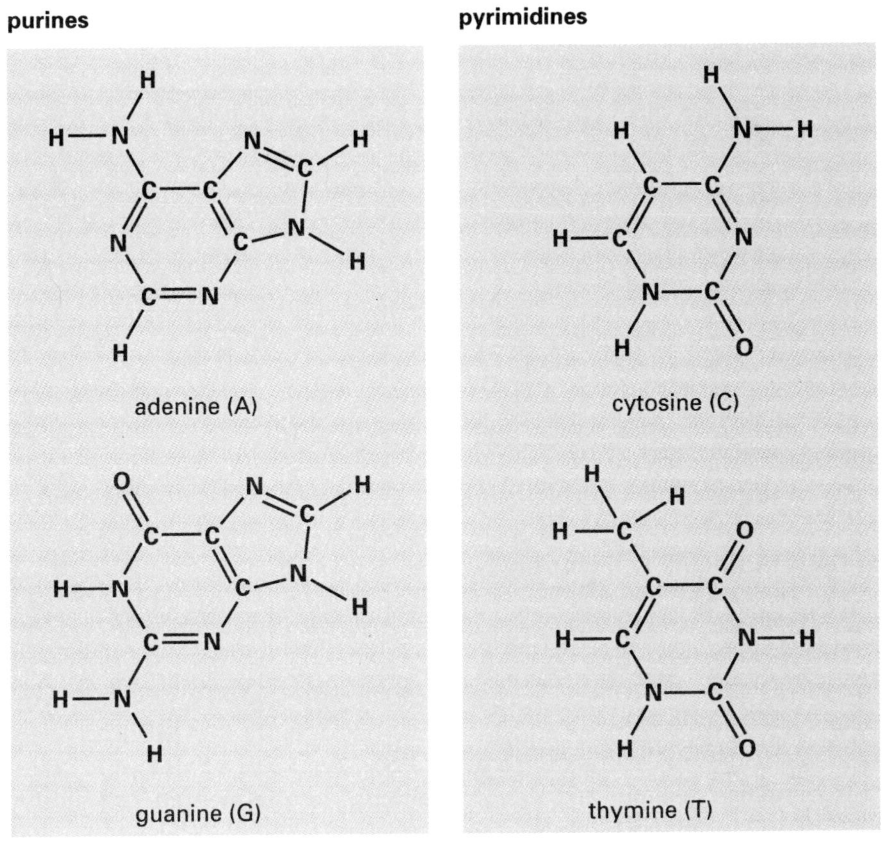

purines   **pyrimidines**

Figure 4.2. The Four bases found in DNA.

2′-deoxyguanosine-5′-triphosphate (dGTP), 2′-deoxycytidine-5′-triphosphate (dCTP), and 2′-deoxythymidine-5′-triphosphate (dTTP).

DNA molecules are very large. The single chromosome of the bacterium *Escherichia coli* is made up of two strands of DNA that are hydrogen-bonded together to form a single circular molecule comprising 9 million nucleotides. Humans have 46 DNA molecules in each cell, each forming one chromosome. We inherit 23 chromosomes from each parent. Each set of 23 chromosomes encodes a complete copy of our **genome** and is made up of 6×10^9 nucleotides (or 3×10^9 **base pairs**—see below). We do not yet know the exact number of genes that encode messenger RNA and therefore proteins in the human genome. The current estimate is in the range of 30,000. Table 4.1 compares the number of predicted messenger RNA genes in the genomes of different organisms. In each organism, there are also approximately 100 genes that code for ribosomal RNAs and transfer RNAs. The role these three types of RNA play in protein synthesis is described in Chapter 8.

Figure 4.3 illustrates the structure of the DNA chain. As nucleotides are added to the chain by the enzyme DNA polymerase (Chapter 5), they lose two phosphate groups. The last (the α phosphate) remains and forms a phosphodiester link between successive deoxyribose residues. The bond forms between the hydroxyl group on the 3′ carbon of the deoxyribose of one nucleotide and the α-phosphate group attached to the 5′ carbon of the next nucleotide. Adjacent nucleotides are hence joined by a 3′–5′ phosphodiester link. The linkage gives rise

Table 4.1. Numbers of Predicted Genes in Various Organisms

Organism	Number of Predicted Genes
Bacterium—*Haemophilus influenzae*	1,709
Yeast—*Saccharomyces cerevisiae*	6,241
Fruit fly—*Drosophila melannogaster*	13,601
Worm—*Caenorhabdites elegans*	18,424
Plant—*Arabidopsis thaliana*	25,498
Human—*Homo sapiens*	~30,000

to the sugar–phosphate backbone of a DNA molecule. A DNA chain has polarity because its two ends are different. In the first nucleotide in the chain, the 5′ carbon of the deoxyribose is phosphorylated but otherwise free. This is called the 5′ end of the DNA chain. At the other end is a deoxyribose with a free hydroxyl group on its 3′ carbon. This is called the 3′ end.

The DNA Molecule Is a Double Helix

In 1953 Rosalind Franklin used X-ray diffraction to show that DNA was a helical (i.e., twisted) polymer. James Watson and Francis Crick demonstrated, by building three-dimensional models, that the molecule is a double helix (Fig. 4.4). Two hydrophilic

IN DEPTH 4.1 Gene Number and Complexity

As the genomes of more and more organisms were sequenced, the most surprising feature to emerge was just how few genes supposedly "complex" organisms possess. The first eukaryotic genome to be sequenced was that of the lowly budding yeast, *Saccharomyces cerevisiae,* the simple unicellular fungus that we use to make bread and beer. *S. cerevisiae* has about 6000 genes. The fruit fly, *Drosophila melanogaster,* a much more complex organism with a brain, nervous and digestive systems, and the ability to fly and navigate, on the other hand, has 13,600 genes, or roughly twice the number in a yeast. Even more surprising was the case of the human genome. Prior to the completion of the Human Genome Project predictions of the number of human genes were in the order of 100,000. Surely this complex vertebrate that could send a spaceship to Mars and write *War and Peace* would need vastly more genes than the fruit fly. In the event, the number of human genes turned out to be much lower than expected, about 30,000, only twice as many genes as in *Drosophila.*

If complex biological and social achievements are not the result of having more genes, where do they come from? Two factors help the human genome generate a more complex organism. First, having more DNA per gene means that more DNA can be used in **enhancer sequences** (page 120), allowing more subtle control of where, when, and to what extent a gene is expressed. Second, there is not a straightforward one-to-one relationship between genes and proteins. **Alternative splicing** (page 118) allows the cell to "cut and paste" a messenger RNA molecule in different ways to produce many different proteins from the same gene. Estimates are that something like 50% of human genes show alternative splicing with the pattern of splicing (the range of proteins produced) varying from tissue to tissue. *Drosophila* genes also show alternative splicing but those of yeast, which contain few **introns** (page 98), do not.

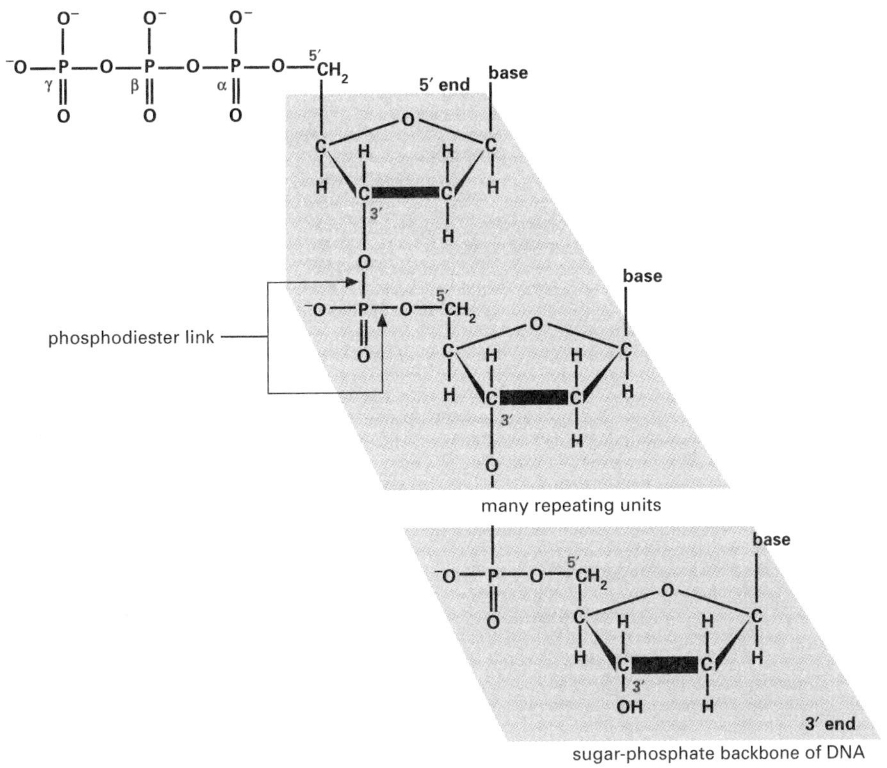

Figure 4.3. The Phosphodiester link and the sugar–phosphate backbone of DNA.

sugar–phosphate backbones lie on the outside of the molecule, and the purines and pyrimidines lie on the inside of the molecule. There is just enough space for one purine and one pyrimidine in the center of the double helix. The Watson–Crick model showed that the purine guanine (G) would fit nicely with the pyrimidine cytosine (C), forming three hydrogen bonds. The purine adenine (A) would fit nicely with the pyrimidine thymine (T), forming two hydrogen bonds. Thus A always pairs with T, and G always pairs with C. The three hydrogen bonds formed between G and C produce a relatively strong base pair. Because only two hydrogen bonds are formed between A and T, this weaker base pair is more easily broken. The difference in strengths between a G–C and an A–T base pair is important in the initiation and termination of RNA synthesis (page 108). The two chains of DNA are said to be antiparallel because they lie in the opposite orientation with respect to one another, with the 3′-hydroxyl terminus of one strand opposite the 5′-phosphate terminus of the second strand. The sugar–phosphate backbones do not completely conceal the bases inside. There are two grooves along the surface of the DNA molecule. One is wide and deep—the major groove—and the other is narrow and shallow—the minor groove (Fig. 4.4). Proteins can use the grooves to gain access to the bases (page 123).

The Two DNA Chains Are Complementary

A consequence of the base pairs formed between the two strands of DNA is that if the base sequence of one strand is known, then that of its partner can be inferred. A **G** in one strand

cytosine-guanine
base pair (C≡G)

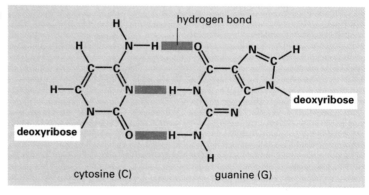

thymine-adenine
base pair (T=A)

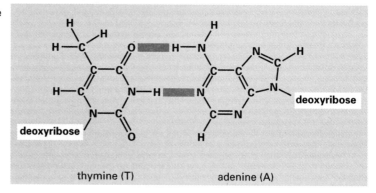

schematic
representation of
the double- helical
structure of DNA

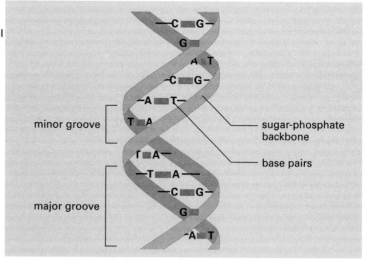

Figure 4.4. The DNA double helix is held together by hydrogen bonds.

will always be paired with a **C** in the other. Similarly an **A** will always pair with a **T**. The two strands are therefore said to be **complementary.**

Example 4.2 **Erwin Chargaff's Puzzling Data**

In a key discovery of the 1950s, Erwin Chargaff analyzed the purine and pyrimidine content of DNA isolated from many different organisms and found that the amounts of A and T were always the same, as were the amounts of G and C. Such an identity was inexplicable at the time, but helped James Watson and Francis Crick build their double-helix model in which every A on one strand of the DNA helix has a matching T on the other strand, and every G on one strand has a matching C on the other.

Different Forms of DNA

The original Watson–Crick model of DNA is now called the B-form. In this form, the two strands of DNA form a right-handed helix. If viewed from either end, it turns in a clockwise direction. B-DNA is the predominant form in which DNA is found. Our genome, however, also contains several variations of the B-form double helix. One of these, Z-DNA, so-called because its backbone has a zig-zag shape, forms a left-handed helix and occurs when the DNA sequence is made of alternating purines and pyrimidines. Thus the structure adopted by DNA is a function of its base sequence.

✳ DNA AS THE GENETIC MATERIAL

Deoxyribonucleic acid carries the genetic information encoded in the sequence of the four bases—adenine, guanine, cytosine, and thymine. The information in DNA is transferred to its daughter molecules through replication (the duplication of DNA molecules) and subsequent cell division. DNA directs the synthesis of proteins through the intermediary molecule RNA. The DNA code is transferred to RNA by a process known as transcription (Chapter 6). The RNA code is then translated into a sequence of amino acids during protein synthesis (Chapter 8). This is the central dogma of molecular biology: DNA makes RNA makes protein.

Retroviruses such as human immunodeficiency virus, the cause of AIDS, are an exception to this rule. As their name suggests, they reverse the normal order of data transfer. Inside the virus coat is a molecule of RNA plus an enzyme that can make DNA from an RNA template by the process known as **reverse transcription.**

✳ PACKAGING OF DNA MOLECULES INTO CHROMOSOMES

Eukaryotic Chromosomes and Chromatin Structure

A human cell contains 46 chromosomes (23 pairs), each of which is a single DNA molecule bundled up with various proteins. On average, each human chromosome contains about 1.3×10^8 base pairs (bp) of DNA. If the DNA in a human chromosome were stretched as

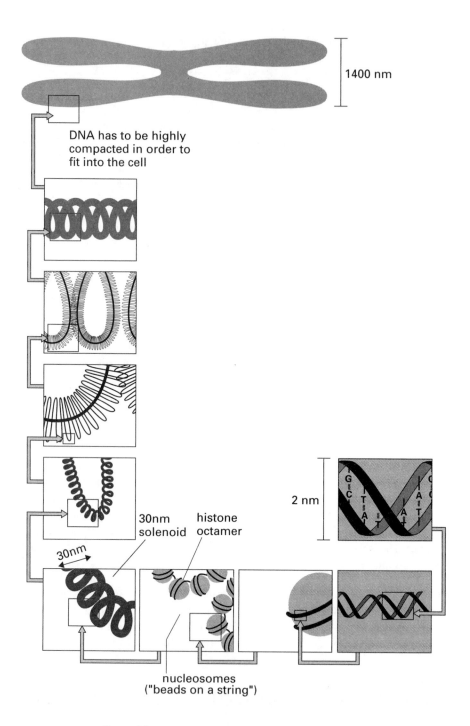

DNA has to be highly
compacted in order to
fit into the cell

1400 nm

2 nm

30nm
solenoid

histone
octamer

30nm

nucleosomes
("beads on a string")

Figure 4.5. How DNA is packaged into chromosomes.

far as it would go without breaking it would be about 5 cm long, so the 46 chromosomes in all represent about 2 m of DNA. The nucleus in which this DNA must be contained has a diameter of only about 10μm, so large amounts of DNA must be packaged into a small space. This represents a formidable problem that is dealt with by binding the DNA to proteins to form chromatin. As shown in Figure 4.5, the DNA double helix is packaged at both small and larger scales. In the first stage, shown on the right of the figure, the DNA double helix with a diameter of 2 nm is bound to proteins known as **histones.** Histones are positively charged because they contain high amounts of the amino acids arginine and lysine (page 185) and bind tightly to the negatively charged phosphates on DNA. A 146 bp length of DNA is wound around a protein complex composed of two molecules each of four different histones—H2A, H2B, H3, and H4—to form a **nucleosome.** Because each nucleosome is separated from its neighbor by about 50 bp of linker DNA, this unfolded chromatin state looks like beads on a string when viewed in an electron microscope. Nucleosomes undergo further packaging. A fifth type of histone, H1, binds to the linker DNA and pulls the nucleosomes together helping to further coil the DNA into chromatin fibers 30 nm in diameter, which are referred to as 30-nm solenoids. The fibers then form loops with the help of a class of proteins known as nonhistones, and this further condenses the DNA (panels on left-hand side of Fig. 4.5) into a higher order set of coils in a process called **supercoiling.**

In a normal interphase cell about 10% of the chromatin is highly compacted and visible under the light microscope (page 57). This form of chromatin is called heterochromatin and is the portion of the genome where no RNA synthesis is occurring. The remaining interphase chromatin is less compacted and is known as euchromatin.

Chromatin is in its most compacted form when the cell is preparing for mitosis, as shown at the top left of Figure 4.5. The chromatin folds and condenses further to form the 1400-nm-wide chromosomes we see under the light microscope. Because the cell is to divide, the DNA has been replicated, so that each chromosome is now formed by two chromatids, each one a DNA double helix. This means the progeny cell, produced by division of the progenitor cell, will receive a full set of 46 chromosomes. Figure 4.6 is a photograph of human chromosomes as they appear at cell division.

Prokaryotic Chromosomes

The chromosome of the bacterium *E. coli* is a single circular DNA molecule of about 4.5×10^6 base pairs. It has a circumference of 1 mm, yet must fit into the 1-μm cell, so like eukaryotic chromosomes it is coiled, supercoiled, and packaged with basic proteins that are similar to eukaryotic histones. However, an ordered nucleosome structure similar to the "beads on a string" seen in eukaryotic cells is not observed in prokaryotes. Prokaryotes do not have nuclear envelopes so the condensed chromosome together with its associated proteins lies free in the cytoplasm, forming a mass that is called the **nucleoid** to emphasize its functional equivalence to the eukaryotic nucleus.

Plasmids

Plasmids are small circular minichromosomes found in bacteria and some eukaryotes. They are several thousand base pairs long and are probably tightly coiled and supercoiled inside the cell. Plasmids often code for proteins that confer resistance to a particular antibiotic. In Chapter 7 we describe how plasmids are used by scientists and genetic engineers to artificially introduce foreign DNA molecules into bacterial cells.

Viruses

Viruses (page 11) rely on the host cell to make more virus. Once viruses have entered cells, the cells' machinery is used to copy the viral genome. Depending on the virus type, the genome may be single- or double-stranded DNA, or even RNA. A viral genome is packaged within a protective protein coat. Viruses that infect bacteria are called **bacteriophages.** One of these, called lambda, has a fixed-size DNA molecule of 4.5×10^4 base pairs. In contrast, the bacteriophage M13 can change its chromosome size, its protein coat expanding in parallel to accommodate the chromosome. This makes M13 useful in genetic engineering (Chapter 7).

IN DEPTH 4.2 DNA—A Gordian Knot

At the start of his career Alexander the Great was shown the Gordian Knot, a tangled ball of knotted rope, and told that whoever untied the knot would conquer Asia. Alexander cut through the knot with his sword. A similar problem occurs in the nucleus, where the 46 chromosomes form 2 m of tangled, knotted DNA. How does the DNA ever untangle at mitosis? The cell adopts Alexander's solution—it cuts the rope. At any place where the DNA helix is under strain, for instance, where two chromosomes press against each other, an enzyme called **topoisomerase II** cuts one chromosome double helix so that the other can pass through the gap. Then, surpassing Alexander, the enzyme rejoins the cut ends. Topoisomerases are active all the time in the nucleus, relieving any strain that develops in the tangled mass of DNA.

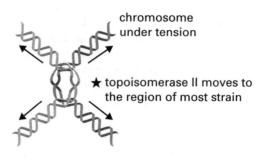

chromosome under tension

★ topoisomerase II moves to the region of most strain

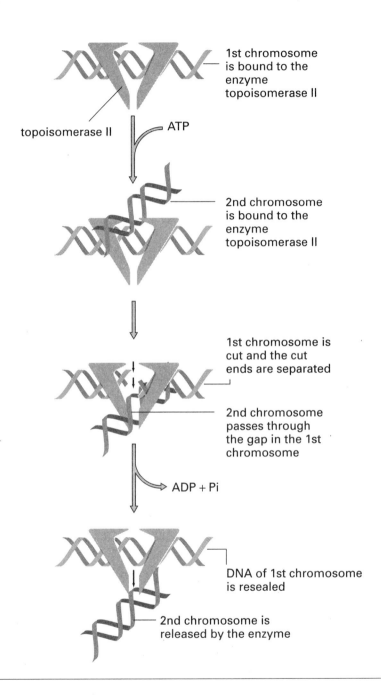

1st chromosome is bound to the enzyme topoisomerase II

topoisomerase II

ATP

2nd chromosome is bound to the enzyme topoisomerase II

1st chromosome is cut and the cut ends are separated

2nd chromosome passes through the gap in the 1st chromosome

ADP + Pi

DNA of 1st chromosome is resealed

2nd chromosome is released by the enzyme

❊ THE GENETIC CODE

Proteins are linear polymers of individual building blocks called amino acids (page 37). The sequence of bases along the DNA strand determines the sequence of the amino acids in proteins. There are 20 different amino acids in proteins but only 4 different bases in DNA (A, T, C, and G). Each amino acid is specified by a **codon,** a group of three bases. Because there are 4 bases in DNA, a three-letter code gives 64 (4 × 4 × 4) possible codons. These 64 codons form the genetic code—the set of instructions that tells a cell the order in which amino acids are to be joined to form a protein. Despite the fact that the sequence of codons on DNA determines the sequence of amino acids in proteins, the DNA helix does not itself play a role in protein synthesis. The **translation** of the sequence from codons into amino acids occurs through the intervention of members of a third class of molecule—messenger RNAs (mRNA) (Fig. 4.7). Messenger RNA acts as a template, guiding the assembly of amino acids into a polypeptide chain. Messenger RNA uses the same code as the one used in DNA with one difference: In mRNA the base uracil (U) (Fig. 2.13 on page 34) is used in place of thymine (T). When we write the genetic code, we usually use the RNA format, that is, we use U instead of T.

The code is read in sequential groups of three, codon by codon. Adjacent codons do not overlap, and each triplet of bases specifies one particular amino acid. This discovery

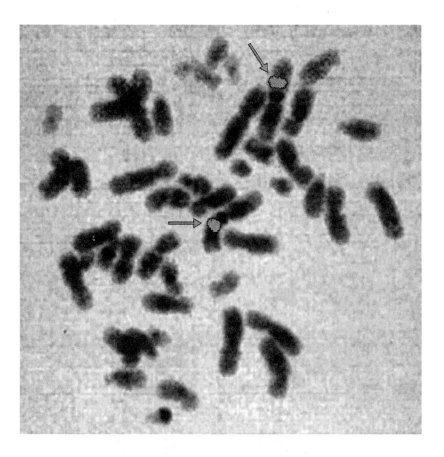

Figure 4.6. A spread of human chromosomes (at metaphase—see page 403). The green signal reveals the gene called *FMO3*, which when mutated causes trimethylaminuria (fish odor syndrome) (Medical Relevance 4.1). There are two copies of the gene, one inherited from each parent, indicated by the arrows. The *FMO3* gene is located on the long arm of chromosome 1, the longest human chromosome.

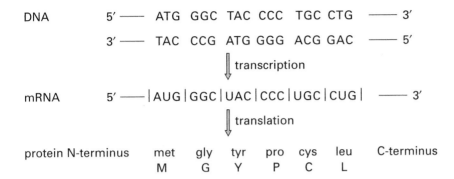

DNA 5' —— ATG GGC TAC CCC TGC CTG —— 3'
3' —— TAC CCG ATG GGG ACG GAC —— 5'

transcription

mRNA 5' ——|AUG|GGC|UAC|CCC|UGC|CUG| —— 3'

translation

protein N-terminus   met   gly   tyr   pro   cys   leu   C-terminus
                       M   G   Y   P   C   L

The 3' to 5' DNA strand is transcribed into an mRNA molecule that is translated, according to the genetic code, into protein. Note that mRNA is made in the 5' to 3' direction and proteins are synthesized from the N terminus. The 3-and 1- letter amino acid codes are shown.

Figure 4.7. The central dogma of molecular biology.

was made by Sydney Brenner, Francis Crick, and their colleagues by studying the effect of various mutations (changes in the DNA sequence) on the bacteriophage T4, which infects the common bacterium *E. coli*. If a mutation caused either one or two nucleotides to be added or deleted, from one end of the T4 DNA, then a defective polypeptide was produced, with a completely different sequence of amino acids. However, if three bases were added or deleted, then the protein made often retained its normal function. These proteins were found to be identical to the original protein, except for the addition or loss of one amino acid.

The identification of the triplets encoding each amino acid began in 1961. This was made possible by using a cell-free protein synthesis system prepared by breaking open *E. coli* cells. Synthetic RNA polymers, of known sequence, were added to the cell-free system together with the 20 amino acids. When the RNA template contained only uridine residues (poly-U) the polypeptide produced contained only phenylalanine. The codon specifying this amino acid must therefore be UUU. A poly-A template produced a polypeptide of lysine, and poly-C one of proline: AAA and CCC must therefore specify lysine and proline, respectively. Synthetic RNA polymers containing all possible combinations of the bases A, C, G, and U, were added to the cell-free system to determine the codons for the other amino acids. A template made of the repeating unit CU gave a polypeptide with the alternating sequence leucine–serine. Because the first amino acid in the chain was found to be leucine, its codon must be CUC and that for serine UCU. Although much of the genetic code was read in this way, the amino acids defined by some codons were particularly hard to determine. Only when specific **transfer RNA** molecules (page 163) were used was it possible to demonstrate that GUU codes for valine. The genetic code was finally solved by the combined efforts of several research teams. The leaders of two of these, Marshall Nirenberg and Gobind Khorana, received the Nobel prize in 1968 for their part in cracking the code.

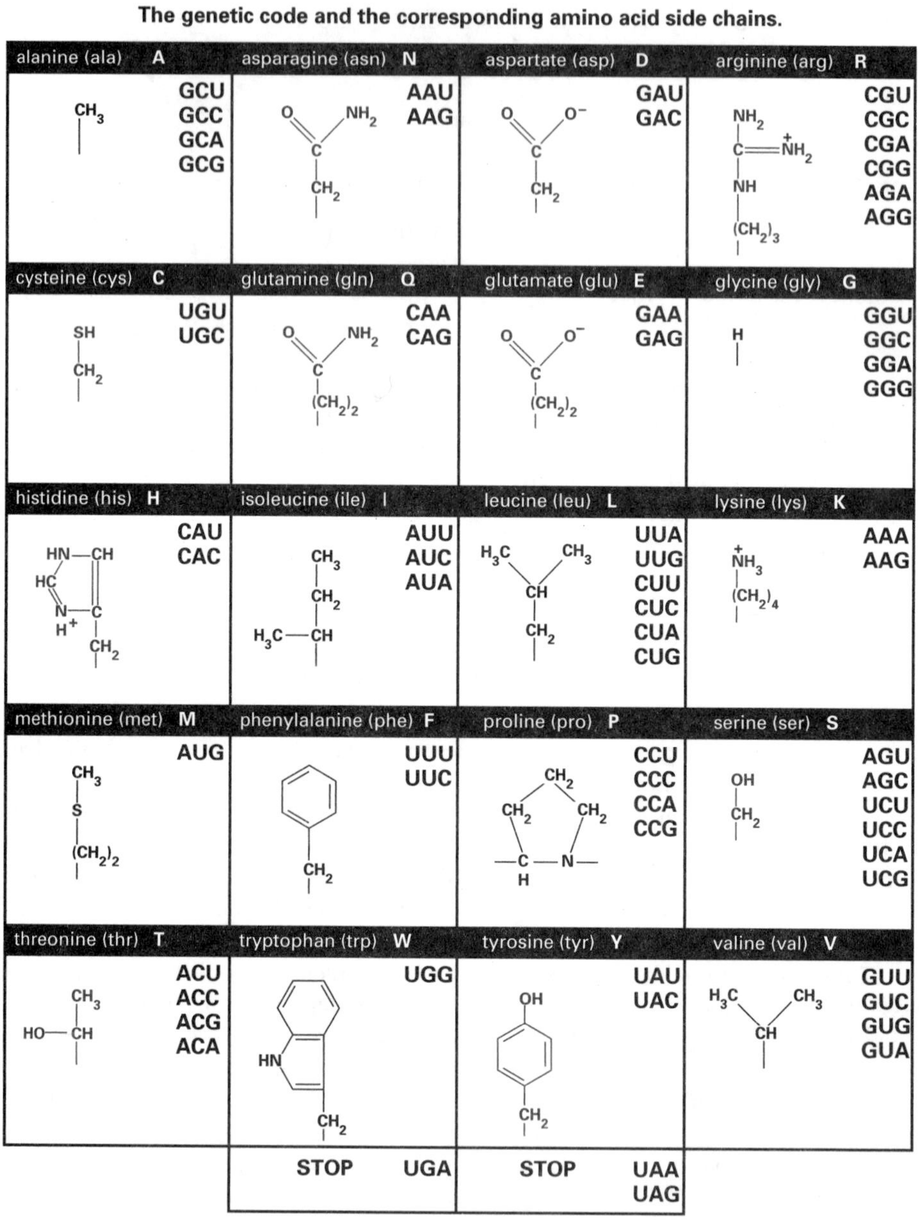

The genetic code and the corresponding amino acid side chains.

Figure 4.8. The genetic code and the corresponding amino acid side chains. The three- and one-letter amino acid abbreviations are shown. Hydrophilic side chains are shown in green, hydrophobic side chains in black. The significance of this distinction is discussed in Chapter 9.

Amino Acid Names Are Abbreviated

To save time we usually write an amino acid as either a three-letter abbreviation, for example, glycine is written as gly and leucine as leu, or as a one-letter code, for example, glycine is G and leucine is L. Figure 4.8 shows the full name, and the three- and one-letter abbreviations, used for each of the 20 amino acids found in proteins.

The Code Is Degenerate But Unambiguous

To introduce the terms *degenerate* and *ambiguous,* consider the English language. English shows considerable degeneracy, meaning that the same concept can be indicated using a number of different words—think, for example, of *lockup, cell, pen, pound, brig,* and *dungeon.* English also shows ambiguity, so that it is only by context that one can tell whether *cell* means *a lockup* or *the basic unit of life.* Like the English language the genetic code shows degeneracy, but unlike language the code is unambiguous.

The 64 codons of the genetic code are shown in Figure 4.8 together with the side chains of the amino acids for which each codes. Amino acids with hydrophilic side chains are shown in green while those with hydrophobic side chains are in black. The importance of this distinction will be discussed in Chapter 9. Sixty-one codons specify an amino acid, and the remaining three act as **stop signals** for protein synthesis. Methionine and tryptophan are the only amino acids coded for by single codons. The other 18 amino acids are encoded by either 2, 3, 4, or 6 codons and so the code is degenerate. No triplet codes for more than one amino acid and so the code is unambiguous. Notice that when two or more codons specify the same amino acid, they usually only differ in the third base of the triplet. Thus mutations can arise in this position of the codon without altering amino acid sequences. Perhaps degeneracy evolved in the triplet system to avoid a situation in which 20 codons each meant one amino acid, and 44 specified none. If this were the case, then most mutations would stop protein synthesis dead.

Start and Stop Codons and the Reading Frame

The order of the codons in DNA and the amino acid sequence of a protein are colinear. The **start signal** for protein synthesis is the codon AUG specifying the incorporation of methionine. Because the genetic code is read in blocks of three, there are three potential **reading frames** in any mRNA. Figure 4.9 shows that only one of these results in the synthesis of the correct protein. When we look at a sequence of bases, it is not obvious which of the reading frames should be used to code for protein. As we shall see later

```
      met    his    glu    tyr
    A U G|C U A|G A A|U A C ... reading frame 1

        cys    stop    asn
    A|U G C|U A G|A A U|A C ... reading frame 2

         ala    arg    ile
    A U|G C U|A G A|A U A|C ... reading frame 3
```

Figure 4.9. Reading frames—the genetic code is read in blocks of three.

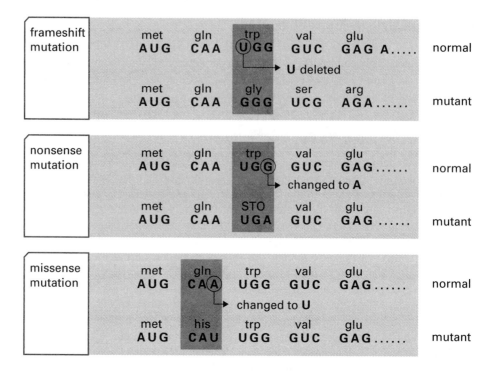

Figure 4.10. Mutations that alter the sequence of bases.

(page 168), the ribosome scans along the mRNA until it encounters an AUG. This both defines the first amino acid of the protein and the reading frame used from that point on. A mutation that inserts or deletes a nucleotide will change the normal reading frame and is called a **frameshift mutation** (Fig. 4.10).

The codons UAA, UAG, and UGA are stop signals for protein synthesis. A base change that causes an amino acid codon to become a **stop codon** is known as a **nonsense mutation** (Fig. 4.10). If, for example, the codon for tryptophan UGG changes to UGA, then a premature stop signal will have been introduced into the messenger RNA template. A shortened protein, usually without function, is produced.

The Code Is Nearly Universal

The code shown in Figure 4.8 is the one used by organisms as diverse as *E. coli* and humans for their nuclear-encoded proteins. It was originally thought that the code would be universal. However, several mitochondrial genes use UGA to mean tryptophan rather than *stop*. The nuclear code for some protists uses UAA and UAG to code for glutamine rather than *stop*.

Missense Mutations

A mutation that changes the codon from one amino acid to that for another is a **missense mutation** (Fig. 4.10). As shown in Figure 4.8, the second base of each codon shows the most consistency with the chemical nature of the amino acid it encodes. Amino acids with charged, hydrophilic side chains usually have A or G—a purine—in the second position. Those with hydrophobic side chains usually have C or U—a pyrimidine—in that position. This has implications for mutations of the second base. Substitution of a purine for a pyrimidine is very likely to change the chemical nature of the amino acid side chain significantly and can therefore seriously affect the protein. Sickle cell anemia is an example of such a mutation. At position 6 in the β-globin chain of hemoglobin the mutation in DNA changes a glutamate residue encoded by GAG to a valine residue encoded by GTG (GUG in RNA). The shorthand notation for this mutation is E6V, meaning that the glutamate (E) at position 6 of the protein becomes a valine (V). This change in amino acid alters the overall charge of the chain, and the hemoglobin tends to precipitate in the red blood cells of those affected. The cells adopt a sickle shape and therefore tend to block blood vessels, causing painful cramplike symptoms and progressive damage to vital organs.

SUMMARY

1. DNA, the cell's database, contains the genetic information necessary to encode RNA and protein.

2. The information is stored in the sequence of four bases. These are the purines, adenine and guanine, and the pyrimidines, cytosine and thymine. Each base is attached to the 1′-carbon atom of the sugar deoxyribose. A phosphate group is attached to the 5′-carbon atom of the sugar. The base + sugar + phosphate is called a nucleotide.

3. The enzyme DNA polymerase joins nucleotides together by forming a phosphodiester link between the hydroxyl group on the 3′ carbon of deoxyribose of one nucleotide and the 5′-phosphate group of another. This gives rise to the sugar–phosphate backbone structure of DNA.

4. The two strands of DNA are held together in a double-helical structure because guanine hydrogen bonds with cytosine and adenine hydrogen bonds with thymine. This means that if the sequence of one strand is known, that of the other can be inferred. The two strands are complementary in sequence.

5. DNA binds to histone and nonhistone proteins to form chromatin. DNA is wrapped around histones to form a nucleosome structure. This is then folded again and again. This packaging compresses the DNA molecule to a size that fits into the cell.

6. The genetic code specifies the sequence of amino acids in a polypeptide. The code is transferred from DNA to mRNA and is read in groups of three bases (a codon) during protein synthesis. There are 64 codons; 61 specify an amino acid and 3 are the stop signals for protein synthesis.

Medical
Relevance
4.1 **Trimethylaminuria**

Trimethylaminuria is an inherited disorder characterized by an unpleasant body odor similar to rotting fish. Mutations in the gene *FMO3,* located on human chromosome 1 (Fig. 4.6) produce a defective version of a protein called flavin-containing monooxygenase 3 (FMO3). One of the functions of this protein, which is found in the liver, is to convert a chemical called trimethylamine (very smelly) to trimethylamine *N*-oxide (nonsmelly). FMO3 carries out this reaction by adding an oxygen onto the nitrogen of trimethylamine. Defective FMO3 protein cannot carry out this reaction. Affected individuals therefore excrete large amounts of trimethylamine in their breath, sweat, and urine. Trimethylamine is the chemical that gives rotting fish its distinctive and unpleasant smell, and this gives rise to the alternative name for trimethylaminuria—fish odor syndrome.

Trimethylamine is derived from our diet in two ways. Fish contain lots of trimethylamine *N*-oxide (page 208). The bacteria that live in our intestines remove the oxygen from the nitrogen of the *N*-oxide to produce trimethylamine. However, one cannot avoid trimethylamine simply by avoiding fish. The phospholipids in the cell membranes of food include phosphatidylcholine, and this is hydrolyzed by our digestive enzymes to release free choline. Gut bacteria then break down the choline further, producing trimethylamine together with acetaldehyde (also called ethanal).

Trimethylaminuria is a difficult disorder to live with. Because of their body odor, sufferers often experience rejection and social isolation. A diet very low in choline, and other chemicals that contain a trimethylamine group, can help to reduce the problem.

Analysis of the faulty genes in different families has revealed different mutations in the *FMO3* gene. Some families carry the missense mutation P153L in which a CCC that codes for a proline as the 153rd amino acid in the normal protein is mutated to CTC (leu), a change that turns out to be critical for the operation of the protein. Other families carry a nonsense mutation in which the GGA (gly) at amino acid 148 in the normal protein, is mutated to TGA, a stop codon. Another family carries a frameshift mutation in which the sequence . . . GGA AAG CGT **G**TC CTG G . . . , which specifies amino acids 188gly lys arg val leu192 in the normal protein, is mutated, by the deletion of the highlighted (bold) G, to . . . GGA AAG CGT TCC TGG . . . , thus generating the amino acid sequence 188gly lys arg **ser trp**192. The frameshift mutation changes not only the amino acid sequence, but also introduces a stop codon further down the mRNA. Because different mutations can give rise to trimethylaminuria, the development of a general diagnostic test for carriers of the faulty gene (who also have one good copy and are therefore carriers who do not themselves have the disorder; page 407) is not possible.

In contrast, a disorder such as sickle cell anemia is caused, in all individuals, by the same mutation (page 81). One test will diagnose the defective gene and prospective parents can be advised and counseled.

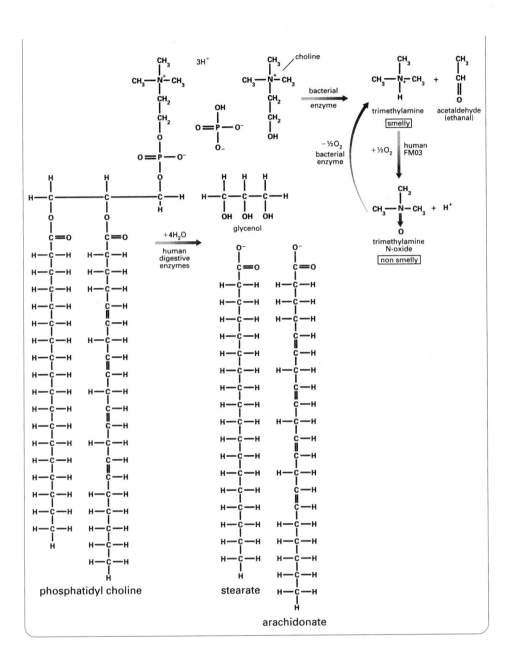

FURTHER READING

DiGuilo, M. 1997. The origin of the genetic code. *Trends Biochem. Sci.* 22: 49–50.

Dolphin, C. T., Janmohamed, A., Smith, R. L., Shephard, E. A., and Phillips, I. R. 1997. Missense mutation in flavin-containing monooxygenase 3 gene, *FMO3,* underlies fish-odour syndrome. *Nature Genetics* 17: 491–494.

Maddox, B. 2002. *Rosalind Franklin: The Dark Lady of DNA*. New York: Harper Collins.

Roca, J. 1995. The mechanisms of DNA topoisomerases. *Trends Biochem. Sci.* 20: 156–160

Watson, J. D., and Crick, F. H. C. 1953. A structure for deoxyribose nucleic acid. *Nature* 171: 737.

Woodcock, C. L., and Horowitz, R. A. 1995. Chromatin organization re-viewed. *Trends Cell Biol.* 5: 272–277.

✹ REVIEW QUESTIONS

For each question, choose the ONE BEST answer or completion.

1. Which base is not found in the DNA of most organisms?
   A. Adenine
   B. Guanine
   C. Cytosine
   D. Uracil
   E. Thymine

2. The bonds that join the nucleotides to form a single strand of DNA are
   A. peptide bonds.
   B. hydrogen bonds.
   C. phosphodiester links.
   D. van der Waals bonds.
   E. disulfide bonds.

3. DNA is packaged into chromatin with the help of the protein called
   A. DNA polymerase I.
   B. histone.
   C. DNA ligase.
   D. topoisomerase I.
   E. RNA.

4. Histone proteins are rich in the amino acids
   A. arginine and asparagine.
   B. leucine and isoleucine.
   C. aspartate and glutamate.
   D. arginine and lysine.
   E. histidine and glutamine.

5. The 20 amino acids found in proteins are coded for by
   A. 59 codons.
   B. 60 codons.
   C. 61 codons.

    D. 63 codons.
    E. 64 codons.

6. Methionine is coded for by
    A. one codon only.
    B. two different codons.
    C. three different codons.
    D. four different codons.
    E. six different codons.

7. A missense mutation is one that
    A. creates a stop codon in the mRNA.
    B. changes the codon of one amino acid to that of another.
    C. changes the reading frame of the mRNA.
    D. creates a break in the DNA strand.
    E. has no effect on the amino acid sequence.

ANSWERS TO REVIEW QUESTIONS

1. *D.* Uracil, which hydrogen bonds with adenine, is present in RNA but is absent from the DNA of most organisms.

2. *C.* The deoxyribose residues of adjacent nucleotides are joined by phosphodiester links.

3. *B.* Histones help to package DNA into chromatin.

4. *D.* Histones bind to negatively charged DNA because they are rich in the positively charged amino acids arginine and lysine.

5. *C.* Of the 64 possible permutations of a triplet of four types of base, 61 code for the 20 amino acids, leaving 3 (UGA, UAG, and UAA) to code for *stop*.

6. *A.* Only AUG codes for methionine, which is the first amino acid found in all newly synthesized proteins of most species.

7. *B.* A missense mutation changes the codon of one amino acid to that of another. A nonsense mutation is one that creates a stop codon and a frameshift mutation is one that changes the reading frame of the mRNA.

5

DNA AS A DATA STORAGE MEDIUM

❋ INTRODUCTION

The genetic material DNA must be faithfully replicated every time a cell divides to ensure that the information encoded in it is passed unaltered to the progeny cells. DNA molecules have to last a long time compared to RNA and protein. The sugar–phosphate backbone of DNA is a very stable structure because there are no free hydroxyl groups on the sugar—they are all used up in bonds, either to the base or to phosphate. The bases themselves are protected from chemical attack because they are hidden within the DNA double helix. Nevertheless, chemical changes—**mutations**—do occur in the DNA molecule, and cells have had to evolve mechanisms to ensure that mutation is kept to a minimum. Repair systems are essential for both cell survival and to ensure that the correct DNA sequence is passed on to daughter cells. This chapter describes how new DNA molecules are made during chromosome duplication and how the cell acts to correct base changes in DNA.

❋ DNA REPLICATION

During replication the two strands of the double helix unwind. Each then acts as a template for the synthesis of a new strand. This process generates two double-stranded daughter DNA molecules, each of which is identical to the parent molecule. The base sequences of the new strands are complementary in sequence to the template strands upon which they were built. This means that G, A, C, and T in the old strand cause C, T, G, and A, respectively, to be placed in the new strand.

Cell Biology: A Short Course, Second Edition, by Stephen R. Bolsover, Jeremy S. Hyams,
Elizabeth A. Shephard, Hugh A. White, Claudia G. Wiedemann
ISBN 0-471-26393-1 Copyright © 2004 by John Wiley & Sons, Inc.

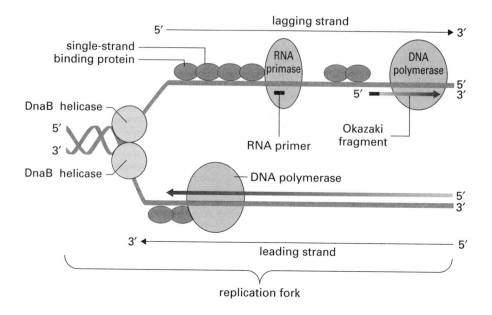

<u>Figure 5.1.</u> DNA replication. The helicases, and the replication fork, are moving to the left.

The DNA Replication Fork

Replication of a new DNA strand starts at specific sequences known as **origins of replication**. The small circular chromosome of *Escherichia coli* has only one of these, whereas eukaryotic chromosomes, which are much larger, have many. At each origin of replication, the parental strands of DNA untwist to give rise to a structure known as the replication fork (Fig. 5.1). This unwinding permits each parental strand to act as a template for the synthesis of a new strand. The structure of the double helix and the nature of DNA replication pose a mechanical problem. How do the two strands unwind and how do they stay unwound so that each can act as a template for a new strand?

✳ PROTEINS OPEN UP THE DNA DOUBLE HELIX DURING REPLICATION

The DNA molecule must be opened up before replication can proceed. The helix is a very stable structure, and in a test tube the two strands separate only when the temperature reaches about 90°C. In the cell the combined actions of several proteins help to separate the two strands. Much of our knowledge of replication comes from studying *E. coli,* but similar systems operate in all organisms, prokaryote and eukaryote. The proteins this bacterium uses to open up the double helix during replication include DnaA, DnaB, DnaC, and single-strand binding proteins.

DnaA Protein

Several copies of the protein DnaA bind to four sequences of nine base pairs within the *E. coli* origin of replication (*ori C*). This causes the two strands to begin to separate (or "melt") because the hydrogen bonds in DNA are broken near to where the DnaA protein binds. The DNA is now in the **open complex** formation and has been prepared for the next stage in replication, which is to open up the helix even further.

Anthrax, Cipro, and the Need to Unwind

The DNA helix must be able to rotate if it is to relieve the rotational stress caused by the formation of the replication fork. This is achieved by an enzyme called topoisomerase I, which breaks a phosphodiester link in one strand of the helix. The DNA helix can now rotate around the remaining unbroken strand to relieve the stress. The phosphodiester link is then reformed. The repeated operation of topoisomerase 1 allows progressive unwinding of the two template strands as replication proceeds.

Topoisomerase I should not be confused with topoisomerase II (page 74), which cuts *both* strands of a DNA double helix to allow a second chromosome to pass through the gap. Topoisomerases II play an essential role during DNA replication in prokaryotes. When the chromosome

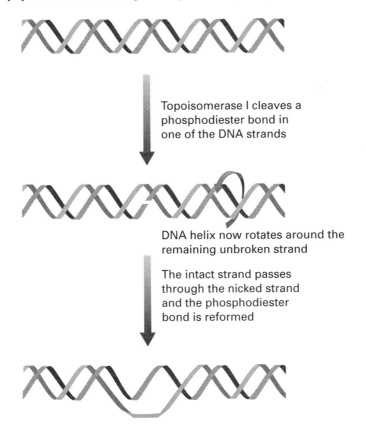

Topoisomerase I cleaves a phosphodiester bond in one of the DNA strands

DNA helix now rotates around the remaining unbroken strand

The intact strand passes through the nicked strand and the phosphodiester bond is reformed

unwinds to form the replication fork, the region of DNA in front of the fork overwinds; the twists cannot dissipate because the prokaryotic chromosome is circular. Topoisomerase II relieves this stress and allows DNA replication to proceed by performing the same action as described on page 000, except that both the DNA double helix that is cut and then resealed, and the double helix that is passed through the gap, belong to the same circular chromosome.

Concerns that a terrorist organization might release large amounts of anthrax spores have caused several governments to stockpile large amounts of the antibiotic Cipro. This works by inhibiting the prokaryotic form of topoisomerase II (sometimes called gyrase) and hence preventing cell replication.

DnaB and DnaC Proteins

DnaB is a **helicase**. It moves along a DNA strand, breaking hydrogen bonds, and in the process unwinds the helix (Fig. 5.1). Two molecules of DnaB are needed, one for each strand of DNA. One DnaB attaches to one of the template strands and moves in the $5'\rightarrow3'$ direction; the second DnaB attaches to the other strand and moves in the $3' \rightarrow 5'$ direction. The unwinding of the DNA double helix by DnaB is an ATP-dependent process. DnaB is escorted to the DNA strands by another protein DnaC. However, having delivered DnaB to its destination, DnaC plays no further role in replication.

Single-Strand Binding Proteins

As soon as DnaB unwinds the two parental strands, they are engulfed by single-strand binding proteins. These proteins bind to adjacent groups of 32 nucleotides. DNA covered by single-strand binding proteins is rigid, without bends or kinks. It is therefore an excellent template for DNA synthesis (Fig. 5.1). Single-strand binding proteins are sometimes called helix-destabilizing proteins.

✤ BIOCHEMISTRY OF DNA REPLICATION

The synthesis of a new DNA molecule is catalyzed by DNA polymerase III. Its substrates are the four deoxyribonucleoside triphosphates, dATP, dCTP, dGTP, and dTTP. DNA polymerase III catalyzes the formation of a phosphodiester link (Fig. 4.3 on page 69) between the 3'-hydroxyl group of one sugar residue and the 5'-phosphate group of a second sugar residue (Fig. 5.2a). The base sequence of a newly synthesized DNA strand is dictated by the base sequence of its parental strand. If the sequence of the template strand is 3' CATCGA 5', then that of the daughter strand is 5' GTAGCT 3'.

DNA polymerase III can only add a nucleotide to a free 3'-hydroxyl group and therefore synthesizes DNA in the 5' to 3' direction. The template strand is read in the 3' to 5' direction. However, the two strands of the double helix are antiparallel. They cannot be synthesized in the same direction because only one has a free 3'-hydroxyl group, the other has a free 5'-phosphate group. No DNA polymerase has been found that can synthesize DNA in the 3' to 5' direction, that is, by attaching a nucleotide to a 5' phosphate, so the synthesis of the two daughter strands must differ. One strand, the **leading strand**, is synthesized continuously while the other, the **lagging strand**, is synthesized discontinuously. DNA polymerase III can synthesize both daughter strands, but must make the lagging strand as a series of short 5' to 3' sections. The fragments of DNA, called **Okasaki fragments** after Reiji Okasaki who discovered them in 1968, are then joined together by DNA polymerase and ligase (Fig. 5.1).

DNA Synthesis Requires an RNA Primer

DNA polymerase III cannot itself initiate the synthesis of DNA. An enzyme called **primase** is needed to catalyze the formation of a short stretch of RNA complementary in sequence to the DNA template strand (Fig. 5.1). This RNA chain, the primer, is needed to prime (or start) the synthesis of the new DNA strand. DNA polymerase III catalyzes the formation of a phosphodiester link between the 3'-hydroxyl group of the RNA primer and the 5'-phosphate group of the appropriate deoxyribonucleotide. Several RNA primers are made along the length of the lagging strand template. Each is extended in the 5' to 3' direction by DNA

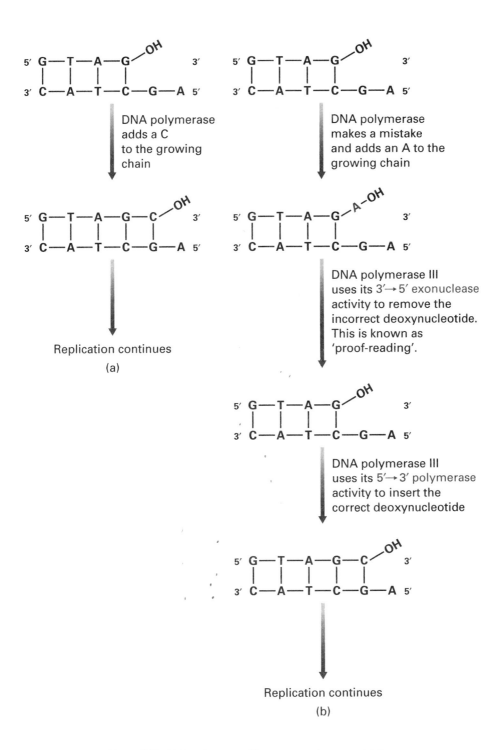

Figure 5.2. DNA polymerase III can correct its own mistakes.

polymerase III until it reaches the $5'$ end of the next RNA primer (Fig. 5.1). In prokaryotes the lagging strand is primed about every 1000 nucleotides whereas in eukaryotes this takes place each 200 nucleotides.

RNA Primers Are Removed

Once the synthesis of the DNA fragment is complete, the RNA primers must be replaced by deoxyribonucleotides. In prokaryotes the enzyme **DNA polymerase I** removes ribonucleotides using its $5' \rightarrow 3'$ **exonuclease** activity and then uses its $5' \rightarrow 3'$ polymerizing activity to incorporate deoxyribonucleotides. In this way, the entire RNA primer gets replaced by DNA. Synthesis of the lagging strand is completed by the enzyme **DNA ligase**, which joins the DNA fragments together by catalyzing formation of phosphodiester links between adjacent fragments.

Eukaryotic organisms probably use an enzyme called **ribonuclease H** to remove their RNA primers. This enzyme breaks phosphodiester links in an RNA strand that is hydrogen-bonded to a DNA strand.

The Self-Correcting DNA Polymerase

The human genome consists of about 3×10^9 base pairs of DNA. DNA polymerase III makes a mistake about every 1 in 10^4 bases and joins an incorrect deoxyribonucleotide to the growing chain. If unchecked, these mistakes would lead to a catastrophic mutation rate. Fortunately, DNA polymerase III has an built-in proofreading mechanism that corrects its own errors. If an incorrect base is inserted into the newly synthesized daughter strand, the enzyme recognizes the change in shape of the double-stranded molecule, which arises through incorrect base pairing, and DNA synthesis stops (Fig. 5.2*b*). DNA polymerase III then uses its $3'$ to $5'$ exonuclease activity to remove the incorrect deoxyribonucleotide and replace it with the correct one. DNA synthesis then proceeds. DNA polymerase III hence functions as a self-correcting enzyme.

Example 5.1   The Meselson–Stahl Experiment

In 1958 Matthew Meselson and Franklin Stahl designed an ingenious experiment to test whether each strand of the double helix does indeed act as a template for the synthesis of a new strand. They grew the bacterium *Escherichia coli* in a medium containing the heavy isotope ^{15}N that could be incorporated into new DNA molecules. After several cell divisions they transferred the bacteria, now containing "heavy" DNA, to a medium containing only the lighter, normal, isotope ^{14}N. Any newly synthesized DNA molecules would therefore be lighter than the original parent DNA molecules containing ^{15}N. The difference in density between the heavy and light DNAs allows their separation using very high speed centrifugation. The results of this experiment are illustrated in the figure. DNA isolated from cells grown in the ^{15}N medium had the highest density and migrated the furthest during centrifugation. The lightest DNA was found in cells grown in the ^{14}N medium for two generations, whereas DNA from bacteria grown for only one generation in the lighter ^{14}N medium had a density half way between these two. This is exactly the pattern expected if each strand of the double helix acts as a template for the synthesis of a new strand. The two heavy parental strands separated during replication, with each acting as a template for a newly synthesized light strand, which remained bound to the heavy strand in a double helix. The resulting DNA was therefore of intermediate density. Only in the second round of DNA replication, when the light strands created during the first round of replication were allowed to act as templates for the construction of complementary light strands, did DNA double helices composed entirely of ^{14}N-containing building blocks appear.

The two scientists were awarded the Nobel prize for this discovery that DNA replication is "semiconservative," meaning that the results are not completely new but are half new and half old.

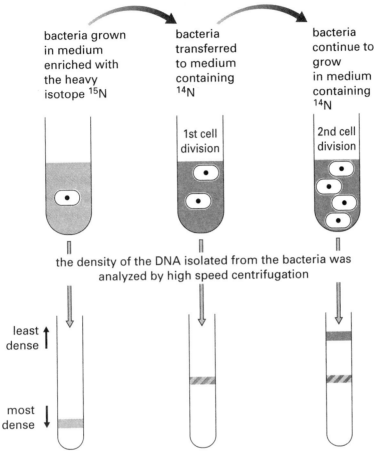

bacteria grown in medium enriched with the heavy isotope ^{15}N

bacteria transferred to medium containing ^{14}N

bacteria continue to grow in medium containing ^{14}N

1st cell division

2nd cell division

the density of the DNA isolated from the bacteria was analyzed by high speed centrifugation

least dense

most dense

This experiment demonstrated that the two strands of the parental DNA unwind and each acts as template for a new daughter strand

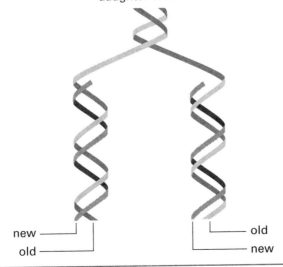

new
old

old
new

✳ DNA REPAIR

Deoxyribonucleic Acid can be damaged by a number of agents, which include oxygen, water, naturally occurring chemicals in our diet, and radiation. Because damage to DNA can change the sequence of bases, a cell must be able to repair alterations in the DNA code if it is to survive and pass on the DNA database unaltered to its progeny cells.

Spontaneous and Chemically Induced Base Changes

The most common damage suffered by a DNA molecule is **depuration**—the loss of an adenine or guanine because the bond between the purine base and the deoxyribose sugar to which it is attached is hydrolyzed by water (Fig. 5.3). Within each human cell about 5000–10,000 depurations occur every day.

Deamination is a less frequent event; it happens about 100 times a day in every human cell. Collision of water molecules with the bond linking the amine group to cytosine sets off a spontaneous deamination that produces uracil (Fig. 5.3). Cytosine base pairs with guanine, whereas uracil pairs with adenine. If this change were not corrected, then a CG base pair would mutate to a UA base pair the next time the DNA strand was replicated.

Ultraviolet light or chemical carcinogens such as benzopyrene, found in cigarette smoke, can also disrupt the structure of DNA. The absorption of ultraviolet light can cause two adjacent thymine residues to link and form a thymine dimer (Fig. 5.4). If uncorrected, thymine dimers create a distortion in the DNA helix known as a **bulky lesion**. This inhibits normal base pairing between the two strands of the double helix and blocks the replication process. Ultraviolet light has a powerful germicidal action and is widely used to sterilize equipment. One of the reasons why bacteria are killed by this treatment is because the formation of large numbers of thymine dimers prevents replication.

Repair Processes

If there were no way to correct altered DNA, the rate of mutation would be intolerable. **DNA excision** and **DNA repair enzymes** have evolved to detect and to repair altered DNA. The role of the repair enzymes is to cut out (excise) the damaged portion of DNA and then to repair the base sequence. Much of our knowledge on DNA repair has been derived from studies on E. coli, but the general principles apply to other organisms such as ourselves. Repair is possible because DNA comprises two complementary strands. If the repair mechanisms can identify which of the two strands is the damaged one, it can then be repaired as good as new by rebuilding it to be complementary to the undamaged one.

Two types of excision repair are described in this section: base excision repair and nucleotide excision repair. The common themes of each of these repair mechanisms are: (1) An enzyme recognizes the damaged DNA, (2) the damaged portion is removed, (3) DNA polymerase inserts the correct nucleotide(s) into position (according to the base sequence of the second DNA strand), and (4) DNA ligase joins the newly repaired section to the remainder of the DNA strand.

Base excision–repair is needed to repair DNAs that have lost a purine (depuration), or where a cytosine has been deaminated to uracil (U). Although uracil is a normal constituent of RNA, it does not form part of undamaged DNA and is recognized and removed by the

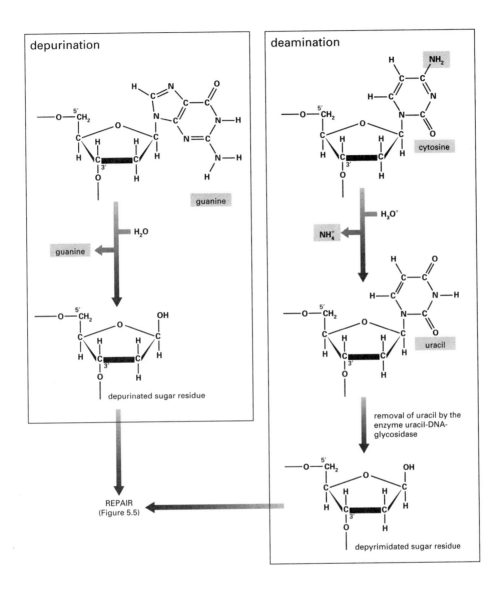

Figure 5.3. Spontaneous reactions corrupt the DNA database.

repair enzyme **uracil–DNA glycosidase** (Fig. 5.3). This leaves a gap in the DNA where the base had been attached to deoxyribose. There is no enzyme that can simply reattach a C into the vacant space on the sugar. Instead, an enzyme called AP endonuclease recognizes the gap and removes the sugar by breaking the phosphodiester links on either side (Fig. 5.5). When DNA has been damaged by the loss of a purine (Fig. 5.3), AP endonuclease also removes the sugar that has lost its base. The AP in the enzyme's name means *ap*yrimidinic (without a pyrimidine) or *ap*urinic (without a purine).

The repair process for reinserting a purine, or a pyrimidine, into DNA is now the same (Fig. 5.5). DNA polymerase I replaces the appropriate deoxyribonucleotide into

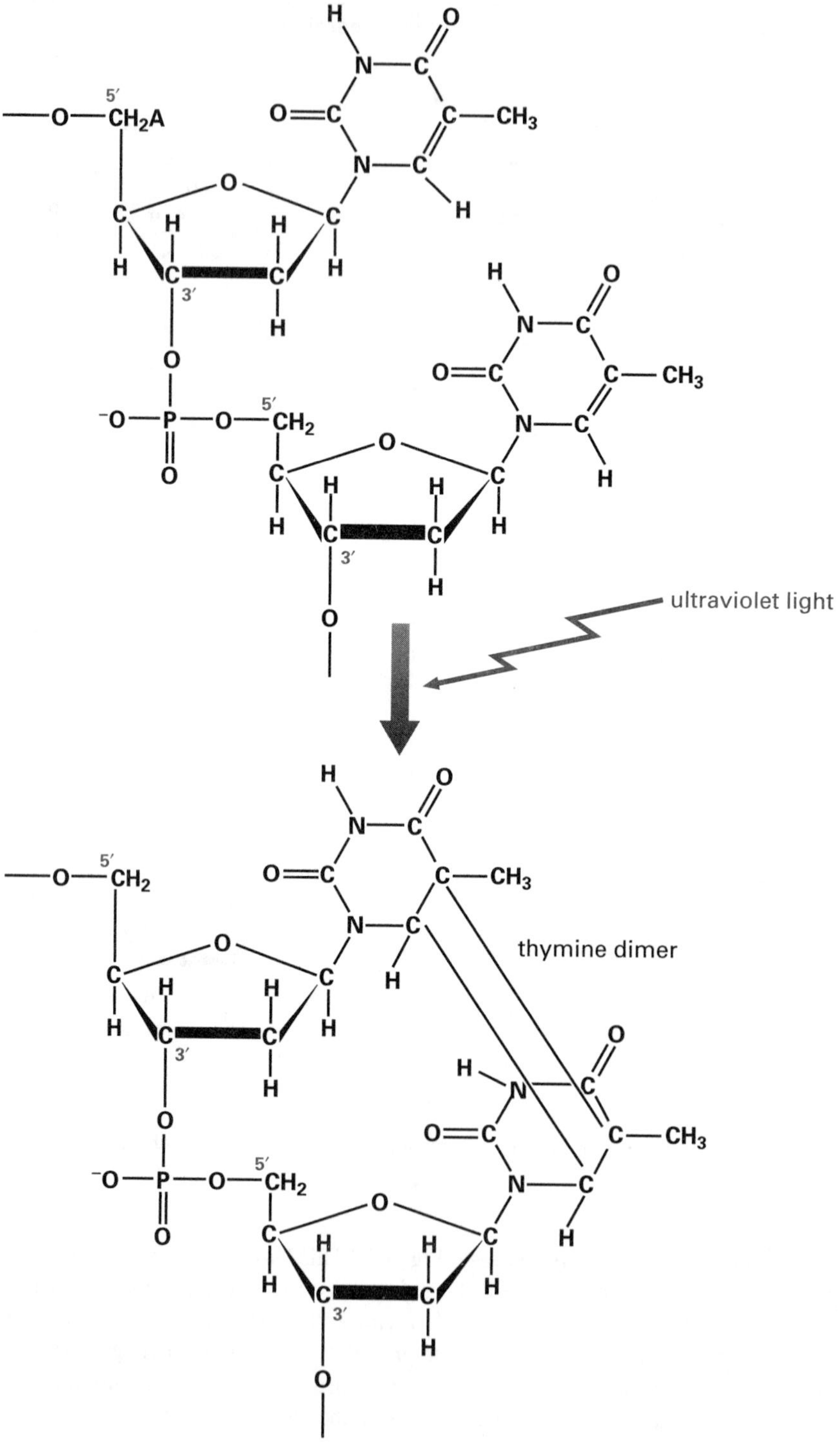

<u>Figure 5.4.</u> Formation of a thymine dimer in DNA.

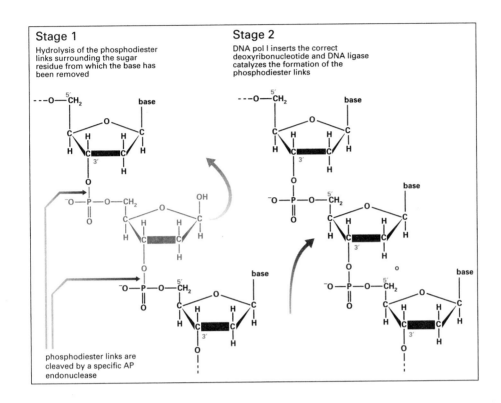

Figure 5.5. DNA repair.

position. **DNA ligase** then seals the strand by catalyzing the reformation of a phosphodiester link.

Example 5.2 A Curious Code

A bacteriophage called PBS2, which infects the bacterium *Bacillus subtilis,* is unusual in that it uses uracil in place of thymine in its DNA. This would be expected to have the unfortunate result—for the virus—that as soon as the bacteriophage DNA entered the bacterial cytoplasm, bacterial repair enzymes would begin cutting the uracils out. PBS2 gets around this problem by quickly making a protein called uracil glycosylase inhibitor, which binds irreversibly to the bacterium's uracil DNA glycosylase and inactivates it. Once this DNA repair enzyme is inactivated, the next problem is to persuade the bacterium to make lots of uracil nucleotides rather than thymine ones. Thymine is normally made from uracil, so all PBS2 does is to produce a second protein that inhibits this conversion, ensuring a good supply of dUTP for replication of its DNA.

Nucleotide excision–repair is required to correct a thymine dimer. The thymine dimer, together with some 30 surrounding nucleotides, is excised from the DNA. Repairing damage of this bulky type requires several proteins because the exposed, undamaged, DNA strand must be protected from nuclease attack while the damaged strand is repaired by the actions of DNA polymerase I and DNA ligase.

> **Medical Relevance 5.2**
>
> **Bloom's Syndrome and Xeroderma Pigmentosum**
>
> DNA helicases are essential proteins required to open up the DNA helix during replication. In Bloom's syndrome, mutations give rise to a defective helicase. The result is excessive chromosome breakage, and affected people are predisposed to many different types of cancers when they are young.
>
> People who suffer from the genetic disorder known as xeroderma pigmentosum are deficient in one of the enzymes for excision repair. As a result, they are very sensitive to ultraviolet light. They contract skin cancer even when they have been exposed to sunlight for very short periods because thymine dimers produced by ultraviolet light are not excised from their genomes.

❋ GENE STRUCTURE AND ORGANIZATION IN EUKARYOTES

Introns and Exons—Additional Complexity in Eukaryotic Genes

Genes that code for proteins should be simple things: DNA makes RNA makes protein, and a gene codes for the amino acids of a protein by the three-base genetic code. In prokaryotes, indeed, a gene is a continuous series of bases that, read in threes, code for the protein. This simple and apparently sensible system does not apply in eukaryotes. Instead, the protein-coding regions of almost all eukaryotic genes are organized as a series of separate bits interspersed with noncoding regions. The protein-coding regions of the split genes are called **exons**. The regions between are called **introns**, short for intervening sequences. In Figure 5.6 we show the structure of the β-globin gene, which contains three exons and two introns. Introns are often very long compared to exons. As happens in prokaryotes, messenger RNA complementary to the DNA is synthesized, but then the introns are spliced out before the mRNA leaves the nucleus (page 118). This means that a gene is much longer

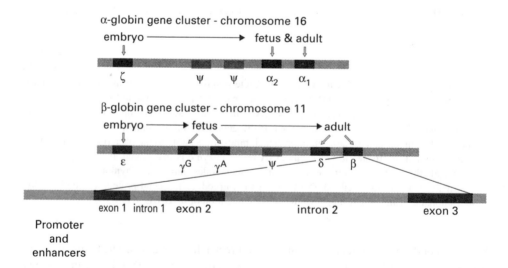

Figure 5.6. The human α- and β-globin gene family clusters. Adults only express α, β, and δ, and of these the expression of δ is very low. The exon/intron boundaries of the β-globin gene are indicated at the bottom.

than the mRNA that ultimately codes for the protein. The name exon derives from the fact that these are the regions of the gene that, when transcribed into mRNA, exit from the nucleus.

In fact there is an evolutionary rationale to this apparently perverse arrangement. As we will see (page 200), a single protein is often composed of a series of **domains,** with each domain performing a different role. The breaks between exons usually correspond to domain boundaries. During evolution, reordering of exons has created new genes that have some of the exons of one gene, and some of the exons of another, and hence generates novel proteins composed of new arrangements of domains, each of which still does its job.

The Major Classes of Eukaryotic DNA

We do not yet fully understand the construction of our nuclear genome. Only about 1.5% of the human genome codes for exonic sequences (i.e., makes protein) with about 23.5% coding for introns, tRNA genes, and ribosomal RNA (rRNA) genes. Most protein-coding genes occur only once in the genome and are called single-copy genes.

IN DEPTH 5.1 Genome Projects

A major goal of biology in the latter decades of the twentieth century was to determine the base sequence of the human genome. This was published in 2001. The goal of biology in the twenty-first century is to determine what all the genes do. This is a difficult task, which is made easier by a comparative genomics approach. Humans share many genes with other organisms. Genes such as those encoding proteins important in metabolism, development, and the control of cell devision are especially conserved through evolution.

The genome of the pufferfish, *Fugu rubripes rupribes,* has about the same number of genes as the human genome, but is only one seventh of the size because it has less extragenic DNA. The pufferfish genome has been sequenced and now can be compared with the human genome to help identify human protein-coding genes.

The genomes of other organisms are also being sequenced. The mouse is a very good model to compare to the human genome. The mouse is a mammal but has a rapid generation time. Mutant mouse models are proving very useful in defining the function of human genes. *Caenorhabditis elegans* is a tiny nematode worm with just 959 body cells. It is also transparent. Many mutant strains of C. *elegans* now exist, and it is a good model for analyzing genes important in development and aging. The zebrafish (*Danio rerio*) is a vertebrate whose embryo is transparent. It is thus easy to spot developmental mutants. A comparison of the mutant gene with that of the human genome will help to identify genes important in human development.

Sophisticated databases have been created to store and analyze base sequence information from the genome projects. Computer programs analyze the data for exon sequences and compare the sequence of one genome to that of another. In this way sequences encoding related proteins (proteins that share stretches of similar amino acids) can be identified. Some important programs, which can be easily accessed through the Internet (see CBASC website) are BLASTN for the comparison of a nucleotide sequence to other sequences stored in a nucleotide database and BLASTP, which compares an amino acid sequence to protein sequence databases.

Many genes have been duplicated at some time during their evolution. Mutation over the succeeding generations causes the initially identical copies to diverge in sequence and produce what is known as a gene family. Members of a gene family usually have a related function—the immunoglobulin gene family, for example, makes antibodies. Different members of a family sometimes encode proteins that carry out the same specialized function but at different times during development. These genes then generate related proteins or **isoforms,** which are often distinguished by placing a Greek letter after the protein name, for example, hemoglobin α and hemoglobin β. Figure 5.6 shows the α- and β-globin gene families. The β-globin gene cluster is on human chromosome 11, and the α-globin gene cluster is on human chromosome 16. Hemoglobin is composed of two α globins and two β globins (page 204). The gene clusters encode proteins produced at specific times during development: from embryo to fetus to adult. The different globin proteins are produced at different stages of gestation to cope with the different oxygen transport requirements during development. The duplication of genes and their subsequent divergence allows the expansion of the gene repertoire, the production of new protein molecules, and the elaboration of ever more specialized gene functions during evolution.

Some sections of DNA are very similar in sequence to other members of their gene family but do not produce mRNA. These are known as **pseudogenes**. There are two in the α-globin gene cluster (ψ in Fig. 5.6). Pseudogenes may be former genes that have mutated to such an extent that they can no longer be transcribed into RNA. Some pseudogenes have arisen because an mRNA molecule has inadvertently been integrated into the genome, copied back into mRNA by an enzyme called **reverse transcriptase** found in some viruses (page 71). Such pseudogenes are immediately recognizable because some or all of their introns were spliced out before the integration occured. Some have the poly-A tail characteristic of intact mRNA (page 118). These are called processed pseudogenes.

Sometimes DNA that encodes RNA is repeated as a series of copies that follow one after the other along the chromosome. Such genes are said to be **tandemly repeated** and include the genes that code for ribosomal RNAs (about 250 copies/cell), transfer RNAs (50 copies/cell), and histone proteins (20–50 copies/cell). The products of these genes are required in large amounts.

This still leaves about 75% of our genome with no very clearly understood function. A large proportion of this so-called **extragenic DNA** is made up of **repetitious DNA** sequences that are repeated many times in the genome. Some sequences are repeated more than a million times and are called **satellite DNA**. The repeating unit is usually several hundred base pairs long, and many copies are often lined up next to each other in tandem repeats. Most of the satellite DNA is found in a region called the **centromere**, which plays a role in the physical movement of the chromosomes that occurs at cell division (page 403), and one theory is that it has a structural function.

Our genome also contains **minisatellite DNA** where the tandem repeat is about 25 bp long. Minisatellite DNA stretches can be up to 20,000 bp in length and are often found near the ends of chromosomes, a region called the **telomere**. **Microsatellite DNA** has an even smaller repeat unit of about 4 bp or less. Again the function of these repeated sequences is unknown, but microsatellites, because their number varies between different individuals, have proved very useful in DNA testing (page 146).

Other extragenic sequences, known as LINEs (long interspersed nuclear elements) and SINEs (short interspersed nuclear elements) occur in our genome. There are about 50,000 copies of LINEs in a mammalian genome. SINEs probably arose by the insertion of small

RNAs into the genome. The most abundant SINE in the human genome is known as the Alu repeat. It occurs about one million times.

GENE NOMENCLATURE

One of the great difficulties that has arisen out of genome-sequencing projects is how to name the genes and the proteins they encode. This has not been easy and a number of committees have been set up to deal with this problem. In general, each gene is designated by an abbreviation, written in capitalized italics. For example, the human gene for flavin-containing monooxygenase is designated *FMO*. Because there is more than one *FMO* gene, we assign a number to identify the specific gene to which we are referring. The gene which when mutated gives rise to trimethylaminuria (page 82) is called *FMO3*. The protein encoded by the *FMO3* gene is written in normal capitals, as FMO3. Similarly, cytochrome P-450 genes (page 249) are abbreviated to *CYP*. *CYP3A4* is a gene that belongs to the *CYP3* family. This family has several members so we must include additional information in the gene name to specify precisely the member of the *CYP3* gene family we are referring to. The protein name is written as CYP3A4.

SUMMARY

1. During replication each parent DNA strand acts as the template for the synthesis of a new daughter strand. The base sequence of the newly synthesized strand is complementary to that of the template strand.

2. Replication starts at specific sequences called origins of replication. The two strands untwist and form the replication fork. Helicase enzymes unwind the double helix, and single-strand binding proteins keep it unwound during replication. DNA polymerase III synthesizes the leading strand continuously in the 5' to 3' direction. The lagging strand is made discontinuously in short pieces in the 5' to 3' direction. These are joined together by DNA ligase.

3. DNA polymerase is a self-correcting enzyme. It can remove an incorrect base using its 3'- to 5'-exonuclease activity and then replace it.

4. DNA repair enzymes can correct mutations. Uracil in DNA, resulting from the spontaneous deamination of cytosine, is removed by uracil–DNA glycosidase. The depyrimidinated sugar is cleaved from the sugar–phosphate backbone by AP endonuclease, and DNA polymerase then inserts the correct nucleotide. The phosphodiester link is reformed by DNA ligase.

5. Protein-coding genes are split into exons and introns. Only exons code for protein. The human genome has a large amount of DNA whose function is not obvious. This includes much repetitive DNA, whose sequence is multiplied many times.

6. Protein-coding genes may be found in repeated groups of slightly diverging structure called gene families, either close together or scattered over the genome. Some of the family members have lost the ability to operate—they are pseudogenes.

FURTHER READING

Brenner, S., Elgar, G., Sandford, R., Macrae, A., Venkatesh, B., and Aparicio, S. 1993. Characterization of the pufferfish (Fugu) genome as a compact model vertebrate genome. *Nature* 366: 265–268.

Friedberg, E. C. 2001. How nucleotide excision repair protects against cancer. *Nature Rev. Cancer* 1: 22–33.

Radman, M., and Wagner, R. 1988. The high fidelity of DNA duplication. *Sci. Am.* 259: 40–46.2.

Scharer, O. D., and Jiricny, J. 2001. Recent progress in the biology, chemistry and structural biology of DNA glycosylases. *Bioessays* 23: 270–281.

✿ REVIEW QUESTIONS

For each question, choose the ONE BEST answer or completion.

1. When a DNA strand with the sequence 3′ GACTAACGATGC 5′ is used as a template on which a daughter DNA sequence is formed, the daughter strand has the sequence
   A. 3′ CTGATTGCTACG 5′
   B. 5′ GCATCGTTAGTC 3′
   C. 5′ CTGATTGCTACG 3′
   D. 3′ CGTAGCAATCAG 5′
   E. 3′ CGUAGCAAUCAG 5′

2. Okazaki fragments are
   A. short peptide sequences.
   B. RNA primers.
   C. newly synthesized stretches of DNA produced during replication of the lagging template strand.
   D. degraded fragments of DNA.
   E. the lengths of DNA between origins of replication.

3. Following a spontaneous DNA deamination of cytosine that generates uracil, repair proceeds in the following order:
   A. Removal of the uracil from its ribose, removal of the sugar to leave a gap, addition of an appropriate new deoxyribonucleotide, reformation of the phosphodiester links.
   B. Removal of the sugar from the sugar–phosphate backbone, removal of the hydrogen-bonded uracil, addition of an appropriate new deoxyribonucleotide, reformation of the phosphodiester links.
   C. Removal of the uracil from its ribose, insertion of cytosine which hydrogen bonds to guanosine, reformation of the bond between cytosine and ribose.
   D. Addition of an appropriate new deoxyribonucleotide to the opposite side of the DNA helix, rotation of the helix by topisomerase 1, reformation of the phosphodiester links.
   E. Transamination of uracil to regenerate cytosine, checking of the conversion by self-correcting DNA polymerase.

4. Which of the following do not play a role in replication?
   A. RNA primers
   B. Deoxynucleotides
   C. Helicase
   D. Stop codons
   E. DNA polymerase

5. During DNA replication, the new DNA is formed on the lagging strand in a series of independent short sections because

    A. DNA polymerase III can synthesize DNA in both the 3′ to 5′ and 5′ to 3′ directions but makes more mistakes when working in the 5′ to 3′ direction.

    B. DNA polymerase III can only operate in the 5′ to 3′ direction.

    C. Thymine dimers are excised from the growing strand, leaving it cut into short fragments.

    D. The hydrogen bonding between adenine and thymine is relatively weak.

    E. DnaA causes the DNA strands to melt.

6. The human genome comprises

    A. protein-coding genes.

    B. genes encoding mRNAs.

    C. sequences that do not encode genes.

    D. genes for ribosomal RNAs and transfer RNAs.

    E. all of the above.

ANSWERS TO REVIEW QUESTIONS

1. *C.* 5′ CTGATTGCTACG 3′—DNA is always synthesized in the 5′ to 3′ direction, and the sequence of the newly synthesized strand is complementary to that of the template strand.

2. *C.* Okazaki fragments are newly synthesized stretches of DNA produced during replication of the lagging template strand.

3. *A.* Answers B, C, and E are more or less topologically and chemically plausible but are not the way repair proceeds. Answer D is nonsense.

4. *D.* Stop codons are used to halt protein synthesis during translation of mRNA. None of the DNA transcription machinery takes any special notice of bases that will generate a stop codon in mRNA; they are copied in exactly the same way as the rest of the genome.

5. *B.* Because DNA polymerase can only proceed in the 5′ to 3′ direction, the lagging strand must be replicated as a series of short sections called Okazaki fragments.

6. *E.* All are correct statements; the genome represents all the information on the DNA.

6

TRANSCRIPTION AND THE CONTROL OF GENE EXPRESSION

Transcription (or RNA synthesis) is the process whereby the information held in the nucleotide sequence of DNA is transferred to RNA. The three major classes of RNA are ribosomal RNA (rRNA), transfer RNA (tRNA), and messenger RNA (mRNA). All play key roles in protein synthesis. Genes encoding mRNAs are known as protein-coding genes. A gene is said to be **expressed** when its genetic information is transferred to mRNA and then to protein. Two important questions are addressed in this chapter: how is RNA synthesized, and what factors control how much is made?

✳ STRUCTURE OF RNA

Ribonucleic acid is a polymer made up of monomeric nucleotide units. RNA has a chemical structure similar to that of DNA, but there are two major differences. First, the sugar in RNA is ribose instead of deoxyribose (Fig. 6.1). Second, although RNA contains the two purine bases adenine and guanine and the pyrimidine cytosine, the fourth base is different. The pyrimidine uracil (U) replaces thymine (Fig. 6.1). The building blocks of RNA are therefore the four ribonucleoside triphosphates adenosine 5′-triphosphate, guanosine 5′-triphosphate, cytidine 5′-triphosphate, and uridine 5′-triphosphate. These four nucleotides are joined together by phosphodiester links (Fig. 6.2). Like DNA the RNA chain has direction. In the first nucleotide in the chain, the 5′ carbon of the ribose is phosphorylated and is available for formation of phosphodiester links. This is called the 5′ end of the RNA chain. At the

Cell Biology: A Short Course, Second Edition, by Stephen R. Bolsover, Jeremy S. Hyams,
Elizabeth A. Shephard, Hugh A. White, Claudia G. Wiedemann
ISBN 0-471-26393-1 Copyright © 2004 by John Wiley & Sons, Inc.

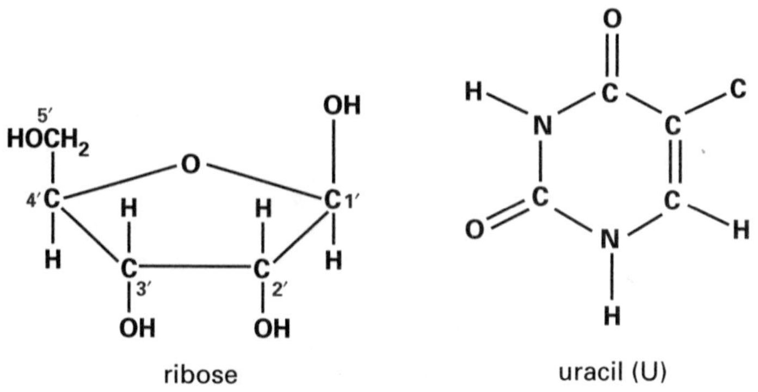

Figure 6.1. RNA contains the sugar ribose and the base uracil in place of deoxyribose and thymine.

other end is a ribose with a free hydroxyl group on its 3′ carbon. This is called the 3′ end. RNA molecules are single-stranded along much of their length, although they often contain regions that are double stranded due to intramolecular base pairing.

✺ RNA POLYMERASE

In any gene only one DNA strand acts as the template for transcription. The sequence of nucleotides in RNA depends on their sequence in the DNA template. The bases T, A, G, and C in the DNA template will specify the bases A, U, C, and G, respectively, in RNA. DNA is transcribed into RNA by the enzyme RNA polymerase. Transcription requires that this enzyme recognize the beginning of the gene to be transcribed and catalyze the formation of phosphodiester links between nucleotides that have been selected according to the sequence within the DNA template (Fig. 6.2).

✺ GENE NOTATION

Figure 6.3 shows the notation used in describing the positions of nucleotides within and adjacent to a gene. The nucleotide in the template strand at which transcription begins is designated with the number +1. Transcription proceeds in the **downstream** direction, and nucleotides in the transcribed DNA are given successive positive numbers. Downstream sequences are drawn, by convention, to the right of the transcription start site. Nucleotides that lie to the left of this site are called the **upstream** sequences and are identified by negative numbers.

✺ BACTERIAL RNA SYNTHESIS

Escherichia coli genes are all transcribed by the same RNA polymerase. This enzyme is made up of five subunits (polypeptide chains). The subunits are named α (there are two of these), β, β', and σ. Each of the subunits has its own job to do in transcription. The role of the sigma (σ) factor is to recognize a specific DNA sequence called the **promoter,** which lies just upstream of the gene to be transcribed (Fig. 6.3). *E. coli* promoters contain two important regions. One centered around nucleotide -10 usually has the sequence TATATT.

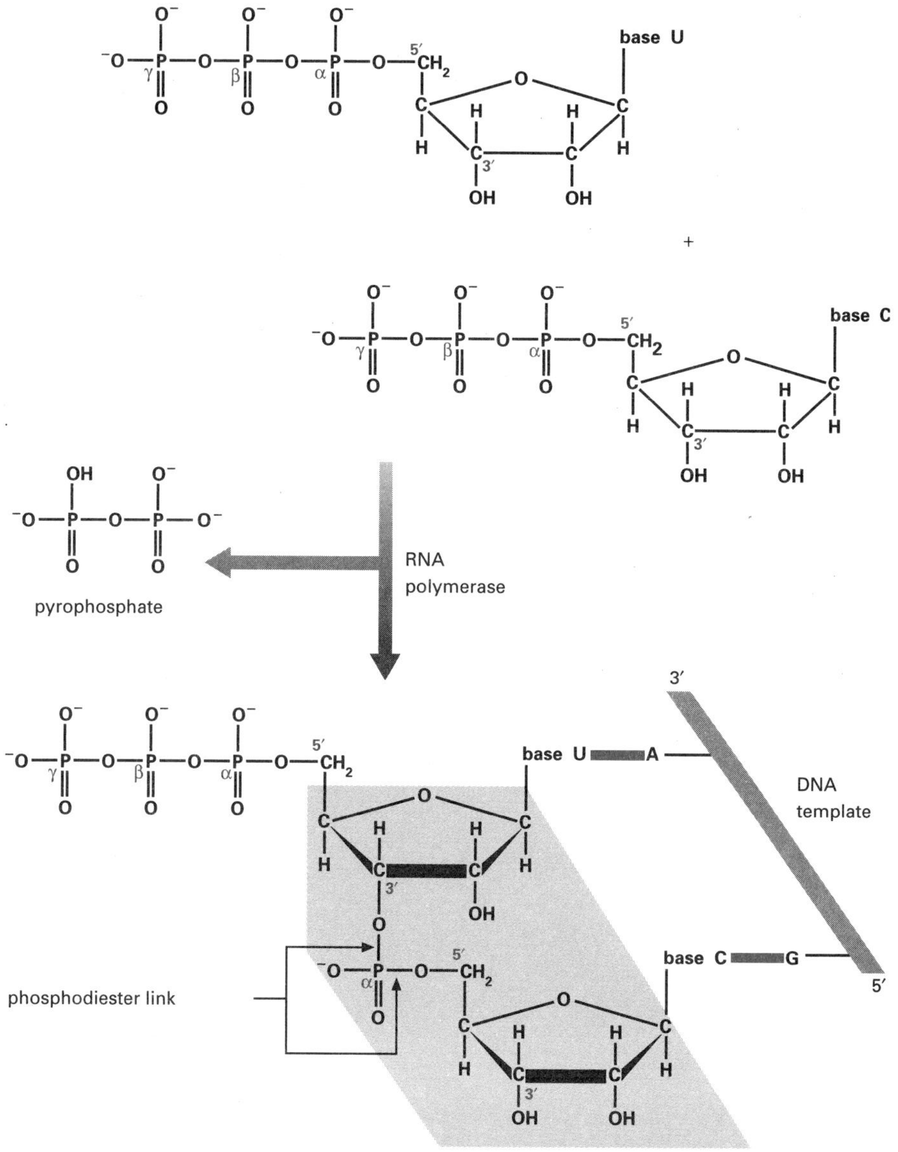

base U

+

base C

$^-O-P-O-P-O^-$
pyrophosphate

RNA polymerase

base U — A
3'

DNA template

base C — G
5'

phosphodiester link

base U

base C

Figure 6.2. Synthesis of an RNA strand.

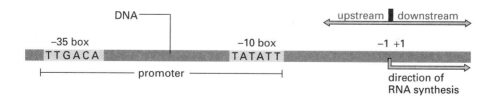

DNA

upstream ▮ downstream

−35 box

TTGACA

−10 box

TATATT

−1 +1

promoter

direction of RNA synthesis

Figure 6.3. Numbering of a DNA sequence.

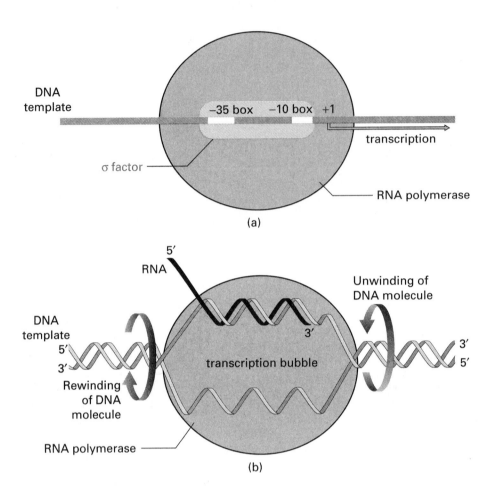

Figure 6.4. (a) RNA polymerase binds to the promoter. (b) DNA helix unwinds and RNA polymerase synthesizes an RNA molecule.

This sequence is called the −10 box (or the Pribnow box). The second, centered near nucleotide −35 often has the sequence TTGACA. This is the −35 box.

On binding to the promoter sequence (Fig. 6.4a), the σ factor brings the other subunits (two of α plus one each of β and β′) of RNA polymerase into contact with the DNA to be transcribed. This forms the **closed promoter complex.** For transcription to begin, the two strands of DNA must separate, enabling one strand to act as the template for the synthesis of an RNA molecule. This formation is called the **open promoter complex.** The separation of the two DNA strands is helped by the **AT**-rich sequence of the −10 box. There are only two hydrogen bonds between the bases adenine and thymine; thus it is relatively easy to separate the two strands at this point. DNA unwinds and rewinds as RNA polymerase advances along the double helix, synthesizing an RNA chain as it goes. This produces a **transcription bubble** (Fig. 6.4b). The RNA chain grows in the 5′ to 3′ direction, and the template strand is read in the 3′ to 5′ direction (Fig. 6.4).

When the RNA chain is about 10 bases long, the σ factor is released from RNA polymerase and plays no further role in transcription. The β subunit of RNA polymerase binds ribonucleotides and joins them together by catalyzing the formation of phosphodiester links as it moves along the DNA template. The β' subunit helps to keep the RNA polymerase attached to DNA. The two α subunits are important as they help RNA polymerase to assemble on the promoter (see discussion of the *lac* operon below).

RNA polymerase has to know when it has reached the end of a gene. *Escherichia coli* has specific sequences, called terminators, at the ends of its genes that cause RNA polymerase to stop transcribing DNA. A terminator sequence consists of two regions rich in the bases G and C that are separated by about 10 bp. This sequence is followed by a stretch of A bases. Figure 6.5 shows how the terminator halts transcription. When the GC-rich regions are transcribed, a hairpin loop forms in the RNA with the first and second GC-rich regions aligning and pairing up. Formation of this structure within the RNA molecule causes the transcription bubble to shrink because where the template DNA strand can no longer bind to the RNA molecule it reconnects to its sister DNA strand. The remaining interactions between the adenines in the DNA template and the uracils in the RNA chain have only two hydrogen bonds per base pair and are therefore too weak to maintain the transcription bubble. The RNA molecule is then released, transcription terminates, and the double helix reforms. This type of transcription termination is known as rho-independent termination.

Some *E. coli* genes contain different terminator sites. These are recognized by a protein, known as rho, which frees the RNA from the DNA. In this case transcription is terminated by a process known as rho-dependent termination.

❋ CONTROL OF BACTERIAL GENE EXPRESSION

Many bacterial proteins are always present in the cell in a constant amount. However, the amount of other proteins is regulated by the presence or absence of a particular nutrient. To grow and divide and make the most efficient use of the available nutrients, bacteria have to adjust quickly to changes in their environment. They do this by regulating the production of proteins required for either breakdown or synthesis of a particular compound. Gene expression in bacteria is controlled mainly at the level of transcription. This is because bacterial cells have no nuclear envelope, and RNA synthesis and protein synthesis are not separate but occur simultaneously. This is one reason why bacteria lack the more sophisticated control mechanisms that regulate gene expression in eukaryotes.

Each bacterial promoter usually controls the transcription of a cluster of genes coding for proteins that work together on a particular task. This collection of related genes is called an **operon** and is transcribed as a single mRNA molecule called a polycistronic mRNA. As shown in Figure 6.6, translation of this mRNA produces the required proteins because there are several start and stop codons for protein synthesis along its length. Each start and stop codon (page 79) specifies a region of RNA that will be translated into one particular protein. The organization of genes into operons ensures that all the proteins necessary to metabolize a particular compound are made at the same time and hence helps bacteria to respond quickly to environmental changes.

The three major factors involved in regulating how much RNA is made are (1) nucleotide sequences within or flanking a gene, (2) proteins that bind to these sequences, and (3) the environment. The human intestine contains many millions of *E. coli* cells that must respond very quickly to the sudden appearance of a particular nutrient. For instance, most foods do not contain the disaccharide lactose (Fig. 6.7), but milk contains large amounts. Within minutes of our drinking a glass of milk, *E. coli* in our intestines start to produce the enzyme β-galactosidase that cleaves lactose to glucose and galactose (Fig. 6.7). In general,

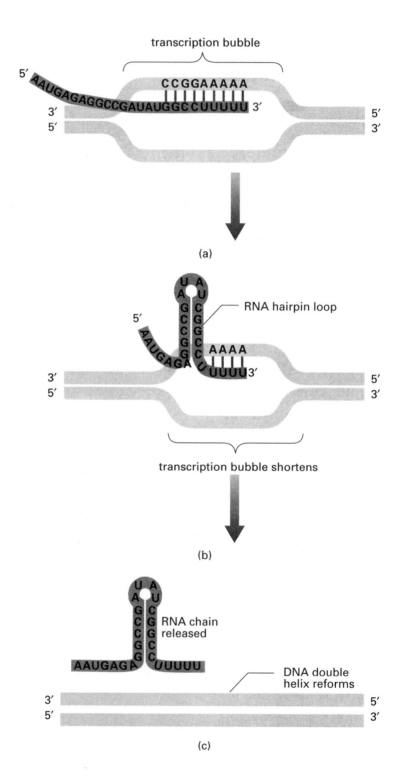

Figure 6.5. Transcription termination in *Escherichia coli*.

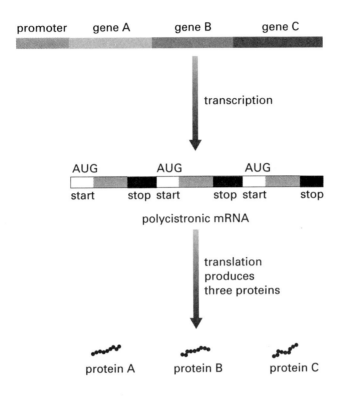

Figure 6.6. A bacterial operon is transcribed into a polycistronic mRNA.

the substrates of β-galactosidase are compounds like lactose that contain a β-galactoside linkage and are therefore called β-galactosides.

lac, an Inducible Operon

β-Galactosidase is encoded by one of the genes that make up the lactose *(lac)* operon, which is shown in Figure 6.8. The operon contains three protein-coding genes called *lac z*, *lac y*, and *lac a*. β-Galactosidase is encoded by the *lac z* gene. As noted before, gene names are always italicized, while the protein product is always in standard type. *Lac y* encodes β-galactoside permease, a carrier (page 316) that helps lactose get into the cell. The *lac a* gene codes for transacetylase. This protein is thought to remove compounds that have a structure similar to lactose but that are not useful to the cell.

In the absence of β-galactoside compounds like lactose, there is no need for *E. coli* to produce β-galactosidase or β-galactoside permease, and the cell contains only a few molecules of these proteins. The *lac* operon is said to be inducible because the rate of transcription into RNA increases greatly when a β-galactoside is present. How is the transcription of the *lac z*, *lac y*, and *lac a* genes switched on and off? A repressor protein (the product of the *lac i* gene) binds to a sequence in the *lac* operon known as the operator. The operator lies next to the promoter so that, when the repressor is bound, RNA polymerase is unable to bind to the promoter. In the absence of a β-galactoside, the *lac* operon spends most of its time in the state shown in Figure 6.8a. The repressor is bound to the operator, RNA polymerase cannot bind, and no transcription occurs. Only for the small fraction of time

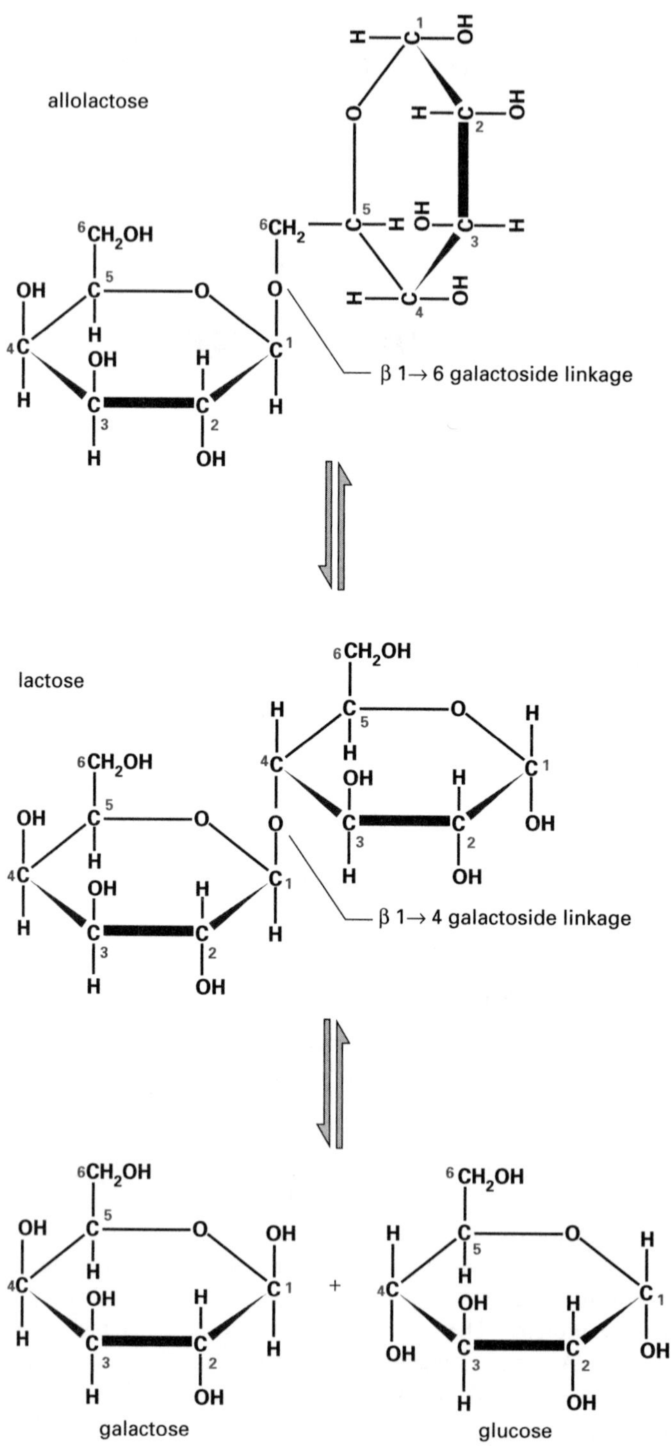

Figure 6.7. Reactions catalyzed by β-galactosidase.

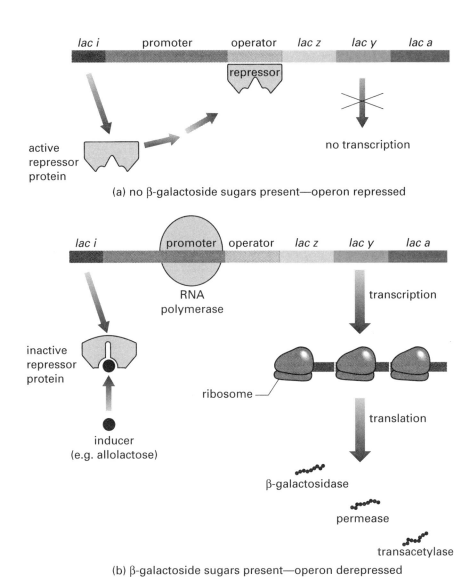

Figure 6.8. Transcription of the *lac* operon requires the presence of an inducer.

that the operator is unoccupied by the repressor can RNA polymerase bind and generate mRNA. Thus in the absence of a β-galactoside, only very small amounts of β-galactosidase, β-galactoside permease, and transacetylase are synthesized.

If lactose appears, it is converted to an isomer called allolactose. This conversion is carried out by β-galactosidase (Fig. 6.7); as we have seen, a small amount of β-galactosidase is made even when β-galactoside is absent. The repressor protein has a binding site for allolactose and undergoes a conformational change when bound to this compound (Fig. 6.8b). This means that the repressor is no longer able to bind to the operator. The way is then clear for RNA polymerase to bind to the promoter and to transcribe the operon. Thus in a short time the bacteria produce the proteins necessary for utilizing the new food source. The concentration of the substrate (lactose in this case) determines whether or not mRNA is synthesized. The *lac* operon is said to be under negative regulation by the repressor protein.

The transcription of the *lac* operon is controlled not only by the repressor protein but also by another protein, the **catabolite activator protein** (**CAP**). If both glucose and lactose are present, it is more efficient for the cell to use glucose as the carbon source because the utilization of glucose requires no new RNA and protein synthesis, all the proteins necessary being already present in the cell. Only in the absence of glucose, therefore, does *E. coli* transcribe the *lac* operon at a high rate. This control operates through an intracellular messenger molecule called **cyclic adenosine monophosphate (cyclic AMP)** (Fig. 6.9). When glucose concentrations are low, the concentration of cyclic AMP increases. Cyclic AMP binds to CAP, and the complex then binds to a sequence upstream of the *lac* operon promoter (Fig. 6.10) where it has a remarkable effect. The DNA surrounding CAP bends by about 90°. Both α subunits of RNA polymerase are now able to make contact with CAP at the same time so that the affinity of RNA polymerase for the *lac* promoter is increased. The result is that now the *lac z, lac y,* and *lac a* genes can be transcribed very efficiently. The *lac* operon is said to be under positive regulation by the CAP–cAMP complex.

To recap, the control of the *lac* operon is not simple. Several requirements need to be met before it can be transcribed. The repressor must not be bound to the operator, and the CAP–cyclic AMP complex and RNA polymerase must be bound to their respective DNA binding sites. These requirements are only met when glucose is absent and a β-galactoside, such as the sugar lactose, is present.

Other compounds such as isopropylthio-β-D-galactoside (IPTG) (Fig. 6.11) can bind to the repressor but are not metabolized. These **gratuitous inducers** are very useful in DNA research and in biotechnology. Chapter 7 deals with this and with some of the industrial applications of the *lac* operon.

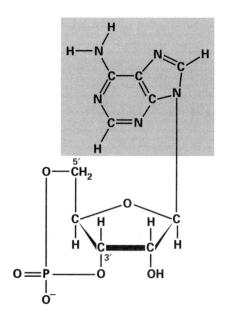

Figure 6.9. Cyclic adenosine monophosphate, also called cyclic AMP or just cAMP.

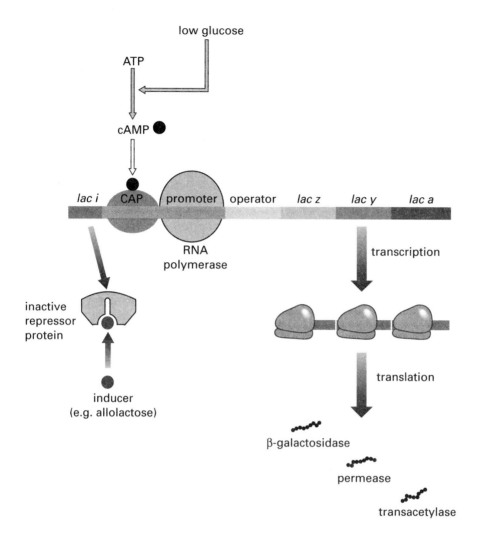

Figure 6.10. For efficient transcription of the *lac* operon, both cAMP and a β-galactoside sugar must be present.

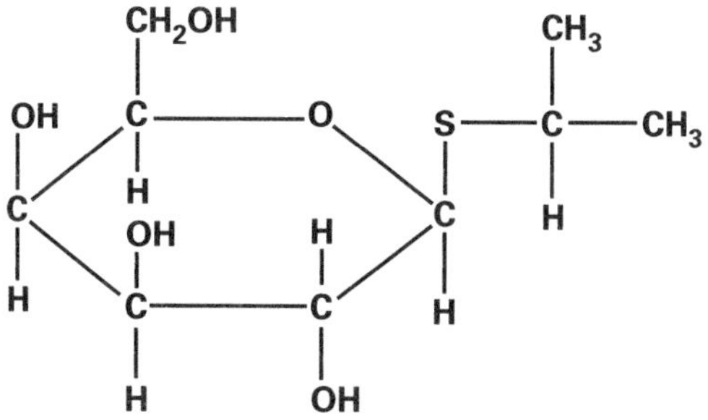

CH₂OH and structure labels:

IPTG (isopropyl thio-β-D-galactoside)

Figure 6.11. Isopropylthio-β-ᴅ-galactoside (IPTG), which can bind to the *lac* repressor protein but which is not metabolized.

Example 6.1 **Quorum Sensing: Squids That Glow in the Dark**

The bacterium *Vibrio fischeri* lives free in seawater but is also found at high densities in the light-emitting organs of the nocturnal squid *Euprymna scolopes,* where it synthesizes an enzyme called luciferase that generates light. This phenomenon is called bioluminescence. When living free in seawater *V. fischeri* synthesizes almost no luciferase; indeed, there would be little point in doing so because the light emitted by a single bacterium would be too dim for anything to see. *V. fischeri* only begins making lots of luciferase when the density of bacteria is high—just as it is in the squid light organs. The word *quorum* is defined in Webster's Dictionary as "the number of . . . members of a body that when duly assembled is legally competent to transact business." Thus *quorum sensing* is a good description of *V. fischeri's* behavior. How does it work?

Luciferase is a product of the *lux A* and *lux B* genes in a bacterial operon called the *lux* operon. A region in the *lux* operon promoter binds a transcription factor, LuxR, which is only active when it has bound a small, uncharged molecule called N-acyl-HSL (also called VAI for *V. fischeri* autoinducer). N-acyl-HSL in turn is made by an enzyme called N-acyl-HSL synthase (or VAI synthase) that is encoded by the gene *lux I,* which is part of the *lux* operon. In free-living *V. fischeri* the *lux* operon is transcribed at a low level. Small amounts of N-acyl-HSL are made, which immediately leak out of the cell into the open sea without binding to luxR. When the bacterium is concentrated in the squid's light organs, then some N-acyl-HSL binds to LuxR, increasing transcription of the *lux* operon. This makes more luciferase—but it also makes more N-acyl-HSL synthase. The concentration of N-acyl-HSL therefore rises—so transcription of the *lux* operon increases further. This means that the genes for N-acyl-HSL synthase, luciferase, and the enzymes that produce the substrate for luciferase are now transcribed at a high rate. This **autoinduction** of the *lux* operon by N-acyl-HSL is a form of **positive feedback** (page 303). When luciferase carries out its reaction, the squid will luminesce at the intensity of moonlight so that as it glides at night over the coral reefs of Hawaii it does not cast a dark shadow; thus its prey are less likely to notice its presence.

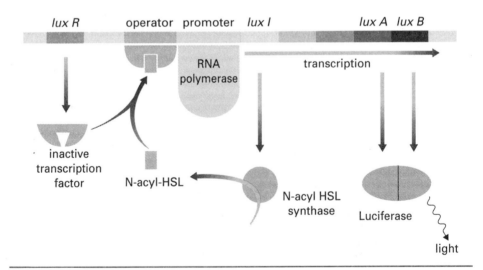

trp, a Repressible Operon

Operons that code for proteins that synthesize amino acids are regulated in a different way from the *lac* operon. These operons are only transcribed if the amino acid is not present, and transcription is switched off if there is already enough of the amino acid around. In this

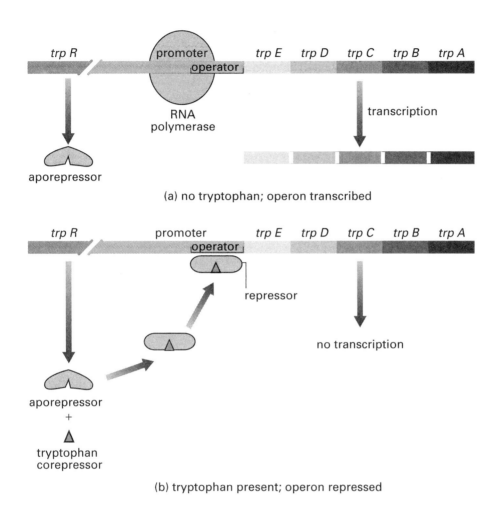

Figure 6.12. Transcription of the *trp* operon is controlled by the concentration of the amino acid tryptophan. (*a*) No tryptophan, operon transcribed. (*b*) Tryptophan present, operon repressed.

way the cell carefully controls the concentration of free amino acids. The tryptophan (*trp*) operon is made up of five structural genes encoding enzymes that synthesize the amino acid tryptophan (Fig. 6.12). This is a **repressible operon.** The cell regulates the amount of tryptophan produced by preventing transcription of the *trp* operon mRNA when there is sufficient tryptophan about. As with the *lac* operon, the transcription of the *trp* operon is controlled by a regulatory protein. The gene *trp R* encodes an inactive repressor protein that is called an aporepressor. Tryptophan binds to this to produce an active repressor complex and is therefore called the **corepressor.** The active repressor complex binds to the operator sequence of the *trp* operon and prevents the attachment of RNA polymerase to the *trp* promoter sequence. Therefore, when the concentration of tryptophan in the cell is high, the active repressor complex will form, and transcription of the *trp* operon is prevented. However, when the amount of tryptophan in the cell decreases, the active repressor complex cannot be formed. RNA polymerase binds to the promoter, transcription of the *trp* operon proceeds, and the enzymes needed to synthesize tryptophan are produced. This is an example of negative feedback (page 303).

Many other operons are regulated by similar mechanisms in which specific regulatory proteins interact with specific small molecules.

EUKARYOTIC RNA SYNTHESIS

Eukaryotes have three types of RNA polymerase. **RNA polymerase I** transcribes the genes that code for most of the ribosomal RNAs. All messenger RNAs are synthesized using **RNA polymerase II.** Transfer RNA genes are transcribed by **RNA polymerase III.** This last enzyme also catalyzes the synthesis of several small RNAs including the 5S ribosomal RNA. The chemical reaction catalyzed by these three RNA polymerases, the formation of phosphodiester links between nucleotides, is the same in eukaryotes and bacteria.

Messenger RNA Processing

A newly synthesized eukaryotic mRNA undergoes several modifications before it leaves the nucleus (Fig. 6.13). The first is known as capping. Very early in transcription the 5′-terminal triphosphate group is modified by the addition of a guanosine via a 5′-5′-phosphodiester link. The guanosine is subsequently methylated to form the **7-methyl guanosine cap.** The 3′ ends of nearly all eukaryotic mRNAs are modified by the addition of a long stretch of adenosine residues, the **poly-A tail** (Fig. 6.13). A sequence AAUAAA is found in most eukaryotic mRNAs about 20 bases from where the poly-A tail is added and is probably a signal for the enzyme poly-A polymerase to bind and to begin the polyadenylation process. The length of the poly-A tail varies, it can be as long as 250 nucleotides. Unlike DNA, RNA is an unstable molecule, and the capping of eukaryotic mRNAs at their 5′ ends and the addition of a poly-A tail to their 3′ end increases the lifetime of mRNA molecules by protecting them from digestion by nucleases.

Many eukaryotic protein-coding genes are split into exon and intron sequences. Both the exons and introns are transcribed into mRNA. The introns have to be removed and the exons joined together by a process known as RNA splicing before the mRNA can be used to make protein. Removal of introns takes place within the nucleus. Splicing is complex and not yet fully understood. It has, however, certain rules. Within an mRNA the first two bases following an exon are always GU and the last two bases of the intron are AG. Several small nuclear RNAs (snRNAs) are involved in splicing. These are complexed with a number of proteins to form a structure known as the spliceosome. One of the snRNAs is complementary in sequence to either end of the intron sequence. It is thought that binding of this snRNA to the intron, by complementary base pairing, brings the two exon sequences together, which causes the intron to loop out (Fig. 6.13). The proteins in the spliceosome remove the intron and join the exons together. Splicing is the final modification made to the mRNA in the nucleus. The mRNA is now transported to the cytoplasm for protein synthesis.

As well as removing introns, splicing can sometimes remove exons in a process called **alternative splicing.** This allows the same gene to give rise to different proteins at different times or in different cells. For example, alternative splicing of the gene for the molecular motor **dynein** produces motors that transport different types of cargo (page 390).

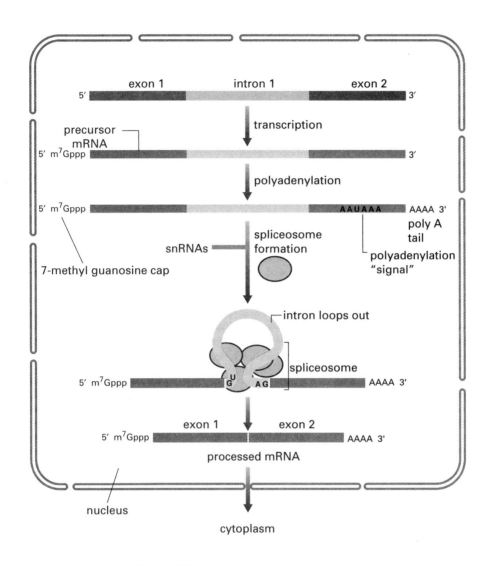

Figure 6.13. mRNA processing in eukaryotes.

❋ CONTROL OF EUKARYOTIC GENE EXPRESSION

Since most eukaryotes are multicellular organisms with many cell types, gene expression must be controlled so that different cell lineages develop differently and remain different. A brain cell is quite different from a liver cell because it contains different proteins even though the DNA in the two cell types is identical. During development and differentiation, different sets of genes are switched on and off. Hemoglobin, for example, is only expressed in developing red blood cells even though the globin genes are present in all types of cell. Genetic engineering technology (Chapter 7) has made the isolation and manipulation of eukaryotic genes possible. This has given us some insight into the extraordinarily complex processes that regulate transcription of eukaryotic genes and allow a fertilized egg to develop into a multicellular, multitissue adult.

Unlike the situation in bacteria, the eukaryotic cell is divided by the nuclear envelope into nucleus and cytoplasm. Transcription and translation are therefore separated in space and in time. This means that the expression of eukaryotic genes can be regulated at more than one place in the cell. Although gene expression in eukaryotes is controlled primarily by regulating transcription in the nucleus, there are many instances in which expression is controlled at the level of translation in the cytoplasm or by altering the way in which the primary mRNA transcript is processed.

The interaction of RNA polymerase with its promoter is far more complex in eukaryotes than it is in bacteria. This section describes how the transcription of a gene, encoding mRNA, is transcribed by RNA polymerase II. In contrast to bacterial RNA polymerase, RNA polymerase II cannot recognize a promoter sequence. Instead, other proteins known as transcription factors bind to the promoter and guide RNA polymerase II to the beginning of the gene to be transcribed.

The promoter sequence of most eukaryotic genes encoding mRNAs contains an AT-rich region about 25 bp upstream of the transcription start site. This sequence, called the TATA box, binds a protein called the transcription factor IID (TFIID), one of whose subunits is called the TATA-binding protein, or TBP (Fig. 6.14a). Several other transcription factors (TFIIA, TFIIB, TFIIE, TFIIF, and TFIIH) then bind to TFIID and to the promoter region (Fig. 6.14b). TFIIF is the protein that guides RNA polymerase II to the beginning of the gene to be transcribed. The complex formed between the TATA box, TFIID, the other transcription factors, and RNA polymerase is known as the transcription preinitiation complex. Note that the proteins with the prefix TFII are so named because they are *t*ranscription *f*actors that help RNA polymerase II to bind to promoter sequences.

Although many gene promoters contain a TATA box, some do not. These TATA-less genes usually encode proteins that are needed in every cell and are hence called housekeeping genes. The promoters of these genes contain the sequence GGGGCGGGGC, called the GC box. A protein called Sp1 binds to the GC box and is then able to recruit TATA-binding protein to the DNA even though there is no TATA box for the latter to bind to. TATA-binding protein then recruits the rest of the transcription preinitiation complex so that transcription can proceeed.

In either case, transcription begins when the carboxy-terminal domain of RNA polymerase II is phosphorylated. This region is rich in the amino acids serine and threonine each of which contains an OH group in their side chain. When these OH groups are phosphorylated (pages 190, 252), RNA polymerase II breaks away from the preinitiation complex and proceeds to transcribe DNA into mRNA (Fig. 6.14c).

Although the formation of the initiation protein complex is sometimes enough to produce a few molecules of RNA, the binding of other proteins to sequences next to the gene greatly increases the rate of transcription producing much more mRNA. These proteins are also called transcription factors, and the DNA sequences to which they bind are called enhancers, so named because their presence enhances transcription. **Enhancer sequences** often lie upstream of a promoter, but they have also been found downstream. Enhancer sequences and the proteins that bind to them play an important role in determining whether a particular gene is to be transcribed. Some transcription factors bind to a gene to ensure that it is transcribed at the right stage of development or in the right tissue. Figure 6.15 shows how one gene can be transcribed in skeletal muscle but not

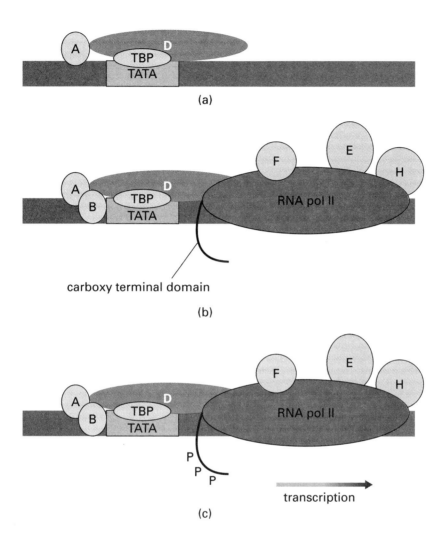

Figure 6.14. RNA polymerase is guided to the promoter by TFII accessory proteins.

in the liver simply because of the presence or absence of proteins that bind to enhancer sequences.

Glucocorticoids Cross the Cell Membrane to Activate Transcription

Glucocorticoids are steroid hormones produced by the adrenal cortex that increase the transcription of several genes important in carbohydrate and protein metabolism. Because they are uncharged and relatively nonpolar, steroid hormones can pass through the plasma membrane by simple diffusion to enter the cytosol. Here they encounter a class of transcription factors that have a binding site for steroid hormones and are therefore called **steroid hormone receptors** (Fig. 6.16). In the absence of glucocorticoid hormone, its receptor remains

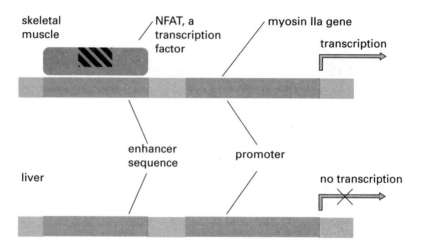

Figure 6.15. Tissue-specific transcription.

in the cytosol and is inactive because it is complexed to two molecules of an inhibitor protein known as Hsp90. However, when the glucocorticoid hormone enters the cell and binds to its receptor, the Hsp90 protein is displaced. The targeting sequence (page 215) that targets the receptor to the nucleus is uncovered, and the glucocorticoid receptor–hormone complex can now move into the nucleus. Here, two molecules of the complex bind to a

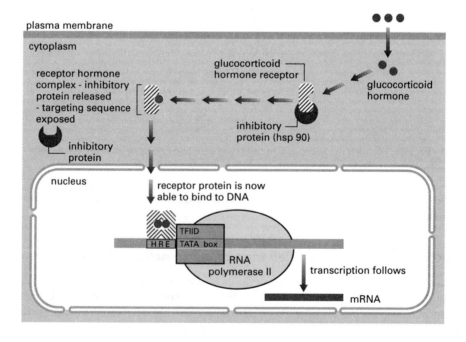

Figure 6.16. The steroid hormone receptor acts to increase gene transcription in the presence of hormone.

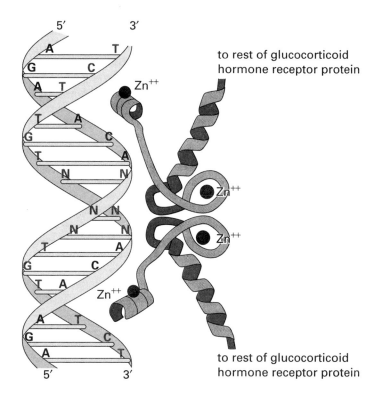

Figure 6.17. The palindromic HRE binds to dimerized steroid hormone receptor.

15-bp sequence known as the hormone response element (HRE) that lies upstream of the TATA box. The HRE is an enhancer sequence. The glucocorticoid receptor–hormone complex interacts with the preinitiation complex bound to the TATA box, and the rate at which RNA polymerase transcribes genes containing the HRE is increased.

Figure 6.17 shows why the glucorticoid hormone receptor binds to DNA as a dimer. The HRE is a palindrome—the sequence on both the top and bottom strands is the same when read in the 5′ to 3′ direction. Each strand of the HRE has the 6-bp sequence 5′ AGAACA 3′ that is known as the core recognition motif. This is the sequence to which a single receptor molecule binds. Because the HRE contains two recognition motifs, it binds two molecules of glucocorticoid receptor. The two 6-bp sequences are separated by 3 base pairs, which presumably are there to provide sufficient space for the receptor homodimer to fit snugly on the double helix. The glucocorticoid receptor is unaffected by the identity of these three base pairs, so the nucleotides are simply indicated as N in Fig. 6.17.

Chemicals that are released from one cell and that alter the behavior of other cells are called **transmitters.** Glucocorticoids are an example of transmitters that alter gene transcription. The cells of multicellular organisms turn the transcription of genes on and off in response to many extracellular chemicals. Unlike steroid hormones, most of these transmitters cannot enter the cell and must activate **intracellular messenger** systems inside the cell that in turn carry the signal onward from the plasma membrane to the nucleus (Chapter 16).

Example 6.2 Glucocorticoid Hormones Can Repress Transcription:
Rheumatoid Arthritis

Glucocorticoid treatment has been shown to give some relief to individuals who suffer from the debilitating disorder rheumatoid arthritis. Collagenase, an enzyme that digests collagen, is generated in the joints of these patients, causing destruction of the extracellular matrix and therefore chronic inflammation. Transcription of the collagenase gene is controlled by an enhancer sequence called the AP1 site. For transcription to occur, the enhancer must be occupied by a transcription factor called AP1, which, like the active glucocorticoid receptor, is a dimer. Glucocorticoid hormones inhibit transcription of the collagenase gene by an ingenious mechanism. The glucocorticoid hormone, on entering the cell interacts with the glucocorticoid receptor as shown in Figure 6.16. The receptor–hormone complex moves to the nucleus and binds to the proteins that would otherwise dimerize to form AP1. The heterodimer is unable to activate transcription of the collagenase gene. By depleting the pool of AP1 subunits, the glucocorticoid receptor–hormone complex prevents transcription of the collagenase gene.

Medical Relevance 6.1   **St. John's Wort and the Pill**

When we ingest plant products and other chemicals foreign to us, the body responds by increasing the amounts of a group of proteins known as the cytochromes P450 (CYPs). This remarkable response, found largely in the liver, is our built-in detoxication system. The foreign chemical signals the appropriate *CYP* gene to activate, transcription takes place, and more CYP protein is produced. The CYP protein then metabolizes the foreign chemical so that it will be cleared from the body quickly and efficiently through the urine or feces. To the liver, a medicinal drug is just another foreign chemical, and the cytochrome P450 system is used to destroy it.

St. John's wort is a herbal remedy, taken by many as a natural antidepressant. One of the active components is the chemical hyperforin, which has been shown to decrease the effectiveness of several therapeutic drugs including the oral contraceptive pill, the HIV protease inhibitor indinavir, and immunosuppressants such as cyclosporin. This is because each of these drugs is metabolized by a cytochrome P450 called CYP3A4, and the gene for CYP3A4 is activated by hyperforin. Because the activation of the *CYP3A4* gene results in more CYP3A4 protein, the prescribed drug, in a person also taking St. John's wort, is metabolized more rapidly, cleared from the body quickly, and is therefore less effective.

The transcription of the *CYP3A4* gene is increased because hyperforin passes to the nucleus and binds to a receptor called the pregnane X receptor (PXR). The receptor–hyperforin complex is then able to dimerize with another receptor called RXR. The dimer binds to an enhancer sequence of the *CYP3A4* gene, interacts with the transcription preinitiation complex, and triggers RNA polymerase II to begin transcription of CYP3A4 mRNA.

This activation of the *CYP3A4* gene by St. John's wort means that this herbal remedy should be taken with caution if a patient has been prescribed a therapeutic drug. CYP3A4 is responsible for the metabolism of about 80% of currently prescribed drugs.

SUMMARY

1. DNA is transcribed into RNA by the enzyme RNA polymerase. The three types of RNA are ribosomal RNA (rRNA), transfer RNA (tRNA), and messenger RNA (mRNA). Uracil, adenine, cytosine, and guanine are the four bases in RNA.

2. In bacteria, RNA polymerase binds to the promoter sequence just upstream of the start site of transcription. The enzyme moves down the DNA template and synthesizes an RNA molecule. RNA synthesis stops once the enzyme has transcribed a terminator sequence.

3. Bacterial genes encoding proteins for the same metabolic pathway are often clustered into operons. Some operons are induced in the presence of the substrate of their pathway, for example, the *lac* operon. Others are repressed in the presence of the product of the pathway, for example, the *trp* operon.

4. Eukaryotic mRNAs are modified by the addition of a 7-methyl-guanosine cap at their 5' end. A poly-A tail is added to their 3' end. Intron sequences are removed, and the exon sequences joined together by the process known as splicing. The fully processed mRNA is then ready for transport to the cytoplasm and protein synthesis.

5. In eukaryotes, there are three RNA polymerases—RNA polymerases I, II, and III. RNA polymerase II needs the help of the TATA-binding protein and other transcription factors to become bound to a promoter. This group of proteins is called the transcription preinitiation complex, and this is sufficient to make a small number of RNA molecules. However, to make a lot of RNA in response to a signal, such as a hormone, other proteins bind to sequences called enhancers. These proteins interact with the initiation complex and increase the rate of RNA synthesis.

FURTHER READING

Latchman, D. 2002. *Gene Regulation, A Eukaryotic Perspective.* 4th ed. Cheltenham: Nelson Thornes Ltd.

Lloyd, G., Landini, P., and Busby, S. 2001. Activation and repression of transcription initiation in bacteria. *Essays Biochem.* 37: 17–31.

Roberts, G. C., and Smith, C. W. 2000. Alternative splicing: Combinatorial output from the genome. *Curr. Opin. Chem. Biol.* 6: 375–383.

Tjian, R. 1995. Molecular machines that control genes. *Sci. Am.* 272: 38–45.

Quorum sensing: Hastings, J. W., and Greenberg, E. P. 1999. Quorum sensing: The explanation of a curious phenomenon reveals a common characteristic of bacteria. *J. Bacteriol.* 188: 2667–2668.

St. John's wort: Moore, L. B., Goodwin, B., Jones, S. A., Wisely, G. B., Serabjit-Singh, C. J., Willson, T. M., Collins, and Kliewer, J. L., 2000. S. A. St. John's wort induces hepatic drug metabolism through activation of the pregnane X receptor. *Proc Natl Acad Sci U S A* 13: 7500–7502.

✳ REVIEW QUESTIONS

For each question, choose the ONE BEST answer or completion.

1. A DNA strand of sequence 3′ GTCAAGGATATTCGAT 5′ would be transcribed in the nucleus to form a polymer of sequence
   A. 3′ GTCAAGGATATTCGAT 5′.
   B. 5′ GUCAAGGAUAUUCGAU 3′.
   C. 5′ CAGUUCCUAUAAGCUA 3′
   D. 3′ AUCGAAUAUCCUUGAC 5′
   E. NH_3^+-Gln-Phe-Leu-COOH

2. RNA polymerase from *E. coli* is a multisubunit protein composed of α, β, β', and σ. The promoter sequence is recognized and bound by
   A. α.
   B. β.
   C. β'.
   D. a β:β' dimer.
   E. σ.

3. A rho-independent transcription terminator sequence in prokaryotic DNA has
   A. a GC-rich region followed by a string of uracils.
   B. a GC-rich region followed by a string of adenines.
   C. several stop codons.
   D. the sequences TTGACA and TATATT separated by about 17 bases.
   E. the sequence AAUAAA and a poly-A tail up to 250 nucleotides long, separated by about 20 bases.

4. The *lac* operon is transcribed when
   A. lactose is present and glucose is absent.
   B. cAMP concentrations in the cell are high.
   C. The cAMP–CAP protein is bound to the *lac* promoter region.
   D. the *lac* repressor is bound to allolactose or a similar shaped molecule.
   E. all of the above.

5. The *trp* operon is transcribed when
   A. tryptophan concentrations in the cell are high.
   B. the *trp* repressor is bound to tryptophan or a similar shaped molecule.
   C. tryptophan is bound to its aporepressor.
   D. the appropriate corepressor is absent.
   E. all of the above.

6. Before a newly synthesized eukaryotic mRNA attaches to the ribosome, it undergoes
   A. capping.
   B. polyadenylation.
   C. splicing.
   D. transport to the cytosol.
   E. all of the above.

7. The preinitiation complex of proteins forms at a eukaryotic protein-coding gene promoter in order to
   A. align RNA polymerase I.
   B. align RNA polymerase II.

C. align RNA polymerase III.

D. align the helicase DnaB.

E. inhibit transcription.

ANSWERS TO REVIEW QUESTIONS

1. **C.** 5′ CAGUUCCUAUAAGCUA 3′. Transcription generates RNA (not protein, which is made at the next step, translation, outside the nucleus in eukaryotes). RNA is synthesized in the 5′ to 3′ direction, and the sequence of the newly synthesized strand is complementary to that of the template DNA strand.

2. **E.** It is the σ factor that binds the promoter sequence and then brings the other subunits onto the DNA.

3. **B.** The terminator sequence in DNA comprises a GC-rich region followed by a string of adenines. Concerning other answers: (C) stop codons exert their effect during translation, not during transcription. (D) The sequences TTGACA and TATATT, the −35 and −10 boxes, are found in the promoter region of prokaryotic genes, not at the end. (E) Processed eukaryotic RNA terminates with the sequence AAUAAA and a poly-A tail up to 250 nucleotides long, separated by about 20 bases. The presence of uracil should have been enough to tell you that this was not DNA!

4. **E.** The *lac* operon is transcribed when lactose is present, generating allolactose, which binds to the repressor, preventing its binding to the promoter, AND if glucose is absent, causing an increase in the cAMP concentration and therefore allowing the cAMP–CAP protein to bind to the *lac* promoter region.

5. **D.** The corepressor is tryptophan itself, and the *trp* operon is only transcribed when tryptophan is absent and therefore needs to be made by the cell. When tryptophan is present at a sufficiently high concentration (viz. A), then it binds to its aporepressor (B, C), which then binds to the promoter, preventing transcription.

6. **E.** All these processes occur before eukaryotic mRNA is translated at the ribosome.

7. **B.** RNA polymerase II transcribes protein-coding genes in eukaryotes. DnaB is part of the machinery of DNA replication that assembles, at the appropriate time, at the origin of replication—not at a promoter.

7

RECOMBINANT DNA AND GENETIC ENGINEERING

Deoxyribonucleic acid is the cell's database. Within the base sequence is all the information necessary to encode RNA and protein. A number of biological and chemical methods now give us the ability to isolate DNA molecules and to determine their base sequence. Once we have the DNA and know the sequence, many possibilities open up. We can identify mutations that cause disease, make a human vaccine in a bacterial cell, or alter a sequence and hence the protein it encodes. The knowledge of the entire base sequence of the human genome, and of the genomes of many other organisms, such as bacteria that cause disease, is set to revolutionize medicine and biology. In future years the power of genetic engineering is likely to impact ever more strongly on industry and on the way we live. This chapter describes some of the important methods involved in recombinant DNA technology at the heart of which is DNA cloning.

❋ DNA CLONING

Since DNA molecules are composed of only four nucleotides, their physical and chemical properties are very similar. Hence it is extremely difficult to purify individual species of DNA by classical biochemical techniques similar to those used successfully for the purification of proteins. However, we can use DNA cloning to help us to separate DNA molecules from each other. A **clone** is a population of cells that arose from one original cell (the **progenitor**) and, in the absence of mutation, all members of a clone will be genetically identical. If a foreign gene or gene fragment is introduced into a cell and the cell then grows and divides

Cell Biology: A Short Course, Second Edition, by Stephen R. Bolsover, Jeremy S. Hyams,
Elizabeth A. Shephard, Hugh A. White, Claudia G. Wiedemann
ISBN 0-471-26393-1 Copyright © 2004 by John Wiley & Sons, Inc.

repeatedly, many copies of the foreign gene can be produced, and the gene is then said to have been cloned. A DNA fragment can be cloned from any organism. The basic approach to cloning a gene is to take the genetic material from the cell of interest, which in the examples we will describe is a human cell, and to introduce this DNA into bacterial cells. Clones of bacteria are then generated, each of which contains and replicates one fragment of the human genetic material. The clones that contain the gene we are interested in are then identified and grown separately. We therefore use a biological approach to isolate DNA molecules rather than physical or chemical techniques.

✺ CREATING THE CLONE

How do we clone a human DNA sequence? The human genome has 3×10^9 base pairs of DNA, and the DNA content of each cell (other than eggs and sperm) is identical. However, each cell expresses only a fraction of its genes. Different types of cells express different sets of genes and thus their mRNA content is not the same. In addition, processed mRNA is shorter than its parent DNA sequence and contains no introns (page 98). Consequently, it is much easier to isolate a DNA sequence by starting with its mRNA. We therefore start the cloning process by isolating mRNA from the cells of interest. The mRNA is then copied into DNA by an enzyme called **reverse transcriptase** that is found in some viruses. As the newly synthesized DNA is complementary in sequence to the mRNA template, it is known as **complementary DNA, or cDNA.** The sample of cDNAs, produced from the mRNA, will represent the product of many different genes.

The way in which a cDNA molecule is synthesized from mRNA is shown in Figure 7.1. Most eukaryotic mRNA molecules have a string of A's at their $3'$ end, the poly-A tail (page 118). A short run of T residues can therefore be used to prime the synthesis of DNA from an mRNA template using reverse transcriptase. The resulting double-stranded molecule is a hybrid containing one strand of DNA and one of RNA. The RNA strand is removed by digestion with the enzyme ribonuclease H. This enzyme cleaves phosphodiester links in the RNA strand of the paired RNA-DNA complex, making a series of nicks down the length of the RNA. DNA polymerase (page 90) is then added. This homes in on the nicks and then moves along replacing ribonucleotides with deoxyribonucleotides. Lastly, DNA ligase (page 92) is used to reform any missing phosphodiester links. In this way a double-stranded DNA molecule is generated by the replacement of the RNA strand with a DNA strand. If the starting point had been mRNA isolated from liver cells, then a collection of cDNA molecules representative of all the mRNA molecules within the liver will have been produced. These DNA molecules have now to be introduced into bacteria.

Introduction of Foreign DNA Molecules into Bacteria

Cloning Vectors.   To ensure the survival and propagation of foreign DNAs, they must be inserted into a vector that can replicate inside bacterial cells and be passed on to subsequent generations of the bacteria. The vectors used for cloning are derived from naturally occurring bacterial plasmids or bacteriophages. Plasmids (page 74) are small circular DNA molecules found within bacteria. Each contains an origin of replication (page 88) and thus can replicate independently of the bacterial chromosome and produce many copies of itself. Plasmids often carry genes that confer antibiotic resistance on the host bacterium. The advantage of this to the scientist is that bacteria containing the plasmid can be selected for in a population of other bacteria simply by applying the antibiotic. Those bacteria with the antibiotic resistance gene will survive, whereas those without it will die. Figure 7.2 shows

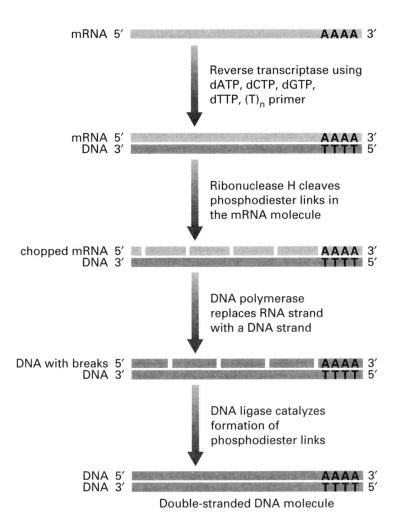

Figure 7.1. Synthesis of a double-stranded DNA molecule.

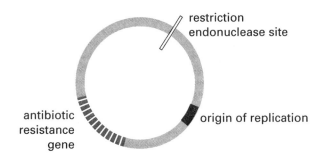

Figure 7.2. A plasmid cloning vector.

the basic components of a typical plasmid cloning vector: an antibiotic resistance gene, a restriction endonuclease site (see next section) at which foreign DNA can be inserted, and an origin of replication so the plasmid can copy itself many times inside the bacterial cell.

Example 7.1   The Cloning Vector pBluescript

The plasmid pBluescript is based on a naturally occurring plasmid that has been engineered to include several valuable features. pBluescript has an origin of replication, and the ampicillin resistance gene for selecting bacterial cells that have taken up the plasmid. The multiple cloning site (MCS) allows the scientist to cut the plasmid with the most appropriate restriction endonuclease for the task at hand.

The MCS lies within the *lac Z* gene, which codes for the enzyme β-galactosidase (page 112). β-Galactosidase converts a substrate known as X-gal to a bright blue product. Cells containing pBluescript without a foreign DNA in the MCS will produce blue colonies when grown on agar plates. However, when the *lac Z* gene is disrupted by insertion of a foreign DNA in the MCS, the cells containing the recombinant plasmids will grow to produce a colony with the normal color of white. This is because the function of the *lac Z* gene is destroyed and β-galactosidase is not produced. This is the basis of a test, called the blue/white assay, to identify colonies containing recombinant plasmids.

Another important feature of pBluescript is the presence of the bacteriophage T7 and T3 promoter sequences, which flank the MCS. These promoter sequences are used to transcribe mRNA from a cDNA cloned into one of the sites within the MCS. By selecting the promoter sequence, and adding the appropriate RNA polymerase (T3 or T7 RNA polymerase), either the sense or the antisense mRNA can be synthesized. Antisense RNAs are used in a number of techniques, for example, in situ hybridization (page 147) to detect cells producing a specific mRNA.

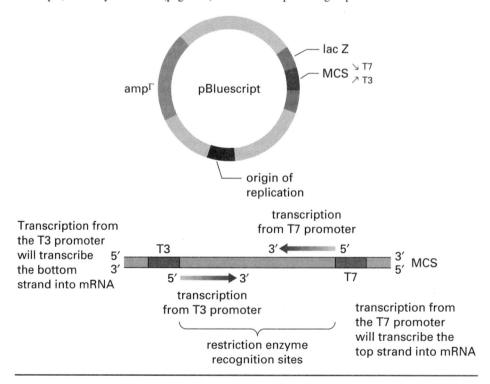

Recognition sites of some common restriction endonucleases

Bacterial species/strain	Enzyme name	Recognition sequences and cleavage sites
Bacillus amyloliquefaciens **H**	*Bam* H1	⇩ GGATCC CCTAGG ⇧
Escherichia coli Ry13	*Eco* R1	⇩ GAATTC CTTAAG ⇧
Providencia stuartii 164	*Pst* 1	⇩ CTGCAG GACGTC ⇧
Serratia marcescens SB	*Sma* H1	⇩ CCCGGG GGGCCC ⇧
Rhodopseudomonas sphaeroides	*Rsa* 1	⇩ GTAC CATG ⇧

Figure 7.3. Recognition sites of some common restriction endonucleases.

Bacteriophages (page 74) are viruses that infect bacteria and utilize the host cell's components for their own replication. The bacteriophage genome is, like a plasmid, circular, although many viruses use RNA rather than DNA as their genetic material. If human DNA is inserted into a bacteriophage, the bacteriophage will do the job of introducing it into a bacterium.

Joining Foreign DNAs to a Cloning Vector. Enzymes known as restriction endonucleases are used to insert foreign DNA into a cloning vector. Each restriction endonuclease recognizes a particular DNA sequence of (usually) 4 or 6 bp. The enzyme binds to this sequence and then cuts both strands of the double helix. Many restriction endonucleases have been isolated from bacteria. The names and recognition sequences of a few of the common ones are shown in Figure 7.3. Restriction endonuclease names are conventionally written in italics because they are derived from the Latin name for the bacterium in which the protein occurs.

Some enzymes such as *Bam* HI, *Eco* RI, and *Pst* I make staggered cuts on each strand. The resultant DNA molecules are said to have **sticky ends** (Fig. 7.4) because such fragments can associate by complementary base pairing to any other fragment of DNA generated by the same enzyme. Other enzymes such as *Sma* HI cleave the DNA smoothly to produce

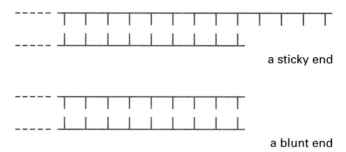

a sticky end

a blunt end

Figure 7.4. Restriction endonulceases generate two types of cut ends in double-stranded DNA.

blunt ends (Fig. 7.4). DNA fragments produced in this way can be joined to any other blunt-ended fragment.

Figure 7.5 illustrates how human DNA is inserted into a plasmid that contains a *Bam* H1 restriction endonuclease site. A short length of synthetic DNA (an oligonucleotide) that includes a *Bam* H1 recognition site is added to each end of the human DNA fragment. Both the human DNA and the cloning vector are cut with *Bam* H1. The cut ends are now complementary and will anneal together. DNA ligase then catalyzes the formation of phosphodiester links between the vector and the human DNA. The resultant molecule is known as a **recombinant plasmid.** If our starting material was mRNA from a sample of liver, we would now have a collection of plasmids each carrying a cDNA from one of the genes that was being transcribed in this organ.

Introduction of Recombinant Plasmids into Bacteria. Figure 7.6 summarizes how recombinant plasmids are introduced into bacteria such as *Escherichia coli.* Bacteria are first treated with concentrated calcium chloride to make the cell wall more permeable to DNA. DNA can now enter these cells, which are said to have been made **competent.** Cells that take up DNA in this way are said to be **transformed.** The transformation process is very inefficient, and only a small percentage of cells actually take up the recombinant molecules. This means that it is extremely unlikely that any one bacterium has taken up two plasmids. The presence of an antibiotic resistance gene in the cloning vector makes it possible to select those bacteria that have taken up a molecule of foreign DNA, since only the transformed cells can survive in the presence of the antibiotic. The collection of bacterial colonies produced after this selection process is a clone library. All the cells of a single colony harbor identical recombinant molecules that began as one mRNA molecule in the original cell sample. Other colonies in the same clone library contain plasmids carrying different DNA inserts. Isolating individual bacterial colonies will produce different clones of foreign DNA. In the example we have described, where the starting DNA material used to produce these clones was a population of cDNA molecules, the collection of clones is called a **cDNA library.**

Selection of cDNA Clones

Having constructed a cDNA library—which may contain many thousands of different clones—the next step is to identify the clones that contain the cDNA of interest. There

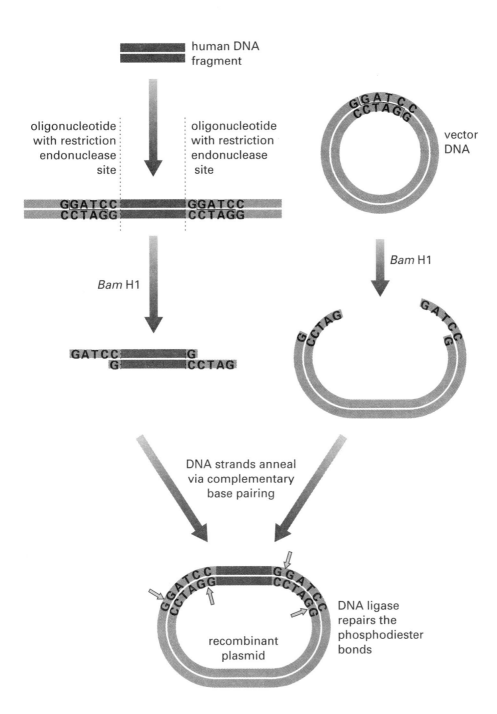

Figure 7.5. Generation of a recombinant plasmid.

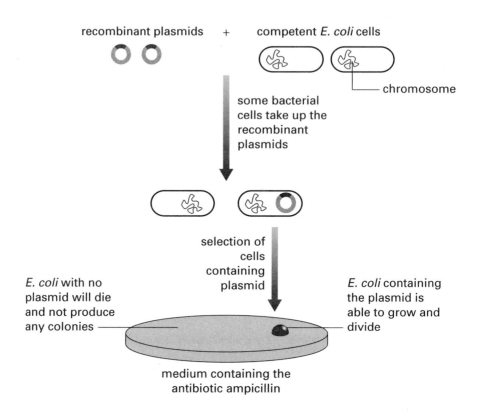

Figure 7.6. Introduction of recombinant plasmids into bacteria.

are many ingenious ways of doing this. We will describe two ways of selecting cDNA clones from a library. One method simply detects the presence of the foreign DNA attached to the plasmid vector, and the second detects the protein encoded for by the foreign DNA. We call the process of selecting specific clones "screening the library".

Preparation of the cDNA Library for Screening.   Bacterial colonies are plated onto agar plates, and the colonies are then replica-plated onto a nylon membrane, which is then treated with detergent to burst (or lyse) the bound cells. If the clone is to be selected by virtue of its DNA sequence (Fig. 7.7), the nylon membrane is processed with sodium hydroxide. This is necessary to break all hydrogen bonds between the DNA strands bound to the nylon membrane and ensures that the DNA is single-stranded. The processed nylon membrane is an exact replica of the DNA contained within each bacterial colony on the agar plate. If the clone is to be selected from the library by detecting the protein encoded by the foreign DNA (Fig. 7.8), then colonies are again replica-plated on to a nylon membrane. This time, however, the nylon membrane is processed to produce an exact copy of the proteins synthesized by each bacterial colony.

Oligonucleotide Probes for cDNA Clones.   If some of the amino acid sequence is known for the protein whose cDNA is to be cloned, an oligonucleotide can be synthesized in vitro (i.e., in a machine rather than in a cell) that has a sequence complementary to

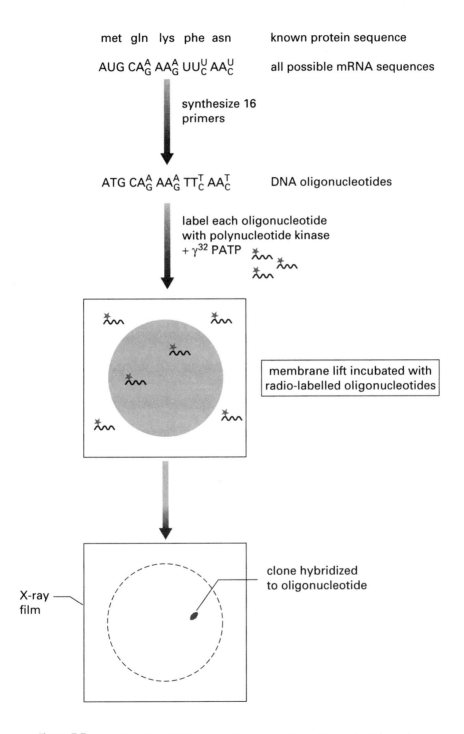

Figure 7.7. Selection of a cDNA clone with a radioactive oligonucleotide probe.

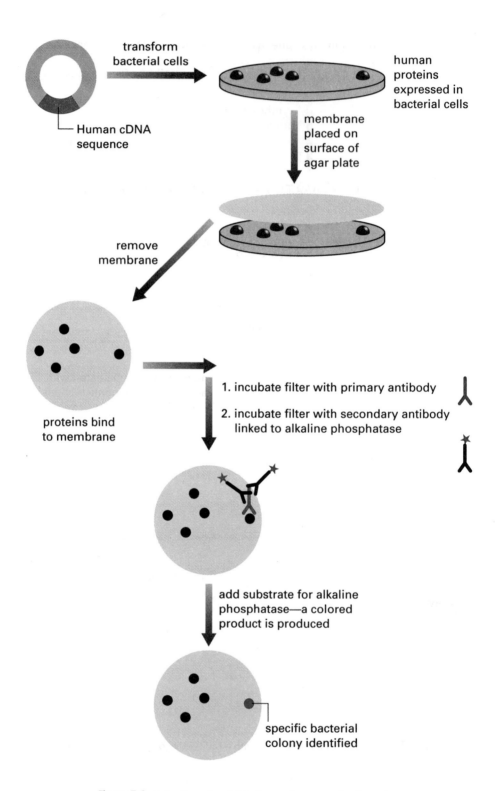

Figure 7.8. Selection of a cDNA clone with an antibody probe.

one of the strands of the cDNA. The first step is to use the genetic code to predict all possible DNA sequences that could code for a short stretch of amino acids within the protein. This strategy is shown in Figure 7.7. The sequence—met gln lys phe asn—can be coded for by 16 possible sequences, because of the redundancy of the genetic code. All 16 oligonucleotide sequences are synthesized. One of the 16 sequences will be complementary to the cDNA we want to select from the library. The oligonucleotides are tagged with a radioactive phosphate group (^{32}P) at their 5′ ends using the enzyme polynucleotide kinase (PNK) and the substrate [γ-^{32}P]ATP, that is, ATP whose γ phosphate is the radioactive isotope ^{32}P. PNK removes the 5′-phosphate group from each oligonucleotide, leaving a 5′-hydroxyl group. The enzyme then transfers the γ(^{32}P) phosphate group of [γ-^{32}P]ATP to the 5′-hydroxyl.

The nylon membrane to which the library DNA is attached is incubated together with the mixture of radiolabeled oligonucleotides. This process is called **hybridization,** a word used whenever two nucleic acid strands associate together by hydrogen bonding. In this case the oligonucleotide complementary in sequence to the clone we want to select will hydrogen bond to the single-stranded DNA on the nylon membrane. Once hybridization is complete, excess oligonucleotide is washed from the nylon membrane, which is now covered with a sheet of X-ray film and placed in a light-tight cassette. The radioactivity in the oligonucleotide will darken the silver grains on the X-ray film—a process known as **autoradiography.** A positive clone will show up as a black spot on the film. Superimposing the X-ray film back onto the original bacterial plate will identify the living bacterial colony that contains the desired foreign DNA clone.

Antibody Probes for cDNA Clones. This method makes use of specific antibodies to detect bacteria expressing the protein product of the DNA to be cloned. For this to work, the foreign DNA must be expressed in the bacterial cells; that is to say, its information must be copied first into mRNA and then into protein. To ensure efficient expression, the plasmid vector contains a bacterial promoter sequence that is used to control transcription of foreign DNA. Such cloning vectors are known as expression vectors. The promoter of the *lac* operon is commonly used in this way. The clone library is plated onto agar plates containing an inducer of the *lac* operon such as IPTG (page 114) to ensure that lots of mRNA and in turn lots of protein is synthesized. Figure 7.8 shows how an antibody, linked to an enzyme (usually alkaline phosphatase), can detect a positive clone by generating a colored product. The pattern of colored spots on the nylon membrane is used to identify the bacterial clones of interest on the original agar plate.

Genomic DNA Clones

The approach described in the previous section permits the isolation of cDNA clones. Complementary DNA clones have many important uses, some of which are described below. However, as a cDNA is a copy of mRNA only, when we want to isolate a gene to investigate its structure and function, we need to create genomic DNA clones. Genes contain exons and introns and have regulatory sequences at their 5′ and 3′ ends and are therefore much larger than cDNAs. The vectors used to clone genes must therefore be able to hold long stretches of DNA. Plasmids used for cDNA cloning cannot do this. A selection of vectors used to clone genes is shown in Table 7.1. Vectors such as the P1 artificial chromosomes (PACs)—based on the bacteriophage P1—can hold about 150,000 bp of DNA. PACs have been used very successfully in the Human Genome Project (page 144). Yeast artificial chromosome

Table 7.1. Vectors Used for Cloning Genomic DNA

Genomic DNA Cloning Vector	Size of Insert (kb)
Bacteriophage λ	9–23
Cosmid	30–44
PAC (P1 artificial chromosome)	130–150
BAC (bacterial artificial chromosome)	Up to 300
YAC (yeast artificial chromosome)	200–600

(YAC) vectors can hold between 200,000 and 600,000 million base pairs of foreign DNA. The choice of genomic vector is governed by the size of DNA the scientist needs to clone. PAC, Bacterial artificial chromosome (BAC), and YAC vectors are needed if an entire gene sequence is to be represented in a single clone. PAC, BAC, and YAC vectors contain all the sequences needed to produce a minichromosome in the appropriate cell type: PACs and BACs in bacterial cells and YACs, as their name implies, to replicate foreign sequences in a yeast cell. The YAC vector therefore contains sequences that will allow the DNA to be processed by the host yeast cell as if it were a normal chromosome and to allow replication alongside the other chromosomes in the cell. Thus a YAC vector has sequences that specify the yeast centromere (page 403), telomeres (the ends of the new chromosome), and the yeast origin of replication. A YAC vector also contains selectable marker genes so that only those cells that have been transformed by a correctly constructed YAC chromosome will survive.

To generate the large DNA fragments needed for genomic cloning, the chromosomal DNA is incubated with a restriction endonuclease for a very short time. Not all the recognition sites for that enzyme are cleaved, and large fragments of DNA are hence produced by partial digestion. Genomic DNA fragments are joined to genomic cloning vectors in the same way we join cDNAs to cDNA cloning vectors. In the example shown in Figure 7.9, human DNA has been introduced into the genome of a bacteriophage known as lambda (λ). This particular vector can accommodate up to 23,000 bp of foreign DNA in its genome. Bacteria are then infected, generating a collection of bacteria called a genomic DNA library.

To select the genomic DNA sequence of interest, the library is plated onto a layer (or lawn) of cultured bacteria so that many copies of the recombinant bacteriophage can be produced. A single λ bacteriophage infects a single *E. coli*. The recombinant bacteriophages then multiply inside the host cells. The cells die and lyse, and the bacteriophages spread to the surrounding layer of bacteria and infect them. These cells lyse, in turn, and the process

Example 7.2   **Cloning a Receptor Protein cDNA**

Glutamate is one of the most important transmitters in the brain. The gene coding for the glutamate receptor, the protein on the surface of nerve cells that, upon binding glutamate, allows an influx of sodium ions (page 372), remained uncloned for a number of years. Success came with the use of a very clever cloning strategy, based on the function of the receptor. mRNA was isolated from brain cells and used as the template for the production of cDNA molecules. These were inserted into a plasmid expression vector. Following the introduction of these cDNAs into bacteria, a cDNA library representative of all the mRNAs in the brain was produced. The many thousands of cDNA clones in the library were then divided into pools. Each of the many pools was then injected into a frog egg (*Xenopus* oocyte), which transcribed the cDNAs into RNA and translated the RNA into protein. To see which of the oocytes had been injected with the cDNA for the glutamate receptor, these cells were whole cell patch clamped (page 321). Glutamate was applied to the oocytes, and the oocyte whose injected pool had included the cDNA for the glutamate receptor responded with an inward current of sodium ions indicating the presence of glutamate receptors in the plasma membrane.

The pool of cDNAs giving this response was further divided into smaller pools. Each of these was rescreened for the presence of glutamate receptor activity. This was followed by several rounds of rescreening. For each round a further subdivision was made of the cDNAs into pools containing fewer and fewer cDNA molecules. Eventually each pool contained only a single cDNA so that the cDNA for the glutamate receptor could be identified. A number of other receptors have now been isolated using the same strategy in which a functional assay is used to identify the cDNA encoding the receptor.

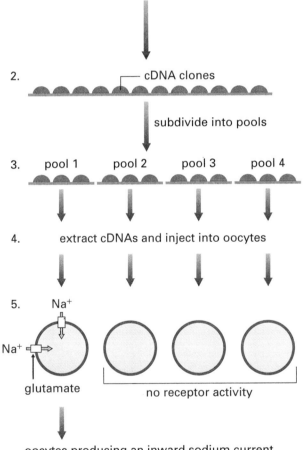

1. mRNA from brain used to make cDNA

2. cDNA clones

subdivide into pools

3. pool 1   pool 2   pool 3   pool 4

4. extract cDNAs and inject into oocytes

5. Na⁺

Na⁺

glutamate   no receptor activity

oocytes producing an inward sodium current
on glutamate application identify the pool containing
the cDNA coding for the glutamate receptor

repeat steps 4 and 5 on pool 1 cDNAs, each
time using smaller cDNA subpools until a
single cDNA giving a positive response in
the receptor assay has been identified

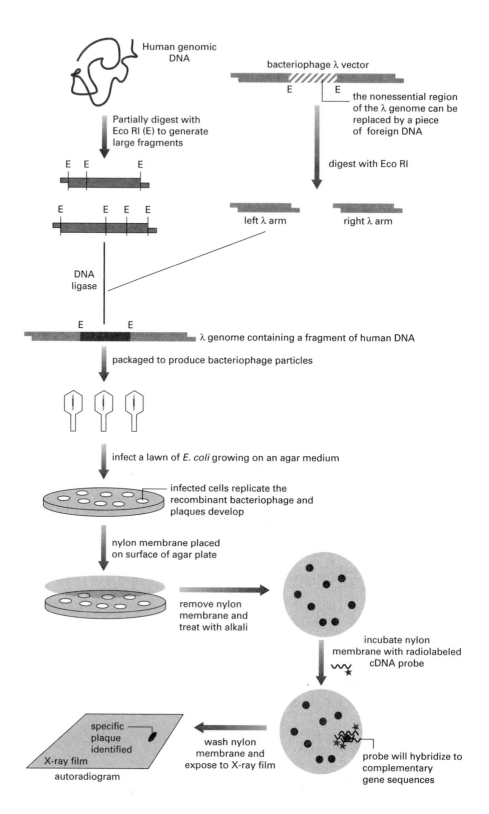

Figure 7.9. Generation and selection of genomic DNA clones.

is repeated. The dead cells give rise to a clear area on the bacterial lawn called a plaque. Each plaque contains many copies of a recombinant bacteriophage that can be transferred to a nylon membrane (Fig. 7.9). Specific DNA clones are selected by incubating the nylon membrane with a radiolabeled cDNA probe complementary to the genomic sequence being searched for. This produces a radioactive area on the nylon membrane that is identified by autoradiography. The use of a cDNA sequence as a **gene probe** makes the task of isolating the corresponding genomic sequence much easier.

A radioactive cDNA probe can be synthesized using the method called random priming. The cDNA clone is heated so that the two strands will separate. Each strand will act as the template for the synthesis of a new DNA strand. A mixture of random hexamers, six nucleotide sequences, containing all possible combinations of the four bases (A, T, C, G) is added to the denatured cDNA along with DNA polymerase and the four deoxynucleotides dATP, dTTP, dCTP, and dGTP. The hexamers will hydrogen-bond to their corresponding sequences on the cDNA templates and prime the synthesis of new DNA strands. If, for example, radiolabeled [α-^{32}P]dATP is included in the reaction, the newly synthesized DNA strands will be radioactive.

✳ USES OF DNA CLONES

DNA Sequencing

The ability to determine the order of the bases within a DNA molecule has been one of the greatest technical contributions to molecular biology. DNA is made by the polymerization of the four deoxynucleotides dATP, dGTP, dCTP, and dTTP. These are joined together when DNA polymerase catalyzes the formation of a phosphodiester link between a free 3'-hydroxyl on the deoxyribose sugar moiety of one nucleotide and a free 5'-phosphate group on the sugar residue of a second nucleotide. However, the artificial dideoxynucleotides ddATP, ddGTP, ddCTP, and ddTTP have no 3'-hydroxyl on their sugar residue (Fig. 7.10), and so if they are incorporated into a growing DNA chain, synthesis will stop. This is the basis of the dideoxy chain termination DNA sequencing technique devised by Frederick Sanger and for which he was awarded a Nobel prize in 1980.

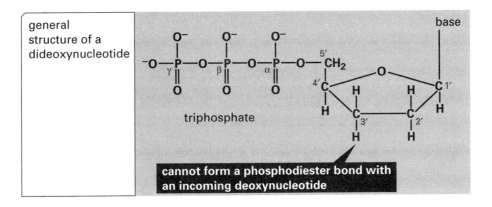

Figure 7.10. A dideoxynucleotide.

This technique is illustrated in Figure 7.11. A cloned piece of DNA of unknown sequence is first joined to a short oligonucleotide whose sequence is known. The DNA is then made single-stranded so that it can act as the template for the synthesis of a new DNA strand. All DNA synthesis requires a primer; in this case a primer is provided that is complementary in sequence to the oligonucleotide attached to the template DNA. Four separate mixtures are prepared. Each contains the DNA template, the primer (which has been radiolabeled), DNA polymerase, and the four deoxynucleotides. The mixtures differ in that each also contains a low concentration of one of the four dideoxynucleotides ddATP, ddGTP, ddCTP, or ddTTP. When a molecule of dideoxynucleotide is joined to the newly synthesized chain, DNA synthesis will stop.

Let us follow what happens in the tube containing ddTTP in Figure 7.11. The first base that DNA polymerase encounters in the DNA template to be sequenced is an A. Since the tube contains much more dTTP than ddTTP, DNA polymerase will add a dTTP to most of the primer molecules. However, a small fraction of the primers will have a ddTTP added to them instead of dTTP since DNA polymerase can use either nucleotide as a substrate. The next base encountered is a G. DNA polymerase is unable to attach dCTP to the ddTTP since there is no OH group on the 3′ carbon of the sugar, and so DNA synthesis is terminated. The majority of strands, however, had a dTTP added, and for these DNA polymerase can proceed, building the growing strand. No problems are encountered with the next six bases. However, the eighth base in the template strand is another A, and once again a small fraction of the growing strands will have ddTTP added instead of dTTP. In the same way as before, these strands can grow no further. This process will be repeated each time an A occurs on the template strand. When the reaction is over, the tube will contain a mixture of DNA fragments of different length, each of which ends in a ddTTP. Similarly, each of the other three tubes will contain a mixture of DNA chains of different length, each of which ends in either ddCTP, ddATP, or ddGTP. To determine the sequence of the newly synthesized chains, each of the four samples is loaded onto a polyacrylamide gel. Because the primer used was radiolabeled, all the new DNA chains will carry a radioactive tag so that, after electrophoresis, the pattern of DNA fragments on the gel can be detected by autoradiography. Each terminated reaction will show up as a black band on the X-ray film. The smallest DNA molecules are that fraction in the tube containing ddTTP whose growth was blocked after the first base, T. These move farthest and produce the band at the bottom of the T lane. DNA molecules one nucleotide larger were produced in the tube containing ddCTP—their growth was blocked after the second base, C. These move almost as far, but not quite, producing the band at the bottom of the C lane. Reading bands up from the bottom of the gel therefore tells us the sequence in which bases were added to the unknown strand: T, C, and so on. Because the new chain is complementary in sequence to its template strand, the sequence of the template strand can be inferred.

The Human Genome Project required the process of DNA sequencing to be automated. Instead of using radioactivity and one reaction tube for each dideoxynucleotide, we now use dideoxynucleotides tagged with fluorescent dyes. Each of the four dideoxynucleotides ddATP, ddGTP, ddCTP, and ddTTP is tagged with a different fluorescent dye. This means that all four of the termination reactions can be carried out in a single reaction tube and loaded into the same well on the polyacrylamide gel. As the reaction product drips from the bottom of the gel, the fluorescence intensity of each of the four colors corresponding to the four dideoxynucleotides is monitored, and this information is transferred straight to a computer where the data are analyzed. An example of a DNA trace produced using fluorescently tagged dideoxynucleotides is shown on the CBASC website. Each peak represents a terminated

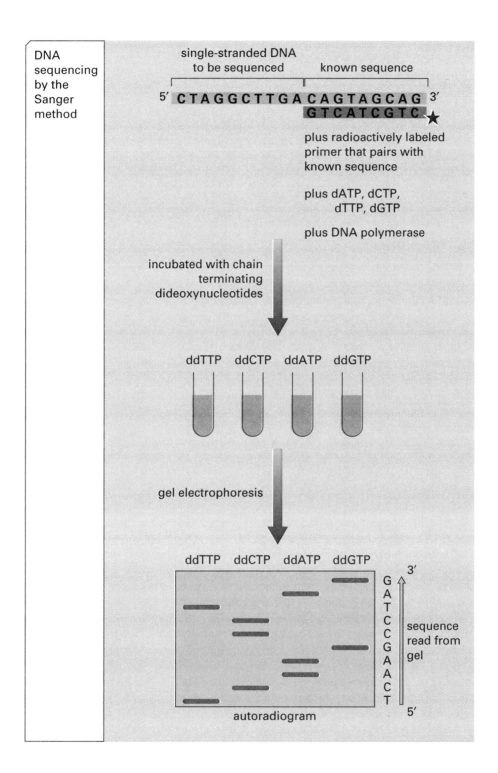

Figure 7.11. DNA sequencing and the dideoxy chain termination method.

DNA product; so by reading the order of the peaks, the DNA sequence is determined. In this example G is yellow, A is green, T is red, and C is blue.

To determine the base sequence of the entire human genome, the DNA was cut by partial restriction endonuclease digestion (page 140) to give large fragments of about 150,000 bp. These were cloned into a vector such as PAC (Table 7.1). The aim was to create a library of clones that overlapped one another. Each clone was sequenced, as described above, and their base sequences compared. Because the clones overlapped, it was possible, by comparing their sequences, to line up the position of each clone with respect to its neighbors. This required the development of sophisticated computer programs, and the creation of a large database of information to order the 3×10^9 base pairs that comprise the human genome.

Southern Blotting

In 1975 Ed Southern developed an ingenious technique, now known as Southern blotting, which can be used to detect specific genes (Fig. 7.12). Genomic DNA is isolated and digested with one or more restriction endonucleases. The resultant fragments are separated according to size by agarose gel electrophoresis. The gel is soaked in alkali to break hydrogen bonds between the two DNA strands and then transferred to a nylon membrane. This produces an exact replica of the pattern of DNA fragments in the agarose gel. The nylon membrane is incubated with a cloned DNA fragment tagged with a radioactive label (page 143). The gene probe is heated before adding to the nylon membrane to make it single-stranded so it will base pair, or hybridize, to its complementary sequences on the nylon membrane. As the gene probe is radiolabeled, the sequences to which it has hybridized can be detected by autoradiography.

Mutations that change the pattern of DNA fragments—for instance, by altering a restriction endonuclease recognition site or deleting a large section of the gene—can easily be detected by Southern blotting. This technique is therefore useful in determining whether an individual carries a certain genetic defect or if a fetus is homozygous for a particular disorder. All that is needed is a small DNA sample from white blood cells or, in the case of a fetus, from the amniotic fluid in which it is bathed, or by removing a small amount of tissue from the chorion villus that surrounds the fetus in the early stages of pregnancy.

Forensic laboratories use Southern blotting to generate DNA fingerprints from samples of blood or semen left at the scene of a crime. A DNA fingerprint is a person-specific Southern blot. The gene probe used in the test is a sequence that is repeated very many times within the human genome—a microsatellite sequence (page 100). Everyone carries a different number of these repeated sequences, and because they lie adjacent to each other on the chromosome they are called VNTRs (variable number tandem repeats). When genomic DNA is digested with a restriction endonuclease and then analyzed by Southern blotting, a DNA pattern of its VNTRs is produced. Unless they are identical twins, it is extremely unlikely that two individuals will have the same DNA fingerprint profile. It has been estimated that if eight restriction endonucleases are used, the probability of two people who are not identical twins generating the same pattern is one in 10^{30}.

A special type of Southern blot, called a zoo blot, can be used to reveal genes that are similar in different species. Such genes, which have been conserved through evolution, are likely to code for crucial proteins. A probe generated from a genomic DNA library is used to probe genomic DNA from many different species. A probe that hybridizes with DNA from a number of species is likely to represent all or part of such a conserved gene.

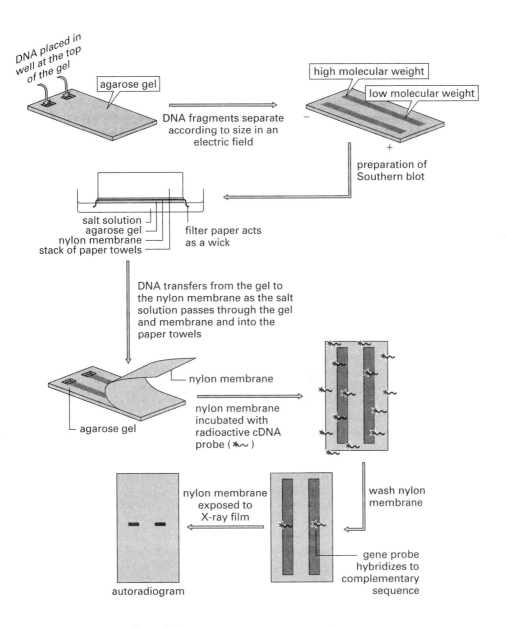

Figure 7.12. The technique of Southern blotting.

In situ Hybridization

It is possible, using the technique of **in situ hybridization,** to identify individual cells ex-pressing a particular mRNA. To do this we need to synthesize an antisense RNA molecule—an RNA that is complementary in sequence to the mRNA of interest. In a test tube, the appropriate strand of the cloned cDNA is copied into the antisense RNA using RNA poly-merase. The RNA is then labeled with a modified nucleotide that can subsequently be detected using an antibody and a color reaction. The cDNA must first be cloned into an

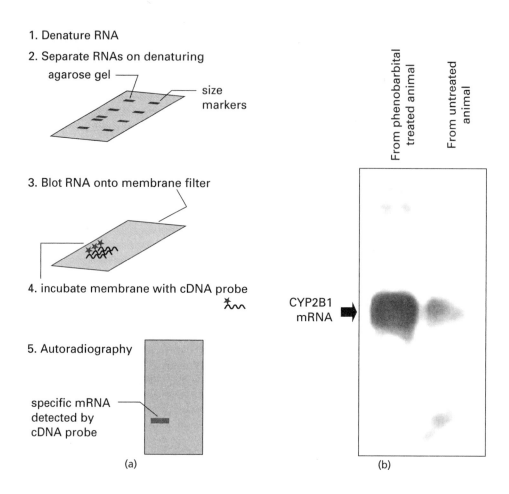

1. Denature RNA

2. Separate RNAs on denaturing
   agarose gel
   size markers

3. Blot RNA onto membrane filter

4. incubate membrane with cDNA probe

5. Autoradiography

specific mRNA detected by cDNA probe

(a)

From phenobarbital treated animal

From untreated animal

CYP2B1 mRNA

(b)

Figure 7.13. (a). The technique of northern blotting. (b). A northern blot reveals that the *CYP2B1* gene is upregulated in animals given phenobarbital.

expression vector that contains a promoter sequence to which RNA polymerase can bind. A thin tissue section, attached to a glass microscope slide, is incubated with the antisense RNA. The antisense RNA will hybridize, in the cell, to its complementary mRNA partner. Excess antisense RNA is washed off the slide, leaving only the hybridized probe. The color reaction is now carried out so that the cells expressing the mRNA of interest can be identified by bright-field microscopy (page 3).

Northern Blotting

Figure 7.13*a* shows a blotting technique that can determine the size of an mRNA and tell us about its expression patterns. RNA is denatured by heating to remove any intramolecular double-stranded regions and then electrophoresed on a denaturing agarose gel. The RNA is transferred to a nylon membrane (as described on page 136 for the transfer of DNA). The nylon membrane is incubated with a radiolabeled, single-stranded cDNA probe

Table 7.2. Blotting Techniques

	What Is Probed	Nature of the Probe	Book Page for Description
Southern blot	DNA	DNA	146
Northern blot	RNA	cDNA or RNA	148
Western blot	Protein	Antibody	166

(page 143), or an antisense RNA probe (page 147). Following hybridization, excess probe is washed off and the nylon membrane exposed to X-ray film. The mRNA is visualized on the autoradiogram (page 139) because it hybridized to the radioactive probe. By analogy with Southern blotting, this technique is called **northern blotting** (Table 7.2). Figure 7.13*b* shows a northern blot for a cytochrome P450 (page 249) mRNA known as CYP2B1. The gene for this protein is activated in the liver by the barbiturate phenobarbital and hence a lot of CYP2B1 mRNA is made.

Production of Mammalian Proteins in Bacteria

The large-scale production of proteins using cDNA-based expression systems has wide applications for medicine and industry. It is increasingly being used to produce polypeptide-based drugs, vaccines, and antibodies. Such protein products are called **recombinant** because they are produced from a recombinant plasmid. For a mammalian protein to be synthesized in bacteria its cDNA must be cloned into an expression vector (as described on page 139). Insulin was the first human protein to be expressed from a plasmid introduced into bacterial cells and has now largely replaced insulin from pigs and cattle for the treatment of diabetes. Other products of recombinant DNA technology include growth hormone and factor VIII, a protein used in the treatment of the blood clotting disorder hemophilia. Factor VIII was previously isolated from donor human blood. However, because of the danger of infection from viruses such as HIV, it is much safer to treat hemophiliacs with recombinant factor VIII. It should, in theory, be possible to express any human protein via its cDNA.

Protein Engineering

The ability to change the amino acid sequence of a protein by altering the sequence of its cDNA is known as protein engineering. This is achieved through the use of a technique known as site-directed mutagenesis. A new cDNA is created that is identical to the natural one except for changes designed into it by the scientist. This DNA can then be used to generate protein in bacteria, yeast, or other eukaryotic cell lines.

The first use of protein engineering is to study the protein itself. A comparison of the catalytic properties of the normal and mutated form of an enzyme helps to identify amino acid residues important for substrate and cofactor binding sites (Chapter 11). This technique was also used to identify the particular charged amino acid residues responsible for the selectivity of ion channels (page 437). Now scientists are using protein engineering to generate new proteins as tools, not only for scientific research but for wider medical and industrial purposes.

Subtilisin is a protease and is one of the enzymes used in biological washing powder. The natural source of this enzyme is *Bacillus subtilis,* an organism that grows on pig feces. To produce, from this source, the 6000 tons of subtilisin used per year by the soap powder industry is a difficult and presumably unpleasant task. The cDNA for subtilisin has been isolated and is now used by industry to synthesize the protein on a large-scale in *E. coli*. The wild-type (natural) form of subtilisin is, however, prone to oxidation because of a methionine present at position 222 in the protein. Its susceptibility to oxidation makes it an unsuitable enzyme for a washing powder that must have a long shelf life and be robust enough to withstand the rigours of a washing machine with all its temperature cycles. Scientists therefore changed the codon for methionine (AUG) to the codon for alanine (GCG). When the modified cDNA was expressed in *E. coli,* the resulting enzyme was found to be active and not susceptible to oxidation. This was excellent news for the makers of soap powder. However, it is always necessary to check the kinetic parameters of a "new" protein produced from a modified cDNA. For subtilisin (met$_{222}$) the K_M (page 244) is 1.4×10^{-4} M while for subtilisin (ala$_{222}$) the K_M is 7.3×10^{-4} M. This means that at micromolar concentrations of dirt the modified enzyme will bind less dirt than the wild-type one, but the dirt concentrations caked onto our clothes are well above micromolar. The product turnover number, k_{cat} (page 242), is 50 s^{-1} for subtilisin (met$_{222}$) and 40 s^{-1} for subtilisin (ala$_{222}$): The mutant enzyme is slightly slower, but not by much. By changing a met to an ala, a new enzyme has been produced that can do a reasonable job and is stable during storage and in our washing machines.

Green fluorescent protein is found naturally in certain jellyfish. Protein engineering has now created a palette of proteins with different colors (blue, cyan, green, and yellow). However, the great advantage of these proteins to biologists is that **chimeric proteins** (proteins composed of two parts, each derived from a different protein) incorporating a fluorescent protein are intrinsically fluorescent. This means that our protein of interest can be imaged inside a living cell using a fluorescence microscope (page 6). The fluorescent part of the chimeric protein tells us exactly where our protein is targeted in the cell and if this location changes in response to signals.

Figure 7.14 illustrates how this approach can be used to determine what concentration of glucocorticoid drug is required to cause the glucocorticoid receptor to move to the nucleus. The plasmid, like many plasmids designed for convenience of use, contains a multiple cloning site (MCS), sometimes called a polylinker, which is a stretch of DNA that contains several restriction endonuclease recognition sites. A convenient restriction endonuclease is used to cut the plasmid (in this case, for green fluorescent protein) and the cDNA for the glucocorticoid receptor is inserted. The plasmid also contains a promoter sequence, derived from a virus, that will drive the expression of the DNA into mRNA in mammalian cells that have been infected with the plasmid (or **transfected**). The chimeric protein, synthesized from the mRNA, will have one part coded for by the cDNA of interest—the other part being the fluorescent protein. The plasmid is grown up in bacteria and then used to transfect mammalian cells. In the absence of glucocorticoid the protein, and therefore the green fluorescence, is in the cytosol. When enough glucocorticoid is added, it binds to the chimeric protein, which then moves rapidly to the nucleus.

Polymerase Chain Reaction

A technique called the polymerase chain reaction (PCR) has revolutionized recombinant DNA technology. It can amplify DNA from as little material as a single cell and from very old tissue such as that isolated from Egyptian mummies, a frozen mammoth, and insects trapped in ancient amber. A simple mouth swab can yield enough cheek cell DNA to determine carriers of a particular recessive genetic disorder. PCR is used to amplify DNA

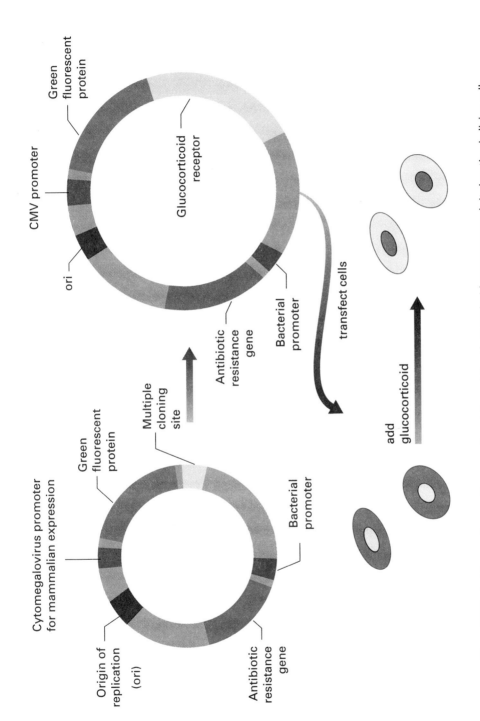

Figure 7.14. A chimera of green fluorescent protein and the glucocorticoid receptor reveals its location in living cells.

from fetal cells or from small amounts of tissue found at the scene of a crime. The tool that makes PCR possible is thermostable DNA polymerase, an enzyme that survives and can even function at extremely high temperatures that would denature (page 206) most enzymes. Thermostable DNA polymerases are isolated from bacteria that live in extremely hot deep-sea volcanic environments.

Figure 7.15 shows how PCR uses a thermostable DNA polymerase and two short oligonucleotide DNA sequences called primers. Each primer is complementary in sequence to a short length of one of the two strands of DNA to be amplified. The primers direct DNA polymerase to copy each of the template strands. The DNA duplex is heated to 90°C to separate the two strands (step 1). The mixture is cooled to 60°C to allow the primers to anneal to their complementary sequences (step 2). At 72°C the primers direct the thermostable DNA polymerase to copy each of the template strands (step 3). These three steps, which together constitute one cycle of the PCR, produce twice the number of original templates. The process of template denaturation, primer annealing, and DNA synthesis is repeated many times to yield many thousands of copies of the original target sequence.

Identifying the Gene Responsible for a Disease

Until recently, the starting point for an identification of the gene responsible for a particular inherited disease was a pattern of inheritance in particular families plus a knowledge of the tissues affected. It is very difficult to find the gene responsible for a disease when the identity of the normal protein is unknown. Very often, the first clue is the identification of other genes that are inherited along with the malfunctioning gene and that are therefore likely to lie close on the same chromosome (this is **linkage,** page 408). In the past, **chromosome walking** was then used to identify the disease gene. This is a slow and tedious process that involves isolating a series of overlapping clones from a genomic clone library. One starts by isolating a single clone and then uses this as a probe to find the next clone along the chromosome. The second clone is used to find the third, and this is used to find a fourth clone, and so on. Each successive clone is tested to find out if it might include all or part of the gene of interest (e.g., by using northern blotting, page 148, to see if the gene is expressed in a tissue known to be affected in the disease). Once a candidate gene is identified, one can find out if it was the gene of interest by sequencing it in unaffected individuals and in disease sufferers; if it is the disease gene, its sequence will be different in the affected people. With the publication of the entire human genome, chromosome walking to generate a series of overlapping clones for testing is unnecessary, but identifying the gene that is responsible for a particular inherited condition is even now a time-consuming task.

Reverse Genetics

Because beginning with an inherited defect in function and working toward identification of a gene is even today a time-consuming task, more and more scientists are now working the other way: they take a gene with a known sequence but unknown function and deduce its role. Since we know the complete genome of a number of species, we can sit at a computer and identify genes that look interesting—for example, because their sequence is similar to a gene of known function. The gene of interest can be mutated and reinserted into cells or organisms, and the cells or organisms tested for any altered function. This approach is called **reverse genetics.**

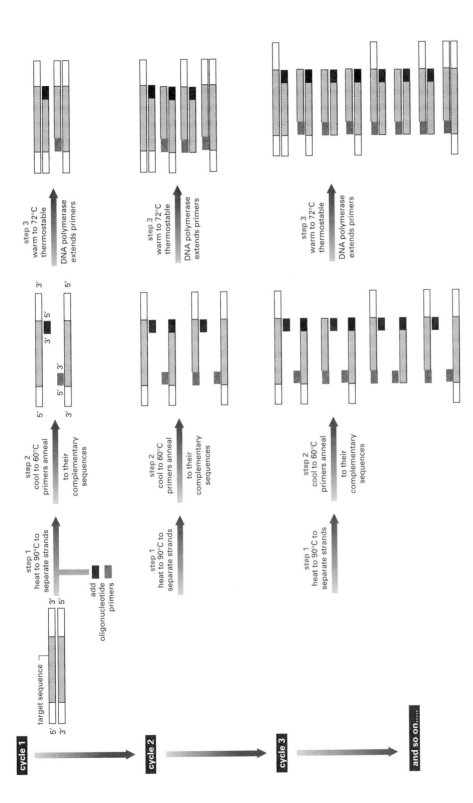

Figure 7.15. Amplification of a DNA sequence using the polymerase chain reaction.

Microarrays and Cancer Classification

Microarrays or gene chips are tiny glass wafers to which cloned DNAs are attached. The principle of microarray technology is to isolate mRNA from a particular cell and to hybridize this to the DNAs on the chip. Because the mRNAs are tagged with a fluorescent dye, the DNAs to which they hybridize on the chip can be detected. Excess mRNA is removed, and fluorescent areas are viewed using a special scanner and microscope. Computer algorithms have been written to analyze the hybridization patterns seen for a particular microarray. The number and type of DNAs used to make a microarray is dependent on the question to be answered.

One of the spin-offs of the Human Genome Project is to identify sets of genes involved in disease. Microarrays are being used to type the mRNAs produced by different cancers in the hope that this will lead to better diagnosis and therefore better treatment. Leukemia, a cancer affecting the blood, is not a single type of disease. Microarray analysis is helping to classify different types of leukemia more precisely by cataloging the mRNAs expressed in different patients. Preliminary studies are encouraging. A microarray of 6817 cDNAs was used to compare the blood mRNAs from patients with acute lymphoblastic leukemias or acute myeloid leukemias. The mRNAs of each patient was passed over the microarray and complementary sequences hybridized. Fifty cDNA/mRNA hybrids were found that could be used to refine the classification of lymphoblastic and myeloid leukemias. In addition, the patterns allowed the classification of lymphoblastic leukemias into T-cell or B-cell classes. This separation of lymphoblastic leukemias into two classes is an important distinction when it comes to deciding on the best treatment for a patient. As more patients with leukemia are investigated using microarrays, it should be possible to design smaller chips, with fewer cDNAs, but with greater prognosis value.

A recent study has shown that the fate of women with breast cancer can be predicted by microarray analysis of their cancer tissue. Several genes were shown to be important for this prediction. If the women had a certain gene expression pattern, then the disease was likely to recur within a 5-year period. However, if the gene pattern was of a second type, the cancer almost never returned. Women that fall into this second class also show no additional benefit from chemotherapy or radiotherapy. Microarray analysis not only offers a survival prognosis for women with breast cancer but has also shown that unpleasant treatment for the disease is unnecessary for those possessing the second type of gene pattern.

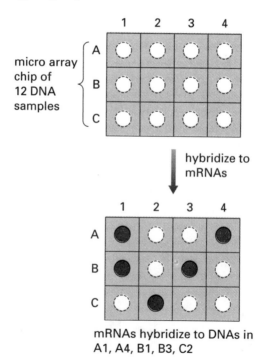

micro array
chip of
12 DNA
samples

hybridize to
mRNAs

mRNAs hybridize to DNAs in
A1, A4, B1, B3, C2

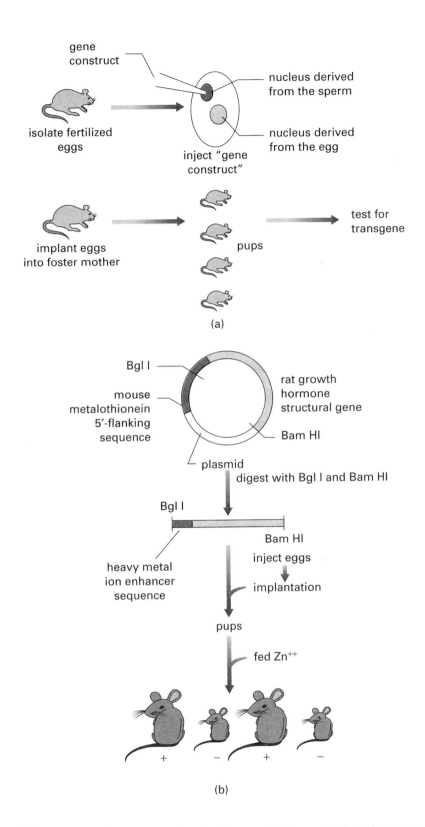

Figure 7.16. (a). Transgenic mouse carrying a foreign gene. (b). the metalothionein gene contains a heavy-metal ion enhancer sequence. The + mice carry the transgene while littermates without the transgene are indicated by −.

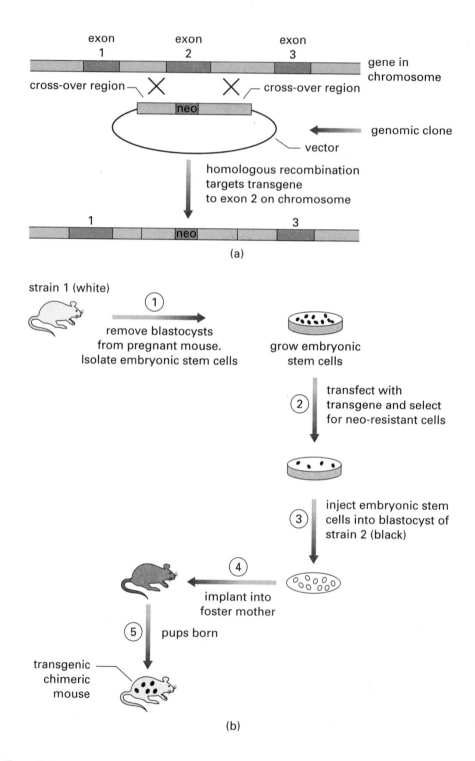

Figure 7.17. Knock-out transgenic mice. (a). Transgene is incorporated into genome of embryonic stem cells by homologous recombination. (b). Genetically modified embryonic stem cells are injected into a blastocyst, which is implanted into a foster mother.

Transgenic Animals

A transgenic animal is produced by introducing a foreign gene into the nucleus of a fertilized egg (Fig. 7.16a). The egg is then implanted into a foster mother and the offspring are tested to determine whether they carry the foreign gene. If they do, a transgenic animal has been produced. The first transgenic mice ever made were used to identify an enhancer sequence that activates the metallothionein gene when an animal is exposed to metal ions in its diet. The 5′ flanking sequence of the metallothionein gene was fused to the rat growth hormone gene (Fig. 7.16b). This DNA construct, the transgene, was injected into fertilized eggs. When the mice were a few weeks old, they were given drinking water containing zinc. Mice carrying the transgene grew to twice the size of their litter mates because the metallothionein enhancer sequence, stimulated by zinc, had increased growth hormone production.

Transgenic farm animals—such as sheep synthesizing human factor VIII in their milk—have been created. This is an alternative to producing human proteins in bacteria.

Transgenic mice are increasingly being used to prove a protein's function. To do this the gene for the protein is knocked out. This is done by either inserting a piece of foreign DNA into the gene, or by deleting the gene from the mouse genome. The consequences of the protein's absence are then established. Figure 7.17 describes the method called **insertional mutagenesis** for knocking out a gene's function. The first step is to isolate a genomic clone containing the gene to be knocked-out. A marker gene, such as the drug resistance gene *neo*, is then inserted into the genomic clone, usually in exon 2 of the gene. This means that the normal, functional product of the gene cannot be synthesized. This construct is the transgene, which is then introduced into embryonic stem (ES) cells. These are cells derived from the inner mass of a blastocyst—that is, a very early embryo—of a white mouse. **Homologous recombination** inside the embryonic stem cells will replace the normal gene with the transgene. Cells in which this rare event happens will survive when grown on neomycin while embryonic stem cells not containing the transgene will die. The genetically modified embryonic stem cells are inserted into the blastocyst cavity of a black mouse and the blastocyst implanted into a foster mother. Transgenic mice will be chimeric and have a mixed color coat because the cells derived from the genetically modified embryonic stem cells will give a white-color coat while the cells from the blastocyst will give a black-color coat. Subsequent breeding will produce a pure white mouse that is homozygous for the knocked-out gene. The effect of knocking-out the gene can then be analyzed.

✳ ETHICS OF DNA TESTING FOR INHERITED DISEASE

The applications of recombinant DNA technology are exciting and far-reaching. However, the ability to examine the base sequence of an individual raises important ethical questions. Would you want to know that you had inherited a gene that will cause you to die prematurely? Some of you might feel fine about this and decide to live life to the full. We suspect most people would not want to know their fate. But what if you have no choice and DNA testing becomes obligatory should you wish to take out life insurance? In the United Kingdom insurance companies are now able to ask for the results of the test for Huntington's disease. This is a fatal degenerative brain disorder that strikes people in their forties. From the insurance company's point of view DNA testing could mean higher premiums according to life expectancy or at worst refusal of insurance cover. There is much ongoing debate on this issue.

On the positive side, we can now test for changes in certain genes that are involved in drug metabolism. Adverse drug reactions are the fourth highest cause of death in the United States. By knowing, for example, which individuals carry mutations in genes coding for cytochromes P450 (page 249), medicines that need to be metabolized by these proteins can be avoided.

Analysis of fetal DNA can inform parents if their child is affected by an inherited disorder. It is now possible, using the PCR, to examine the genotype of a single cell from an eight-cell embryo. If the DNA test is negative, the embryo can be inserted into the mother, and the parents know they have not passed on a defective gene to their child. This combination of DNA testing and in vitro fertilization has given rise to the unfortunate phrase "designer babies".

In the future it may be possible to correct genetic defects before or after birth by replacing mutated genes by a normal copy. The technology to carry out such experiments is very demanding and important ethical questions are raised. In Chapter 20 we will discuss the potential use of gene therapy for the correction of the genetic disorder cystic fibrosis.

IN DEPTH 7.1 Genetically Modified (GM) Plants—Can They Help to Feed the World?

The arguments about the value of GM crops and the damage they cause the environment will continue for a long time. These have largely concerned plants that produce insecticide and plants that are resistant to herbicide. Almost unnoticed in the maelstrom of claim and counterclaim has been the use of genetic engineering to produce nutritionally enhanced crops. Rice is a staple food in many countries but lacks many vital nutrients. About 800 million children suffer from vitamin A deficiency, which can result in blindness and a weakened immune system. In an attempt to overcome this severe nutritional deficiency, a group of Swiss scientists have engineered the rice endosperm to produce provitamin A (beta-carotene). This is converted in the body to vitamin A. Four genes were introduced into the rice endosperm, three genes from the daffodil and one from the bacterium *Erwinia*. These genes code for all the proteins needed to make beta-carotene. The transgenic rice is golden in color because of the presence of large amounts of beta-carotene.

Almost 24% of the world's population, mainly women, are deficient in iron. Rice has been engineered with the gene for ferritin, an iron-binding protein. This rice has higher amounts of iron and has been shown to correct iron deficiency in laboratory animals. The plan is to cross the provitamin-A-enriched, iron-enriched strains with agricultural rice strains to produce a hardy and nutritional crop. Golden rice, enriched in iron, provides a great opportunity to improve the health of billions of people.

SUMMARY

1. DNA sequences can be cloned using reverse transcriptase, which copies mRNA into DNA to make a hybrid mRNA:DNA double-stranded molecule. The mRNA strand is then converted into DNA by the enzymes ribonuclease H and DNA polymerase. The new double-stranded DNA molecule is called complementary DNA (cDNA).

2. Restriction endonucleases cut DNA at specific sequences. DNA molecules cut with the same enzyme can be joined together. To clone a cDNA, it is joined to a cloning

vector—a plasmid or a bacteriophage. Genomic DNA clones are made by joining fragments of chromosomal DNA to a cloning vector. When the foreign DNA fragment has been inserted into the cloning vector, a recombinant molecule is formed.

3. Recombinant DNA molecules are introduced into bacterial cells by the process of transformation. This produces a collection of bacteria (a library) each of which contains a different DNA molecule. The DNA molecule of interest is then selected from the library using either an antibody or a nucleic acid probe.

4. There are many important medical, forensic and industrial uses for DNA clones. These include:
   - Determination of the base sequence of the cloned DNA fragment.
   - In situ hybridization to detect specific cells making RNA complementary to the clone.
   - Southern blotting and genetic fingerprinting to analyze an individual's DNA pattern.
   - Synthesis of mammalian proteins in bacteria or eukaryotic cells.
   - Changing the DNA sequence to produce a new protein.
   - Generation of fluorescent protein chimeras for subsequent microscopy on live cells.
   - The polymerase chain reaction, which lets us produce many copies of a DNA molecule in a test tube.
   - Production of transgenic animals to study gene function.

FURTHER READING

McPherson J. D., et al. 2001. A physical map of the human genome. *Nature* 409: 934–941.

Mullis, K. B. 1990. The unusual origin of the polymerase chain reaction. *Sci. Am.* 262: 56–65.

Venter, J. C., et al. 2001. The sequence of the human genome. *Science* 291: 1304–51.

Golden rice: Beyer, P., Al-Babili, S., Ye, X., Lucca, P., Schaub, P., Welsch, R., and Potrykus, I. 2002. Golden Rice: Introducing the beta-carotene biosynthesis pathway into rice endosperm by genetic engineering to defeat vitamin A deficiency *J. Nutrition* 132: 506S–510S.

Microarrays: Friend, S. H., and Stoughton, R. B. 2002. The magic of microarrays. *Sci. Am.* 286: 44–49.

✿ REVIEW QUESTIONS

For each question, choose the ONE BEST answer or completion.

1. Plasmids used as cloning vectors
   A. are circular molecules.
   B. have an origin of replication.
   C. carry antibiotic resistance genes.
   D. have unique restriction endonuclease cutting sites.
   E. all of the above.

2. A cDNA clone library is a collection of clones representing
  A. the entire genome of the sample tissue.
  B. all protein-coding gene sequences in the genome of the sample tissue.
  C. all intronic sequences of genes expressed in the sample tissue.
  D. the mRNAs expressed in the sample tissue.
  E. all tRNAs being used in the sample tissue.

3. A cDNA clone library can be screened with an antibody if
  A. the plasmid vector contains a convenient promoter such as that of the *lac* operon to increase expression.
  B. the plasmid vector contains a multiple cloning site.
  C. the nylon "colony lift" membrane is first soaked in alkali to denature DNA.
  D. the nylon "colony lift" membrane is first treated with a strong protease to break peptide bonds.
  E. an antibody that recognizes IPTG, and that is linked to an enzyme such as alkaline phosphatase, is available.

4. A radioactive gene probe can be synthesized by the method of random priming if the following factors are present in the incubation mixture:
  A. The DNA template from which the probe will be made, present as single-stranded DNA
  B. The four deoxynucleotides, dATP, dCTP, dGTP, and dTTP, one of which should carry a radioactive tag
  C. Primers for DNA synthesis
  D. DNA polymerase
  E. All of the above

5. To sequence DNA using the automated dideoxy chain termination method, the reaction mixture should contain the DNA to be sequenced plus
  A. the four deoxynucleotides, dATP, dCTP, dGTP, and dTTP.
  B. the four dideoxynucleotides, ddATP, ddCTP, ddGTP, and ddTTP, each carrying a different fluorescent tag.
  C. a DNA primer.
  D. DNA polymerase.
  E. all of the above.

6. The technique called
  A. Southern blotting is used to analyze RNA.
  B. western blotting is used to analyze DNA.
  C. eastern blotting is used to analyze RNA.
  D. Southern blotting is used to analyze DNA.
  E. northern blotting is used to analyze protein.

7. Each cycle in the polymerase chain reaction involves the following steps:
  A. Polymerization then annealing then denaturation
  B. Denaturation then polymerization then annealing
  C. Denaturation then annealing then polymerization
  D. Denaturation then chimera formation then polymerization
  E. Denaturation then polymerization then sequencing

ANSWERS TO REVIEW QUESTIONS

1. *E.* All these statements are true. Plasmids used as cloning vectors are circular, have an origin of replication so that they can replicate in the host bacterium, carry antibiotic genes so that transformed cells can be selected for, and have unique restriction enzyme cutting sites to allow genes of interest to be inserted into the plasmid.

2. *D.* cDNA means DNA that is complementary to the mRNA found in a tissue sample. cDNAs are generated using a poly-T primer that binds to processed mRNA, but which will not bind to tRNA or rRNA, which have no poly-A sequence.

3. *A.* Addition of the gratuitous inducer *lac* operon inducer IPTG should ensure good production of the protein, which can then be recognized by the antibody. Considering the other answers: (B) The multiple cloning site makes insertion of the gene of interest easier, but has no implications regarding the method to be used to then screen the library. (C) If one is using an antibody to recognize protein, the presence of DNA is largely irrelevant. (D) The antibody is used to recognize the particular protein coded by the gene of interest. It would therefore be highly counterproductive to destroy the proteins with a protease. (E) There is no reason why one should want to use an antibody that recognizes IPTG.

4. *E.* The primers bind to the template, allowing DNA polymerase to generate the probe using the deoxynucleotides provided. Because one of the deoxynucleotides carries a radioactive tag, the resulting probe is radioactive and can therefore be used, for example, to reveal the position of complementary DNA on a Southern blot.

5. *E.* The primer binds to the DNA to be sequenced. DNA polymerase then begins work using the deoxynucleotides, but each time it instead incorporates a dideoxynucleotide; synthesis of that strand stops so that the DNA of that specific length is flagged with a fluorophore of the color corresponding to the base encountered.

6. *D.* Southern blotting is used to analyze DNA, northern blotting is used to analyze RNA, and western blotting is used to analyze protein (Table 7.2). There is no eastern blotting (although there is a technique called SouthWestern blotting!).

7. *C.* PCR proceeds by cycles of denaturation (separation of the DNA strands), annealing (of primers onto the strands), and polymerization to form double helices.

8

MANUFACTURING PROTEIN

The genetic code (page 75) dictates the sequence of amino acids in a protein molecule. The synthesis of proteins is quite complex, requiring three types of RNA. Messenger RNA (mRNA) contains the code and is the template for protein synthesis. Transfer RNAs (tRNAs) are adapter molecules that carry amino acids to the mRNA. Ribosomal RNAs (rRNAs) form part of the ribosome that brings together all the components necessary for protein synthesis. Several enzymes also help in the construction of new protein molecules. This chapter describes how the nucleotide sequence of an mRNA molecule is translated into the amino acid sequence of a protein.

Figure 8.1 shows the basic mechanism of protein synthesis, also called translation. In the first step, free amino acids are attached to tRNA molecules. In the second step, a ribosome assembles on the mRNA strand to initiate synthesis. In the third step, the ribosome travels along the mRNA. At each codon on the RNA a tRNA binds, bringing the amino acid defined by that codon to be added to the growing polypeptide chain. In the last, fourth, step the ribosome encounters a stop codon and protein synthesis is terminated.

❋ ATTACHMENT OF AN AMINO ACID TO ITS tRNA

Amino acids are not directly incorporated into protein on a messenger RNA template. An amino acid is carried to the mRNA chain by a tRNA molecule. tRNAs are small, about 70–80 nucleotides in length, and are folded into precise three-dimensional structures because of hydrogen bonding between bases in particular stretches of the molecule. This gives rise

Cell Biology: A Short Course, Second Edition, by Stephen R. Bolsover, Jeremy S. Hyams,
Elizabeth A. Shephard, Hugh A. White, Claudia G. Wiedemann
ISBN 0-471-26393-1 Copyright © 2004 by John Wiley & Sons, Inc.

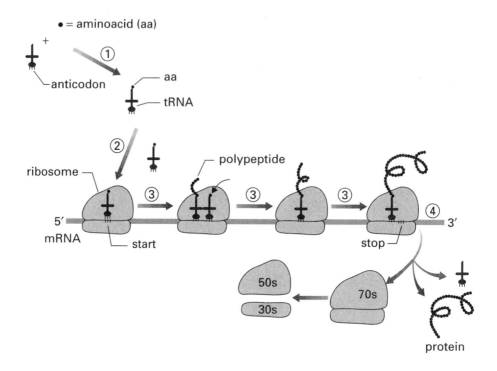

Figure 8.1. Overview of protein synthesis

to four double-stranded regions, and it is these that give tRNA its characteristic cloverleaf structure when drawn in two dimensions as in Figure 8.2.

Each tRNA molecule has an amino acid attachment site at its 3′ end and an **anticodon,** three bases that are complementary in sequence to a codon on the mRNA. The tRNA binds to the mRNA molecule because hydrogen bonds form between the anticodon and codon. For example, the codon for methionine is 5′ AUG 3′ which will base pair with the anticodon 3′ UAC 5′.

Transfer RNA, the Anticodon, and the Wobble

Although 61 codons specify the 20 different amino acids, there are not 61 tRNAs; instead the cell economizes. The codons for some amino acids differ only in the third position of the codon. Figure 4.8 on page 78 shows that when an amino acid is encoded by only two different triplets the third bases will be either C and U, or A and G. For example aspartate is coded by GAC and GAU and glutamine by CAA and CAG. The **wobble** hypothesis suggests that the pairing of the first two bases in the codon and anticodon follows the standard rules—G bonds with C and A bonds with U—but the base pairing in the third position is not as restricted and can wobble. If there is a pyrimidine, U or C, in the third position of the codon, it can fit with any purine, G or A, in the anticodon, and vice versa. Thus, only one tRNA molecule is required for two codon sequences. The anticodon of some tRNAs contains the unusual nucleoside inosine (I), whose base is the purine hypoxanthine (Fig. 2.13, page 34). Inosine can base pair with any of U, C, or A in the third position of the codon. Some

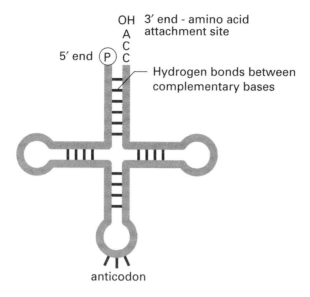

OH 3' end - amino acid
A attachment site
C
5' end (P) C
Hydrogen bonds between
complementary bases

anticodon

Figure 8.2. Transfer RNA (tRNA).

tRNA molecules can therefore base pair with as many as three different codons provided the first two bases of the codon are the same. For example, the tRNA for isoleucine has the anticodon UAI and can therefore base pair with any of AUU, AUC, or AUA.

The attachment of an amino acid to its correct tRNA molecule is illustrated in Figure 8.3. This process occurs in two stages, both catalyzed by the enzyme **aminoacyl tRNA synthase.** During the first reaction, the amino acid is joined, via its carboxyl group, to an adenosine monophosphate (AMP) and remains bound to the enzyme. All tRNA molecules have at their 3' end the nucleotide sequence CCA. In the second reaction aminoacyl tRNA synthase transfers the amino acid from AMP to the tRNA, forming an ester bond between its carboxyl group and either the 2'- or 3'-hydroxyl group of the ribose of the terminal adenosine (A) on the tRNA to form an **aminoacyl tRNA.** This step is often referred to as amino acid activation because the energy of the ester bond can be used in the formation of a lower energy peptide bond between two amino acids. A tRNA that is attached to an amino acid is known as a **charged tRNA.** There are at least 20 aminoacyl tRNA synthases, one for each amino acid and its specific tRNA.

❋ THE RIBOSOME

The ribosome is the cell's factory for protein synthesis. Each ribosome consists of two subunits, one large and one small, each of which is made up of RNA plus a large number of proteins. The ribosomal subunits and their RNAs are named using a parameter, called the **S value.** The S value, or **Svedberg unit,** is a sedimentation rate. It is a measure of how fast a molecule moves in a gravitational field. For example, the bigger a ribosomal subunit, the quicker it will sediment and the larger the S value. Prokaryotic ribosomes, and those found

●●●● **IN DEPTH 8.1 How We Study Proteins in One Dimension**

The technique known as **SDS–PAGE** is widely used to analyze the spectrum of proteins made by a particular tissue, cell type, or organelle. It is also invaluable for assessing the purity of isolated proteins. SDS stands for sodium dodecyl sulfate and PAGE for polyacrylamide gel electrophoresis.

The aim of the technique is to denature the proteins to be analyzed and then to separate them according to their size in an electrical field. To do this we first add a chemical called 2-mercaptoethanol to the protein sample. This will break any **disulfide bonds** (page 191) within a protein or between protein subunits. Next SDS, which is an anionic detergent, is added and the protein sample boiled. SDS coats each protein chain with negative charge. Each individual polypeptide in the sample becomes covered with an overall net negative charge. This means that when placed in an electrical field the SDS-coated proteins will separate according to their size because the smaller proteins move most quickly toward the positive electrode or **anode.** Polyacrylamide provides the matrix through which the proteins move during electrophoresis. The monomeric form, acrylamide, is poured into a mold. A solid but porous gel forms as the acrylamide polymerizes. The shape of the mold is such that wells are formed in the gel into which the protein sample can be loaded for electrophoresis.

When electrophoresis is complete, the proteins are stained by incubating the gel in a solution of Coomassie brilliant blue. Each protein band stains blue and is detectable by eye. However, if the amount of protein is very low, a more sensitive detection system is needed such as a silver stain. Proteins of known molecular mass are also electrophoresed on the gel. By comparison with the standard proteins, the mass of an unknown protein can be determined.

If we want to follow the fate of a single protein in a complex mixture of proteins, we combine SDS–PAGE with a technique called western blotting. The name western blotting is, like northern blotting, a play on the name of Dr. Ed Southern who devised the technique of Southern blotting to analyze DNA (Table 7.2).

The protein mixture is separated by SDS–PAGE. A nylon membrane is then placed up against the polyacrylamide gel and picks up the proteins, so that the pattern of protein spots on the original polyacrylamide gel is preserved on the nylon membrane. The nylon membrane is then incubated with an antibody specific for the protein of interest. This antibody, the primary antibody, will seek out and bind to its partner protein on the nylon membrane. A second antibody is added that will bind to the primary antibody. To be able to detect the specific protein of interest on the membrane, the secondary antibody is attached to an enzyme. In the figure shown, the enzyme used was horseradish peroxidase. A substrate is added and is converted by the enzyme into a colored product. The protein of interest is seen as a colored band on the nylon membrane. The same enzyme-linked secondary antibody can be used in other laboratories or at other times for western blotting of many different proteins because the specificity is determined by the unlabeled primary antibody.

Part A of the figure shows the Coomassie brilliant-blue-stained pattern of proteins isolated from the endoplasmic reticulum of liver. The leftmost lane is from a phenobarbital-treated animal while the middle lane is from an untreated control animal. The dark bands indicate the presence of protein. The spectrum of proteins is very similar in the two samples, except that a band with a relative molecular mass (M_r) of about 52,000 is much darker in the sample from the treated animal. This tells us that drug treatment has caused an increase in the production of a protein with this relative molecular mass. Western blotting (part B) using an antiCYP2B1

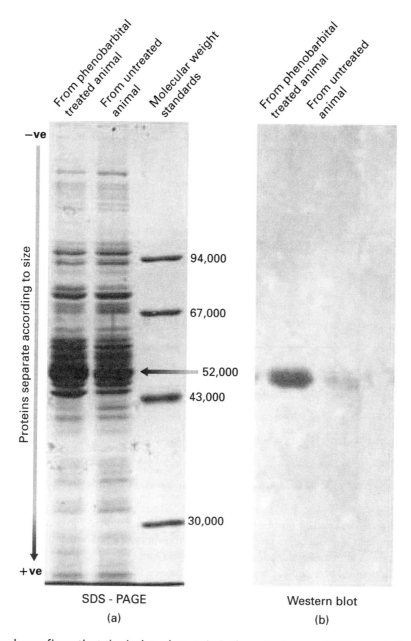

SDS - PAGE
(a)

Western blot
(b)

antibody confirms that the induced protein is the cytochrome P450 protein known as CYP2B1. The *CYP2B1* gene is activated by phenobarbital to produce more CYP2B1 protein to metabolize and clear the drug from the body (page 249).

We already showed how northern blotting revealed that transcription of the *CYP2B1* gene is increased after phenobarbital treatment (Fig. 7.13 on Page 148). The western blot shown here demonstrates that, as expected, the amount of CYP2B1 protein is increased as well.

Sodium dodecyl sulfate (SDS) is a major constituent of hair shampoo, where it is usually called by its alternative name of sodium lauryl sulfate.

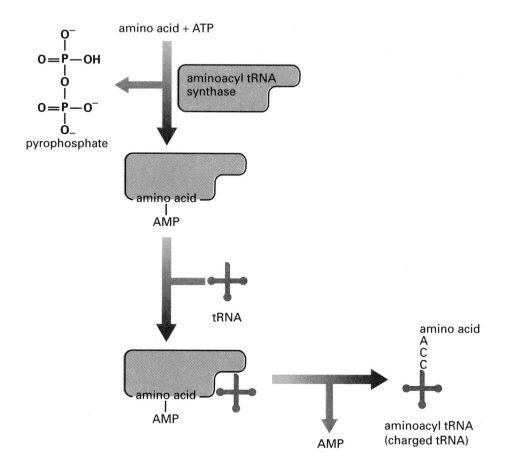

Figure 8.3. Attachment of an amino acid to its tRNA.

inside mitochondria and chloroplasts, are 70S when fully assembled and comprise a larger, 50S subunit and a smaller 30S one. Eukaryotic ribosomes are 80S when fully assembled and comprise a larger, 60S subunit and a smaller 40S one.

The formation of a peptide bond (page 38) between two amino acids takes place on the ribosome. The ribosome has binding sites for the mRNA template and for two charged tRNAs (Fig. 8.4). An incoming tRNA with its linked amino acid occupies the **aminoacyl site (A site),** and the tRNA attached to the growing polypeptide chain occupies the **peptidyl site (P site).**

🏵 BACTERIAL PROTEIN SYNTHESIS

Ribosome-Binding Site

For protein synthesis to take place, a ribosome must first attach to the mRNA template. AUG is not only the start codon for protein synthesis; it is used to code for all the other methionines in the protein. How does the ribosome recognize the correct AUG at which to

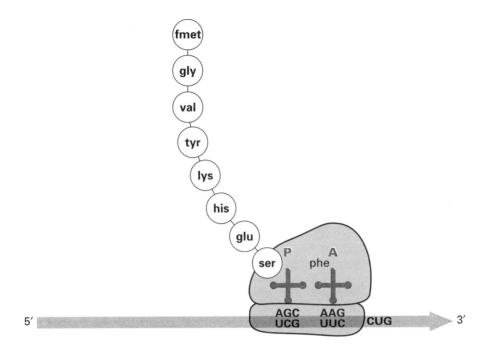

Figure 8.4. The P and A sites on the ribosome.

begin protein synthesis? All bacterial mRNAs have at their 5′ end a stretch of nucleotides called the untranslated (or leader) sequence, which do not code for the protein. These nucleotides are nevertheless essential for the correct placing of the ribosome on the mRNA. Bacterial mRNA molecules usually have a nucleotide sequence similar to 5′ GGAGG 3′ whose center is about 8 to 13 nucleotides upstream of (5′ to) the AUG start codon. This sequence is complementary to a short stretch of sequence, 3′ CCUCC 5′, found at the 3′ end of the rRNA molecule within the 30S ribosomal subunit (Fig. 8.5). The mRNA and the rRNA interact by complementary base pairing to place the 30S ribosomal subunit in the correct position to start protein synthesis. The sequence on the mRNA molecule is called the ribosome-binding site. This is sometimes referred to as the Shine–Dalgarno sequence after the two scientists who found it.

Because the genetic code is read in triplets of three bases, there are three possible reading frames (page 79). The reading frame that is actually used by the cell is defined by the first AUG that the ribosome encounters downstream of the ribosome-binding site.

Chain Initiation

The first amino acid incorporated into a new bacterial polypeptide is always a modified methionine, formyl methionine (fmet) (Fig. 8.6). Methionine first attaches to a specific tRNA molecule, tRNA^{fmet}, and is then modified by the addition of a formyl group that attaches to its amino group. tRNA^{fmet} has the anticodon sequence 5′ CAU 3′ that binds to its complementary codon, the universal start codon AUG.

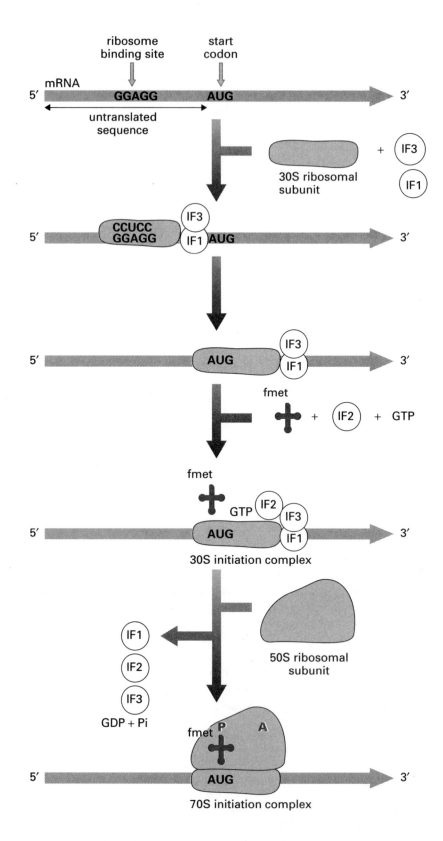

Figure 8.5. Formation of the 70S initiation complex.

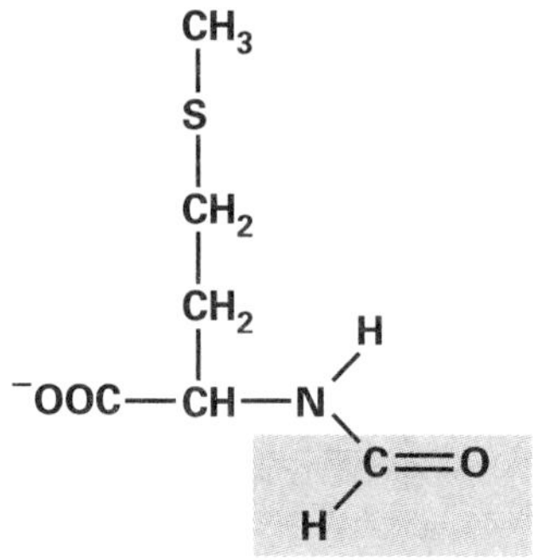

$$CH_3$$
$$|$$
$$S$$
$$|$$
$$CH_2$$
$$|$$
$$CH_2$$
$$|$$
$$^-OOC—CH—N$$

formyl
group

Figure 8.6. Formyl methionine.

Example 8.1 The Irritating Formyl Methionine

White blood cells are strongly attracted by any peptide that begins with formyl methionine: they assume an amoeboid shape (Fig. 1.5 on page 5) and begin to crawl toward the source of the peptide. To a white blood cell, the presence of a peptide beginning with formyl methionine means that there is an infection nearby that needs to be fought. This is because the body's own proteins do not contain formyl methionine: only prokaryotes begin protein synthesis with this modified amino acid.

The 70S Initiation Complex

The initiation phase of protein synthesis involves the formation of a complex between the ribosomal subunits, an mRNA template and tRNA^{fmet} (Fig. 8.5). A 30S subunit attaches to the ribosome-binding site as described above. tRNA^{fmet} then interacts with the AUG initiation codon, and finally the 50S ribosomal subunit attaches. The ribosome is now complete, and the first tRNA and its amino acid are in place in the P site of the ribosome. A 70S initiation complex has been formed, and protein synthesis can begin. The ribosome is orientated so that it will move along the mRNA in the 5′ to 3′ direction, the direction in which the information encoded in the mRNA molecule is read.

Three proteins called initiation factors 1, 2, and 3, together with the nucleotide guanosine triphosphate, are needed to help the 70S initiation complex form. Initiation factors 1 and 3 are attached to the 30S subunit. Initiation factor 3 helps in the recognition of the ribosome-binding site on the mRNA. Initiation factor 2 specifically recognizes tRNA^{fmet} and binds it to the ribosome. When the 50S subunit attaches, the three initiation factors are released and the guanosine triphosphate is hydrolyzed, losing its γ phosphate to become guanosine diphosphate.

Elongation of the Protein Chain

The synthesis of a protein begins when an aminoacyl tRNA enters the A site of the ribosome (Fig. 8.7). The identity of the incoming aminoacyl tRNA is determined by the codon on

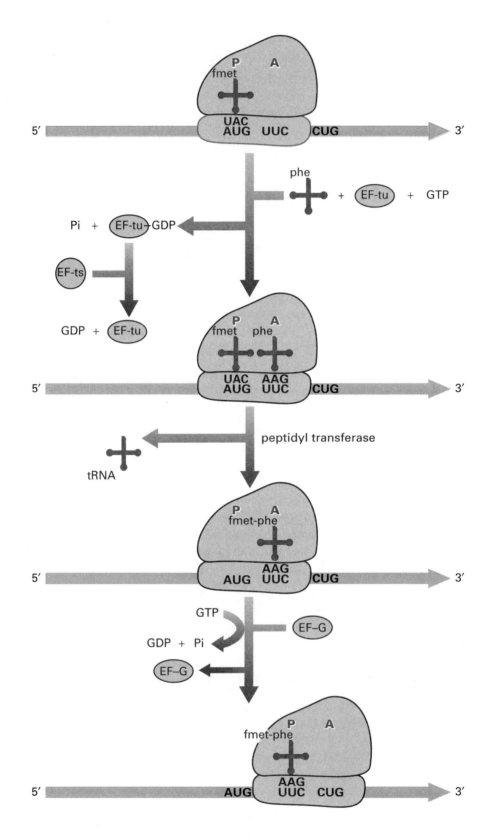

Figure 8.7. Elongation of the protein chain. Pi represents the inorganic phosphate ion HPO_4^{2-}.

the mRNA. If, for example, the second codon is 5′ UUC 3′, then phenylalanyl tRNA^{phe} (whose anticodon is 5′ GAA 3′) will occupy the A site. The P site has, of course, already been occupied by tRNA^{fmet} during the formation of the initiation complex. Now that both the A and P sites are occupied, the enzyme **peptidyl transferase** catalyzes the formation of a peptide bond between the two amino acids (fmet and phe in this example). The dipeptide is attached to the tRNA occupying the A site. Because the tRNA in the P site is no longer attached to an amino acid, it is released from the ribosome and it can be reused. The ribosome now moves along the mRNA to the right so that the tRNA dipeptide now occupies the P site (leaving the A site free for another incoming aminoacyl tRNA). The movement of the ribosome, three nucleotides at a time, relative to the mRNA, is called **translocation,** a term often used in biology to describe movement that is actively driven and precisely controlled. The process of peptide bond formation is followed by translocation, and the whole process is repeated until the ribosome reaches a stop signal and protein synthesis terminates.

Proteins are synthesized beginning at their amino or N terminus (page 184). The first amino acid hence has a free (although formylated) amino group. The last amino acid in the chain has a free carboxyl group and is known as the carboxyl or C terminus (page 185).

Elongation of a polypeptide chain needs the help of three proteins called **elongation factors.** These proteins are needed to speed up the process of protein synthesis. To bind to the A site, the aminoacyl tRNA must form a complex with elongation factor tu and a molecule of guanosine triphosphate (GTP). On hydrolysis of GTP to guanosine diphosphate (GDP) and inorganic phosphate ion (Pi), the aminoacyl tRNA is able to enter the A site. Elongation factor tu is released from the aminoacyl tRNA by the action of a second protein, elongation factor ts. This removes the GDP bound to elongation factor tu, and the protein is recycled. The movement of the ribosome needs the help of a third protein elongation factor G and the energy supplied by the hydrolysis of a second GTP molecule.

Example 8.2 **The Diptheria Bacterium Inhibits Protein Synthesis**

Some bacteria cause disease because they inhibit eukaryotic protein synthesis. Diphtheria was once a widespread and often fatal disease caused by infection with the bacterium *Corynebacterium diptheriae.* This organism produces an enzyme (diphtheria toxin) that inactivates eukaryotic elongation factor 2 (the equivalent of the bacterial elongation factor G). Diphtheria toxin splits the bond between ribose and nicotinamide in NAD$^+$ (page 37), releasing free ribose and attaching the remainder, called ADP-ribose, to elongation factor 2, a process known as ADP ribosylation. The protein is now inactive and is unable to assist in the movement of the ribosome along the mRNA template. Protein synthesis therefore stops in the affected human cells. The advantage of this to the bacterium is that all the amino acids that the host was using to make protein are now available for its own use.

The Polyribosome

More than one polypeptide chain is synthesized from an mRNA molecule at any given time. Once a ribosome has begun translocating along the mRNA, the start AUG codon is free, and another ribosome can bind. A second 70S initiation complex forms. Once this ribosome

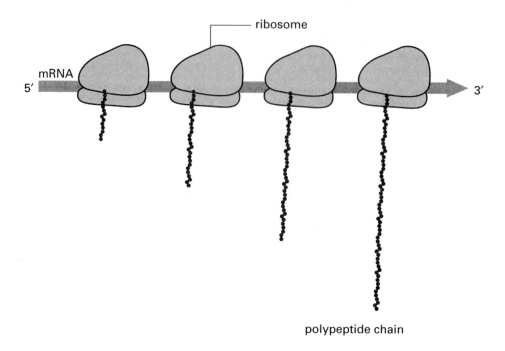

Figure 8.8. Many ribosomes reading one mRNA form a polyribosome.

has moved away, a third ribosome can attach to the start codon. This process is repeated until the mRNA is covered with ribosomes. Each of these spans about 80 nucleotides. The resultant structure, the polyribosome or polysome (Fig. 8.8), is visible under the electron microscope. This mechanism allows many protein molecules to be made at the same time on one mRNA.

Termination of Protein Synthesis

There are three codons, UAG, UAA, and UGA, that have no corresponding tRNA molecule. These codons are called *stop codons*. Instead of interacting with tRNAs, the A site occupied by one of these codons is filled by proteins known as **chain release factors.** In the presence of these factors the newly synthesized polypeptide chain is freed from the ribosome, and the mRNA, tRNA, and the 30S and 50S ribosomal subunits dissociate (Fig. 8.9). Release factor 1 causes polypeptide chain release from UAA and UAG, and release factor 2 terminates chains with UAA and UGA. A third protein, release factor 3, cooperates with the other two to stop protein synthesis. When the A site is occupied by a release factor, the enzyme peptidyl transferase is unable to add an amino acid to the growing polypeptide chain and instead catalyzes the hydrolysis of the bond joining the polypeptide chain to the tRNA. The carboxyl (COOH) end of the protein is therefore freed from the tRNA, and the protein is released.

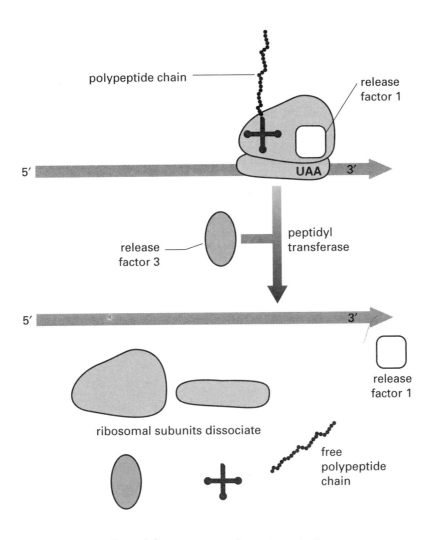

Figure 8.9. Termination of protein synthesis.

The Ribosome Is Recycled

It would be wasteful if a complex structure such as the ribosome were used only once. When protein synthesis is over, the two ribosomal subunits dissociate and can be reused when needed.

✸ EUKARYOTIC PROTEIN SYNTHESIS IS A LITTLE MORE COMPLEX

Elongation of the polypeptide chain and the termination of protein synthesis in eukaryotes does not differ very much from that described for bacteria. However, the initiation of protein synthesis is more complex in eukaryotes. Their proteins always start with methionine instead of the formyl methionine used in bacterial protein synthesis. A special transfer RNA, $tRNA_i^{met}$, is used to initiate protein synthesis from the AUG start codon. The methionine is often removed from the protein after synthesis. Eukaryotic mRNAs do not contain the bacterial Shine–Dalgarno sequence for ribosome binding. Eukaryotic mRNAs have at their 5' end a 7-methyl guanosine cap (page 118). A number of proteins are needed to attach the small, 40S, ribosomal subunit to the cap. Once attached, the 40S subunit slides along the leader sequence of the mRNA looking for the start codon. All eukaryotic mRNAs have a sequence very similar to 5' CCACC 3' adjacent to the initiating AUG codon. This sequence (known as the Kozak sequence after the scientist who noted it) tells the ribosome that it has reached the start AUG codon. The recognition of the AUG codon that specifies the start site for translation also requires the help of at least nine proteins. The $tRNA_i^{met}$ binds to the 40S subunit and then the large 60S ribosomal subunit attaches to form the 80S initiation complex. Protein synthesis can now begin.

Example 8.3   How Hepatitis C Takes Control

The hepatitis C virus shuts down the liver protein synthesis machinery and forces the resources of the infected cell to be diverted into production of viral proteins that will be used to make new virus. It does this by commandeering the cell's ribosomes to its own mRNAs. Hepatitis C mRNAs do not have a 7-methyl guanosine cap at their 5' ends. Instead they have internal ribosome entry sites (IRES), which are specific sequences to which the ribosomes bind tightly. The clamping of the ribosome to the IRES forces initiation of viral protein synthesis at the expense of host cell protein synthesis.

Hepatitis C is a particularly nasty virus because infected people can go on to develop liver cancer, cirrhosis, and other chronic liver disease. In the United States alone, 10,000 people die per year from hepatitis C infection. It is hoped that drugs can be developed that will prevent the host ribosome from binding to the viral IRES and so halt the production of new virus in an infected liver.

Antibiotics and Protein Synthesis

Many antibiotics work by blocking protein synthesis, a property that is extensively exploited in research and medicine. Many antibiotics only inhibit protein synthesis in bacteria and not in eukaryotes. They are therefore extremely useful in the treatment of infections because the invading bacteria will die but protein synthesis in the host organism remains unaffected. Examples are chloramphenicol, which blocks the peptidyl transferase reaction, and tetracycline, which inhibits the binding of an aminoacyl tRNA to the A site of the ribosome. Both of these antibiotics therefore block chain elongation. Streptomycin, on the other hand inhibits the formation of the 70S initiation complex because it prevents $tRNA^{fmet}$ from binding to the P site of the ribosome.

Puromycin causes the premature release of polypeptide chains from the ribosome and acts on both bacterial and eukaryotic cells. This antibiotic has been widely used in the study of protein synthesis. Puromycin can occupy the A site of the ribosome because its structure resembles the 3' end of an aminoacyl-tRNA (Fig. 8.10). However, puromycin does not bind to the mRNA. Puromycin blocks protein synthesis because peptidyl transferase uses it as a substrate and forms a peptide bond between the growing polypeptide and the antibiotic. Once translocation has occurred, the growing polypeptide has no strong attachment to the mRNA and is therefore released from the ribosome.

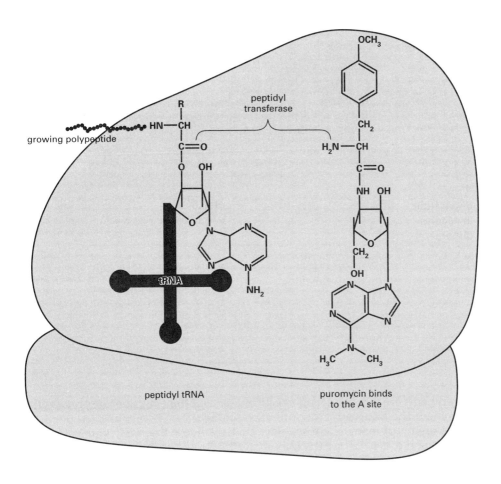

Figure 8.10. The antibiotic puromycin inhibits protein synthesis.

IN DEPTH 8.2 Proteomics

Proteomics is the study of the **proteome**—the complete protein content of a cell. It is the proteins a cell makes that allow it to carry out its specialized functions. Although, for example, a liver and a kidney cell have many proteins in common, they also each possess a unique subset of proteins, which gives to them their own characteristics. Similarly, a cell will need to make different proteins, and proteins in different amounts, according to its metabolic state. The goal of proteomics is to identify all of the proteins produced by different cells and how a particular disease changes a cell's protein profile.

The separation of a cell's protein mixture into individual components is tackled using a technique known as two-dimensional polyacrylamide gel electrophoresis. This produces a pattern of protein spots. These patterns are recorded and serve as templates for the comparison of the proteomes of different cells. Spots that change during a cell's development, or changes that occur in disease, can be easily identified. A protein spot of interest is excised from the polyacrylamide gel and the protein broken into small, overlapping peptide fragments by proteolytic enzymes.

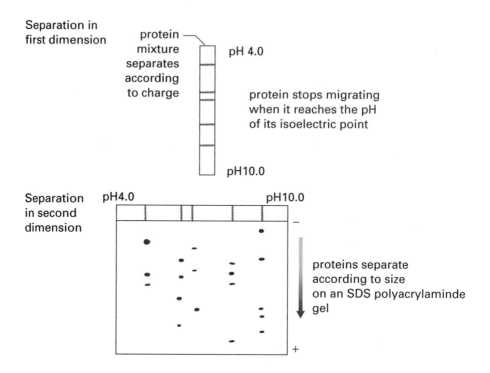

The fragments are fed into a mass spectrometer, and their peptide mass fingerprint determined. The fingerprint identifies the protein.

The dream of many scientists, now that we know the base sequence of the human genome, is to define the human proteome and to determine all of its variations. However, the genome projects use a single automated method to sequence the order of the four bases in DNA, and data are obtained at a relatively rapid pace. In contrast it is very labor intensive to identify proteins and much effort is being made to devise ways of speeding things up. The aim of proteomic research centers is to increase their throughput from about 40 to 100 peptide samples an hour, to the daunting number of 1 million peptides a day.

SUMMARY

1. During protein synthesis, the genetic code in an mRNA is translated into a sequence of amino acids. An amino acid attaches to the 3′ end of a tRNA to form an aminoacyl tRNA. Protein synthesis takes place on the ribosome, that binds both to mRNA and tRNA. It has two tRNA binding sites, the P site and the A site.

2. Initiation of protein synthesis in bacteria involves the binding of the 30S ribosomal subunit to the mRNA. tRNA^{fmet} binds to the initiation codon, and then the 50S ribosomal subunit attaches and the 70S initiation complex is formed.

3. Protein synthesis begins when a second aminoacyl tRNA occupies the A site. Each incoming amino acid is specified by the codon on the mRNA. The anticodon on the tRNA hydrogen bonds to the codon, thus positioning the amino acid on the ribosome.

A peptide bond is formed, by peptidyl transferase, between the amino acids in the P and A sites. The newly synthesized peptide occupies the P site, and another amino acid is brought into the A site. This process of elongation requires a number of proteins (elongation factors); as it continues, the peptide chain grows. When a stop codon is reached, the polypeptide chain is released with the help of proteins known as release factors.

4. More than one ribosome can attach to an mRNA. This forms a polyribosome, and many protein molecules can be made simultaneously from the same mRNA.

5. Many antibiotics fight disease because they inhibit particular steps in protein synthesis.

FURTHER READING

Arnez, J. G., and Moras, D. 1997. Structural and functional considerations of the aminoacylation reaction. *Trends Biochem. Sci.* 22: 211–216.

Lytle, J. R., Wu, L., and Robertson, H. D. 2001. The ribosome binding site of hepatitis C virus mRNA. *J. Virol.* 75: 7629–7636.

Moore, P. B., and Steitz, T. A. 2002. The involvement of RNA in ribosome function. *Nature* 418: 229–235.

Ribas de Pouplana, L., and Schimmel, P. 2001. Aminoacyl-tRNA synthetases: Potential markers of genetic code development. *Trends Biochem. Sci.* 26: 591–596.

❋ REVIEW QUESTIONS

For each question, choose the ONE BEST answer or completion.

1. Which of the following steps is NOT part of translation?
   A. Synthesis of mRNA
   B. Charging of tRNA
   C. Initiation of protein synthesis
   D. Elongation of the polypeptide chain
   E. Termination of protein synthesis

2. In bacterial and eukaryotic mRNAs, the normal start codon for protein synthesis is
   A. GAU.
   B. UGA.
   C. AUG.
   D. AGU.
   E. UAG.

3. The enzyme that attaches the amino acid to its tRNA is
   A. aminoacyl mRNA synthase.
   B. aminoacyl tRNA synthase.
   C. amino acid attachment enzyme.

    D. elongation factor tu.
    E. tRNA polymerase.

4. Bacterial mRNAs usually have the nucleotide sequence 5′ GGAGG 3′, which bonds to
    A. tRNA.
    B. methionine.
    C. the 30S ribosomal subunit.
    D. the 40S ribosomal subunit.
    E. protein initiation factors.

5. The enzyme that catalyzes the formation of a peptide bond between two amino acids is
    A. amino acid transferase.
    B. peptide joinase.
    C. peptidyl transferase.
    D. peptide elongation factor.
    E. puromycin.

6. Which of the following do NOT play a part in the termination of protein synthesis?
    A. The codons AUG, AGG, and GUA
    B. The codons UAA, UAG and UGA
    C. Release factor 1
    D. Release factor 3
    E. Peptidyl transferase

7. The base hypoxanthine in the nucleotide inosine can hydrogen bond with
    A. G, U, and C.
    B. G, U, and A.
    C. G, A, and C.
    D. U, C, and A.
    E. A only.

ANSWERS TO REVIEW QUESTIONS

1. *A.* Synthesis of mRNA is not part of translation. All the others are, in particular, "charging" means attaching an amino acid to a tRNA ready for use in protein synthesis.

2. *C.* AUG, which codes for the amino acid methionine.

3. *B.* The enzyme is called aminoacyl tRNA synthase. Concerning the other answers: (A) Because amino acids are not attached to mRNA, there is no such process in cells. (C) Amino acid attachment enzyme is a plausible name, albeit less specific since it does not say to what the amino acid is attached. However, the name is not used. (D) Elongation factor tu plays a role later in the process of protein synthesis, not during tRNA charging. (E) tRNA polymerase is not a name that is ever used. If it did mean anything, it would mean the enzyme that makes tRNA (there is such an enzyme in eukaryotes: it is called RNA polymerase III), not the one that attaches an amino acid.

4. *C.* The rRNA molecule within the 30S ribosomal subunit has a short stretch of sequence, 3′ CCUCC 5′, at its 3′ end that is complementary in sequence to the 5′ GGAGG 3′ of the mRNA. Hydrogen bonding between these two sequences aligns the ribosome for protein synthesis. Concerning answer D: the 40S ribosomal subunit plays a similar role in eukaryotes as the 30S unit does in prokaryotes. However, eukaryotic mRNAs do not have the prokaryotic GGAGG Shine–Dalgarno sequence.

5. **C.** The enzyme is called peptidyl transferase. Concerning the other answers: A, B, and D are plausible answers, though D is the least plausible since most enzymes have names ending in *ase;* however, they are not the names used. Puromycin is an antibiotic, not a protein (and therefore certainly not an enzyme).

6. **A.** These are codons directing the incorporation of methionine, arginine, and valine, respectively. All the others do play a part in the termination of protein synthesis, in particular, UAA, UAG, and UGA are stop codons, while if the A site is occupied by a release factor, peptidyl transferase catalyzes the hydrolysis of the bond joining the polypeptide chain to the tRNA. The carboxyl (COOH) end of the protein is therefore freed from the tRNA and the completed protein is released.

7. **D.** Hypoxanthine can bond with any of U, C, or A.

9

PROTEIN STRUCTURE

Virtually everything cells do depends upon proteins. We are all, regardless of build, made up of water plus more or less equal amounts of fat and protein. Although the DNA in our cells contains the information necessary to make our bodies, DNA itself is not a significant part of our body mass. Nor is it a chemically interesting molecule, in the sense that one length of DNA is much the same as another in terms of shape and chemical reactivity. The simplicity of DNA arises because it is a polymer made up of only four fairly similar monomers, and this is appropriate because the function of DNA is simply to remain as a record and to be read during transcription. In contrast, proteins made using the instructions in DNA vary enormously in physical characteristics and function and can be considered as constituting the much more complex proteome (page 178). Silk, hair, the lens of an eye, an immunoassay (such as found in a pregnancy test kit), and cottage cheese are all just protein plus more or less water, but they are different because the proteins they contain are different. Proteins carry out almost all of the functions of the living cell including, of course, the synthesis of new DNA. Neither growth nor development would be possible without proteins.

Most proteins have functions that depend on their ability to recognize other molecules by binding. This recognition depends on specific three-dimensional binding sites that make multiple interactions with the **ligand,** the molecule being bound. To do this a protein must itself have a specific three-dimensional structure. Each of the huge number of protein functions demands its own protein structure. Evolution has produced this diversity by using a palette of 20 amino acid monomers, each with its own unique shape and chemical properties, as the building blocks of proteins. Huge numbers of very different structures are therefore possible.

Cell Biology: A Short Course, Second Edition, by Stephen R. Bolsover, Jeremy S. Hyams,
Elizabeth A. Shephard, Hugh A. White, Claudia G. Wiedemann
ISBN 0-471-26393-1 Copyright © 2004 by John Wiley & Sons, Inc.

✻ NAMING PROTEINS

To be able to discuss proteins, we need to give them names. Naming conventions vary between different areas of biology. We have already seen how enzymes such as DNA polymerase are named for the reaction they catalyze. Many proteins have names that describe their structure or their role in cells, such as hemoglobin and connexin. However, the pace with which new proteins are being discovered at the moment means that many are not given proper names but are simply referred to by their size: p38 (page 418) and p53 (page 415) have relative molecular masses of about 38,000 and about 53,000, respectively. Clearly this could cause confusion, so we sometimes add the name of the gene as a superscript: p16^{INK4a} is a protein of relative molecular mass (M_r) of about 16,000 that is the product of the *INK4a* gene.

✻ POLYMERS OF AMINO ACIDS

Translation produces linear polymers of α-amino acids. If there are fewer than around 50 amino acids in a polymer, we tend to call it a *peptide*. More and it is a *polypeptide*. Proteins are polypeptides, and most have dimensions of a few nanometers (nm), although structural proteins like keratin in hair are much bigger. The relative molecular masses of proteins can range from 5000 to hundreds of thousands.

The Amino Acid Building Blocks

The general structure of α-amino acids, the building blocks of polypeptides and proteins, is shown in Figure 9.1. R is the side chain. It is the side chain that gives each amino acid its unique properties.

During the process of translation (page 173), peptidyl transferase joins the amino group of one amino acid to the carboxyl of the next to generate a peptide bond. A generalized polypeptide is shown in Figure 9.1. The backbone, a series of peptide bonds separated by the α carbons, is shown in green. At the left-hand end is a free amino group; this is known as the **N terminal** or **amino terminal.** At the right-hand end is a free carboxyl group; this is

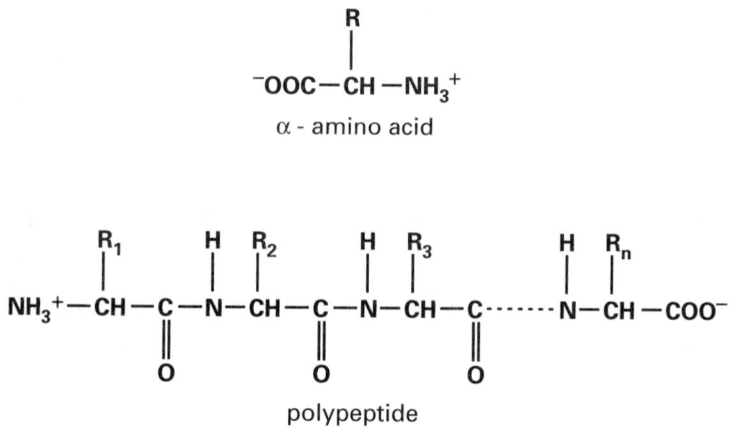

Figure 9.1. α-Amino acids and the peptide bond.

known as the **C terminal** or **carboxy terminal.** Peptides are normally written this way, with the C terminal to the right.

The properties of individual polypeptides are conferred by the side chains of their constituent amino acids. Many different properties are important—size, electrical charge, the ability to participate in particular reactions—but the most important is the affinity of the side chain for water. Side chains that interact strongly with water are hydrophilic. Those that do not are hydrophobic. We have already encountered the 20 amino acids coded for by the genetic code (page 75), but we will now describe each in turn, beginning with the most hydrophilic and ending with the most hydrophobic. Each amino acid has both a three-letter abbreviation and a one-letter code (Fig. 9.2). In the following section we refer to each amino acid by its full name and give the three- and one-letter abbreviations. This will help you to familiarize yourself with the amino acid abbreviations, which are used in other sections of the book.

Figure 9.3 shows four amino acids with hydrophilic side chains. Aspartate (Asp, D) and glutamate (Glu, E) have acidic side chains. At the pH of the cytoplasm, these side chains bear negative charges, which interact strongly with water molecules. As they are usually charged, we generally name them as their ionized forms: aspartate and glutamate rather than aspartic acid and glutamic acid. A polypeptide made entirely of these amino acids is very soluble in water. Lysine (Lys, K) and arginine (Arg, R) have basic side chains. At the pH of the cytoplasm, these side chains bear positive charges, which again interact strongly with water molecules.

Figure 9.4 shows histidine (His, H), whose side chain is weakly basic with a pK_a of 7.0. At neutral pH about half the histidine side chains, therefore, bear a positive charge. The fact that histidine is equally balanced between protonated and unprotonated forms gives it important roles in enzyme catalysis. At neutral pH a polypeptide made entirely of histidine will be very soluble in water.

Figure 9.4 also shows cysteine (Cys, C) whose thiol (—SH) group is weakly acidic with a pK_a (page 25) of about 8. At neutral pH most (about 90%) of cysteine side chains will have their hydrogen attached. Even so the charge on the remaining 10% means that a polypeptide made entirely of cysteine will be soluble in water. Cysteine's thiol group is chemically reactive and has important roles in some enzyme-active sites. Two cysteine thiol groups can form a **disulfide bond** (page 191).

Figure 9.5 shows the three amino acids serine (Ser, S), threonine (Thr, T), and tyrosine (Tyr, Y). These amino acids have hydroxyl (—OH) groups in their side chains that can hydrogen bond with water molecules. A polypeptide composed of these amino acids is soluble in water.

Asparagine (Asn, N) and glutamine (Gln, Q) (Fig. 9.6) are the amides of aspartate and glutamate. They are hydrophilic.

Glycine (Gly, G; Fig. 9.6) has nothing but a hydrogen atom for its side chain. It is relatively indifferent to its surroundings.

The five amino acids shown in Figure 9.7 have side chains of carbon and hydrogen only. These are alanine (Ala, A), valine, (Val, V), leucine (Leu, L), isoleucine (Ile, I), and phenylalanine (Phe, F). The side chains cannot interact with water so they are hydrophobic. A polypeptide composed entirely of these amino acids does not dissolve in water but will dissolve in olive oil.

Tryptophan (Trp, W; Fig. 9.8) is the largest of the amino acids. Its double ring side chain is mainly hydrophobic. Methionine (Met, M; Fig. 9.8) is also hydrophobic: its sulfur atom is in the middle of the chain so cannot interact with water. Last comes proline (Pro, P; Fig. 9.9). Proline is not really an amino acid at all—it is an **imino acid,** but biologists give it

The genetic code and the corresponding amino acid side chains.

alanine (ala) A	asparagine (asn) N	aspartate (asp) D	arginine (arg) R
CH₃ GCU GCC GCA GCG	O NH₂, C, CH₂ — Site for attachment of sugars AAU AAG	O O⁻, C, CH₂ — Negatively charged. Can be phosphorylated GAU GAC	NH₂ C=⁺NH₂ NH (CH₂)₃ — Positively charged CGU CGC CGA CGG AGA AGG
cysteine (cys) C	glutamine (gln) Q	glutamate (glu) E	glycine (gly) G
SH CH₂ — About 10% are deprotonated and hence negatively charged. Forms disulfide bonds UGU UGC	O NH₂, C, (CH₂)₂ CAA CAG	O O⁻, C, (CH₂)₂ — Negatively charged. Can be phosphorylated GAA GAG	H — The smallest side chain GGU GGC GGA GGG
histidine (his) H	isoleucine (ile) I	leucine (leu) L	lysine (lys) K
HN—CH, HC, N—C, H⁺, CH₂ — About 50% are protonated. pK is 7.0. Can be phosphorylated CAU CAC	CH₃ CH₂ H₃C—CH AUU AUC AUA	H₃C CH₃ CH CH₂ CH₂ UUA UUG CUU CUC CUA CUG	⁺NH₃ (CH₂)₄ — Positively charged AAA AAG
methionine (met) M	phenylalanine (phe) F	proline (pro) P	serine (ser) S
CH₃ S (CH₂)₂ AUG	CH₂ (benzene ring) UUU UUC	CH₂ CH₂ CH₂ —C—N— H — Introduces a kink in the polypeptide chain CCU CCC CCA CCG	OH CH₂ — Can be phosphorylated AGU AGC UCU UCC UCA UCG
threonine (thr) T	tryptophan (trp) W	tyrosine (tyr) Y	valine (val) V
CH₃ HO—CH — Can be phosphorylated. Site for attachment of sugars ACU ACC ACG ACA	(indole ring) HN CH₂ — The largest side chain UGG	OH (phenol ring) CH₂ — Can be phosphorylated UAU UAC	H₃C CH₃ CH GUU GUC GUG GUA
	STOP UGA	STOP UAA UAG	

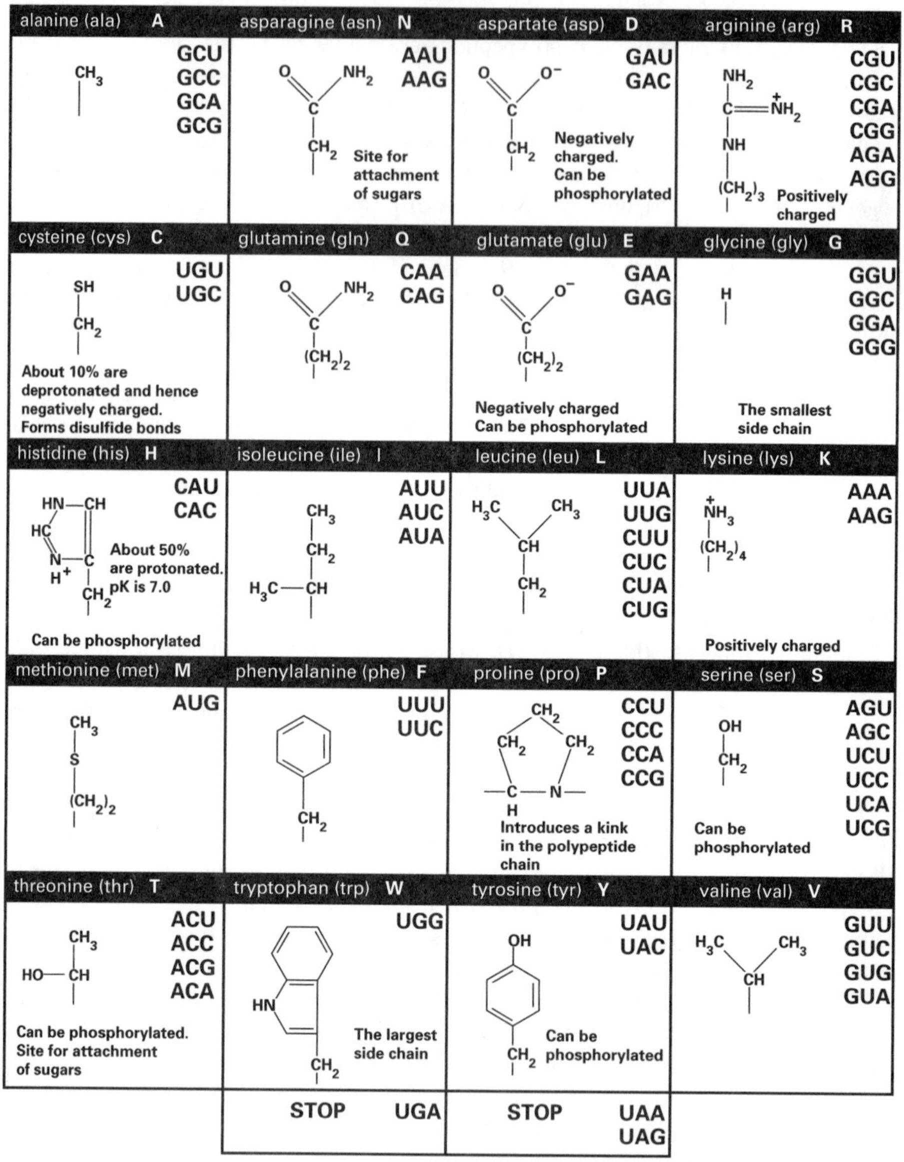

Figure 9.2. The genetic code and the corresponding amino acid side chains, with special properties noted. Hydrophilic side chains are shown in green, hydrophobic side chains in black.

honorary amino acid status. Because the side chain (which is hydrophobic) is connected to the nitrogen atom and therefore to the peptide bond, including proline in the polypeptide chain introduces a kink.

How is it that amino acids with hydrophobic side chains are present in solution in the cytosol and can be picked up by their respective tRNAs? The answer is simple: as free amino acids, they all bear two charges, a positive charge on the amino group and a negative

Asp *D* *Aspartate*

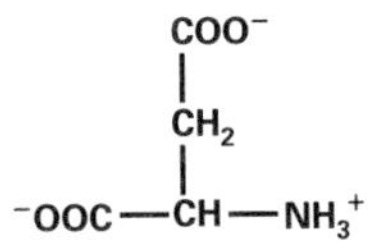

Glu *E* *Glutamate*

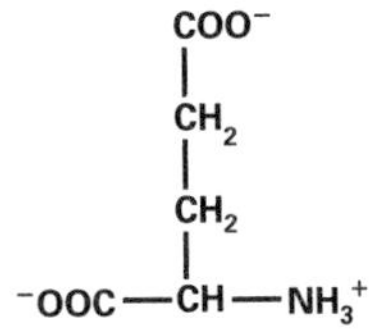

Lys *K* *Lysine*

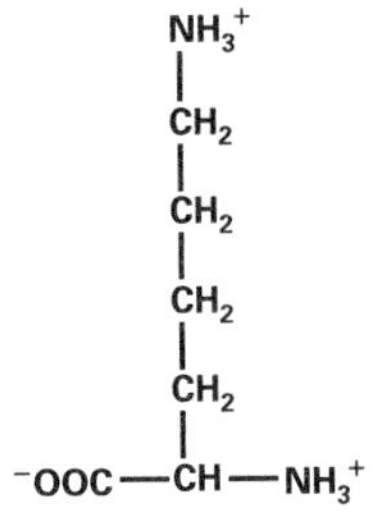

Arg *R* *Arginine*

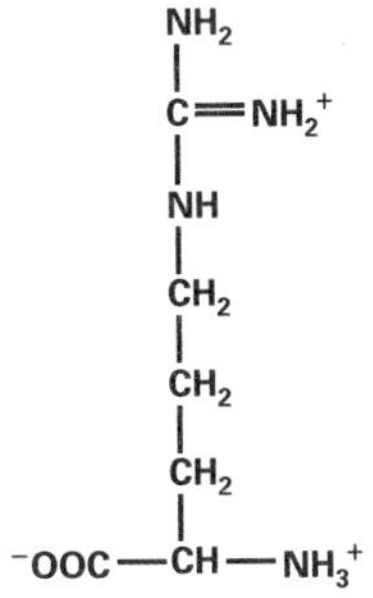

Figure 9.3. Amino acids with hydrophilic side chains.

His *H* *Histidine*

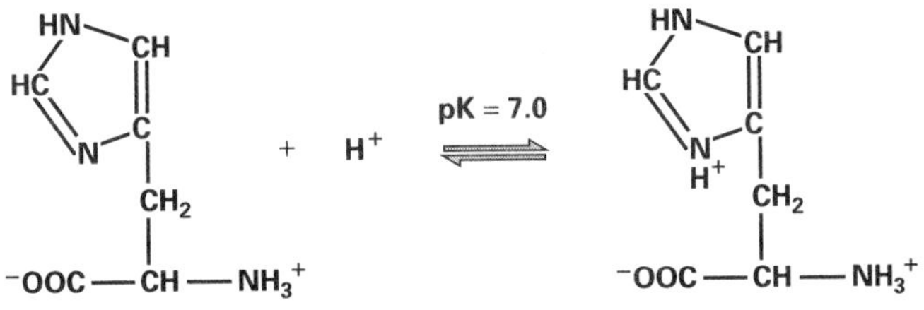

Cys *C* *Cysteine*

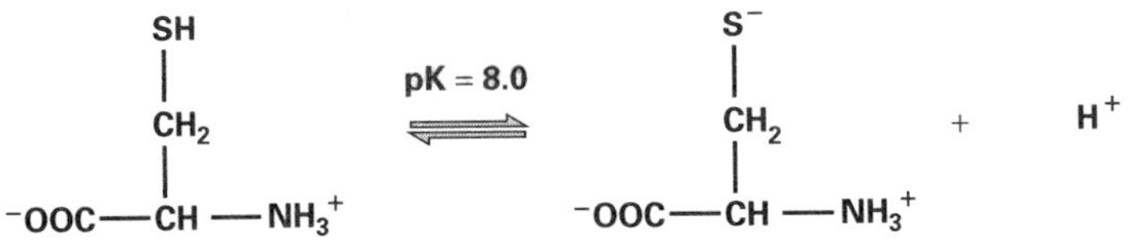

Figure 9.4. Histidine and cysteine.

Ser   S   Serine

Thr   T   Threonine

Tyr   Y   Tyrosine

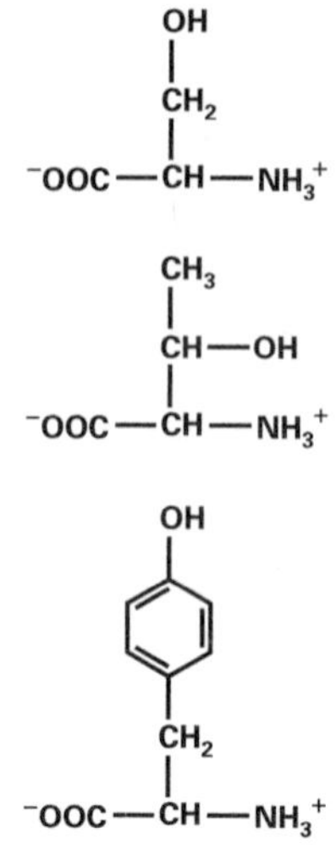

Figure 9.5. Amino acids with hydroxl groups in their side chains.

charge on the carboxyl group. Like all small ions they are soluble in water. Formation of the peptide bond removes these charges. The term *residue* is used to refer to the amino acid when it is part of a peptide chain.

The Unique Properties of Each Amino Acid

Although we have classified side chains on the basis of their affinity for water, their other properties are important (Fig. 9.2).

Asn   N   Asparagine

Gln   Q   Glutamine

Gly   G   Glycine

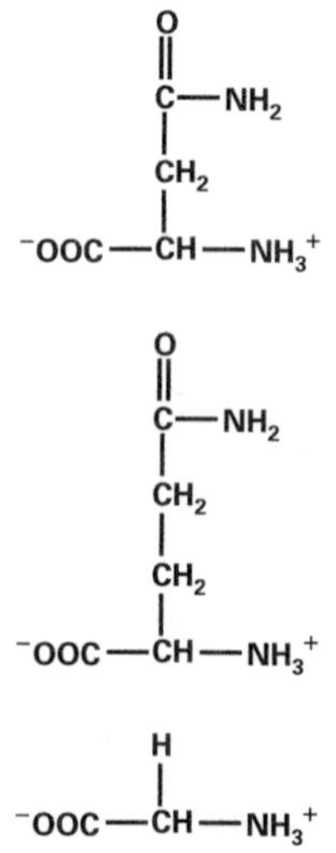

Figure 9.6. Asparagine, glutamine, and glycine.

Ala *A* *Alanine*

Val *V* *Valine*

Leu *L* *Leucine*

Ile *I* *Isoleucine*

Phe *F* *Phenylalanine*

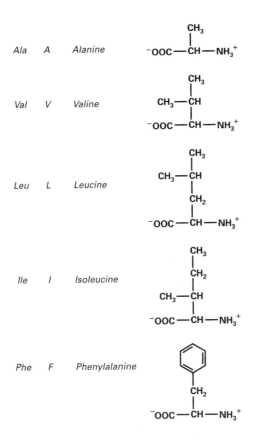

Figure 9.7. Amino acids with simple hydrophobic side chains.

Trp *W* *Tryptophan*

Met *M* *Methionine*

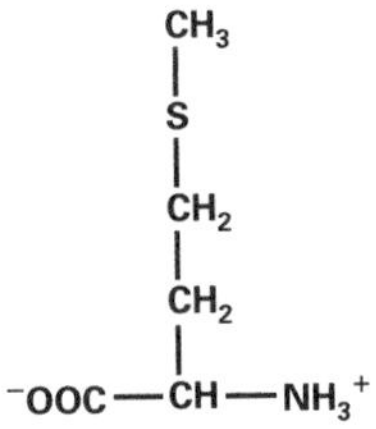

Figure 9.8. Amino acids with more complicated hydrophobic side chains.

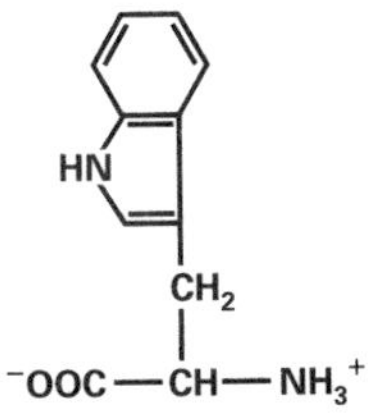

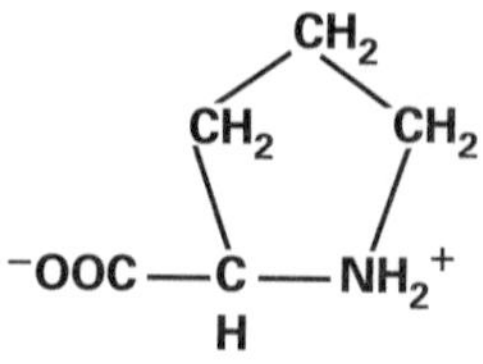

Figure 9.9. Proline, an imino acid.

CHARGE. If a side chain is charged it will be hydrophilic. But charge has other effects too. A positively charged residue such as a lysine will attract a negatively charged residue such as glutamate. If the two residues are buried deep within a folded polypeptide, where neither can interact with water, then it will be very difficult to pull them apart. We call such an electrostatic bond inside a protein a **salt bridge.** Negatively charged residues will attract positively charged ions out of solution, so a pocket on the surface of a protein lined with negatively charged residues will, if it is the right size, form a binding site for a particular positively charged ion like sodium or calcium.

POSTTRANSLATIONAL MODIFICATIONS. After a polypeptide has been synthesized some of the amino acid side chains may be modified. A good example of this is glycosylation, in which chains of sugars may be added to asparagine or threonine side chains (page 225). A very specific posttranslational modification occurs in the connective tissue protein collagen (page 13). After the polypeptide chain has been synthesized, specific prolines and lysines are hydroxylated to hydroxyproline and hydroxylysine. Because hydroxyl groups can hydrogen bond with water, this helps the extracellular matrix to form a hydrated gel (page 13).

Example 9.1 **Hydroxyproline, Hydroxylysine, and Scurvy**

Scurvy, a very unpleasant disease in which blood vessels become leaky, is caused by a lack of vitamin C in the diet. As far as we know, the only use of vitamin C in the body is to participate in the post-translational modification of proline and lysine to generate the hydroxyproline and hydroxylysine in collagen molecules. Without vitamin C collagen, necessary for the physical strength of blood vessels, fails to form properly. In 1747 James Lind demonstrated that citrus fruits (which we now know to be rich in vitamin C) prevented scurvy, a finding that allowed explorers to make long-distance sea voyages that had previously been impossible.

Not all posttransational modifications are so permanent. One very important modification is phosphorylation, the attachment of a phosphate group. We have already seen how sugars can be phosphorylated (page 33), but so can the side chains of the six amino acids serine, threonine, tyrosine, aspartate, glutamate, and histidine (Fig. 9.10). Usually the phosphate group comes from ATP, in which case the enzyme that does the job is called a **kinase.** Phosphate groups carry two negative charges and can therefore markedly alter the balance of electrical forces within a protein. Another group of enzymes called **phosphatases** remove the phosphate groups. Since phosphorylation is readily reversible, it is often used to turn proteins off and on. For example, the transcription factor NFAT (page 222) is only

phosphoesters

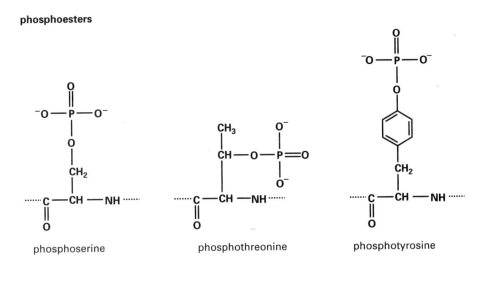

phosphoserine   phosphothreonine   phosphotyrosine

phosphoanhydrides   **phosphoimides**

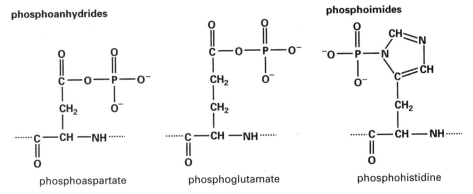

phosphoaspartate   phosphoglutamate   phosphohistidine

Figure 9.10. Six side chains can be phosphorylated.

active in the dephosphorylated state, while insulin receptor substrate number 1 (page 356) is activated by phosphorylation.

DISULFIDE BONDING.   Under oxidizing conditions the thiol (—SH) groups of two cysteine residues can link together to form a *disulfide bond* (or *disulfide bridge*) (Fig. 9.11). Proteins made for use outside the cell often have disulfide bonds that confer additional rigidity on the protein molecule. If all the peptide bonds in a polypeptide are hydrolyzed, any cysteines that were linked by disulfide bonds remain connected in dimers called—very confusingly—cystine molecules.

✤ OTHER AMINO ACIDS ARE FOUND IN NATURE

Ornithine and citrulline (page 293) are α-amino acids that play a vital role in the body—they are used as part of the urea cycle to get rid of ammonium ions, which would otherwise poison us (page 293). However, they are not used as building blocks in the synthesis of polypeptides.

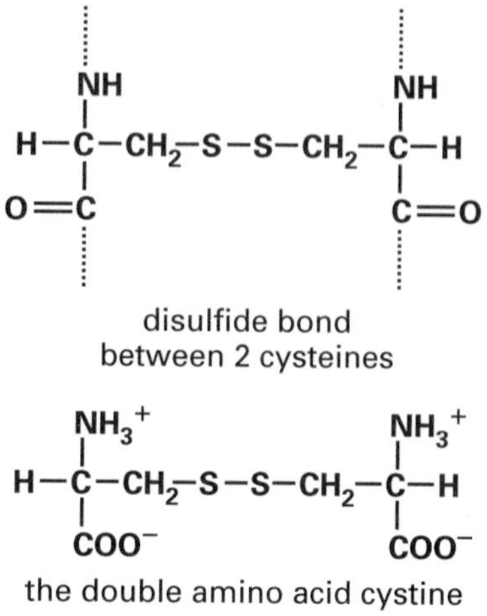

NH   NH
| |
H—C—CH₂—S—S—CH₂—C—H
| |
O=C   C=O

disulfide bond
between 2 cysteines

NH₃⁺   NH₃⁺
| |
H—C—CH₂—S—S—CH₂—C—H
| |
COO⁻   COO⁻

the double amino acid cystine

Figure 9.11. Oxidation of adjoining cysteine residues produces a disulfide bond. Proteolysis of the polypeptide releases cystine.

There are many other nonprotein amino acids. Very early in the evolution of life, the palette of amino acids used to make polypeptides became fixed at the 20 that we have described. Almost the entire substance of all living things on earth is either polypeptide, composed of these 20 monomers, or other molecules synthesized by enzymes that are themselves proteins made of these 20 monomers. Natural selection has directed evolution within the constraints imposed by the palette.

❋ THE THREE-DIMENSIONAL STRUCTURES OF PROTEINS

Proteins are polypeptides folded into specific, complex three-dimensional shapes. Generally, hydrophobic amino acids pack into the interior of the protein while hydrophilic amino acid side chains end up on the surface where they can interact with water. The amino acid side chains pack together tightly in the interior: there are no spaces. It is the three-dimensional structure of a particular protein that allows it to carry out its role in the cell. The shape can be fully defined by stating the position and orientation of each amino acid, and such knowledge lets us produce representations of the protein (Fig. 9.12). However, when discussing protein structure, it is helpful to think in terms of a series of different levels of complexity.

The **primary structure** of a protein is the sequence of amino acids. Lysozyme is an enzyme that attacks bacterial cell walls. It is found in secretions such as tears and in the white of eggs. Lysozyme has the following primary structure:

(NH₂)KVFGRCELAAAMKRHGLDNYRGYSLGNWVCAAKFESNFNTQATNRNTD GSTDYGILQINSRWWCDNGRTPGSRNLCNIPCSALLSSDITASVNCAKKIVSDGDG MNAWVAWRNRCKGTDVQAWIRGCRL(COOH).

Numbering is always from the amino terminal end where synthesis of the protein begins on the ribosome. Lysozyme has four disulfide bonds between four pairs of cysteines. The 129 amino acids of hen egg white lysozyme are shown in linear order in Figure 9.12a

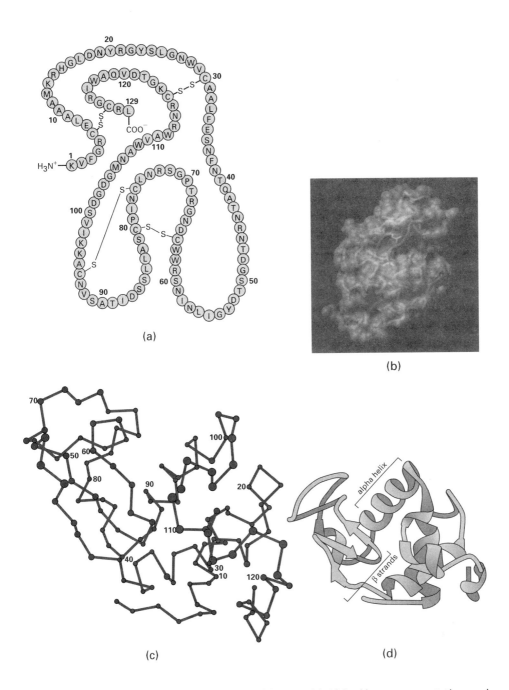

Figure 9.12. Lysozyme: (a) linear map, (b) space filling model, (c) backbone representation, and (d) cartoon.

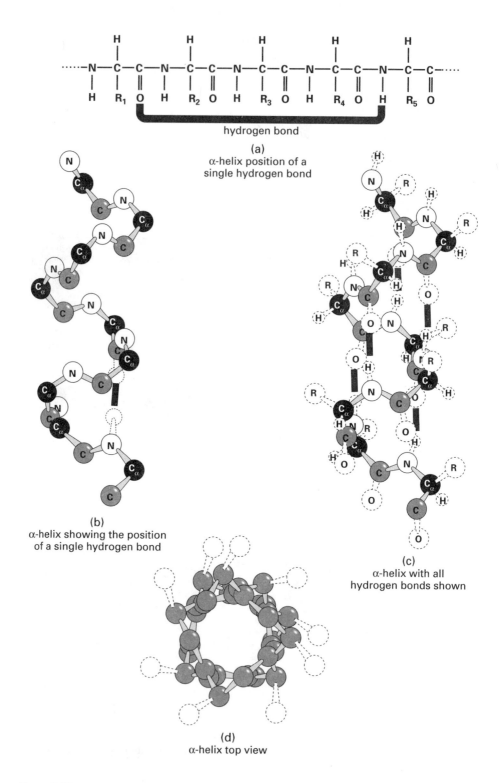

(a)
α-helix position of a
single hydrogen bond

(b)
α-helix showing the position
of a single hydrogen bond

(c)
α-helix with all
hydrogen bonds shown

(d)
α-helix top view

Figure 9.13. The α helix: (a) the nitrogen atom in each peptide bond forms a hydrogen bond with the oxygen of the peptide bond four ahead. (b) An α helix, with one hydrogen bond drawn. (c) An α helix, with all the hydrogen bonds drawn. (d) An end-on view of an α helix.

with the disulfide bonds indicated. Lysozyme was the first enzyme to have its three-dimensional structure fully determined (in 1965). Figure 9.12b shows that structure, with all of the atoms that form the molecule displayed. We see little except an irregular surface. However, if the amino acid side chains are stripped away and the path of the peptide-bonded backbone drawn (Fig. 9.12c), we see that some regions of the protein backbone are ordered in a repeating pattern. Two types of protein backbone organization are common to many proteins. These are named the α **helix** and the β **sheet.** Figure 9.12d redraws the peptide backbone of lysozyme to emphasize these patterns, with the lengths of peptide participating in β sheets represented as arrows. Collectively these repeating patterns are known as **secondary structures.** There are other regions of the protein that do not have any such ordered pattern.

In an α helix the polypeptide chain twists around in a spiral, each turn of the helix taking 3.6 amino acid residues. This allows the nitrogen atom in each peptide bond to form a hydrogen bond with the oxygen four residues ahead of it in the polypeptide chain (Fig. 9.13). All the peptide bonds in the helix are able to form such hydrogen bonds, producing a rod in which the amino acid side chains point outward. Because it introduces a kink into the polypeptide chain, proline cannot participate in an α helix.

In a β sheet lengths of polypeptide run alongside each other, and hydrogen bonds form between the peptide bonds of the strands. This generates a sheet that has the side chains protruding above and below it (Fig. 9.14). Along a single strand the side chains alternate up then down, up then down. Because the actual geometry prevents them from being completely flat, they are sometimes called β pleated sheets. A polypeptide chain can form two types of β sheet: Either all of the strands in the β sheet are running in the same direction (Fig. 9.14a) forming a **parallel** β **sheet** or they can alternate in direction (Fig. 9.14b) making an **antiparallel** β **sheet.** The polypeptide chains in β sheets are fully extended unlike the chain in an α helix.

In structural proteins like the **keratin** in hair or the fibroin in silk, the whole polypeptide chain is ordered into one of these secondary structures. Such fibrous proteins have relatively simple repeating shapes and do not have binding sites for other molecules. Most proteins have regions without secondary structures, the precise folding and packing of the amino acids being unique to the protein, side by side with regions of secondary structure. We have seen that secondary structures such as the α helix and β sheet are formed because of hydrogen bonds involving the peptide bonds of the backbone. The three-dimensional structure is held together by various interactions most of which are individually weak but collectively produce stable molecules. These interactions involve the amino acid side chains that interact with one another, with the backbone, with water molecules, and (if there is one) with a prosthetic group or cofactor (page 205). These interactions are hydrogen bonds, electrostatic interactions, van der Waals forces, hydrophobic interactions, and (in some proteins) disulfide bonds.

Hydrogen Bonds

If a hydrogen attached to an oxygen, nitrogen, or sulfur by a covalent bond gets close to a second electron-grabbing atom, then that second atom also grabs a small share of the electrons to form a hydrogen bond. The atom to which the hydrogen is covalently bonded is called the donor because it is losing some of its share of electrons; the other electron-grabbing atom is the acceptor. In order for a hydrogen bond to form, the donor and acceptor must be within a fixed distance of one another (typically 0.3 nm) with the hydrogen on a straight line between them. In an α helix the nitrogen atom within a peptide bond shares its

•••• IN DEPTH 9.1 Chirality and Amino Acids

A chiral structure is one in which the mirror image of the structure cannot be super-imposed on the structure. The shape of your whole body is not chiral. Your mirror image could be rotated through 180°, so that it faces into the mirror, then step back and be superimposed on you. Your right hand, however, is chiral: Its mirror image is not a right hand any more but a left hand. Molecules are often chiral—this arises if a carbon atom has four different groups attached to it and is therefore asymmetric. The α carbon of α-amino acids is asymmetric for all except glycine. It is possible for there to be more than one asymmetric carbon in a molecule. If there is an asymmetric carbon in a molecule, there will be two different ways that the groups on that carbon can be arranged and so there will be two **optical isomers.** These isomers interact with polarized light differently. Although a different system is used in chemistry, optical isomers of amino acids (and sugars) are denoted L and D. L amino acids are exclusively used in proteins and predominate in the metabolism of amino acids generally. However D amino acids are found in nature. D alanine occurs in bacterial cell walls while some antibiotics such as valinomycin and grami-cidin A contain D amino acids. These molecules are synthesized in entirely different ways from proteins. Among sugars it is the D forms that are mainly used by cells.

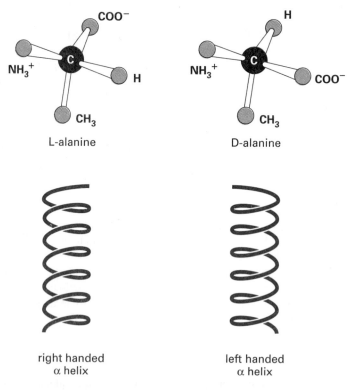

L-alanine   D-alanine

right handed   left handed
α helix α helix

Helices are chiral too. The α helix found in proteins is right handed, like a regular screw thread. Reflect a length of right-handed α helix in a mirror, and you will get a left-handed α helix composed of D amino acids. Because of the actual structures, it turns out that L amino acids fit nicely in a right-handed α helix, while D amino acids fit nicely in a left-handed α helix. (If you doubt the chirality of helices, try making wire models of the two shown in the accompanying figure and see if they can be superimposed.)

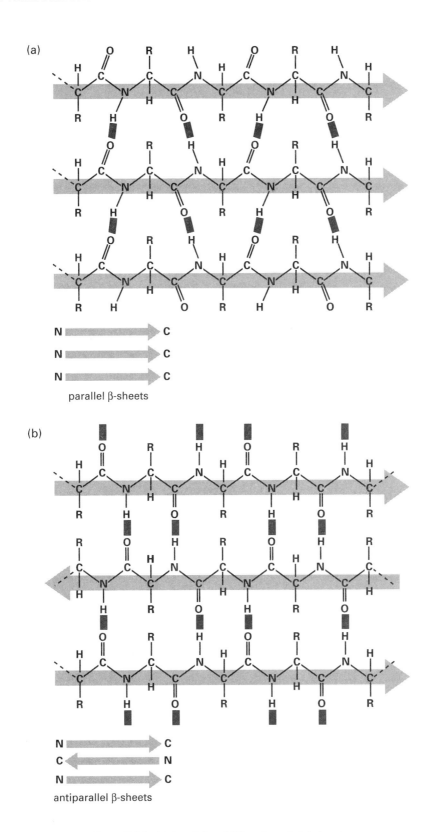

Figure 9.14. (a) Parallel and (b) antiparallel β sheets.

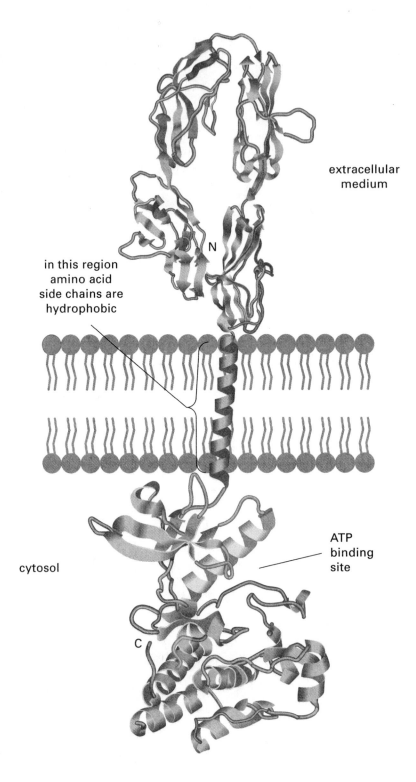

extracellular
medium

in this region
amino acid
side chains are
hydrophobic

N

ATP
binding
site

cytosol

C

Figure 9.15. Cartoon representation of the platelet derived growth factor receptor.

hydrogen atom with the carbonyl group oxygen of the peptide bond four residues ahead of it in the chain.

Electrostatic Interactions

If positive and negatively charged amino acid residues are buried in the hydrophobic interior of a protein, where neither can interact with water, then they will attract each other and the force between them will be stronger than it would be in water. Such an electrostatic bond inside a protein is called a salt bridge.

Polar groups such as hydroxyl and amide groups are dipoles. They have an excess of electrons at one atom and a corresponding deficiency at another, as we have already seen for the water molecule (page 27). The partial charges of dipoles will be attracted to other dipoles and to fully charged groups.

van der Waals Forces

These are relatively weak close-range interactions between atoms. Imagine two atoms sitting close together. At a given instant more of the electrons on one atom may be on one side, and this exposes the positive charge on the nucleus. This positive charge attracts the electrons of the adjacent atom thus exposing its nuclear charge, which would attract the electrons of another atom and so on. The next instant the electrons will have moved so we have a situation of fluctuating attractions between atoms. These forces are important in the close-packed interiors of proteins and membranes and in the specific binding of a ligand to its binding site.

Hydrophobic Interactions

Molecules that do not interact with water (and that are therefore classed as hydrophobic) force the water around them to become more organized: hydrogen-bonded "cages" form. This organization is thermodynamically undesirable and is minimized by the clustering together of such molecules. This is called the **hydrophobic effect.** A polypeptide with hydrophilic and hydrophobic residues will spontaneously adopt a configuration in which the hydrophobic residues are not exposed to water. They can achieve this either by sitting in a lipid bilayer (Fig. 9.15) or by adopting a globular shape in which the hydrophobic residues are clustered in the center of the protein.

Disulfide Bonds

Extracellular proteins often have disulfide bonds between specific cysteine residues. These are strong covalent bonds, and they tend to lock the molecule into its conformation. Although relatively few proteins contain disulfide bonds, those that do are more stable and are therefore easy to purify and study. For this reason many of the first proteins studied in detail, such as the digestive enzymes chymotrypsin and ribonuclease and the bacterial cell wall degrading enzyme lysozyme, have disulfide bonds.

✳ TERTIARY STRUCTURE: DOMAINS AND MOTIFS

The three-dimensional protein structure often has protrusions, clefts, or grooves on the surface where particular amino acids are positioned to form sites that bind ligands and, in the case of enzymes, catalyze reactions within or between ligands. The whole three-dimensional arrangement of the amino acids in the protein is called the tertiary structure.

A tertiary structure is unique to a particular protein. However, common patterns or **motifs** occur in tertiary structures. For example, many proteins with different functions show a β-barrel structure where β sheet is rolled up to form a tube. Green fluorescent protein (page 150) is one example (Fig. 9.16). Often the tertiary structure is seen to divide into discrete regions. The calcium-binding protein **calmodulin** shows this clearly (Fig. 9.17). Its single chain is organized into two **domains,** one shown in green, the other in black, joined by only one strand of the polypeptide chain. In calmodulin the two domains are very similar, and the modern gene probably arose through duplication of an ancestral gene that was half as big. Domains are easier to see than to define. "A separately folded region of a single polypeptide chain" is as good a definition as any.

Domains may be similar or different in both structure and function. The catabolite-activating protein (CAP) of *Escherichia coli* (page 114) binds to a specific sequence of bases on DNA, assisting RNA polymerase to bind to its promoter and initiate transcription of the *lac* operon. It does this only when it has bound cAMP (which in turn only happens when the intracellular glucose concentration is low). One of the domains of CAP has the job of binding cAMP. Another recognizes DNA sequences using a *helix-turn-helix* motif (Fig. 9.18). One of these helices fits into the major groove of the DNA where it can make specific interactions with the exposed edges of the bases.

Proteins that interact with DNA often do so *via* **zinc fingers.** The DNA binding domain of the glucocorticoid receptor (page 121) contains two zinc finger motifs. Each finger is formed by the coordination of four cysteine residues with a Zn^{2+} ion (Fig. 9.19a). The domain contains two α helices (Fig. 9.19b), one in each finger. The first helix, also known as the recognition helix, fits into the major groove of the DNA and makes contacts with specific bases. The second helix is bent at right angles to the first and helps to stabilize the receptor homodimer on DNA by promoting dimerization between the two receptor monomers. This finger can alternatively interact with other proteins that help to activate or repress transcription, such as AP1 subunits (page 124).

Domains usually correspond to exons (page 98) in the gene. It is therefore relatively easy for evolution to create new proteins by mixing and matching domains from existing proteins. The cAMP gated channel (page 350) arose in the ancestor of the vertebrates when a mutation occurred that spliced the cAMP-binding domain from cAMP-dependent protein kinase (page 353) onto the end of the voltage-gated sodium channel (page 330).

Integral membrane proteins have special tertiary structures. The polypeptide chain may cross the membrane once or many times, but each time it does so it has a region of hydrophobic side chains that can be in a hydrophobic environment such as the interior of a lipid bilayer. The commonest membrane-spanning structure is a 22-amino-acid α helix. Figure 9.15 shows an integral membrane protein called the platelet-derived growth factor receptor (page 353). Hydrophobic side chains on the transmembrane α helix interact with the hydrophobic interior of the membrane. The protein is held tightly in the membrane because for it to leave would expose these hydrophobic side chains to water.

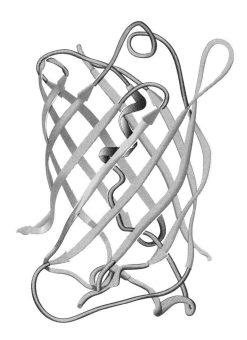

Figure 9.16. Green fluorescent protein molecule comprises a ß barrel and a central α helix.

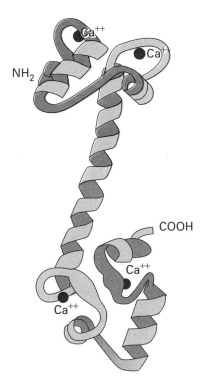

Figure 9.17. Calmodulin is composed of two very similar domains.

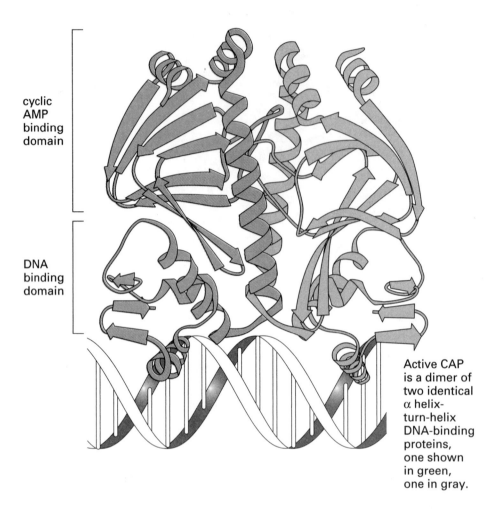

cyclic
AMP
binding
domain

DNA
binding
domain

Active CAP
is a dimer of
two identical
α helix-
turn-helix
DNA-binding
proteins,
one shown
in green,
one in gray.

Figure 9.18. Active catabolite activator protein is a dimer.

Example 9.2 **Recognition by Binding**

Many of the jobs proteins do depend on recognition and binding. Proteins vary greatly in the specificity with which they bind other substances. Histones are at one extreme. About 25% of their amino acids are lysine or arginine, whose side chains are positively charged. Histones therefore bind to DNA—any DNA—through an electrostatic attraction to the negative charges on the phosphates of the DNA backbone.

The glucocorticoid receptor also binds DNA, but it is at the very other extreme of specificity. After binding glucocorticoid hormone, the hormone–receptor complex enters the nucleus and binds to the base sequence 5′ AGAACA 3′. The protein does this using a zinc finger domain (page 200), which inserts an α helix into the major groove of DNA. The receptor works as a dimer so the sequence 5′ AGAACA 3′ must appear twice in the DNA, separated by three nucleotides (page 203). Once bound the receptor binds other proteins that in turn stimulate transcription.

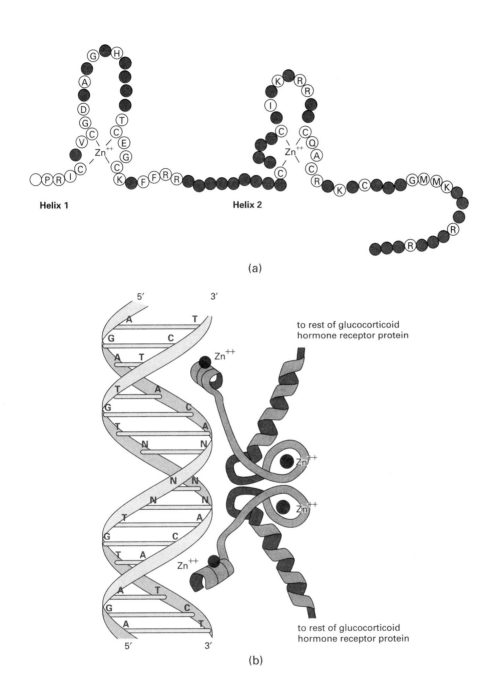

Figure 9.19. (a) Two zinc finger motifs in the glucocorticoid receptor. Each circle is one amino acid residue. Amino acid residues that are important in binding Zn^{2+} or DNA are specified. (b) Cartoon representation of the zinc finger domains of a dimerized pair of glucocorticoid hormone receptors interacting with DNA.

IN DEPTH 9.2 Hydropathy Plotting—The PDGF Receptor

One of the central problems in structural biology is the prediction of the structure of a protein from the primary structure or indeed from the DNA sequence once the gene has been found. Although we have come some way in understanding protein folding, the problem remains unsolved. It is possible, however, to identify proteins that will have similar structures. One of the simpler things that can be looked for is regions of hydrophobic amino acids: if these are 21–22 amino acids long, they are likely to be membrane spanning α helices. The figure shows a **hydropathy plot** for the platelet-derived growth factor (PDGF) receptor. Each amino acid in the protein has a hydrophobicity allotted to it, so that ionized groups like those on aspartate and arginine get a big negative score while groups like those on phenylalanine and leucine get a big positive score. The plot is a running average of the hydrophobicity along the polypeptide chain. The protein begins with a somewhat hydrophobic region: this is the signal sequence (page 215) that directs the growing protein to the endoplasmic reticulum. The rest

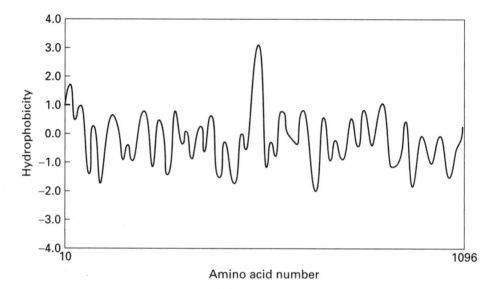

of the protein is neutral or somewhat hydrophilic except for a prominent short hydrophobic region in the center. We can therefore predict that this protein will cross the membrane once at this location. Figure 9.15 shows the full predicted structure of this protein.

✹ QUATERNARY STRUCTURE: ASSEMBLIES OF PROTEIN SUBUNITS

Many proteins associate to form multiple molecular structures that are held together tightly but not covalently by the same interactions that stabilize tertiary structures. For example, gap junction channels (page 55) are formed from six identical monomers. CAP is only active when it dimerizes (Fig. 9.18). Hemoglobin, the protein that carries oxygen in our red blood cells, is formed from four individual polypeptide chains, two α chains and two

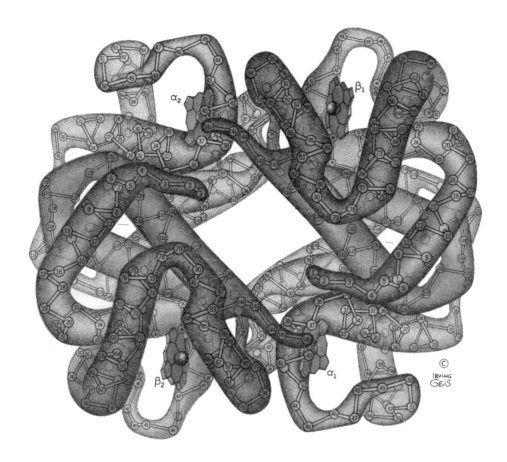

Figure 9.20. Hemoglobin, a tetrameric protein with heme prosthetic groups. (Illustration: Irving Geis. Rights owned by Howard Hughes Medical Institute. Reproduction by permission only.)

β chains (Fig. 9.20). In all these cases, we call the three-dimensional arrangement of the protein subunits as they fit together to form the multimolecular structure the **quaternary structure** of the protein.

✳ PROSTHETIC GROUPS

Even with the enormous variety of structures available, there are some functions that proteins need help with because the 20 protein amino acid side chains do not cover the properties required. Moving electrons in oxidation and reduction reactions and binding oxygen are good examples. Proteins therefore associate with other chemical species that have the required chemical properties. Hemoglobin uses iron-containing heme groups to carry oxygen molecules (Fig. 9.21). The general name for a nonprotein species that is tightly bound to a protein and helps it perform its function is **prosthetic group.**

Some proteins have tightly bound metal ions that are essential to their function. We have already met the zinc fingers of the glucocorticoid receptor and the iron in hemoglobin. Other proteins use magnesium, manganese, or copper.

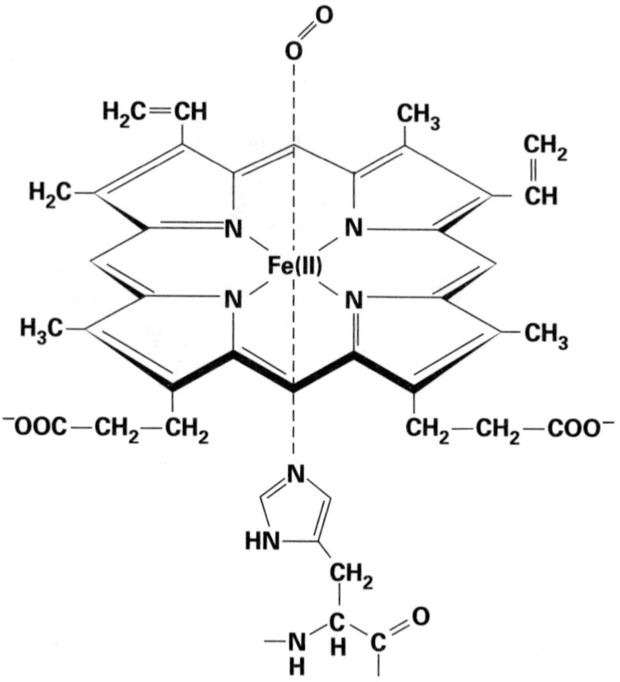

Figure 9.21. Iron-containing prosthetic group heme in the form in which it is found in oxygenated hemoglobin. The oxygen molecule is at the top. (From D. Voet and J. D. Voet, *Biochemistry,* 2nd ed., New York: Wiley, 1995, p. 216. ©1995 John Wiley & Sons, Inc. This material is used by permission of John Wiley & Sons, Inc.)

✳ THE PRIMARY STRUCTURE CONTAINS ALL THE INFORMATION NECESSARY TO SPECIFY HIGHER-LEVEL STRUCTURES

Protein structures are stable and functional over a small range of environmental conditions. Outside this range the pattern of interactions that stabilizes the tertiary structure is disrupted and the molecule **denatures**—activity disappears as the molecule loses its structure. Denaturation may be caused by many factors, which include excessive temperature, change of pH, and detergents. Concentrated solutions of urea (8 mol per liter) have long been used by biochemists to denature proteins. Unlike heat and pH, urea does not cause the protein to precipitate. Physicochemical techniques have shown that in urea solution all of the higher levels of structure are lost and that polypeptide chains adopt random, changing conformations. Reagents such as urea that do this are called **chaotropic reagents.** If the urea is removed (by dialysis or simply by dilution), the protein refolds, regaining its structure and biological activity. This shows that the sequence of amino acids contains all of the information necessary to specify the final structure. The refolding of a urea-denatured protein cannot be random. Even a small protein with 100 amino acids would take some 10^{50} years to try out all of the structural conformations available. The fact that refolding does happen and happens on a time scale of seconds tells us that there must be a folding pathway, and the process is not random. Secondary structures may form first and act as folding units. In the cell folding is assisted by proteins called **chaperones** (page 221).

Medical Relevance 9.1

Protein Folding Gone Awry—Mad Cow Disease

Some years ago one of the dogmas of biology was overthrown—it was long believed that diseases could only be transmitted by structures that contained nucleic acids, that is, by viruses or by microorganisms. It is now clear that a group of brain diseases can be transmitted by a protein. These are the diseases called spongiform encephalopathies: bovine spongiform encephalopathy in cows (mad cow disease), scrapie in sheep, and Creutzfeldt–Jacob disease and kuru in humans. These diseases are rare in humans but have recently increased in farm animals.

The infectious agent is called a **prion** and is a protein. It is coded for by a gene the animals have as part of their genome. In healthy individuals the gene is expressed in the cells of the nervous system and generates an innocuous protein called PrP^c (prion-related protein cellular). PrP^c has a globular region at the C-terminal end but the N-terminal region seems to be unstructured.

Sometimes the same polypeptide chain folds up differently with the disordered N-terminal section instead folding into a structure rich in β sheet. This is called PrP^{sc} (prion-related protein scrapie). The disease arises because PrP^{sc} can cause normal PrP^c protein molecules to fold up into the abnormal form. This keeps happening and lumps of PrP^{sc} form that damage nerve cells. The evidence is that it spread in cattle herds through food that contained recycled meat from infected animals. It is not yet known how one PrP^{sc} molecule triggers the aberrant folding of a normal PrP^c molecule.

Before the problem was fully comprehended, infected animals had been used for human food. There have been a number of cases in people that are thought to have come from eating contaminated beef—these are new variant Creutzfeldt–Jacob disease. It remains to be seen how many people will develop this fatal disease.

•••● IN DEPTH 9.3 Curing Mad Mice with Smelly Fish

If a cell is immersed in a solution more dilute than its own cytosol, it tends to swell, as we have seen in the case of plant cells (page 53). On the other hand, cells immersed in concentrated solutions shrink. The overall strength of a solution is its **osmolarity,** and the movement of water into or out of cells because of solution strength is called **osmosis.** Many fish are exposed to changes in the salt concentration of the water around them, either because they swim from salt water to fresh and back (salmon, eels) or because they live in estuaries. The cells of these fish adjust the osmolarity of their cytosol to match the surrounding water by increasing the concentration of various small, easily synthesized molecules, one of which is urea. However, this introduces a further problem because high concentrations of urea are chaotropic—they cause cellular proteins to unfold.

$$CH_3 - N(CH_3)(CH_3) \rightarrow O$$

Trimethylamine *N*-oxide

The small molecule trimethylamine *N*-oxide, in contrast, helps proteins to fold into their correct shape. One of the most dramatic examples of this is that trimethylamine *N*-oxide, added to mouse nerve cells in culture along with the scrapie agent PrP^{sc}, protects the endogenous PrP^{c} prion protein from being incorrectly refolded to generate PrP^{sc} (page 207). Fish producing urea, therefore, also upregulate a gene encoding a flavin-containing monooxygenase (page 82) and produce trimethylamine *N*-oxide. A battle of opposing forces then ensues, with urea tending to denature the protein and trimethylamine *N*-oxide counteracting this to maintain the protein in its correctly folded state.

When fish die and rot, bacterial action converts trimethylamine *N*-oxide to trimethylamine, which gives rotting fish its distinctive smell (page 82).

SUMMARY

1. Polypeptides are linear polymers of α-amino acids linked by peptide bonds. There are 20 amino acids coded for by the genetic code. They differ in the properties of their side chains, which range from hydrophobic groups to charged and uncharged hydrophilic groups.

2. In addition to their hydrophilicity, side chains of individual amino acids have specific properties of which the most important are charge, the ability to form disulfide bonds, and the ability to undergo posttranslational modification.

3. Phosphorylation (on S, T, Y, D, E, or H) is a posttranslational modification that allows the balance of electrical charges on a protein to be dramatically and reversibly altered.

4. Proteins are polypeptides that have a complex three-dimensional shape.

5. It is convenient to consider protein structure as having three levels. The primary structure is the linear sequence of the amino acid monomers.

6. In some parts of a protein regular, repeated foldings of the polypeptide chain can be seen: these are secondary structures. Secondary structures are held together by hydrogen bonds between the carboxyl oxygens and the hydrogens on the nitrogens of the peptide bonds.

7. There are two common types of secondary structure. In the α helix the chain coils upon itself making a spiral with hydrogen bonds running parallel to the length of the spiral. In β sheets the hydrogen bonds are between extended strands of polypeptide that run alongside one another. The hydrogen bonds are at right angles to the strands and in the plane of the sheet. There are two types—parallel and antiparallel.

8. The final, complex folding of a protein is its tertiary structure. Interactions between side chains stabilize the tertiary structure.

9. Some proteins have a quaternary structure. This is an association of subunits, each of which has a tertiary structure.

FURTHER READING

Branden, C., and Tooze. J. 1999. *Introduction to Protein Structure,* 2nd ed. New York: Garland.

Creighton, T. 1993. *Proteins, Structures and Molecular Properties,* 2nd ed. New York: W. H. Freeman.

Fersht, A. 1999. *Structure and Mechanism in Protein Science.* New York: W. H. Freeman.

McGee, H. 1987. *On Food and Cooking,* London: Unwin.

Perutz, M. 1992. *Protein Structure: New Approaches to Disease and Therapy.* New York: W. H. Freeman.

Tanford, C., and Reynolds, J. 2001. *Nature's Robots—A History of Proteins.* Oxford: Oxford University Press.

Voet D., and Voet J. D. 2003. Biochemistry, 3rd ed. New York: Wiley.

✻ REVIEW QUESTIONS

For each question, choose the ONE BEST answer or completion.

1. Peptide bonds result from
   A. the reaction of amino groups with water.
   B. the reaction of two carboxyl groups and removal of water.
   C. the reaction of an amino group with a carboxyl group and loss of water.
   D. reactions between amino acid side chains.
   E. transfer of a phosphate group from ATP to the amino acid.

2. The amino acids found in proteins are called
   A. α amino acids.
   B. ß amino acids.
   C. γ amino acids.
   D. δ amino acids.
   E. π amino acids.

3. The side chains of the amino acids serine, asparagine, and glutamine are all
   A. normally negatively charged at pH 7.
   B. normally positively charged at pH 7.
   C. able to interact with water.
   D. hydrophobic.
   E. indifferent to their environment.

4. The amino acids making up an α helix
   A. have side chains that point outward—away from the center of the helix.
   B. have side chains that point inward—making up a hydrophobic core for the helix.
   C. have side chains that do not interact with amino acids elsewhere in the protein.
   D. have side chains that are always hydrophobic.
   E. include proline.

5. Disulfide bonds
   A. form between two methionine residues.
   B. form between two cystine residues.
   C. form between two cysteine residues.
   D. are found in cytoplasmic proteins.
   E. attach microtubules to organelles.

6. Integral membrane proteins are helped to locate across the lipid bilayer by
   A. glycosylation.
   B. formation of disulfide bonds.
   C. phosphorylation.
   D. using an α helix made up of amino acids with hydrophilic side chains.
   E. using an α helix made up of amino acids with hydrophobic side chains.

7. Proteins cannot usually be denatured by
   A. concentrated urea solutions.
   B. oxygen.
   C. strong acid.
   D. concentrated ethanol solutions.
   E. heat.

ANSWERS TO REVIEW QUESTIONS

1. **C.** The amino group of one amino acid reacts with the carboxyl group of another with loss of water.

2. **A.** They are α-amino acids meaning that the amino group is on the same carbon as the carboxyl group.

3. **C.** Serine, asparagine, and glutamine all have polar side chains that have no overall charge (so A and B are wrong). These are able to hydrogen bond with water and so are hydrophilic.

4. **A.** The side chains point outward away from the center of the helix. Note that because proline introduces a kink into the polypeptide chain, it cannot participate in an α helix, so E is wrong.

5. **C.** Disulfide bonds are formed between two cysteine residues. Concerning answer B: cystine is the name given to the double amino acid found when proteins containing disulfide bonds are hydrolyzed. If we spoke of a cystine residue in a protein we would therefore mean two cysteines that were already joined by a disulfide bond. Two such residues would not be able to form a further disulfide bond.

6. **E.** Most proteins cross membranes by forming an α helix made up of hydrophobic amino acids. The side chains of amino acid residues in an α helix point out, so the residues need to be hydrophobic to interact with the hydrophobic interior of the bilayer. Such a helix is usually 22 amino acids long, which matches the thickness of a membrane bilayer. Formation of disulfide bonds is unlikely to help proteins adopt a transmembrane configuration, while glycosylation and phosphorylation both make proteins more hydrophilic, making a transmembrane configuration less likely.

7. **B.** Concentrated urea solutions, strong acid, concentrated ethanol solutions, and heat can all denature proteins. In contrast, although oxygen may sometimes oxidize amino acids, such an oxidation does not normally result in loss of structure.

10

INTRACELLULAR PROTEIN TRAFFICKING

An essential feature of all eukaryotic cells is their highly compartmentalized organization, which allows the spatial and temporal separation of the synthesis, modification, degradation, secretion, or take up of molecules. Each of the various organelles within cells is specialized for one or more tasks and therefore needs a specialized set of proteins to carry out its function. In this chapter we will examine how a protein, once synthesized on a ribosome, is precisely and actively moved (**translocated**) to the correct cellular compartment.

✳ THREE MODES OF INTRACELLULAR PROTEIN TRANSPORT

Newly synthesized proteins must be delivered to their appropriate site of function within the cell. The mechanisms and machinery that eukaryotic cells use to accomplish this are highly conserved from yeast to humans. There are three possible ways by which the cell accomplishes the task (Fig. 10.1). *First,* the protein may fold into its final form as it is synthesized and then move through an aqueous medium to its final destination, remaining folded all the way. Delivery of proteins to the nucleus follows this scheme; the proteins are synthesized on cytosolic ribosomes and pass through nuclear pores into the nucleoplasm by a process called **gated transport.** In the *second* form of transport, called **transmembrane translocation,** unfolded polypeptide chains are threaded across one or more membranes to reach their final destination. Proteins destined for the interior of peroxisomes, mitochondria, and chloroplasts are synthesized on cytosolic ribosomes and may fold up completely or

Cell Biology: A Short Course, Second Edition, by Stephen R. Bolsover, Jeremy S. Hyams,
Elizabeth A. Shephard, Hugh A. White, Claudia G. Wiedemann
ISBN 0-471-26393-1 Copyright © 2004 by John Wiley & Sons, Inc.

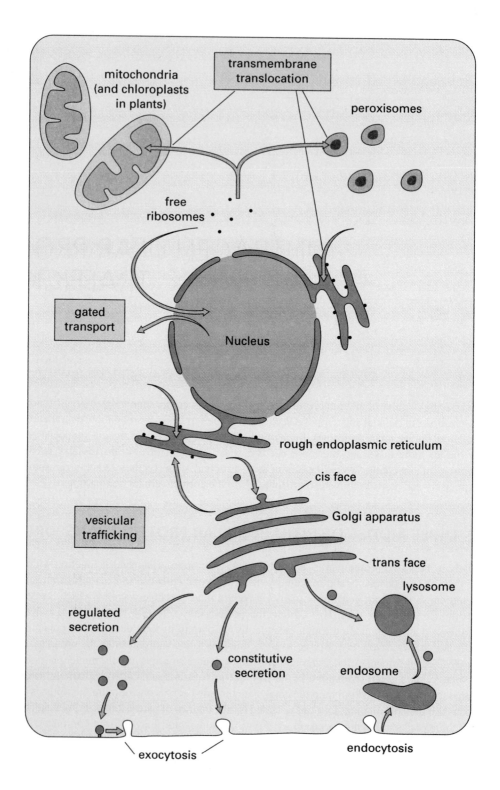

Figure 10.1. The three modes of protein transport.

partially into their final form but are then unfolded in order to be transported across the appropriate membrane. Proteins that are synthesized on the rough endoplasmic reticulum are threaded across the membrane and into the interior of that organelle while they are being synthesized. In the *third* form of transport, small closed bags made of membrane, called **vesicles,** carry newly synthesized protein from the endoplasmic reticulum to the Golgi apparatus and between other compartments in a process called **vesicular trafficking.** The fate of a protein—whether to remain in the cytosol or be sent along one of these alternative transport paths—is determined by sections of the protein itself that act as **sorting signals.** When the protein is first made on the ribosome, it is simply a stretch of polypeptide. The initial sorting decisions must therefore be made on the basis of particular amino acid sequences called **targeting sequences.** For proteins synthesized on the rough endoplasmic reticulum, additional sorting signals such as sugars and phosphate groups can be added by enzymes that modify the chemical structure of the protein. In general, sorting operates by proteins containing a specific sorting signal binding to a receptor protein, which in turn binds to translocation machinery situated in the membrane of the appropriate compartment.

Targeting Sequences

Targeting sequences (also known as **localization sequences**) usually comprise a length of 3–80 amino acids that are recognized in the cytosol by specific receptors that then guide the protein to the correct site and make contact with the appropriate translocation machinery. Once the protein has been imported into the new location the targeting sequence is often removed by enzymes that break the peptide bond between the targeting sequence and the rest of the protein. Some targeting sequences have been characterized better than others. The targeting sequence encoding import into the endoplasmic reticulum consists of about 5–15 mostly hydrophobic amino acids at the N terminus of the protein called the **signal sequence.** The import signal for mitochondria is a stretch of 20–80 amino acids in which positively charged side chains stick out on one side of the helix and hydrophobic side chains stick out on the other, a so-called amphipathic helix. A cluster of about five positively charged amino acids located within the protein sequence targets a protein to the nucleus, while the best known peroxisomal targeting sequence is the C-terminal tripeptide Ser-Lys-Leu-COOH.

Retention

Another class of sorting signal does not activate the transport of a protein out of its present location but is rather a signal to the cell that the protein has reached its final destination and should not be moved. For example, proteins with the motif Lys-Asp-Glu-Leu-COOH (KDEL) at their C terminus are retained within the endoplasmic reticulum.

❋ TRANSPORT TO AND FROM THE NUCLEUS

In contrast to the situation in prokaryotes, RNA transcription and protein synthesis are separated in space and time in eukaryotes. Exchange of material between the nucleus and the cytoplasm is essential for the basic functioning of these cells and must be tightly controlled. RNA and ribosomal subunits that are assembled in the nucleus have to enter the cytoplasm where they are required for protein synthesis. On the other hand, proteins such as histones and transcription factors must enter the nucleus to carry out their functions. The nuclear pore complex mediates trafficking between the nucleus and the cytoplasm. The pore is made from a large number of proteins, and we refer to this type of structure as a multiprotein complex. Transport through the nuclear pore complex is mediated by signals and requires both energy and transporter proteins.

Example 10.1 Holding Calcium Ions in the Endoplasmic Reticulum

One of the functions of the smooth endoplasmic reticulum is to hold calcium ions ready for release into the cytosol when the cell is stimulated. A protein called calreticulin (short for calcium-binding protein of the endoplasmic reticulum) helps hold the calcium ions. Its primary structure is

(NH$_2$)MLLSVPLLLGLLGLLGLAVAEPAVYFKEQFLDGDGWTSRWIESKHKSDFGKFVLSSGKF
YGDEEKDKGLQTSQDARFYALSASFEPFSNKGQTLVVQFTVKHEQNIDCGGGYVKLFP
NSLDQTDMHGDSEYNIMFGPDICGPGTKKVHVIFNYKGKNVLINKDIRCKDDEFTHLYT
LIVRPDNTYEVKIDNSQVESGSLEDDWDFLPPKKIKDPDASKPEDWDERAKIDDPTDSKP
EDWDKPEHIPDPDAKKPEDWDEEMDGEWEPPVIQNPEYKGEWKPRQIDNPDYKGTWIH
PEIDNPEYSPDPSIYAYDNFGVLGLDLWQVKSGTIFDNFLITNDEAYAEEFGNETWGVTK
AAEKQMKDKQDEEQRLKEEEEDKKRKEEEEAEDKEDDEDKDEDEEDEEDKEEDEEED
VPGQAKDEL(COOH)

The first 17 amino acids include 14 hydrophobic ones, shown in black: this is the signal sequence that triggers translocation of the protein into the endoplasmic reticulum. The last 4 amino acids, KDEL, ensure that it remains there. In between these two sorting signals is the functional core of the protein.

The Nuclear Pore Complex

The nuclear pore complex is embedded in the double membrane of the nuclear envelope (Fig. 10.2). The basic structure of the nuclear pore complex, which has been mainly eluci- dated by electron microscopy, is very similar in all eukaryotic cells. It consists of over 50 different components, called nucleoporins, which form eight identical subunits arranged in a circle. Seen from the side the nuclear pore complex comprises three rings. The inner ring sends out radial spokes into the center of the pore and includes transmembrane proteins that anchor the complex to the nuclear envelope. A central transporter, also called the *plug,* is situated in the heart of each nuclear pore complex, forming the gate for macromolecular traffic.

Gated Transport Through the Nuclear Pore

In general, a protein has to display a distinct signal for it to be transported through the nuclear pore. Proteins with a nuclear localization signal are transported in, while proteins with a nuclear export signal are transported out. Mobile transporter proteins, usually mediating either export or import, recognize the appropriate targeting sequence and then interact with structural elements of the nuclear pore. This causes the pore to expand, allowing the active transport of the transporter–protein complex. The precise mechanism by which large molecules move through the nuclear pore is not yet fully understood. However, the discovery of the role played by the GTPase Ran in nuclear transport has given valuable insight into

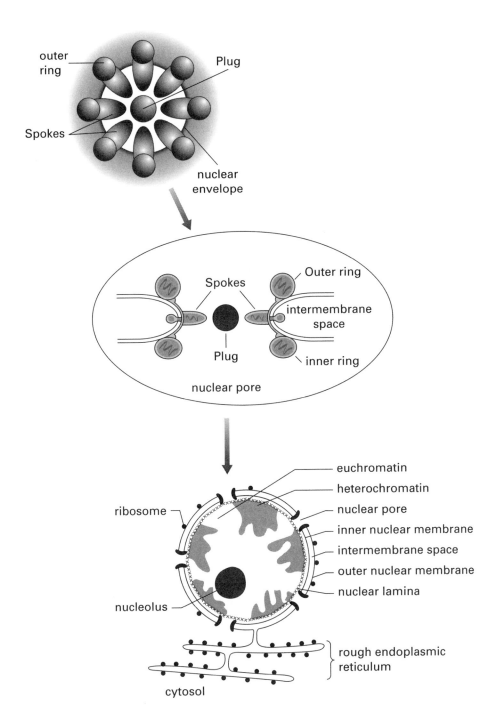

Figure 10.2. The nuclear pore.

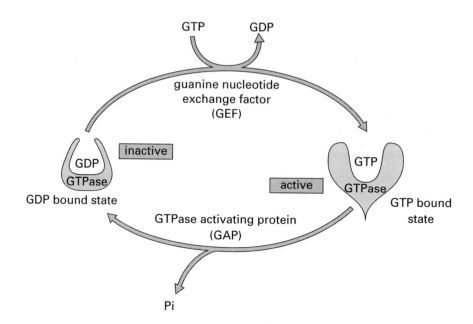

Figure 10.3. The GDP/GTP cycle of a GTPase. Pi represents an inorganic phosphate ion.

how directionality of transport into and out of the nucleus is achieved. GTP hydrolysis by Ran provides the energy for transport. We will now describe this process.

GTPases and the GDP/GTP Cycle

GTPases form a family of proteins that are often involved when cells need to control complex processes. They all share the ability to hydrolyze the nucleotide GTP but otherwise differ markedly in the processes they control and their mode of operation. We have already met one GTPase: the protein EF-tu, a component of the protein synthesis mechanism (page 172). Once a GTPase has hydrolyzed GTP to GDP it adopts an inactive shape and is unable to activate its target process (Fig. 10.3). In contrast, if the protein expels the GDP and binds a molecule of GTP, it then adopts its active form. The cycle between the GDP-bound and GTP-bound state is regulated by effector proteins. GTPase activating proteins or GAPs speed up the rate at which GTPases hydrolyze GTP, and hence the rate at which they inactivate, while guanine nucleotide exchange factors, or GEFs, assist in the exchange of GDP for GTP and therefore help GTPases adopt their active configuration.

GTPases in Nuclear Transport

In the case of nuclear pore transport, the GEFs that operate on Ran are found in the nucleus, while Ran GAPs are cytosolic (Fig. 10.4). Thus nucleoplasmic Ran is predominantly in the GTP bound state (Ran:GTP), while most cytosolic Ran has GDP bound (Ran:GDP).

Figure 10.5 shows how Ran regulates import of proteins into the nucleus. An import transporter binds the nuclear localization sequence on the protein. As long as the transporter remains on the cytosolic side of the nuclear envelope, its cargo will remain bound. However,

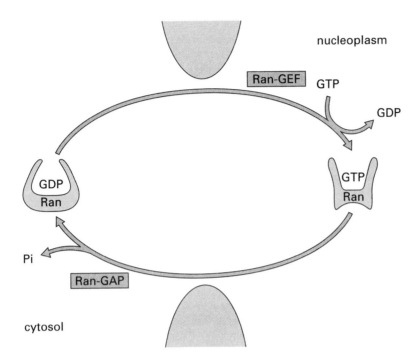

Figure 10.4. Ran GEF and GAP are localized to the nucleoplasm and cytosol, respectively. Pi represents an inorganic phosphate ion.

once the transporter finds itself on the nucleoplasmic side, Ran in its active, GTP-bound state binds and causes the cargo to be released. Now, as long as the transporter remains on the nucleoplasmic side, it will have Ran:GTP bound and will be unable to bind cargo. However, once the transporter finds itself on the cytoplasmic side Ran-GAPs attached to the cytosolic face of the nuclear pore complex will cause Ran to hydrolyze its bound GTP to GDP and Ran will dissociate from the transporter, which is then able to bind more cargo.

The same principle is used when a protein is to be exported from the nucleus (Fig. 10.6). In this case the export receptor can only bind proteins with an export sequence if they also bind Ran:GTP. As long as the transporter remains on the nucleoplasmic side of the nuclear envelope, its cargo will remain bound. However, once the transporter finds itself on the cytoplasmic side Ran GAPs will cause Ran to hydrolyze its bound GTP to GDP and both Ran and the cargo will dissociate from the transporter, which is then unable to bind more cargo until it moves back into the nucleoplasm where Ran:GTP is available.

Although Ran can drive both nuclear import and nuclear export, in each case it is the active GTP-bound form of the protein that binds to the transporter. The GDP-bound form is inactive and cannot bind the transporter.

Some proteins move back and forth between the cytosol and the nucleus by successively revealing and masking nuclear localization sequences. For example, the glucocorticoid receptor (page 121) only reveals its nuclear localization sequence when it has bound glucocorticoid.

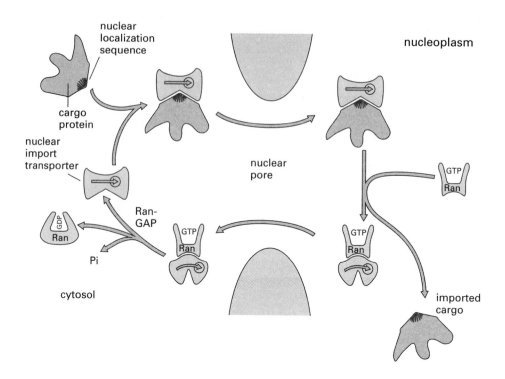

Figure 10.5. Nuclear import. Pi represents an inorganic phosphate ion.

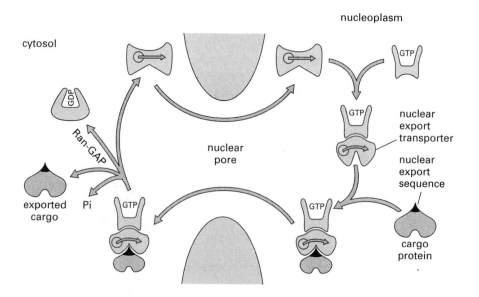

Figure 10.6. Nuclear export. Pi represents an inorganic phosphate ion.

TRANSPORT ACROSS MEMBRANES

Transport to Mitochondria

Mitochondria have their own DNA and manufacture a small number of their own proteins. However, the majority of mitochondrial proteins are coded for by nuclear genes. These are synthesized on free ribosomes and only imported into the mitochondrion posttranslationally. For example, proteins destined for the mitochondrial matrix carry a targeting sequence at their N terminus and are recognized by a receptor protein in the outer membrane. The mitochondrial receptor makes contact with a translocation complex, which unfolds the protein and moves it across both outer and inner membrane simultaneously. Once the protein is translocated, the targeting sequence is cleaved off and the protein refolded.

Chaperones and Protein Folding

A correctly addressed protein may fail to be targeted to an organelle if it folds too soon into its final three-dimensional shape. For example, movement of proteins into the mitochondrial matrix requires that a protein must move through a channel through the outer and inner membranes. This channel is just wide enough to allow an unfolded polypeptide to pass through. Our cells have proteins called chaperones, which as the name indicates, "look after" proteins. Chaperones use energy derived from the hydrolysis of ATP to keep newly synthesized proteins destined for the mitochondrial matrix in an unfolded state. As soon as the protein moves through the channels and into the matrix, the matrix targeting sequence is cleaved. The protein now folds into its correct shape. Some small proteins can fold without help. Larger proteins are helped to fold in the mitochondrial matrix by a chaperone protein called chaperonin, which provides a surface on which another protein can fold. Chaperonin itself does not change shape when helping another protein to fold.

Certain stresses that cells can experience, such as excessive heat, can cause proteins to denature (page 206). The cell responds by making proteins called heat-shock proteins in large amounts. The heat-shock proteins bind to misfolded proteins, usually to a hydrophobic region exposed by denaturation, and help the protein to refold. Like chaperone proteins, the heat-shock proteins are not themselves changed, but instead form a platform on which the denatured protein can refold itself. Heat-shock proteins are found in all cell compartments and also in bacteria.

Transport to Peroxisomes

Most organelles that are bound by a single membrane have their proteins made at the rough endoplasmic reticulum and transported to them in vesicles (Fig. 10.1). Peroxisomes are an exception: their proteins are synthesized on free-floating polyribosomes and then transported to their final destination. Peroxisomal targeting sequences on the protein bind to peroxisome import receptors in the cytosol. The complex of cargo and receptor docks onto the peroxisomal membrane and then crosses the membrane to enter the peroxisome. Here, the protein cargo is released and the import receptor is shuttled back into the cytosol.

Blocking Calcineurin—How Immunosuppressants Work
The drug cyclosporin A is invaluable in modern medicine because it suppresses the immune response that would otherwise cause the rejection of transplanted organs. It does this by blocking a critical stage in the activation of T lymphocytes, one of the cell types in the immune system. T lymphocytes signal to other cells of the immune system by synthesizing and releasing the protein interleukin 2. Transcription of the interleukin 2 gene is activated by a transcription factor called NFAT.

NFAT has a sorting signal that would normally direct it to the nucleus, but in unstimulated cells this is masked by a phosphate group, so NFAT remains in the cytoplasm and interleukin 2 is not made. However, when foreign proteins, for example, any beginning with f-met (page 171), activate the T lymphocyte, they cause the concentration of calcium ions in the cytosol to increase (an increase of calcium concentration is a common feature of cell stimulation to be described in Chapter 16). Calcium activates a **phosphatase** called **calcineurin,** which removes the phosphate group from many substrates including NFAT. NFAT then moves to the nucleus and activates interleukin 2 transcription. The released interleukin 2 activates other immune system cells that attack the foreign body. Cyclosporin blocks this process by inhibiting calcineurin, so that even though calcium rises in the cytoplasm of the T lymphocyte, NFAT remains phosphorylated and does not move to the nucleus.

Medical Relevance 10.3

How Protein Mistargeting Can Give You Kidney Stones

Primary hyperoxaluria type 1 is a rare genetic disease in which calcium oxalate "stones" accumulate in the kidney. Healthy people convert dietary glyoxylate (from plants, see page 297) to the useful amino acid glycine by the enzyme alanine glyoxylate aminotransferase (AGT) (page 250). AGT is located in peroxisomes in liver cells. If glyoxylate cannot be converted to glycine, it is instead oxidized to oxalate and excreted by the kidney, where it tends to precipitate as hard lumps of calcium oxalate. Two thirds of patients with primary hyperoxaluria type 1 have a mutant form of AGT that simply fails to work. However, the other third have an AGT with a single amino acid change (G170R) that works reasonably well, at least in the test tube. However, this amino acid change is enough to make the mitochondrial import system believe that AGT is a mitochondrial protein and import it inappropriately, so that no AGT is available to be transported to the peroxisomes. For the clinician, the mistargeting of AGT in primary hyperoxaluria type 1 poses an unusual problem, namely, how to explain to a patient that the way to cure their kidney stones is to have a liver transplant!

Synthesis on the Rough Endoplasmic Reticulum

Synthesis of proteins destined for import into the endoplasmic reticulum starts on free polyribosomes. When the growing polypeptide chain is about 20 amino acids long, the endoplasmic reticulum signal sequence is recognized by a **signal recognition particle** that is made up of a small RNA molecule and several proteins (Fig. 10.7). The signal recognition particle brings the ribosome to the endoplasmic reticulum membrane where it interacts with a specific receptor—the **signal recognition particle receptor** (or the **docking protein**). This interaction directs the polypeptide chain to a **protein translocator.** Once this has occurred the signal recognition particle and its receptor are no longer required and are released. Protein synthesis now continues; and, as the polypeptide continues to grow, it threads its way through the membrane via the protein translocator, which acts as a channel allowing hydrophilic stretches of polypeptide chain to cross. Once the polypeptide chain has entered the lumen of the endoplasmic reticulum, the signal sequences may be cleaved off by an enzyme called signal peptidase. Some proteins do not undergo this step but instead retain their signal sequences.

The platelet-derived growth factor receptor is an example of an integral membrane protein. It contains a stretch of 22 hydrophobic amino acids that spans the plasma membrane (Fig. 9.15 on page 198). The first part of the polypeptide to be synthesized is an endoplasmic reticulum signal sequence, so the polypeptide begins to be threaded into the lumen of the endoplasmic reticulum. This section will become the extracellular domain of the receptor.

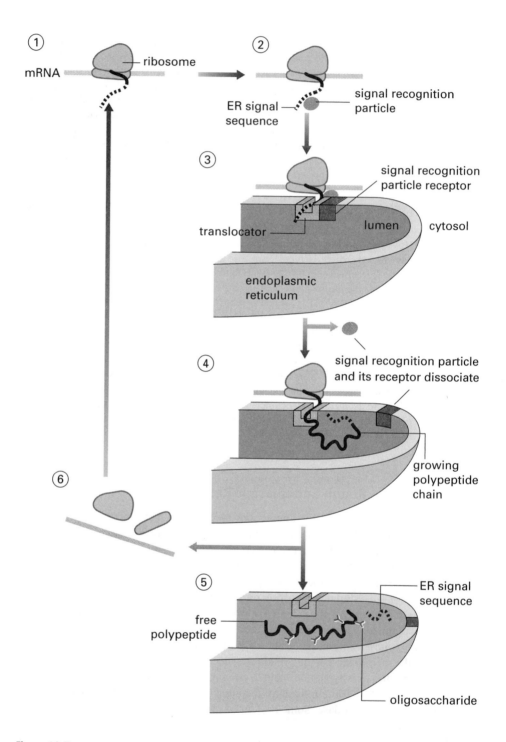

Figure 10.7. Transport of a growing protein across the membrane of the endoplasmic reticulum.

When the stretch of hydrophobic residues is synthesized, it is threaded into the translocator in the normal way but cannot leave at the other end because the amino acid residues do not associate with water. As synthesis continues, therefore, the newest length of polypeptide bulges into the cytosol. Once synthesis stops, this section is left as the cytosolic domain.

If a protein contains more than one hydrophobic stretch, then synthesis of the second stretch reinitiates translocation across the membrane, so that the protein ends up crossing the membrane more than once.

We can identify key amino acid sequences that play a role in protein targeting by making use of protein engineering. If we join the stretch of nucleotides that codes for the endoplasmic reticulum signal sequence to the cDNA that codes for a cytosolic protein, we produce a chimeric DNA molecule (page 150). When this cDNA is transfected (page 150) into cells, it will be transcribed into mRNA and then translated into a protein that is unchanged except that it now has, at its N terminus, the endoplasmic reticulum import sequence. The protein is targeted to the endoplasmic reticulum. However, because the protein has no other sorting signal, it is not recognized by any other receptor protein. It gets secreted from the cell by the constitutive route. This result shows that constitutive exocytosis is the default route for proteins synthesized on the rough endoplasmic reticulum.

Glycosylation: The Endoplasmic Reticulum and Golgi System

Most polypeptides synthesized on the rough endoplasmic reticulum are glycosylated, that is, they have sugar residues added to them, as soon as the growing polypeptide chain enters the lumen of the endoplasmic reticulum. In a process called N-glycosylation a premade oligosaccharide composed of two *N*-acetyl glucosamines, then nine mannoses, and then three glucoses is added to an asparagine residue by the enzyme oligosaccharide transferase. The three glucose residues are subsequently removed, marking the protein as ready for export from the endoplasmic reticulum to the Golgi apparatus.

The stacks of the Golgi apparatus (Fig. 10.8) have distinct polarity that reflects the passage of proteins through the organelle. The asymmetry of the stack is reflected in the

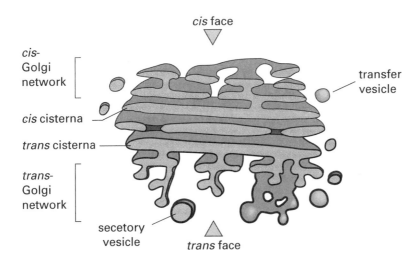

Figure 10.8. The Golgi apparatus.

morphology of the membranes from which it is formed. The **cis** cisternae are made of membranes 5.5 nm thick, like those of the rough endoplasmic reticulum, while the **trans** cisternae are 10 nm thick, like the plasma membrane. Each cisterna is characterized by a central flattened region where the luminal space, as well as the gap between adjacent cisternae, is uniform. The margin of each cisterna is often dilated and is often fenestrated (i.e., has holes through it) as well.

Small, spherical vesicles are always found in association with the Golgi apparatus, especially with the edges of the cis face. These are referred to as transfer vesicles; some of them carry proteins from the rough endoplasmic reticulum to the Golgi stacks, others transfer proteins between the stacks, that is, from *cis* to middle and from middle to *trans*. As proteins move through the Golgi apparatus, the oligosaccharides already attached to them are modified, and additional oligosaccharides can be added. As well as having important functions once the protein has reached its final destination, glycosylations play an important role in sorting decisions at the *trans*-Golgi network.

❋ VESICULAR TRAFFICKING BETWEEN INTRACELLULAR COMPARTMENTS

Most of the single-membrane organelles of the eukaryotic cell pass material between themselves by vesicular traffic, in vesicles that bud off from one compartment to fuse with another (Fig. 10.9). In this way the cargo proteins are never in contact with the cytosol. Two main directions of traffic can be identified (Fig. 10.1). The exocytotic pathway runs from the endoplasmic reticulum through the Golgi apparatus to the plasma membrane. The endocytotic pathway runs from the plasma membrane to the lysosome. This is the route by which extracellular macromolecules can be taken up and processed. If vesicles are to be moved over long distances, they are transported along cytoskeletal highways (page 389).

The Principle of Fission and Fusion

Figure 10.9 illustrates how budding of a vesicle from one organelle, followed by fusion with a second membrane, can transport both soluble proteins and integral membrane proteins to the new compartment. The process retains the "sidedness" of the membrane and the compartment it encloses: the side of an integral membrane protein that faced the lumen of the first compartment ends up facing the lumen of the second compartment, while soluble proteins in the lumen of the first compartment do not enter the cytosol but end up in the lumen of the second compartment, or in the extracellular medium if the target membrane is the plasma membrane.

●●●● IN DEPTH 10.1 Trafficking Movies

Chimeric proteins that contain green fluorescent protein (page 150) and appropriate sorting signals will be processed by the cells machinery just as proteins coded for on its own genes. Thus protein trafficking can be viewed by fluorescence microscopy in live cells that have been transfected with plasmids coding for such chimeras. Movies illustrating many aspects of membrane transport are available on the World Wide Web (see the CBASC website).

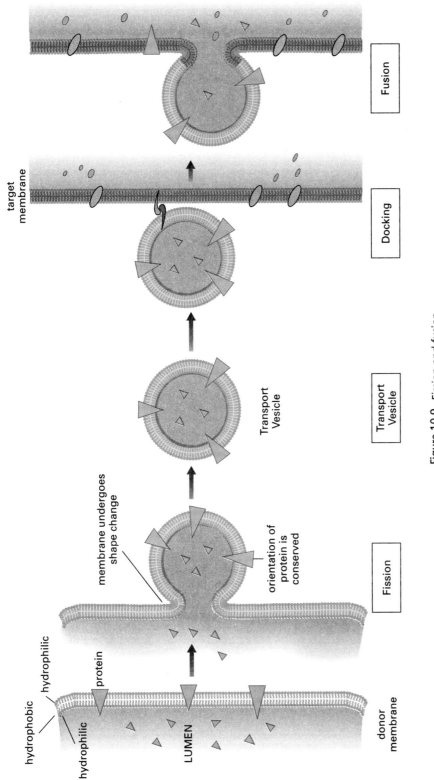

Figure 10.9. Fission and fusion.

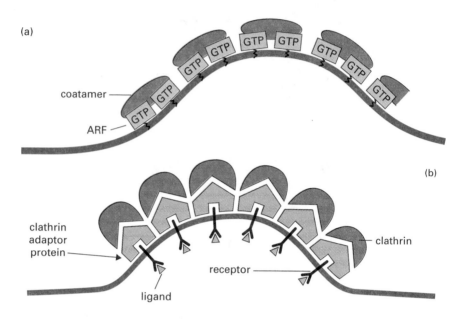

(a)

coatamer

ARF

(b)

clathrin
adaptor
protein

clathrin

receptor

ligand

Figure 10.10. Generation of buds by (a) coatamers and (b) clathrin.

Vesicle Formation

Vesicle formation is the process during which cargo is captured and the lipid membrane is shaped with the help of cytosolic proteins into a bud, which is then pinched off in a process called fission. The ordered assembly of cytosolic proteins into a coat over the surface of the newly forming vesicle is responsible for forcing the membrane into a curved shape (Fig. 10.10). There are two types of coats that serve this function: **coatomer** coats and **clathrin** coats. The coat must be shed before fusion of the vesicle with its target membrane can occur.

Coatomer-Coated Vesicles

Transport along the default pathway uses coatomer-coated vesicles. This is the mechanism used in trafficking between the endoplasmic reticulum and Golgi, between the individual Golgi stacks, and in budding of constitutive secretory vesicles from the *trans* Golgi. The coatomer coat consists of seven different proteins that assemble into a complex. The current model for coatomer coat formation at the Golgi is that a guanine nucleotide exchange factor in the donor membrane exchanges GDP for GTP in the GTPase **ARF.** This causes ARF to adopt its active configuration, in which a fatty acid tail is exposed that anchors ARF into the donor membrane (Fig. 10.10a). Membrane-bound ARF is the initiation site for coatamer assembly and coatomer-coated vesicle formation. The coat is only shed when the vesicle is docking to its target membrane. An ARF-GAP in the target membrane causes the hydrolysis of GTP, and the resulting conformational change causes ARF to retract its hydrophobic tail and become cytosolic and therefore causes the coat to be shed.

Clathrin-Coated Vesicles

Clathrin-coated vesicles mediate selective transport. They are, for example, the means by which protein in the trans Golgi that bears mannose-6-phosphate is collected into a vesicle bound for the lysosome. Clathrin-coated vesicles carry proteins and lipids from the plasma membrane to the endosome, and operate in other places where selective transport is required.

Figure 10.10b illustrates how clathrin generates a vesicle. The process starts when the cargo of interest binds to integral proteins of the donor membrane that are selective receptors for that cargo. **Clathrin adaptor proteins** then bind to cargo-loaded receptors and begin to associate, forming a complex. Lastly, clathrin molecules bind to this complex, forming the coat and bending the membrane into the bud shape.

Even though clathrin can force the membrane into a bud shape, it cannot force the bud to leave as an independent vesicle. One of the best-studied membrane fission events is endocytosis. Here a GTPase called **dynamin** forms a ring around the neck of a budding vesicle. GTP hydrolysis then causes a change in dynamin's shape that mechanically pinches the vesicle off from its membrane of origin. Unlike coatamer, clathrin coats dissociate as soon as the vesicle is formed, leaving the vesicle ready to fuse with the target membrane.

Example 10.2 **Buds in a Test Tube**

If phospholipids are shaken up with aqueous medium, they spontaneously form artificial vesicles called liposomes. If these are then incubated with purified coatomer proteins, ARF, and GTP, bud formation and coated vesicles can be observed by electron microscopy. In contrast neither buds nor vesicles form if liposomes are incubated with clathrin and clathrin adaptor proteins. Clathrin-mediated budding requires ligand to be present in the vesicles, and receptors for that ligand to be present in the membrane of the vesicle. Furthermore even if buds did form, clathrin on its own, unlike coatamer, cannot cause fission. Clathrin-coated buds require another protein such as dynamin to trigger fission.

The Trans-Golgi Network and Protein Secretion

At its *trans* face the Golgi apparatus breaks up into a complex system of tubes and sheets called the *trans*-Golgi network (Fig. 10.8). Although there is some final processing of proteins in the *trans*-Golgi network, most of the proteins reaching this point have received all the modifications necessary to make them fully functional and to specify their final destination. Rather, the *trans*-Golgi network is the place where proteins are sorted into the appropriate vesicles and sent down one of three major pathways: **constitutive secretion, regulated secretion,** or transport to lysosomes.

Vesicles for constitutive or regulated secretion, though functionally different, look very much alike and are directed to the cell surface (Fig. 10.1). When the vesicle membrane comes into contact with the plasma membrane, it fuses with it. At the point of fusion the membrane is broken through, the contents of the vesicle are expelled to the extracellular space, and the vesicle membrane becomes a part of the plasma membrane. This process, by which the contents of a vesicle are delivered to the plasma membrane and following membrane fusion and breakthrough are released to the outside, is called **exocytosis.**

The difference between the constitutive and regulated pathways of secretion by exocytosis is that the former is always "on" (vesicles containing secretory proteins are presented for exocytosis continuously) while the regulated pathway is an intermittent one in which the vesicles containing the substance to be secreted accumulate in the cytoplasm until they receive a specific signal, usually an increase in the concentration of calcium ions in the surrounding cytosol, whereupon exocytosis proceeds rapidly (page 343). After secretion, vesicular membrane proteins are retrieved from the plasma membrane by **endocytosis** and transported to the **endosomes** (Fig. 10.1). From the endosomes vesicles are targeted to the lysosomes, back to the Golgi, or into the pool of regulated secretory vesicles. In a cell that is not growing in size, the amount of membrane area added to the plasma membrane by exocytosis is balanced over a period of minutes by endocytosis of the same area of plasma membrane.

Notice that in exocytosis the membrane of the vesicle becomes incorporated into the plasma membrane; consequently the integral proteins and lipids of the vesicle membrane become the integral proteins and lipids of the plasma membrane. This is the principal, if not the only way, that integral proteins made on the rough endoplasmic reticulum are added to the plasma membrane.

Targeting Proteins to the Lysosome

One of the best-understood examples of sorting in the *trans*-Golgi network is lysosomal targeting (Fig. 10.11). Proteins that are destined for the lysosome are synthesized on the rough endoplasmic reticulum and therefore, like all proteins synthesized here, have a mannose-containing oligosaccharide added. Because they do not have an endoplasmic reticulum retention signal such as KDEL, they are transported to the Golgi apparatus. There proteins

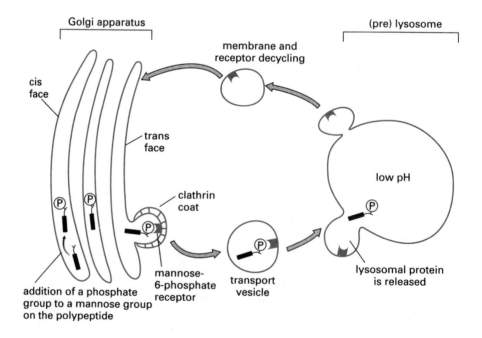

Figure 10.11. Targeting of protein to the lysosome.

destined for the lysosome are modified by phosphorylation of some of their mannose residues to form mannose-6-phosphate (Fig. 2.11 on page 32). Once the proteins reach the trans-Golgi network, specific receptors for mannose-6-phosphate recognize this sorting signal and cause the proteins to be packaged into vesicles that are transported to the lysosome, where they fuse. In the low pH (5.0) environment of the lysosome, the lysosomal protein can no longer bind to its receptor. The phosphate group is removed by a phosphatase. Vesicles containing the receptor bud off from the lysosome and deliver the mannose-6-phosphate receptors back to the trans-Golgi network.

The function of the lysosome is to degrade unwanted materials. To carry out this function, an inactive or primary lysosome fuses with a vesicle containing the material to be digested. This makes a secondary lysosome. The vesicles with which primary lysosomes fuse may be bringing materials in from outside the cell or they may be vesicles made by condensing a membrane around worn out or unneeded organelles in the cells own cytoplasm. The latter are sometimes called autophagic vacuoles. Some materials may not be digestible and remain in the lysosome for the lifetime of the cell. These small dense remnant lysosomes are called residual bodies.

Fusion

Membrane fusion is the process by which a vesicle membrane incorporates its components into the target membrane and releases its cargo into the lumen of the organelle or, in the case of secretion, into the extracellular medium. Different steps in membrane fusion are distinguished. First, the vesicle and the target membrane mutually identify each other. Then, proteins from both membranes interact with one another to form stable complexes and bring the two membranes into close apposition, resulting in the docking of the vesicle to the target membrane. Finally, considerable energy needs to be supplied to force the membranes to fuse, since the low-energy organization—in which the hydrophobic tails of the phospholipids are kept away from water while the hydrophilic head groups are in an aqueous medium—must be disrupted, even if only briefly, as the vesicle and target membranes distort and then fuse.

Each type of vesicle must only dock with and fuse with the correct target membrane, otherwise the protein constituents of all the different organelles would become mixed with each other and with the plasma membrane. Our understanding of the molecular processes leading to membrane fusion is only just beginning to take shape, but our current understanding is that two types of proteins, called **SNARES** and **Rab family** GTPases work together to achieve this. SNARES located on the vesicles (v-SNARES) and on the target membranes (t-SNARES) interact to form a stable complex that holds the vesicle very close to the target membrane (Fig. 10.12). Not all vSNARES can interact with all tSNARES, so SNARES provide a first level of specificity. So far, over 50 members of the Rab family have been identified in mammalian cells, and each seems to be found at one particular site where it regulates one specific transport event, thus controlling which vesicle fuses with which target. For example, the recycling of the mannose-6-phosphate receptor back from the lysosome to the trans-Golgi network (Fig. 10.11) requires Rab9, and yellow fluorescent protein-Rab9 chimeras (page 150) locate to the returning vesicles. GTP hydrolysis by Rabs is thought to provide energy for membrane fusion.

Another important factor in membrane fusion is the lipid composition at the fusion site. In particular phosphoinositides, a group of lipids that we will encounter again in Chapter 16 (page 346), seem to play a crucial role.

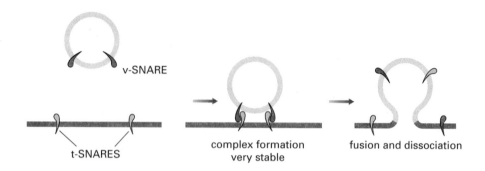

Figure 10.12. SNAREs and vesicle fusion.

Example 10.3 **SNARES, Food Poisoning, and Face-Lifts**

Botulism, food poisoning caused by a toxin released from the anaerobic bacterium *Clostridium botulinum,* is fortunately rare. Botulinum toxin comprises a number of enzymes that specifically destroy those SNARE proteins required for regulated exocytosis in nerve cells. Without these proteins regulated exocytosis cannot occur, so the nerve cells cannot tell muscle cells to contract. This causes paralysis: most critically, paralysis of the muscles that drive breathing. Death in victims of botulism results from respiratory failure.

Low concentrations of botulinum toxin (or "BoTox") can be injected close to muscles to paralyze them. For example, in a "chemical facelift," botulinum toxin is used to paralyze facial muscles, producing an effect variously described as "youthful" and "zombie-like."

SUMMARY

1. The basic mechanisms of intracellular protein trafficking are similar in all eukaryotic cells, from yeast cells to human nerve cells.

2. The final destination of a protein is defined by sorting signals that are recognized by specific receptors. The polypeptide chain itself contains targeting sequences while glycosylation and phosphorylation can add additional sorting signals.

3. Some sorting signals activate translocation of a protein to a new location, while others such as the endoplasmic reticulum retention signal KDEL cause the protein to be retained at its present location.

4. Nuclear proteins are synthesized on free ribosomes and carried through the nuclear pore by Ran-mediated gated transport. Other proteins with a nuclear export signal are carried the other way, again by a Ran-mediated process.

5. Peroxisomal proteins, together with the majority of mitochondrial and chloroplast proteins that are not coded for by mitochondrial or chloroplast genes, are synthesized on free ribosomes and then transported across the membrane(s) of the target organelle.

6. Proteins with an endoplasmic reticulum signal sequence are synthesized on the rough endoplasmic reticulum. The growing polypeptide chain is fed across the membrane as it is synthesized. The signal sequence may then be cleaved off.

7. The default pathway for proteins synthesized on the rough endoplasmic reticulum is to pass through the Golgi apparatus and be secreted from the cell via the constitutive pathway.

8. Glycosylation in the endoplasmic reticulum and then in the Golgi apparatus has two functions: to produce the final, functional form of the protein and to add further sorting signals such as the lysosomal sorting signal mannose-6-phosphate.

9. Vesicles shuttle between most of the single-membrane cellular organelles, the exception being peroxisomes. Budding and then fission of vesicles from the donor membrane can be driven by coatamer family proteins or by clathrin plus dynamin.

10. A tight association of vesicular and target SNARE proteins forces vesicles into close association with the target membrane. Fusion requires the action of a Rab family GTPase.

FURTHER READING

Hammer III, J. A., and Wu, X.S. 2002. Rabs grab motors: Defining the connections between Rab GTPases and motor proteins. Curr. Opin. Cell Biol. 14: 69–75

Jahn, R., and Südhof, T. C. 1999. Membrane fusion and exocytosis. *Annu. Rev. Biochem.* 68: 863–893.

Lin, R. C., and Scheller, R. H. 2000. Mechanisms of synaptic vesicle exocytosis. *Annu. Rev. Cell Dev. Biol.* 16: 19–49.

Segev, N. 2001. Ypt and Rab GTPases: Insight into functions through novel interactions. Curr. Opin. Cell Biol. 13: 500–511.

Zerial, M., and McBride, H. 2001. Rab proteins as membrane organizers. *Nat. Rev. Mol. Cell Biol.* 2: 107–117.

❈ REVIEW QUESTIONS

For each question, choose the ONE BEST answer or completion.

1. Cytosolic ribosomes synthesize proteins that will remain free in the cytosol plus those destined for the interior of
   A. peroxisomes, endoplasmic reticulum, and the Golgi apparatus.
   B. peroxisomes, mitochondria, chloroplasts, and lysosomes.
   C. peroxisomes, mitochondria, chloroplasts, and the nucleus.
   D. mitochondria, chloroplasts, lysosomes, and the nucleus
   E. the Golgi apparatus, lysosomes, and the nucleus.

2. The motif KDEL at the C terminus of a protein marks it for
   A. import into the endoplasmic reticulum and subsequent translocation to the Golgi apparatus.
   B. import into the endoplasmic reticulum and subsequent translocation to the lysosome.

C. import into mitochondria and subsequent translocation to the mitochondrial intermembrane space.

D. retention within the endoplasmic reticulum.

E. retention within the trans-Golgi network.

3. Addition of the appropriate GAP to a solution containing a GTPase, GTP, and GDP only will

A. increase the fraction of the GTPase that is in the GDP-loaded state.

B. increase the fraction of the GTPase that is in the GTP-loaded state.

C. cause an active cycling process in which the GTPase binds GTP, hydrolyses it to GDP, releases the GDP, and binds another GTP.

D. cause an active cycling process in which the GTPase binds GDP, hydrolyses it to GTP, releases the GTP, and binds another GDP.

E. have no detectable effect.

4. Glycosylation of proteins occurs in the

A. peroxisome.

B. mitochondrion.

C. lysosome.

D. endoplasmic reticulum.

E. all these organelles.

5. The lysosomal sorting signal is

A. *N*-acetyl-glucosamine.

B. asparagine.

C. Ran:GTP.

D. ribose-6-phosphate.

E. mannose-6-phosphate.

6. Endocytotic vesicles can only form in the presence of

A. receptors specific for a ligand that is present in the extracellular medium.

B. adaptor proteins.

C. clathrin.

D. dynamin.

E. all of the above.

7. Ribosomes that are located on the rough endoplasmic reticulum are there by virtue of

A. an interaction between a signal recognition particle, a hydrophobic sequence at the N terminus of the protein being synthesized by the ribosome, and a receptor in the membrane of the endoplasmic reticulum.

B. a KDEL sequence at the C terminus of the 60S subunit.

C. an interaction between the sequence CCUCC on the 16S subunit and a receptor in the membrane of the endoplasmic reticulum.

D. mannose-6-phosphate at the 5′ terminus of the 40S subunit.

E. a hydrophobic stretch of RNA at the 5′ terminus of the 40S subunit.

ANSWERS TO REVIEW QUESTIONS

1. *C.* See Figure 10.1.

2. *D.* The KDEL motif does not activate a translocation process; rather it holds proteins bearing the sequence in the endoplasmic reticulum if they have already been translocated there.

3. *A.* GAP means GTPase activator protein, so the GTPase will hydrolyze GTP to GDP more rapidly once the GAP has been added. Concerning other answers: (B) A GEF, a guanine nucleotide exchange factor, would do this. (C) This won't happen because there is no GEF present to cause the GTPase to release its GDP and then bind a GTP. (D) This is nonsense—hydrolysis of GTP yields GDP, not the other way around.

4. *D.* Glycosylation occurs in the endoplasmic reticulum and in the Golgi apparatus.

5. *E.* The lysosomal sorting signal is mannose-6-phosphate. Concerning other answers: (B) It is true that the oligosaccharide that contains the mannose-6-phosphate signal is attached to an asparagine residue, but asparagine by itself is not a lysosomal targeting signal. (C) Ran is concerned with nuclear import/export. (D) Ribose is a pentose sugar so it does not have a number 6 carbon!

6. *E.* All of these are required for endocytotic vesicles to form. Receptors specific for a ligand that is present in the extracellular medium, adaptor proteins, and clathrin are required for the generation of a clathrin-coated invagination of the plasma membrane; dynamin is then required to pinch the invagination off to make an endocytotic vesicle.

7. *A.* All the other answers are more or less nonsensical. In particular, it is important to note that ribosomes are found at the endoplasmic reticulum by virtue of a signal sequence on the protein that they are synthesizing, *not* because the ribosome itself is any different from those that are free in the cytosol, as was implied by answers B through E.

11

HOW PROTEINS WORK

The three-dimensional structures of proteins generate binding sites for other molecules. This reversible binding is central to most of the biological roles of proteins, whether the protein is a gap junction channel that binds a similar channel on another cell (page 55) or a transcription factor that binds to DNA. One special class of proteins, enzymes, have sites that not only bind another molecule but then catalyze a chemical reaction involving that molecule.

How Proteins Bind Other Molecules

Proteins can bind other protein molecules, DNA or RNA, polysaccharides, lipids, and a very large number of other small molecules and inorganic ions and can even bind dissolved gases such as oxygen, nitrogen, and nitric oxide. Binding sites are usually very specific for a particular ligand, although the degree of specificity can vary widely. Usually the binding is reversible so that there is an equilibrium between the free and bound ligand.

A binding site is usually a cleft or pocket in the surface of the protein molecule, which is made up of amino acid side chains appropriately positioned to make specific interactions with the ligand. All of the forces that stabilize tertiary structures of proteins are also used in ligand–protein interaction: Hydrogen bonds, electrostatic interactions, the hydrophobic effect, and van der Waals forces all have their roles. Even covalent bonds may be formed in a few cases—some enzymes form a transient covalent bond with the substrate as part of the mechanism used to effect the reaction.

Cell Biology: A Short Course, Second Edition, by Stephen R. Bolsover, Jeremy S. Hyams,
Elizabeth A. Shephard, Hugh A. White, Claudia G. Wiedemann
ISBN 0-471-26393-1 Copyright © 2004 by John Wiley & Sons, Inc.

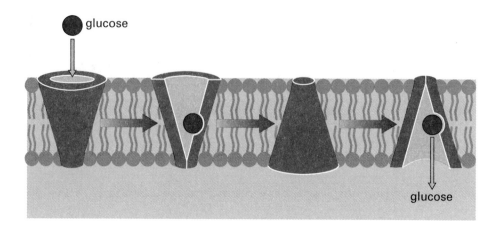

Figure 11.1. The glucose carrier switches easily between two shapes.

Dynamic Protein Structures

It is easy to get the impression that protein structures are fixed and immobile. In fact proteins are always flexing and changing their structure slightly around their lowest energy state. A good term for this is "breathing." Many proteins have two low-energy states in which they spend most of their time, like a sleeper who, though twisting and turning throughout the night, nevertheless spends most time lying on their back or side. An example is the **glucose carrier** (Fig. 11.1). This is a transmembrane protein that forms a tube through the membrane. It is stable in one of two configurations. In one the tube is open to the cytosol; in the other the tube is open to the extracellular medium. By switching between the two states, the glucose carrier carries glucose into and out of the cell.

Allosteric Effects

The glucose carrier is able to bind a ligand—the glucose molecule—in either of its low-energy conformations. In contrast, the *lac* repressor (page 113) can only bind its ligand, the operator region of the *lac* operon, in one conformation. On its own the protein predominantly adopts this conformation so transcription is prevented as it binds to the DNA. When the *lac* repressor binds allolactose (a signal that lactose is abundant), it is locked into a second, inactive form that cannot bind to the DNA (Fig. 6.8 on page 113). Transcription is no longer repressed, although the cAMP–CAP complex is additionally required if transcription is to proceed at a high rate. This type of interaction, in which the binding of a ligand at one place affects the ability of the protein to bind another ligand at another location, is called **allosteric** and is usually a property of proteins with a quaternary structure (i.e., with multiple subunits).

Hemoglobin (Fig. 9.20 on page 205) is an example of a protein where allosteric effects play an important role. Each heme prosthetic group, one on each of the four subunits, can bind an oxygen molecule. We can get an idea of what one subunit on its own can do by looking at myoglobin (Fig. 11.2a), a related molecule that moves oxygen within the cytoplasm. Myoglobin has just one polypeptide chain and one heme. The green line in Figure 11.2b shows the oxygen-binding curve for myoglobin. Starting from zero oxygen, the first small increase in oxygen concentration produces a large amount of binding to myoglobin; the

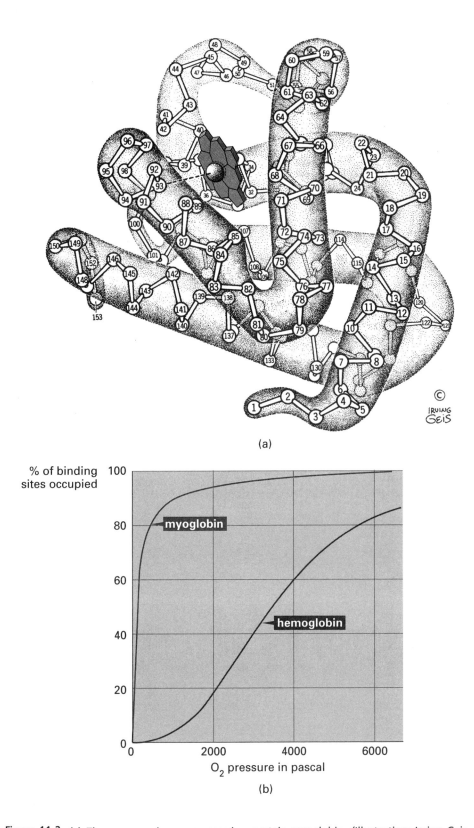

(a)

% of binding sites occupied

100

80 **myoglobin**

60

40 **hemoglobin**

20

0

0   2000   4000   6000

O_2 pressure in pascal

(b)

Figure 11.2. (a) The monomeric oxygen-carrying protein myoglobin. (Illustration: Irving Geis. Rights owned by Howard Hughes Medical Institute. Reproduction by permission only.) (b) Oxygen binding of myoglobin (in green) and hemoglobin (in black) as oxygen pressure increases.

next increase in oxygen produces a slightly smaller amount of binding, and so on, until myoglobin is fully loaded with oxygen. A curve of this shape is known as hyperbolic. The black line in Figure 11.2b shows the oxygen-binding curve for hemoglobin. Starting from zero oxygen, the first small increase in oxygen concentration produces hardly any binding to hemoglobin. The next increase in oxygen produces much more binding so the curve gets steeper before leveling off again as the hemoglobin becomes fully loaded. This behavior is called cooperative. The explanation for this behavior is that the hemoglobin subunits can exist in one of two states, only one of which has a high affinity for oxygen. The way that the four subunits fit together means that they all must be in one form or the other. When oxygen concentration is low, most of the hemoglobin molecules have their subunits in the low-affinity form. As oxygen increases, it begins to bind to the hemoglobin—little at first as most of the hemoglobin is in the low-affinity form and only a little in the high-affinity form. As oxygen binds, more molecules switch to the high-affinity form as the low- and high-affinity forms are in equilibrium. Eventually virtually all of the molecules have made the switch to the high-affinity form. This produces the curve shown in Figure 11.2b. This cooperative oxygen binding makes hemoglobin an effective transporter as it will load up with oxygen in the lungs but will release it readily in the capillaries of the tissues where the oxygen concentration is low. Myoglobin would release little of its bound oxygen at the oxygen concentrations typical of respiring tissues.

Some enzymes show cooperative behavior caused by an allosteric effect that causes binding of one substrate molecule to make it easier for the other substrate to bind. The degree of cooperativity can be altered by the binding of other molecules (called effectors) that act to switch the enzyme on or off.

Chemical Changes That Shift the Preferred Shape of a Protein

Proteins can change conformation as a result of environmental changes, by binding a ligand or by having a particular group attached covalently to them. Anything that changes the pattern of electrostatic interactions within a protein will alter the relative energy of its states. If, for example, a protein contains histidine residues, merely changing the pH will do this. In solutions with pH greater than 7 most of the histidine residues in a protein will be

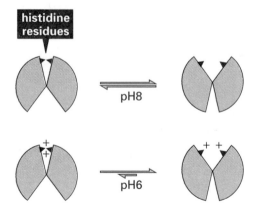

Figure 11.3. A pH change that alters the charge on histidine will alter the balance of forces within a protein. At pH 6, the structure on the right will predominate.

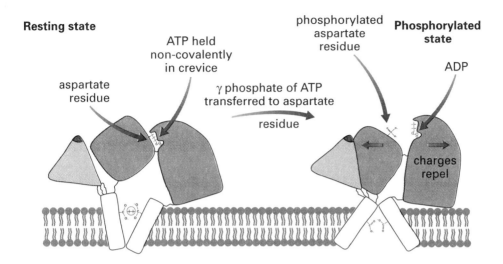

Resting state

aspartate
residue

ATP held
non-covalently
in crevice

γ phosphate of ATP
transferred to aspartate
residue

phosphorylated
aspartate
residue

**Phosphorylated
state**

ADP

charges
repel

Figure 11.4. Phosphorylation changes the charge pattern, and hence the balance of forces within
the calcium ATPase, forcing a change of shape.

uncharged (page 187). In solutions with a pH less than 7, most of the histidine residues will
bear a positive charge. A protein conformation that had two histidine residues close together
(Fig. 11.3) would therefore be stable in alkaline conditions, but in acid conditions the two
residues would each bear a positive charge and repel, destabilizing the conformation.

Another mechanism that is used to change the conformation of proteins is the addition of
a negatively charged phosphate group by a class of enzymes called protein kinases. Proteins
can be phosphorylated on serine, threonine, tyrosine, aspartate, glutamate, or histidine. The
calcium ATPase (Fig. 11.4) is a transmembrane protein that forms a tube that can be open
to the cytosol or to the extracellular medium. In its resting state the tube is open to the
cytosol, while ATP is held by noncovalent interactions in a crevice in a cytosolic domain.
Phosphorylation transfers the γ phosphate of ATP to a nearby aspartate residue. Repulsion
between the phosphoaspartate and the remaining ADP means that the lowest energy state
of the protein is now one in which the tube is open to the extracellular medium. This
mechanism is used to force calcium ions out of the cytoplasm into the extracellular fluid.
The calcium ATPase is described in more detail on page 318.

The movements produced in a protein by phosphorylation are small, on the order of
nanometers or less. When these are repeated many times and amplified by lever systems,
they can produce movements of micrometers or even meters. The beating of a flagellum
(page 386) or the kicking of a leg are both produced by phosphorylation-induced protein
shape changes.

✸ ENZYMES ARE PROTEIN CATALYSTS

Life depends on complex networks of chemical reactions. These are mediated by enzymes.
Enzymes are catalysts of enormous power and high specificity. Consider a lump of sugar.
It is combustible but quite difficult to set alight. A chemical catalyst would speed up its
combustion, and we would end up with heat, a little light, carbon dioxide, and water.
Swallowed and digested, the sucrose is broken down in many steps to carbon dioxide and

water by the action of at least 22 different enzymes, and the energy released is used to drive other reactions in the body.

At a basic level, a reaction carried out by an enzyme can be expressed as

$$E + S \rightleftharpoons ES \rightleftharpoons EP \rightleftharpoons E + P$$

where E is the enzyme, which binds the **substrate** S to form the complex ES. The term *substrate* is used for a ligand that binds to an enzyme and that is then transformed to the product, P. The region of the protein where the substrate binds and the reaction occurs is called the active site. This binding is specific, often highly so. The enzyme β-galactosidase (page 109) is moderately specific and will split not only lactose but also any other disaccharide that has a glycosidic bond to β-galactose. By contrast **glycogen phosphorylase kinase** (page 305) acts with absolute specificity on a single substrate, another enzyme called **glycogen phosphorylase**—none of the thousands of other proteins in the cell can substitute. In general, the specificity of an enzyme is conferred by the shape of the active site and by particular amino acid side chains that interact with the substrate. Binding of the substrate produces the enzyme–substrate complex ES; the catalytic function of the protein then converts the substrate to product, still bound to the enzyme in the complex EP. Finally the product dissociates from the enzyme. Chemical engineers use catalysts made of many materials, and within cells there are catalysts called ribozymes that are made of RNA. However, only proteins, with their enormous repertoire of different shapes can produce catalysts of high selectivity.

The **catalytic rate constant** k_{cat} (also known as the **turnover number**) of an enzyme gives us an idea of the enormous catalytic power of most enzymes. It is defined as the number of molecules of substrate converted to product per molecule of enzyme per unit time (equally it is moles of substrate converted per mole of enzyme per unit time). Many enzymes have k_{cat} values around 1000 to 10,000 per second. The reciprocal of k_{cat} is the time taken for a single event. Thus if k_{cat} is $10,000 \text{ s}^{-1}$, one substrate molecule will be converted every tenth of a millisecond. Some enzymes achieve very much higher rates. Catalase, an enzyme found in peroxisomes, has a k_{cat} of $4 \times 10^7 \text{s}^{-1}$, and so it takes only 25 ns to split a molecule of hydrogen peroxide into oxygen and water.

The Initial Velocity of an Enzyme Reaction

The basic experiment in the study of enzyme behavior is measurement of the appearance of product as a function of time (Fig. 11.5). This is often called an enzyme assay. The product appears most rapidly at the very beginning of the reaction. As the reaction progresses, the rate at which the product appears slows down and eventually becomes zero when the system has reached equilibrium. All reactions are in principle reversible; thus the observed overall rate of the reaction is actually the difference between the rate at which the product is being formed and the rate at which the product is being broken down again in the reverse reaction. In many enzyme reactions the equilibrium lies strongly toward the product.

●●●● **IN DEPTH 11.1 What to Measure in an Enzyme Assay**

In an enzyme assay we measure the appearance of product as a function of time. In principle we could do this by starting identical reactions in a series of test tubes and stopping the reaction in each test tube at a different time after the start and then measuring the amount of product in each tube.

However, if either the substrate or the product has a property that can be measured while the reaction is proceeding, then the whole assay can be performed in one test tube. Many enzyme assays are done by using changes in the absorption of light when the substrate is converted to product. Other optical properties can be used. In their original work Michaelis and Menten studied an enzyme that breaks the disaccharide sucrose down into glucose and fructose. The mixture of glucose plus fructose rotated polarized light differently from sucrose, and they used this property to follow the course of the reaction. If there is no convenient optical property, then others may be available; for example, one can monitor the progress of a reaction in which the polysaccharide glycogen (page 30) is hydrolyzed by digestive enzymes by measuring the resulting fall in viscosity.

We can simplify the analysis of enzyme reactions if we consider only the start of the reaction, when there is no product present and so the back reaction can be neglected. The rate of reaction at time zero (the **initial velocity** v_0, sometimes called the initial rate) is found by plotting a graph of product concentration as a function of time and measuring the slope at time zero (Fig. 11.5). In practice the slope is measured over the first 5% of the total reaction. Initial velocity, v_0, is conveniently expressed as the rate at which the concentration of the product increases, that is, moles per liter per second.

Enzymes must be assayed under controlled conditions because temperature, pH, and other factors alter the activity. Most are such very effective catalysts that they must be assayed under conditions where the concentration of the enzyme is always very much less than the concentration of the substrate. Otherwise the reaction would be over in a fraction of a second.

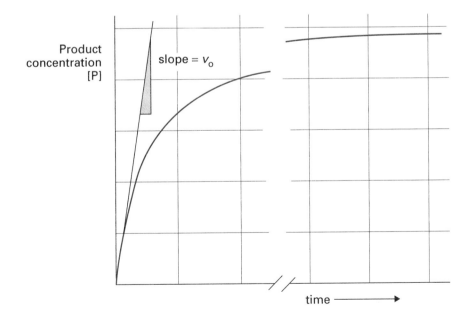

Figure 11.5. Definition of v_0, the initial velocity of a chemical reaction.

Enzymes are studied for many reasons. An understanding of how they achieve their catalytic excellence and specificity is of fundamental interest and has many practical applications as we increasingly use them in industrial processes and we even seek to design enzymes for particular tasks. Measurements of enzyme concentrations and properties allow us to study processes within cells and organisms. The starting point of any study of an enzyme is to determine its activity, measured as k_{cat}, and how tightly it can bind its substrate—its substrate affinity. Substrate affinity is measured by another constant called the **Michaelis constant, K_M**. We will now explain how these constants are measured.

Effect of Substrate Concentration on Initial Velocity

Let us consider a series of experiments designed to see how the initial velocity of an enzyme reaction varies with the concentration of substrate (which is always much greater than that of the enzyme). Each of the smaller graphs in Figure 11.6 shows the result of one of these experiments. As we increase the substrate concentration, we find that at first the velocity increases with each increase in substrate concentration but that, as the substrate

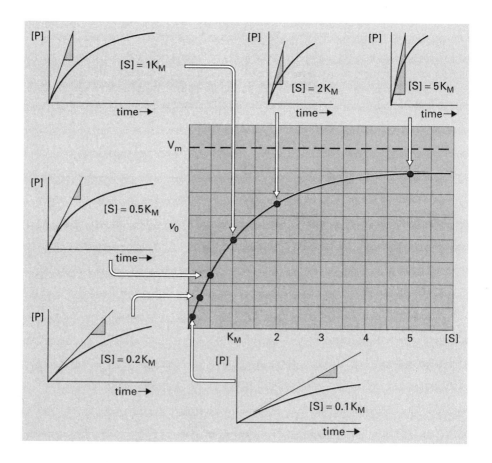

Figure 11.6. v_0 measured in a number of reaction tubes (with [E] constant and always less than [S]) forms a hyperbolic curve when plotted as a function of substrate concentration.

concentration becomes larger, the increases in rate produced get smaller and smaller. We have already met the name of a curve of this type: it is a hyperbola. The initial velocity approaches a maximum value that is never exceeded. Reactions that show this sort of dependence on substrate concentration are said to show **saturation kinetics**.

How can we explain this? The reaction sequence can be simplified to

$$E + S \rightleftharpoons ES \rightarrow E + P$$

(we are using initial velocities so we can ignore any back reaction). The enzyme and substrate must collide in solution, and the substrate must bind at the enzyme's active site to form the ES complex. The chemical reaction then takes place within the ES complex, and finally the product is released. In experiments with higher and higher substrate concentrations, there is ever more ES present, and this increasing ES gives an increasing rate of product release. At very high substrate concentrations virtually all of the enzyme is present as ES, and the observed rate is limited by the ES $\rightarrow$ E + P step. Thus, as the substrate concentration increases, the reaction rate levels off as it approaches a **maximal velocity** called V_m or V_{max}. We can define V_m as the limiting initial velocity obtained as the substrate concentration approaches infinity. It is the product of the catalytic rate constant k_{cat} and the amount of enzyme present, that is, $V_m = k_{cat} [E_{total}]$. Having defined V_m, we now define the Michaelis constant, K_M, as that substrate concentration that gives an initial velocity equal to half V_m.

The plot of v_0 against [S] gives a hyperbolic curve that is described by the equation

$$v_0 = \frac{V_m[S]}{K_M + [S]}$$

where [S] is the substrate concentration. This is called the **Michaelis–Menten equation** after Maud Menten and Leonor Michaelis who propounded a general theory of enzyme action in 1913. The Michaelis constant K_M is the substrate concentration that gives an initial velocity numerically equal to half of V_m. Another way of looking at this is to say that if the enzyme is saturated with substrate, then the rate of reaction will be V_m, while at a substrate concentration giving an initial velocity of $1/2 V_m$ the enzyme is half saturated with substrate. A small value of K_M means that the enzyme has a high affinity for the substrate.

The Effect of Enzyme Concentration

As the concentration of an enzyme increases, so does the initial velocity, and this relationship is linear: if you double the enzyme concentration, you double the initial velocity. This fact means that we can measure how much of an enzyme is present by measuring the rate of the reaction it catalyzes.

●●●● IN DEPTH 11.2 Determination of V_m and K_M

The initial rate of a reaction depends on V_m and K_M according to the Michaelis–Menten equation:

$$v_0 = \frac{V_m[S]}{K_M + [S]}$$

Thus the information necessary to estimate V_m and K_M is contained within data such as that shown in Figure 11.6. However, it is not immediately obvious how one would go about this: The curve flattens off well before V_m is reached so that guessing V_m by eye is near to impossible. Nowadays many computer programs are available that will fit the hyperbolic Michaelis–Menten equation to the raw data. Before this, biologists used mathematical transformations of the Michaelis–Menten equation that yielded straight-line relationships. The best known is called the Lineweaver–Burk plot after its inventors. It simply inverts the Michaelis–Menten equation to get

$$\frac{1}{v_0} = \frac{K_M + [S]}{V_m[S]} = \frac{K_M}{V_m[S]} + \frac{[S]}{V_m[S]}$$

hence

$$\frac{1}{v_0} = \frac{K_M}{V_m} \times \frac{1}{[S]} + \frac{1}{V_m}$$

This function can be recognized as having the form of the equation of a straight line $y = mx + c$, where y is $1/v_0$ and x is $1/[S]$. Thus if one plots $1/v_0$ on the vertical axis and $1/[S]$ on the horizontal axis, one gets a straight line where the y intercept is $1/V_m$, the x intercept is $-1/K_M$, and the slope is K_M/V_m.

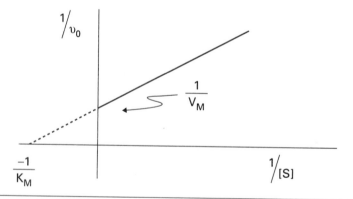

Table 11.1. Values of k_{cat}/K_M for a Number of Enzymes

Enzyme	$k_{cat}(s^{-1})$	K_M (mol liter^{-1})	k_{cat}/K_M (liter mol^{-1} s^{-1})
Trimeric G protein (page 344)	0.02	6×10^{-7}	3.3×10^4
Lysozyme (page 192)	0.5	6×10^{-6}	8.3×10^3
Aminoacyl tRNA synthase (page 165)	7.6	9×10^{-4}	8.4×10^3
Intestinal ribonuclease (page 199)	7.9×10^2	7.9×10^{-3}	1×10^5
Acetylcholinesterase (page 368)	1.4×10^4	9×10^{-5}	1.6×10^8
Catalase (page 60)	4×10^7	1.1	4×10^7

The Specificity Constant

The Michaelis constant K_M reflects the affinity of an enzyme for its substrate; k_{cat} reflects the catalytic ability of an enzyme. The ratio of these, k_{cat}/K_M, is the specificity constant, which is a measure of how good the enzyme is at its job. A high specificity constant means that a reaction goes fast (k_{cat} is big) and the enzyme does not need a high concentration of substrate (K_M is small). When an enzyme has relatively low specificity, that is, it can work on a number of different substrates, the substrate that has the largest specificity constant is the preferred substrate for the enzyme.

A reaction cannot go faster than the rate at which enzyme and substrate actually collide. In some enzymes this rate of collision is the factor limiting the overall rate, and such enzymes are said to be "diffusion limited" and are considered to have reached a perfection of biological design. Diffusion-limited values of k_{cat}/K_M can be calculated to be in the range 10^8–10^{10} liter per mole per second. Table 11.1 shows k_{cat}, K_M, and k_{cat}/K_M for a number of enzymes.

> **Example 11.1** Speed Isn't Everything
>
> In this chapter we emphasize enzymes as tools for performing a reaction that is required in the cell. For this type of enzyme, the faster the reaction occurs, the better, although in many cases an "optimized" enzyme must be controlled so that it only operates at maximum rate when lots of product is needed.
>
> For other enzymes, slowness is a virtue. We will meet trimeric G proteins in Chapter 16. Like the GTPases that we have already met (EF-tu, Ran, and Rab), these are active when they have GTP bound, and switch to an inactive state when they hydrolyze the GTP to GDP. While they are active, they turn on their target processes. In these enzymes, therefore, a slow rate of reaction allows them to act as timer switches. They are activated when GDP is ejected and GTP bound. They then remain active until they hydrolyze the GTP. For the typical trimeric G protein listed in Table 11.1, k_{cat} is 0.02 per second, so the "timer" is set to 50 s (the reciprocal of 0.02 s^{-1}).

✳ ENZYME CATALYSIS

A catalyst speeds the rate of reaction but does not alter the position of the equilibrium E + S ⇌ E + P. Reaction speed is increased by lowering the activation energy barrier between the reactants and products. Figure 11.7 illustrates this effect. The top graph shows the total energy of the system as the reaction proceeds, the horizontal axis representing the progress

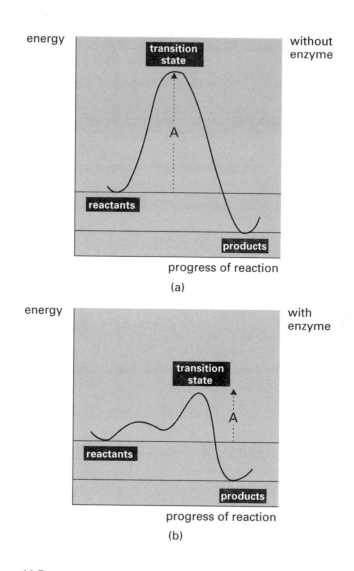

Figure 11.7. Catalysts act by reducing the activation energy (A) of a reaction.

Medical Relevance 11.1

Measurement of Enzyme Concentrations in Blood

Because the initial velocity of a reaction is a linear function of the enzyme concentration, we can measure the concentration of an enzyme by measuring the rate of the reaction it catalyzes. Clinicians make use of this to measure the amount of an enzyme present in the blood. If enzymes that are normally found within cells appear in the blood in more than trace amounts, this indicates that cells are being damaged and their contents are leaking out. The presence of enzymes characteristic of a particular organ tells the clinician which tissue is affected. The digestive enzyme trypsin is stored in an inactive form called trypsinogen in the cells of the pancreas. Trypsinogen is activated once it is secreted into the lumen of the gut. Trypsinogen can be detected in the blood by converting it to the active enzyme and measuring its activity. The presence of trypsinogen in the blood of a fetus is a sign that the cells of the pancreas are dying, and this is in turn an early indicator of cystic fibrosis (page 426).

of the reaction so that at the left we have the reactants, on the right we have the products, and in the middle we have the transition state that the system has to pass through for the reaction to occur. Because the transition state is of higher energy than the reactants, the reaction will only occur if the reactants have additional energy (e.g., as thermal motion) equal to the **activation energy.** The bottom graph shows the same reaction in the presence of a catalyst. Although the energy of the reactants and products has not changed, the presence of the catalyst has reduced the energy of the transition state and has therefore reduced the activation energy. A greater proportion of all the reactant molecules in solution will possess this smaller amount of extra energy, so more can undergo the reaction: the catalyst has speeded up the rate at which the reaction occurs. Each enzyme has evolved to reduce the activation energy of a particular reaction. Clearly the binding of the substrate is as central to catalysis as it is to specificity. The active site contains precisely positioned amino acid side chains that promote the reaction. Daniel Koshland proposed that the enzyme's active site was shaped not so much to fit the substrate but to fit a molecule that was halfway between substrate and product, and therefore to reduce the activation energy of the reaction by stabilizing, and hence reducing the energy of, the transition state. He proposed that the enzyme might actually change shape when it bound the substrate to promote catalysis. This is called induced fit and many enzymes have been shown to do this.

✱ COFACTORS AND PROSTHETIC GROUPS

Enzymes that need to perform reactions that are outside the repertoire of the 20 amino acid side chains recruit other chemical species to help them do the job. The aminotransferases provide a good example. These enzymes are central to amino acid metabolism (page 292) and catalyze the interconversion of amino acids and oxo-acids by moving an amino group (Fig. 11.8). Pyridoxal phosphate (derived from vitamin B_6) is bound by the protein and accepts the amino group from the amino acid donor, which then leaves the enzyme converted to a new oxo-acid. The oxo-acid substrate then binds, accepts the amino group from the pyridoxal phosphate, and leaves having been converted to an amino acid. Pyridoxal phosphate ends up exactly as it started and remains bound to the enzyme ready for another cycle. These helper chemicals are called cofactors when, like pyridoxal phosphate, they are not tightly bound to the enzyme. We already know the term for a molecule that is very tightly bound to a protein and helps it perform its job—it is a prosthetic group (page 205). These terms overlap and usage varies.

The prosthetic group of hemoglobin and myoglobin, heme, (Fig. 9.21 on page 206) binds oxygen. It is a versatile molecule that is also found in electron transfer proteins called cytochromes. Cytochromes found in the inner mitochondrial membrane play important roles in the electron transport chain (page 265). Other cytochromes, the cytochromes P450 that we have met earlier (pages 124, 158, 166), are found in the membranes of the endoplasmic reticulum and mitochondria. There are at least 50 different cytochrome P450s in humans. Some of these proteins carry out very specific metabolic reactions in the process of converting cholesterol into the different steroid hormones our body utilizes. Other cytochromes P450 are less specific in their reactions. Their role is to detoxicate the millions of foreign chemicals to which we are exposed on a daily basis. However, all cytochromes P450 share one property in common: they all insert a single atom of oxygen, derived from molecular oxygen, into a substrate. Subsequent rearrangements often result in the oxygen being in the form of a hydroxyl group in the final reaction product. This transformation converts a

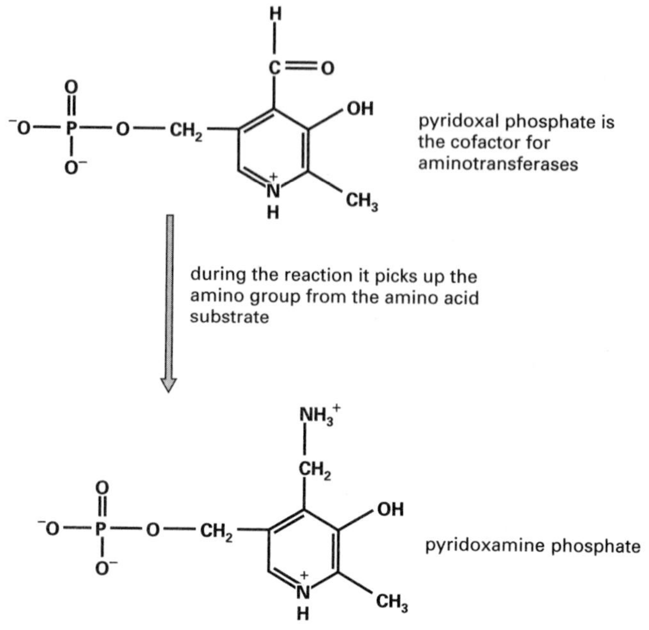

pyridoxal phosphate is
the cofactor for
aminotransferases

during the reaction it picks up the
amino group from the amino acid
substrate

pyridoxamine phosphate

Alanine glyoxylate aminotransferase catalyzes the reaction:

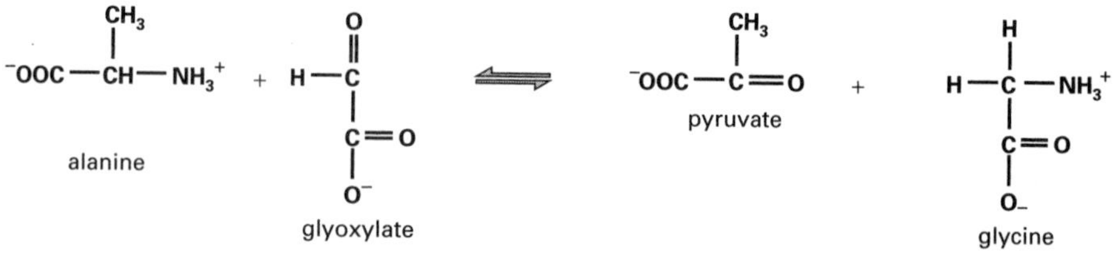

alanine

glyoxylate

pyruvate

glycine

the reaction is in two stages:

E-pyridoxal phosphate   +   alanine   ⇌   E-pyridoxamine phosphate   +   pyruvate

E-pyridoxamine phosphate + glyoxylate   ⇌   E-pyridoxal phosphate   +   glycine

Figure 11.8. Aminotransferases use a cofactor that participates in the reaction but ends up un-changed. E represents the enzyme molecule.

hydrophobic molecule into one that is more hydrophilic. After hydroxylation other molecules are coupled to the hydroxyl group to make the foreign chemical even more hydrophilic and hence allow it to be rapidly excreted in the urine and feces. Unfortunately, in very rare cases, the addition of an oxygen atom into a molecule can increase its reactivity and, instead of being rendered harmless by the action of a cytochrome P450, it is transformed into a very dangerous chemical that can damage DNA. If the damage causes mutation or strand breakage, then cancer can result. Polycyclic aromatic hydrocarbons, found in tobacco smoke, and the chemical aflatoxin, found in the mold that grows on peanuts, are converted into chemical carcinogens by the action of certain cytochromes P450.

✿ ENZYMES CAN BE REGULATED

In this chapter we have discussed the catalytic role of enzymes in isolation. In fact enzymes are like other proteins in showing the complex behavior described in Chapter 9 such as multiple states, multiple binding sites, quaternary structure, and phosphorylation. For instance, some important enzymes have quaternary structures and show cooperativity between the active sites. Such enzymes will not follow the Michaelis–Menten equation, but instead the curve of initial velocity against substrate concentration will have the same S-shaped or sigmoid curve seen in the binding of oxygen to hemoglobin (Fig. 11.2).

A good example of an allosteric enzyme is to be found in the processes that allow us to use sugars as metabolic fuels. The pathway called **glycolysis** (page 284) converts glucose and fructose to pyruvate with a net production of two ATP molecules. The pyruvate is then passed to the mitochondria, which use it to produce more ATP.

The first committed step in this pathway is the addition of a second phosphate group to monophosphorylated fructose. The γ phosphate of ATP is transferred to the number 1 carbon of fructose-6-phosphate. Before this step, sugars can be converted to glycogen or metabolized in other ways, but after this step the sugar is committed to be broken down to pyruvate. The step is catalyzed by the enzyme **phosphofructokinase** (Fig. 11.9). ATP serves as an allosteric inhibitor of this enzyme so that it operates slowly when ATP concentrations are high and there is no need for more ATP to be produced. When there is a drain on ATP, levels of AMP increase (page 259). AMP competes with ATP for the regulatory site and causes the enzyme to switch to the active, high-affinity conformation. Thus as the ATP concentration is reduced and AMP levels increase, the pathway is more and more activated until it is turned on to full speed. We very often see this type of control system, where the product of a pathway or process feeds back to allosterically inhibit an enzyme at the start of that pathway. They are an example of negative feedback (page 303).

It is not only enzymes involved in metabolism that are regulated by ligand binding. Cdk1, the enzyme that triggers mitosis (page 411), is an allosterically regulated enzyme. On its own, in solution, it spends the vast majority of time in an inactive conformation. In order to assume the active conformation, Cdk1 must bind a protein ligand called cyclin B (Fig. 11.10). In addition to this allosteric regulation, Cdk1 is also regulated by phosphorylation: for most of the time the catalytic site is blocked by two phosphate groups within the catalytic domain,

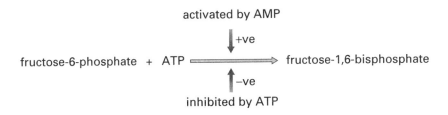

Figure 11.9. Phosphofructokinase is regulated by the binding of ATP or AMP at a regulatory site that is separate from the active site. Binding of ATP inhibits while binding of AMP activates.

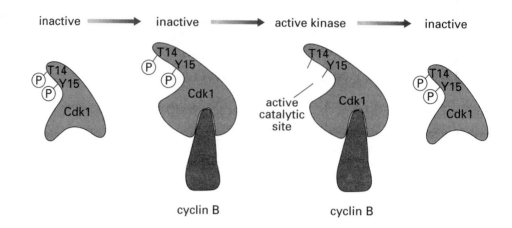

Figure 11.10. Control of Cdk1 by cyclin B and by phosphorylation.

at threonine 14 and tyrosine 15. In order to become an active enzyme, Cdk1 must lose these two phosphates while cyclin B is bound. This complex control means that Cdk1 can act as a checkpoint. If the phosphate groups have been removed from T14 and Y15, AND cyclin B is present at a high enough concentration, THEN it is safe to proceed into mitosis.

Cdk1 is an example of an enzyme that is turned off by phosphorylation. Other enzymes are turned on by phosphorylation. Examples include RNA polymerase II (page 120) and glycogen phosphorylase (page 305).

IN DEPTH 11.3 Rapid Reaction Techniques

Enzyme measurements are usually carried out with low concentrations of the enzyme because higher concentrations would give initial velocities too fast to measure. Initial velocity measurements are always made long after the enzyme–substrate complex has formed. It would be very interesting to be able to observe the actual formation of enzyme–substrate complexes and indeed, more generally, to observe the formation of protein–ligand complexes. These reactions are rapid, occurring on a millisecond time scale or less.

Hartridge and Roughton devised the method illustrated in the accompanying diagram for measuring the rate at which hemoglobin combined with oxygen. A ram pushes two syringes, one containing hemoglobin and one containing oxygenated solution. The solutions mix and pass down a tube, so that the further one looks to the right in the tube, the more time the oxygen has had to bind to the hemoglobin. The association of oxygen with hemoglobin can be monitored by an optical change, the same color change that causes arterial blood to be bright red while venous blood is a dark bluish red. Simply moving an optical detector from left to right along the tube showed the concentration of oxygen–hemoglobin complex after different reaction times. However, all the time the measurements are being made, the hemoglobin solution is running out of the pipe at the right-hand end, so that one can only use this approach when large quantities of the protein are available.

The second figure shows a less wasteful method called stopped flow. Here two solutions are mixed and passed into an observation chamber and from it into a syringe. The plunger of this "stopping syringe" moves back until it hits a stopping plate and the flow stops. Observations are made on the solution as it ages in the observation chamber, using a high-speed recording device that is triggered by the stopping syringe as it hits the plate. Stopped flow allows observations of reactions down to about 0.1 ms after mixing. For example, the rate at which calmodulin (page 200) binds calcium can be measured by filling one syringe with calmodulin, the other with a calcium solution. In this case the reaction can be followed because one tyrosine residue on calmodulin (Tyr138) increases its fluorescence when calmodulin binds calcium.

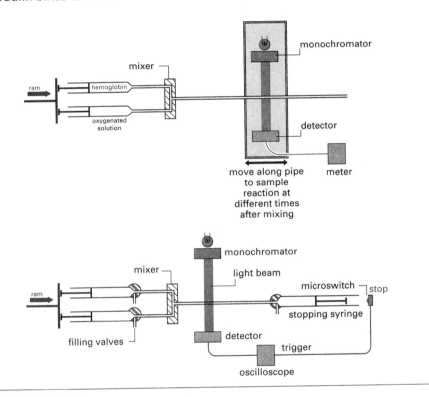

SUMMARY

1. Most protein functions arise from their ability to specifically bind other molecules in a reversible yet specific fashion. Shape changes, triggered by the binding of other molecules, mediate protein movements and function. Allostery is a special case where a shape change induced by a ligand at one site changes the affinity at another site.

2. Enzymes are highly specific biological catalysts. The turnover number or catalytic rate constant, k_{cat}, is the maximum number of substrate molecules that can be converted to product per molecule of enzyme per unit time.

3. The initial velocity (i.e., the rate at the start when product is absent) of many enzymes shows a hyperbolic dependence on substrate concentration. At high substrate concentrations the initial velocity approaches a limiting value V_m, as the enzyme is saturated with substrate. The substrate concentration that gives an initial velocity equal to half V_m, is the Michaelis constant K_M. This indicates the enzyme's affinity for the substrate.

4. The initial velocity is related to the substrate concentration by the Michaelis–Menten equation:

$$v_0 = \frac{V_m[S]}{K_M + [S]}$$

5. The initial velocity is directly proportional to the enzyme concentration.

6. Some enzymes use cofactors to carry out reactions that require different properties from those of the side chains of the 20 amino acids found in proteins.

7. Enzyme activity within cells is modulated by a variety of methods that include phosphorylation and allosteric effects.

FURTHER READING

Cornish-Bowden, A. 1995. *Fundamentals of Enzyme Kinetics,* rev. ed. London: Portland Press.

Fersht, A. 1999. *Structure and Mechanism in Protein Science.* New York: W. H. Freeman.

Voet, D., Voet, J. D. 2003. *Biochemistry,* 3rd ed. New York: Wiley.

REVIEW QUESTIONS

For each question, choose the ONE BEST answer or completion.

1. Conformational or shape changes in proteins allow
   A. conversion of chemical energy into kinetic (movement) energy.
   B. cooperativity between subunits in a quaternary structure.
   C. allosteric control of enzymes.
   D. proteins to carry other molecules across membranes.
   E. all of the above.

2. The turnover or catalytic constant (k_{cat}) of an enzyme is
   A. the number of molecules of substrate converted to product per mole of enzyme per unit time.
   B. the time taken for a single catalytic event.
   C. the number of substrate molecules in a mole.
   D. the number of moles of substrate converted to product per mole of enzyme per unit time.
   E. the initial velocity of the reaction when half the enzyme molecules have bound substrate.

3. The oxygen-binding curve for hemoglobin is S-shaped: starting from zero oxygen, the first small increase in oxygen concentration produces hardly any binding to hemoglobin. The next increase in oxygen produces much more binding so that the curve gets steeper before leveling off again as the hemoglobin becomes fully loaded. Hemoglobin can show this behavior because
   A. it contains heme prosthetic groups.
   B. it contains magnesium, in contrast to the iron in myoglobin.
   C. it comprises two α chains of high oxygen affinity and two β chains of low oxygen affinity.
   D. its four subunits can exist in one of two states, only one of which has a high affinity for oxygen. The way that the four subunits fit together means that they all must be in one form or the other. When oxygen concentration is low, most of the hemoglobin molecules have their subunits in the low-affinity form. As oxygen binds, more molecules switch to the high-affinity form.
   E. its four subunits can exist in one of two states, only one of which has a high affinity for oxygen. The way that the four subunits fit together means that they all must be in one form or the other. When oxygen concentration is low, most of the hemoglobin molecules have their subunits in the high-affinity form. As oxygen binds, more molecules switch to the low-affinity form.

4. Which of the following statements about enzymes is false?
   A. They bind their substrate.
   B. They lower the activation energy for the reaction they catalyze.
   C. They alter the equilibrium constant for the reaction they catalyze.
   D. They stabilize intermediates along the pathway of the reaction they catalyze.
   E. They can change shape to promote the reaction.

5. Protein kinases alter the conformation of their target protein by adding
   A. H^+ ions.
   B. phosphate groups.
   C. histidine residues.
   D. oxygen.
   E. heme groups.

6. A prosthetic group is

    A. a highly charged domain or motif that alters the shape of a protein.

    B. a nonprotein molecule that is loosely attached to a protein and helps it to perform its function.

    C. a nonprotein molecule that is tightly bound to a protein and helps it to perform its function.

    D. a steroid molecule attached to a histidine residue.

    E. a region of a protein that can accept a phosphate group.

7. Phosphofructokinase

    A. produces a product that is inevitably broken down to pyruvate.

    B. uses ATP as a substrate in its catalytic site.

    C. binds ATP at a regulatory site.

    D. works more slowly if ATP concentrations increase.

    E. all of the above.

ANSWERS TO REVIEW QUESTIONS

1. *E.* All these statements about shape changes in proteins are true.

2. *D.* Note that this statement is equivalent to "k_{cat} is the number of molecules of substrate converted to product per molecule of enzyme per unit time" but is not equivalent to the statement in answer A, which is false.

3. *D.* Because the subunits begin at low affinity but switch to the high-affinity form as oxygen binds, the curve for oxygen binding starts out shallow and then gets steeper. Answer E would produce exactly the opposite effect. Of the other answers: (A) is true, but irrelevant—myoglobin also contains heme prosthetic groups, but does not show cooperative behavior. (B) is false: both myoglobin and hemoglobin contain iron. It is chlorophyll, in plants, that uses magnesium. (C) is false: the oxygen affinities of all four subunits are similar and, in any case, a difference in affinity would not cause the binding curve to be S-shaped.

4. *C.* No catalyst, whether inorganic or biological, can alter the equilibrium constant of a reaction. (If one did, it would allow the generation of a perpetual motion machine.)

5. *B.* Concerning the other answers: (A) there are no enzymes that add H^+, those groups that can be protonated (e.g., $—NH_2$, histidine residues) accept H^+ rapidly without the requirement for any catalyst. (C) Amino acids are not added to proteins once they have been synthesized on the ribosome, although processing of synthesized proteins can often involve the removal of one or more amino acids, as, for example, when the endoplasmic reticulum signal sequence is removed (page 223). (D) Those proteins that do bind oxygen, such as hemoglobin, do so spontaneously, without the need for a catalyst.

6. *C.* Both cofactors and prosthetic groups are nonprotein molecules that help a protein to perform its function, but cofactors are only loosely attached while prosthetic groups are attached tightly.

7. *E.* ATP is a substrate for phosphofructokinase, but it also acts at an independent regulatory site, reducing the activity of the enzyme.

12

ENERGY TRADING WITHIN THE CELL

In the nonliving world complex things degrade naturally to simpler things: Gradients of temperature or concentration disappear, chemical reactions approach equilibrium, and uniformity triumphs. Living things do not appear to follow these trends. Cells are complex and divide to make other complex cells: A fertilized egg differentiates to make a whole complex organism. Living things must obey the laws of thermodynamics. The escape from the behavior of nonliving systems is allowed because living systems take matter and energy from the environment and use it to grow, to reproduce, and to repair themselves. Living systems are open systems while nonliving systems are closed.

A chemical reaction that has reached equilibrium can do no work. A good definition of death is the state at which all of the chemical reactions in a cell/organism have reached equilibrium. In a living organism the concentrations of metabolites are often very far from the equilibrium concentrations and yet are more or less constant: This is said to be a *steady state*. Cells can do this because they are open systems taking energy and matter from their environment.

We can use an analogy with the world of economics. It is unlikely that people would spontaneously repair our houses, or feed us, or give us this book, but we can drive these otherwise unlikely processes by spending money. In a similar way, cells can drive otherwise unlikely processes by using up one of four **energy currencies** that are then replaced using energy taken from the outside world.

Cell Biology: A Short Course, Second Edition, by Stephen R. Bolsover, Jeremy S. Hyams,
Elizabeth A. Shephard, Hugh A. White, Claudia G. Wiedemann
ISBN 0-471-26393-1 Copyright © 2004 by John Wiley & Sons, Inc.

✵ CELLULAR ENERGY CURRENCIES

The scientific way of saying that a process will proceed (although a catalyst may be necessary to achieve a reasonable reaction speed) is to say that the change in Gibbs free energy, expressed as ΔG, is negative. One reaction that we will meet again is the hydrolysis of glucose-6-phosphate to yield glucose and a phosphate ion:

$$\text{Glucose-6-phosphate} + H_2O \rightarrow \text{Glucose} + HPO_4^{2-} \qquad \Delta G = -19 \text{ kJmol}^{-1}$$

Since this reaction has a negative ΔG, it can proceed, releasing 19 kJ of energy for every mole of glucose-6-phosphate hydrolyzed. Another reaction we will meet again is the hydrolysis of nucleotides, the building blocks of DNA and RNA (Chapter 4). Simply losing the terminal phosphate from the nucleotide ATP releases 30 kJmol^{-1}:

$$\text{Adenosine triphosphate} + H_2O \rightarrow \text{Adenosine diphosphate} + HPO_4^{2-} + H^+$$
$$\Delta G = -30 \text{ kJmol}^{-1}$$

The reverse of these reactions will of course not proceed. For instance, cells need to phosphorylate glucose to make glucose-6-phosphate but cannot use the reaction

$$\text{Glucose} + HPO_4^{2-} \rightarrow \text{Glucose-6-phosphate} + H_2O \qquad \Delta G = +19 \text{ kJmol}^{-1}$$

The reaction will not proceed because it has a positive ΔG. Crucially, though, an unfavorable (positive ΔG) reaction can occur if it is tightly coupled to a second reaction that has a negative free-energy change (negative ΔG) so the overall change for the reactions put together is negative. Thus cells phosphorylate glucose by carrying out the following reaction:

$$\text{Glucose} + \text{Adenosine triphosphate} \rightarrow \text{Glucose-6-phosphate}$$
$$+ \text{Adenosine diphosphate} + H^+ \quad \Delta G = -11 \text{ kJmol}^{-1}$$

Adenosine triphosphate, or ATP, has given up the energy of its hydrolysis to drive an otherwise energetically unfavorable reaction forward. We call ATP a cellular currency to draw an analogy with money in human society. Just as we can spend money to cause someone to do something they would not otherwise do, such as give us food or build us a house, the cell can spend its energy currency to cause processes that would otherwise not occur. However, the analogy is not exact because energy currencies are not hoarded. There is a continuous *turnover* of ATP to ADP and back again. ATP is therefore not an energy store but simply a way of linking reactions. It can be thought of as a truck that carries metabolic energy to where it is needed and that returns empty to be refilled. The number of trucks is small but the amount moved can be large. An average person hydrolyzes about 50 kg of ATP per day but makes exactly the same amount from ADP and inorganic phosphate. We will see how this happens in this chapter. The cell has a number of energy currencies of which four—NADH, ATP, the hydrogen ion gradient across the mitochondrial membrane, and the sodium gradient across the plasma membrane—are the most important. We will now discuss each of these in turn.

Reduced Nicotinamide Adenine Dinucleotide (NADH)

This, the most energy rich of the four currencies, is shown in Figure 12.1a. NADH is a strong reducing agent. It will readily react to allow two hydrogen atoms to be added to molecules, in the general reaction $NADH + H^+ + X \rightarrow NAD^+ + H_2X$. NAD^+ is shown in Figure 12.1b. Addition of hydrogen atoms to molecules, or the removal of oxygen atoms, is called **reduction.** The opposite of reduction is **oxidation,** the addition of oxygen. Because oxygen atoms tend to take more than their fair share of electrons in any bonds they make, addition of oxygen means removal of electrons, and vice versa, so the most general definition of reduction is the addition of electrons, with oxidation being defined as the removal of electrons. We will later see NADH acting to reduce complex cell chemicals like pyruvate and acetoacetate. However, when it is acting as an energy currency, NADH simply passes its two hydrogen atoms to oxygen, making water. This releases a lot of energy: Every mole of NADH that is used in this way releases 206 kJ of energy.

Medical Relevance 12.1	**NAD^+, Pellagra, and Chronic Fatigue Syndrome**

A small fraction of our total body NAD^+ is lost from the body each day. New NAD^+ is synthesized from the vitamin niacin, which in turn can be synthesized from the essential amino acid tryptophan. If we don't get enough niacin or tryptophan, we develop the disease called pellagra. In pellagra parts of our bodies that use a lot of energy, such as the brain, begin to fail. Pellagra killed about 100,000 Americans in the first half of the twentieth century before Joseph Goldberger of the U.S. Public Heath Service showed that it could be prevented and cured by a varied diet. Niacin is now added to all flour, and pellagra is almost unknown in Western countries.

Since chronic fatigue syndrome is manifested as a lack of energy, some people have suggested that NAD^+ itself, given as a dietary supplement, might reenergize the patients. There is no scientific evidence for this idea, but nevertheless many chronic fatigue syndrome sufferers buy and take NAD^+ in the hope that it may help.

Nucleoside Triphosphates (ATP plus GTP, CTP, TTP, and UTP)

Adenosine triphosphate, the second most energy rich of the four currencies, is shown in Figure 12.2. In earlier chapters we met many chemical processes in the cell that are driven by ATP hydrolysis. When one mole of ATP is hydrolyzed, 30 kJ of energy are released. The γ phosphate is easily transferred between nucleotides in reactions such as this one:

$$ATP + GDP \rightleftharpoons ADP + GTP$$

So as far as energy is concerned, we can regard GTP, CTP, TTP, and UTP as equivalent to the most commonly used nucleotide energy currency, ATP. Another easy, reversible reaction is this one:

$$2\,ADP \rightleftharpoons AMP + ATP$$

in which one ADP transfers its β phosphate to another, so that it itself is left as AMP while converting the other ADP to ATP. This conversion is at equilibrium in cells and is important in maintaining the supply of ATP when ATP is used in reactions.

NADH and ATP take part in so many reactions within the cell that they are often called **coenzymes,** meaning molecules that act as second substrates for many enzymes as they do

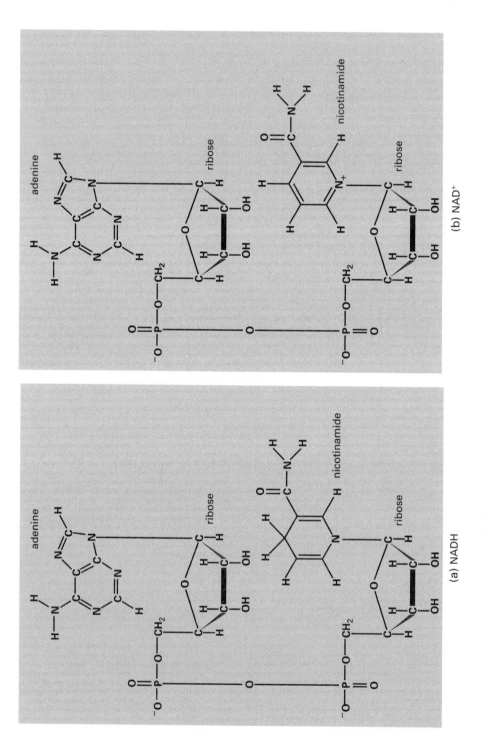

(a) NADH

(b) NAD⁺

Figure 12.1. Nicotinamide adenine dinucleotide, NADH, shown in (*a*), is a strong reducing agent and energy currency. The oxidized form, NAD⁺, is shown in (*b*).

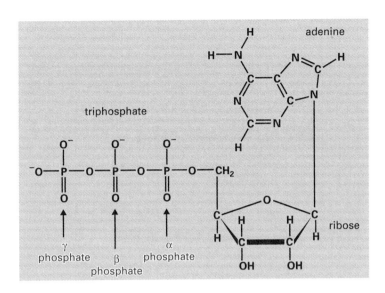

Figure 12.2. Adenosine triphosphate, an energy currency.

their particular jobs. The term is easy to confuse with cofactor (page 249). A cofactor is a chemical species that is loosely associated with an enzyme, helps it carry out its function, and, although it may undergo reactions, ends up in the same state that it began. The concept of coenzymes is very different. Coenzymes are *bona fide* substrates that are converted to products (e.g., NAD^+ and ADP) by the enzyme.

The Hydrogen Ion Gradient Across the Mitochondrial Membrane

The endosymbiotic theory states that mitochondria are derived from bacteria that evolved to live in eukaryotic cells (page 12). The bacterial cytosol is usually about 0.6 pH unit more alkaline than the world outside; that is, H^+ ions are four times more concentrated outside than inside. If they could move freely across the bacterial cell membrane, H^+ ions would rush in down this gradient. Furthermore, there is a voltage difference across the membrane: The inside is about 140 mV more negative than the extracellular medium. Transmembrane voltages are always referred to in terms of the internal voltage relative to that outside: in this case, -140 mV. The transmembrane voltage attracts the positively charged H^+ ions into the bacterium. Any combination of a concentration gradient and a voltage gradient is called an electrochemical gradient. For hydrogen ions at the bacterial membrane the electrochemical gradient is large and inward. Should H^+ ions be allowed to rush into the bacterium, they would release energy: about 17 kJ for every mole that enters.

Figure 12.3 is a representation of a mitochondrion inside a eukaryotic cell. This is not an accurate picture of what a mitochondrion looks like (see Fig. 3.6 on page 59) but rather emphasizes the topology and the function. In the center is shown the mitochondrion with its two membranes. In the very middle is a volume equivalent to bacterial cytosol that is called the mitochondrial matrix. Next comes the inner mitochondrial membrane, then the outer mitochondrial membrane. The intermembrane space is the small space between the two mitochondrial membranes. The green region is the cytosol, bound by

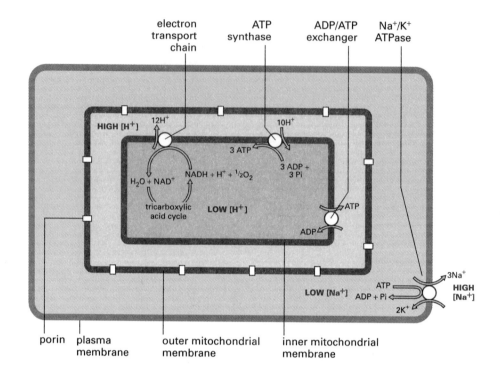

electron transport chain ATP synthase ADP/ATP exchanger Na⁺/K⁺ ATPase

Figure 12.3. Sites in the cell where the energy currencies are interconverted.

the plasma membrane. An integral membrane protein of the outer mitochondrial membrane, called porin, forms a hole or **channel** that lets through all ions and molecules of $M_r < 10,000$, so the ionic composition of the intermembrane space is the same as that of cytosol. However, there is a large electrochemical gradient across the inner mitochondrial membrane. When H^+ ions move in from the intermembrane space to the mitochondrial matrix, they release 17 kJ mol^{-1} of energy, exactly as in the mitochondrion's proposed bacterial ancestors.

The Sodium Gradient Across the Plasma Membrane

Unlike bacteria, most eukaryotic cells do not have an H^+ electrochemical gradient across their plasma membranes. Rather, it is sodium ions that are more concentrated outside the cell than inside (Fig. 12.3). Typically, the sodium concentration in the cytosol is about 10 mmol $liter^{-1}$ while the concentration in the extracellular medium is about 150 mmol $liter^{-1}$. This chemical gradient is supplemented by a voltage gradient. The cytosol is between 70 and 90 mV more negative than the extracellular medium, that is, the transmembrane voltage of the plasma membrane is between -70 and -90 mV. There is therefore a large inward electrochemical gradient for sodium ions. If sodium ions are allowed to rush down this gradient, they release energy—approximately 15 kJ for every mole of Na^+ entering the cytosol.

✳ ENERGY CURRENCIES ARE INTERCONVERTIBLE

A company that buys raw materials in the United States and Mexico, spending dollars and pesos, and then sells products in Europe and Japan, receiving euros and yen, simply converts from euros and yen to dollars and pesos to pay its bills. In the same way, cells convert from the energy currency in which they are in credit to the energy currency they are using up.

Exchange Mechanisms Convert Between the Four Energy Currencies

The cell has mechanisms that transfer energy between the four currencies. The conversions are summarized in Figure 12.4. In a typical animal cell oxidation of fuel molecules in the mitochondria (by the Krebs cycle, page 283) tops up the supply of NADH. The cell then converts this energy currency into the other three. All the interconversions are reversible. Figure 12.3 shows where each conversion mechanism is located.

Consider what happens if a few sodium ions move out of the extracellular fluid and into the cytosol of a eukaryotic cell—for example, when a nerve cell transmits the electrical signal called an action potential (Chapter 15). The sodium gradient has been slightly depleted: the cell holds less of this energy currency than it did before. However, the cell still has plenty of energy in the form of ATP that it can convert into energy as a sodium gradient. It does this using the **sodium/potassium ATPase.** This protein is located in the plasma membrane. Its function is to move $3Na^+$ ions out of the cell and to move $2K^+$ ions into the cell. For this to happen ATP is hydrolyzed to ADP thus giving up energy that is used to push Na^+ out of the cytosol to its higher energy state in the extracellular medium.

The cell has now used some ATP, but it still holds plenty of energy in another currency—the H^+ gradient across the inner mitochondrial membrane. The enzyme **ATP synthase** is located in the inner mitochondrial membrane. The protein interconverts the two energy currencies: as H^+ ions move into the mitochondrion they give up energy that is then used by ATP synthase to make ATP from ADP and inorganic phosphate.

The energy of the H^+ gradient is now depleted. However, the cell still has plenty of energy in its NADH account. The **electron transport chain** allows interconversion of energy as NADH to energy as H^+ gradient. NADH is used to reduce molecular oxygen to

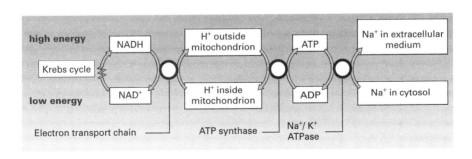

Figure 12.4. Energy flow between the currencies in a normal animal cell.

water, releasing energy that is used to push H^+ ions out of the mitochondrion to their higher energy state in the cytosol.

Each of the three energy conversion systems is a protein structure called a **carrier** because it carries solute across the membrane. The sodium/potassium ATPase carries sodium and potassium ions, while both ATP synthase and the electron transport chain carry H^+ ions. There are many carriers in the cell with a wide variety of functions, some of which we will meet later in this book. The three that convert between the energy currencies are vital and are evolutionarily ancient.

Example 12.1 The Fuel That Makes Bacteria Swim

Although by no means all bacteria are motile, some, such as *Escherichia coli* and *Salmonella typhimurium,* can swim. Their small size makes the process very different from swimming as we experience it. If you, the reader, dive into a swimming pool, your momentum will carry you some distance because your mass is great relative to the density of the water. For a bacterium whose mass is extremely small, water is a highly viscous medium. A bacterium swimming in water experiences roughly what a human swimmer in treacle might feel! Despite this, motile bacteria can achieve speeds of up 100 μm s^{-1} by the use of a structure called the bacterial flagellum. This consists of a rigid helical filament, 20–40 nm in diameter and up to 10 μm long, composed of a single protein, flagellin. The bacterial flagellum operates like a boat's propeller, pushing the bacterium along as it is turned by the motor at its base.

The flagellar motor consists of a series of rings that allow the motor to rotate within the complex layers of membranes and cell wall that make up the bacterial cell surface. Like ATP synthase, the flagellar motor allows H^+ ions to flow in across the plasma membrane. The energy released as the H^+ ions flow down their electrochemical gradient is used to turn the rotor of the motor and hence the flagellum at up to 100 Hz (=100 times per second).

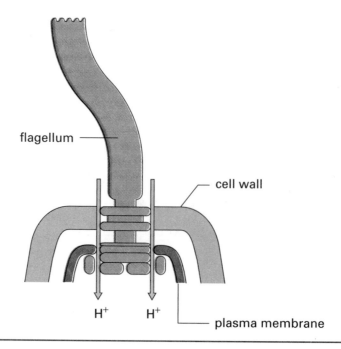

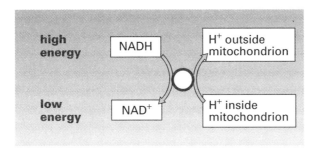

Figure 12.5. Currency conversion: the electron transport chain converts between NADH and the H^+ gradient.

Electron Transport Chain

The reason we breathe is to supply oxygen to the electron transport chain, which uses the oxygen to oxidize NADH in the overall chemical reaction:

$$NADH + H^+ + \frac{1}{2}O_2 \rightleftharpoons NAD^+ + H_2O$$

This reaction releases energy, 206 kJ for every mole of NADH used. The energy is used to carry about $12H^+$ ions from their low-energy state in the mitochondrial matrix to their high-energy state in the cytosol. Figure 12.5 summarizes the reaction in terms of energy currencies. The circle symbolizes the linkage between the energy released in the conversion of NADH to NAD^+ and the energy used to drive H^+ out of the mitochondrial matrix. If the electron transport chain simply allowed NADH to reduce oxygen to water, then the reaction's energy would be released as heat. Instead, the enzymatic function of the electron transport chain is tightly coupled to its function as a carrier that moves H^+ ions. The energy of NADH is thus converted to the energy of the H^+ gradient.

The electron transport chain has six components: four large protein complexes, each containing many polypeptide chains, embedded in the inner membrane, plus two mobile electron carriers. The electron transport process has been the subject of intensive investigation for decades and this has led to some rather complex nomenclature. Some aspects of the process remain unclear.

Electrons are mostly carried by prosthetic groups built into the protein complexes. We have already seen the heme prosthetic group in hemoglobin and myoglobin. The heme-containing proteins of the electron transport chain are, like hemoglobin, colored red or brown and are called *cytochromes* from the Latin for cell and color. The iron in a reduced cytochrome is Fe^{2+} while in an oxidized cytochrome it is Fe^{3+}. The exact colors of the cytochromes vary because the proteins in which they are buried provide different environments. These spectral differences are shown by different letters: cytochromes a, b, c with subscript numbers for more subtle differences: cytochrome c_1, a_3. Other prosthetic groups are iron–sulfur clusters that vary in detail but are essentially iron ions coordinated to sulfurs in the thiol groups of cysteine side chains or free sulfides. Flavins are also present as electron-carrying prosthetic groups and the final complex also contains copper.

The first of the mobile electron carriers is called *ubiquinone* or coenzyme Q. This is chemically a quinone (hence "Q"), which is fairly hydrophobic, and its hydrophobicity is increased by a long hydrocarbon tail (Fig. 12.6). It carries both hydrogen ions and electrons.

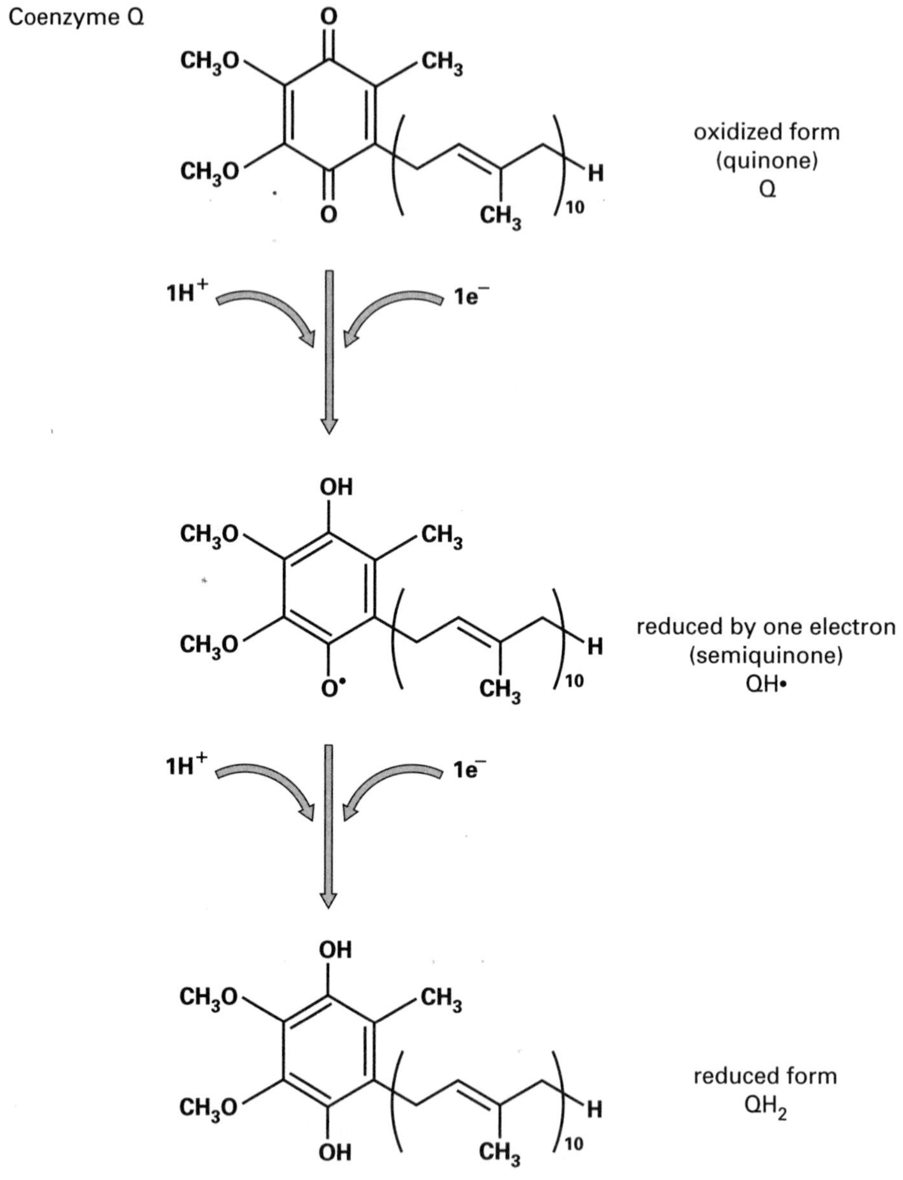

Figure 12.6. Coenzyme Q carries two hydrogen atoms.

The second electron carrier is a small protein found in the mitochondrial intermembrane space called *cytochrome c*. It has a relative molecular mass of 12,270, which is just large enough to be retained by the mitochondrion (the channel porin in the outer membrane passes solute of $M_r = 10,000$ or less). Cytochrome c is soluble but associates with the inner mitochondrial membrane.

We have said that oxidation is loss of electrons, and reduction gain of electrons. Electron carriers that cycle between the reduced state and the oxidized state normally move either one

Overview of the mitochondrial electron transport chain

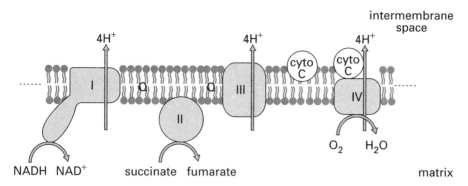

Figure 12.7. The electron transport chain comprises four large multimolecular complexes in the inner mitochondrial membrane, three of which are hydrogen ion carriers.

or two electrons. NADH and a cofactor called $FADH_2$ each carry two electrons. Reduced cytochromes carry single electrons as do the iron–sulfur clusters. Flavins and coenzyme Q can carry either one or two electrons.

Figure 12.7 shows an overview of the electron transport chain. **Complex I** (sometimes called NADH dehydrogenase or NADH-Q oxidoreductase) accepts electrons from NADH, thus oxidizing it, and uses these to reduce coenzyme Q (Fig. 12.8): In doing so it moves four hydrogen ions outward from the matrix to the mitochondrial intermembrane space. The reaction can be summarized as

$$NADH + Q + 5H^+_{matrix} \rightarrow NAD^+ + QH_2 + 4H^+_{intermem}$$

The transport of the hydrogen ions is accomplished using coenzyme Q as it picks up a hydogen ion when it picks up an electron. Complex I has a bound coenzyme Q that collects two hydrogen ions from the matrix when it accepts two electrons from iron–sulfur clusters (which received them from a bound flavin which accepted them from NADH). The bound coenzyme QH_2 formed then passes its electrons to iron–sulfur clusters, which finally reduce a mobile coenzyme Q, which again collects two hydrogen ions from the matrix. These are released to the intermembrane space when this mobile carrier is reoxidized. The machinery that produces this directionality is not yet clear.

Complex II (also known as succinate-Q reductase complex or succinate dehydrogenase) is the only one of the four complexes that is not a hydrogen ion pump. As we will see in Chapter 13 (page 284) complex II is part of the Krebs cycle—it is the enzyme that oxidizes succinate to fumarate using FAD, which becomes $FADH_2$. It passes its electrons to mobile coenzyme Q. Other mitochondrial enzymes that generate $FADH_2$ share this property of being able to reduce coenzyme Q. Because hydrogen ions are not moved across the membrane in this process one gets less ATP formed from the electrons on $FADH_2$ than from those on NADH.

Complex III (Q-cytochrome c oxidoreductase or cytochrome reductase) accepts electrons from reduced coenzyme Q (QH_2) and uses them to reduce the other mobile electron

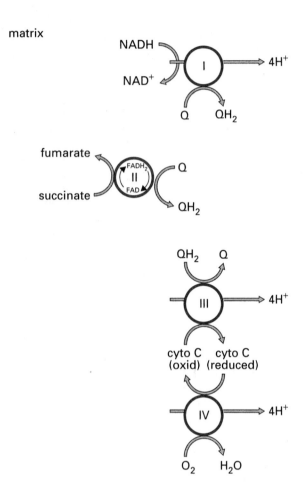

Figure 12.8. Overview of the operation of the mitochondrial electron transport chain.

carrier cytochrome c. This process results in the movement of four hydrogen ions from the matrix to the intermembrane space. Complex II marks a change from two electrons being carried to one. It contains hemes, bound coenzyme Q, and iron–sulfur clusters.

Cytochrome c moves the electrons to complex IV (cytochrome c oxidase). This protein complex reduces oxygen to water and moves four hydrogen ions from the matrix to the intermembrane space. It contains two types of heme and copper ions. Oxygen binds to Fe^{2+} in a heme just as it binds to the heme in myoglobin or hemoglobin. The overall reaction is

$$4\text{Cyto } c_{\text{reduced}} + 8\text{H}^+_{\text{matrix}} + \text{O}_2 \rightarrow 4\text{Cyto } c_{\text{oxidized}} + 2\text{H}_2\text{O} + 4\text{H}^+_{\text{intermem}}$$

The electron transport chain is very similar in all organisms. In eukaryotes the complexes are found in the inner mitochondrial membrane, while in prokaryotes the complexes are found in the plasma membrane.

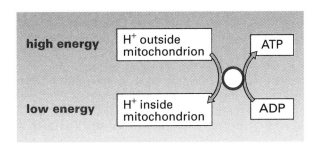

Figure 12.9. Currency conversion: ATP synthase interconverts the H^+ gradient and ATP.

ATP Synthase

The energy stored in the hydrogen ion gradient can be used to drive the synthesis of ATP from ADP and inorganic phosphate. The conversion is accomplished by a marvelous "nanomachine" called ATP synthase. At its most basic, the operation is a simple one and is summarized in Figure 12.9. Once again, the circle symbolizes the linkage between the enzyme and carrier functions. ATP synthase is a protein complex that is also known as mitochondrial ATPase (because it can carry out the reverse reaction, hydrolyzing ATP and driving H^+ out of the mitochondria) or F_1F_0 ATPase (because it can be prepared in two fragments F_1 and F_0) while others call it complex V.

In most mitochondria ATP synthase is the only route by which the large electrochemical gradient for H^+ generated by the electron transport chain can be relieved. If the supply of ADP runs out, then ATP synthase will stop moving H^+ in. If this happens, the electron transport chain will stop too because it cannot keep pushing H^+ out against a greater and greater electrochemical gradient. Another way of stopping ATP synthase is to block its operation with the antibiotic oligomycin; again, if this is done, the electron transport chain will stop too.

The coupling between ATP synthesis and electron transport can be broken if hydrogen ions can be carried back across the membrane by an alternative route. Weak organic acids such as 2,4-dinitrophenol can do this. The dinitrophenol is uncharged in the intermembrane space and in this form is sufficiently hydrophobic to diffuse into the membrane. If it enters the matrix, it will tend to lose its hydrogen ion as the hydrogen ion concentration is much lower. The result is movement of hydrogen ions down their concentration gradient. Such molecules are called "uncouplers." In the presence of an uncoupler electron transport (and so oxidation of fuels) will carry on regardless of the availability of ADP or a working ATP synthase.

●●●● **IN DEPTH 12.1 Brown Fat**

Triacylglycerols are stored within specialized fat cells in the body. Most fat cells are composed of a droplet of lipid surrounded by a thin layer of cytoplasm with a nucleus and a few mitochondria. The resulting tissue is white in color and simply releases or stores fatty acids in response to the needs of the organism. This is the kind of fat that is typically found around our kidneys and under the skin.

A second kind of fat is found in babies. Brown fat cells not only have stored triacylglycerols but are also rich in mitochondria, the cytochromes of the mitochondria giving the brown color. Brown fat is a heat-generating tissue. A channel selective for H^+ called **thermogenin** is found in the inner mitochondrial membrane. As fast as the electron transport chain pushes H^+ ions out of the mitochondrial matrix, they flow through thermogenin back down their electrochemical gradient into the matrix. In other cells this flux would only occur through ATP synthase and would be tightly coupled to the production of ATP from ADP. The presence of thermogenin uncouples the phosphorylation of ADP from the flow of electrons to oxygen so that the electron transport chain can work flat out even though there is no ADP available, as long as the cell contains triacylglycerols, which can be oxidized to regenerate NADH (see β oxidation, page 290). This generates a lot of heat and helps the infant maintain body temperature. Large blocks of brown fat are found in animals that hibernate.

A similar uncoupling mechanism is used by some plants to generate heat. Some arum lilies rely on carrion-eating flies for pollination. Uncoupled mitochondria at the base of the flower generate sufficient heat to evaporate the evil-smelling odorants used to attract the flies.

Sodium/Potassium ATPase

This is a single protein in the plasma membrane, with a carrier action that is tightly linked to an enzymatic one. Under normal conditions, it hydrolyzes ATP. The energy released drives sodium ions up their electrochemical gradient out of the cell. The Na^+/K^+ ATPase also moves potassium ions the other way, into the cytosol. For every ATP hydrolyzed, three Na^+ ions are moved out and two K^+ ions are moved in. Figure 12.10 summarizes the reaction in terms of energy currencies.

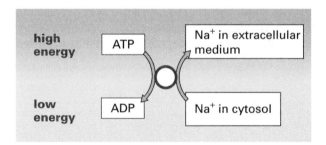

Figure 12.10. Currency conversion: the sodium–potassium ATPase interconverts ATP and the Na^+ gradient.

ADP/ATP Exchanger

ATP synthase makes ATP inside the mitochondrion. For the ATP to be available to the rest of the cell, there needs to be a mechanism to enable it to leave the mitochondrion for use in the cytosol. This job is performed by another carrier, the ADP/ATP exchanger. This protein has no enzymatic action; it simply moves ADP in one direction across the mitochondrial inner membrane and ATP in the opposite direction. In most eukaryotic cells the carrier operates in the direction shown in Figure 12.3. Carriers such as the Na^+/K^+ ATPase together with many synthetic processes use up ATP in the cytosol, producing ADP. Then ADP enters the mitochondria by the ADP/ATP exchanger and is reconverted to ATP by ATP synthase. ATP then leaves the mitochondrion with the help of the ADP/ATP exchanger.

Example 12.2 Chemicals That Interfere with Energy Conversion Are Highly Toxic

The electron transport chain, ATP synthase, and the sodium/potassium ATPase together run the energy currency market and are vital to the cell. Chemicals that interfere with them are very toxic. Rotenone, the most widely used rat poison, blocks complex I of the electron transport chain while cyanide blocks complex IV. Digitalis, from foxgloves, blocks the Na^+/K^+ ATPase.

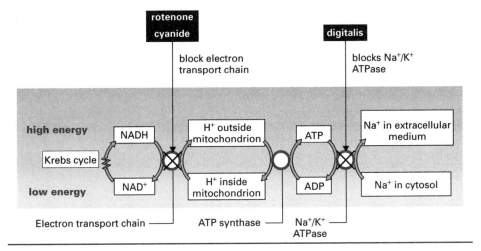

Photosynthesis

The fuels that our cells oxidize to give NADH and $FADH_2$ must of course come from some-where! Ultimately that somewhere is the fixation of carbon dioxide into more complex molecules, which is in turn driven by light energy captured by **photosynthesis.** Photosynthesis also generates the oxygen in the atmosphere. Some 10^{10} tons of carbon are fixed every year.

The **dark reactions** that fix carbon will be described in Chapter 13 (page 302). However, the overall effect can be summarized as:

$$6CO_2 + 18ATP + 12NADPH + 12H_2O \rightarrow C_6H_{12}O_6 + 18ADP + 18Pi + 12NADP^+ + 6H^+$$

●●●● IN DEPTH 12.2 ATP Synthase, Rotary Motor, and Synthetic Machine

ATP synthase can be seen in the electron microscope as tiny beads attached to the the matrix face of the inner mitochondrial membrane by a thinner region. We now know that this enzyme is a molecular motor. The beadlike head or F_1 fragment is attached to proteins that span the membrane—the F_0 part.

Most of the bulk of F_1 is composed of three α and three β subunits arranged alternately. The β subunits bind ATP or ADP and Pi. They are joined to the membrane-bound F_0 part by a central stalk composed of γ and ε subunits. The F_0 part has 10–14 subunits (called c subunits) forming a ring embedded in the membrane and an a subunit, which binds to the outside of the ring of c subunits. The head of the F_1 part is connected to the base by the b subunit, which connects to the a subunit in the membrane.

We have described how the splitting of ATP can drive a conformational change in the protein that has bound it and that this conformational change can produce movement (page 241). The converse of this is true—a conformational change can be produced by movement and this conformational change can drive a reaction.

The stalk of the F_1 part (the γ subunit) is not symmetrical. It fits into the center of the circle of alternating α and β subunits and makes different interactions with each β subunit. The β subunits have three conformations possible—a T conformation, which binds ATP very strongly, an L conformation, which binds ADP and Pi but cannot release them, and an O or open conformation, which can release bound nucleotides. The T conformation has been shown experimentally to catalyze the conversion of ADP and Pi to bound ATP with an equilibium constant near 1 (at equilibrium the concentrations of substrates and products are 1 : 1). Rotation of the stalk (γ subunit) alters the conformation of the β subunits (which do not rotate because the b subunit holds the head stationary).

Imagine that all three β subunits have bound nucleotide. One is in the T conformation, one has the L conformation, and one is O. The T conformation subunit has ADP+Pi interconverting to ATP, and the O and L conformations will have bound ADP and Pi. As the stalk rotates 120°, it converts the T conformation subunit into O so that the ATP can escape and more ADP + Pi can bind. The same movement converts the previous O state subunit into L, trapping the ADP and Pi, and the subunit in the L state converts into T. One third of a rotation generates one ATP, so a full rotation of the stalk generates three as each of the three β subunits cycles from T to O to L and back to T.

This sounds too good to be true! Experiments using the reverse reaction indicate it is true. Isolated F_1 parts were attached to a glass surface so the stalk was sticking upward. The γ subunit (the stalk) was linked to a fluorescently labeled actin filament. This could be seen using a fluorescence microscope, and the movies acquired are published on the Web (see CBASC website). Addition of ATP caused the filament to rotate, and it could be demonstrated that the hydrolysis of a single ATP caused a 120° rotation!

This has not told us how hydrogen ion flow causes the rotation. This remains speculation, but Howard Berg and George Oster have proposed an appealing hypothesis. In the membrane F_0 has a ring of c subunits with the single a subunit attached. The a subunit has two half length hydrogen ion channels. Each c subunit has an aspartate residue in the center that is exposed to the hydrophobic interior of the membrane. If this residue has lost its hydrogen ion, the negative charge is very unfavorable in the hydrophobic environment of the membrane interior. Normally this residue will have its hydrogen attached as the uncharged form can be tolerated in the hydrophobic environment of the membrane interior.

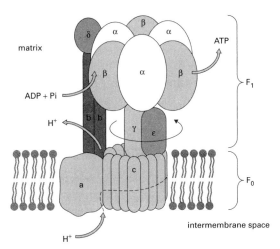

The a subunit makes close contact with two c subunits so their aspartates are exposed to the two half channels. One channel opens toward the matrix and one toward the intermembrane space. The channels provide a hydrophilic environment in which the aspartic acids can lose their hydrogen ions. Once its aspartate has ionized, a c subunit is stuck—it cannot rotate away from the channel on the a subunit. It is much more likely that an aspartate will pick up a hydrogen ion again if it is in contact with the half channel leading to the intermembrane space as the hydrogen ion concentration is around four times higher than it is on the matrix side, and the large transmembrane voltage acts to pull H^+ toward the matrix. Similarly a c subunit contacting the half channel leading to the matrix is much more likely to be ionized. Molecular movements ensure that as soon as a c subunit becomes uncharged, it will be able to move into the hydrophobic membrane. As all of the c subunits are linked in a ring, this movement is a rotation of the ring. The adjacent c subunit, which was in contact with the half channel to the matrix, is very likely to have a charged aspartate. Movement of this into the membrane is unfavorable so the only allowed rotation is movement toward the other channel on the a subunit—the one contacting the hydrogen-ion-rich intermembrane space. It is likely to become uncharged and can rotate into the membrane repeating the process. The ring of c subunits acts to carry hydrogen ions from the intermembrane space back to the matrix. The rotation has to be clockwise because of the design of the a subunit, and the hydrogen ion concentration difference across the membrane. The size of the ring seems to vary with the species, but 10 c subunits are found in ATP synthase from yeast. Ten subunits would use 10 hydrogen ions for one rotation, and one rotation generates 3 ATP so 3.33 hydrogen ions are used per ATP generated.

The experiment described above tends to confirm this theory as the rotation of the stalk of the immobilized F_1 part is anticlockwise, and this is driven by the hydrolysis of ATP so the rotation should be clockwise for ATP synthesis.

The understanding of how cells generate and interconvert their energy currencies is one of the triumphs of twentieth-century biological science. In 1997 the Nobel prize for chemistry was awarded to Jens Skou, Paul Boyer, and John Walker for working out the structure and mechanism of ATP synthase.

The F_0 component of ATP synthase is evolutionarily related to the motor that drives the bacterial flagellum (Example 12.1 on page 264): both generate rotational force from the H^+ gradient. Which came first is a matter for speculation.

The process requires 3 ATP and 2 NADPH for every carbon dioxide fixed. These are regenerated by the **light reactions** of photosynthesis. In plants both light and dark reactions occur in the chloroplast, which, like the mitochondrion, is thought to have originated through endosymbiosis (page 12). Figure 3.6 (on page 59) shows the structure of a chloroplast. It has an outer membrane that has porins, and so is permeable to many molecules, and an inner membrane that is impermeable. The infolding of the inner membrane has been carried further in chloroplasts than in mitochondria and has produced discrete membrane structures called **thylakoids.** Within the thylakoids is the thylakoid space or lumen, and outside the thylakoids is the **stroma.** The dark reactions take place in the stroma and the light reactions in the thylakoid membranes.

When a molecule absorbs a photon of light of an appropriate wavelength (determined by the electronic structure of the molecule), an electron jumps to a higher energy level. Normally it drops back to its normal level, and the energy of the photon is lost as heat. In photosynthesis, however, this high-energy electron is captured by another molecule and transferred through a chain of electron carriers rather as we have seen in mitochondrial electron transport. This allows hydrogen ions to be pumped across the thylakoid membrane, setting up a hydrogen ion electrochemical gradient with the inside of the thylakoid space having a positive voltage, and a lower pH, compared to the stroma. An ATP synthase, very similar to that found in mitochondria, then allows hydrogen ions to flow back from the thylakoid space into the stroma, driving the synthesis of ATP.

Like electron transport in mitochondria, photosynthetic electron transport takes place in large multiprotein complexes that span the thylakoid membrane and uses mobile carriers to shuttle electrons between the complexes. There are actually two different photosystems called photosystem I and photosystem II, which work together. The fundamental pigment involved is **chlorophyll.** Chlorophyll is rather like heme (Fig. 9.21 on page 206) but differs in detail and importantly has a magnesium ion rather than iron at the center and has a long hydrocarbon tail called phytol. Two types of chlorophyll are present: chlorophyll a and chlorophyll b. Structural differences give them different colors, which means they absorb light of different wavelengths.

Most of the chlorophyll molecules are **antenna chlorophylls,** which act to harvest light and pass the energy to a pair of reaction center chlorophylls. Antenna chlorophylls are supplemented by other pigments such as carotenoids, which absorb different wavelengths of light to enable efficient use of most of the light energy reaching the plant.

Let us start with the reaction center chlorophylls in PSII. These chlorophylls absorb light at 680 nm and so the reaction center is called P680. As soon as an electron in one of these is raised to a higher energy (either by energy transfer from antenna chlorophylls or by direct light absorbtion), it is immediately captured by an adjacent molecule (called pheophytin). This generates a charge pair: the chlorophyll is +ve and the pheophytin −ve. The complex structure prevents the charges neutralizing by rapidly transferring the electron from pheophytin to a quinone called **plastoquinone,** which is very similar in structure to ubiquinone (coenzyme Q) in mitochondria (Fig 12.6). This leaves a positively charged reaction center chlorophyll $P680^+$. Also at the reaction center is a cluster of four manganese ions together with bound water. An electron is abstracted from the bound water—this is done four times resulting in oxidation of two molecules of water to give a molecule of O_2. The net result is a movement of electrons from water to plastoquinone, which is reduced to plastoquinol (QH_2). During this process hydrogen ions are pumped from the stroma into the thylakoid space.

Plastoquinol passes its electrons to a copper protein called plastocyanin. This process is carried out by a large membrane protein complex called cytochrome *bf,* which is analogous to complex III in the mitochondrial electron transport chain. The other photosystem now comes into play. Here again we have a pair of reaction center chlorophyll a molecules called P700 (as they absorb light of wavelength 700 nm) at the center of a complex of other chlorophyll molecules and accessory pigments. The electron lifted to a higher energy level is captured by another chlorophyll and immediately passed to plastoquinone. This leaves a +ve charge at the reaction centre (P700$^+$). This charge is neutralized by an electron from plastocyanin, which came from photosystem II.

The captured electron passes from plastoquinone to an iron–sulfur cluster, which in turn passes electrons to a protein called **ferredoxin.** Ferredoxin can pass electrons to an enzyme called ferredoxin-NADP$^+$ reductase, which reduces NADP$^+$ to NADPH. The process has used light energy to oxidize water to oxygen and to drive electrons through a series of carriers to produce NADPH and the movement of hydrogen ions from the stroma to the thylakoid space. This hydrogen ion gradient is used by ATP synthase to make ATP from ADP and inorganic phosphate. The ATP and NADPH are used in the dark reactions to fix carbon dioxide into complex molecules.

If there is no NADP$^+$ available to accept the electrons from ferredoxin, photosynthesis would have to stop were it not for an alternate route. Ferredoxin can also pass its electrons back to the cytochrome *bf* complex, resulting in a cyclic flow of electrons. As the electrons pass along the chain of carriers, hydrogen ions are pumped into the thylakoid space. This process, which allows ATP synthesis without production of either NADPH or oxygen, is called cyclic photophosphorylation.

All Carriers Can Change Direction

In a normal animal cell, the primary source of energy is the Krebs cycle (page 283). This regenerates NADH from NAD$^+$, making at the same time a small amount of ATP. Because the NADH currency is always being topped up, while the others are being used, the direction of operation of the energy conversion systems is usually that shown in Figure 12.3. However, all the carriers are reversible.

Yeast cells in a wine barrel, or muscle cells in the leg of a sprinter, are anaerobic; there is no oxygen available. In this situation cells can make NADH, but the electron transport chain cannot drive H$^+$ out of the mitochondria because there is no molecular oxygen waiting to be reduced by NADH. Instead, the cell's energy needs are met by anaerobic glycolysis (page 286), which makes ATP. Figure 12.11 shows how the cell maintains the amounts of energy currencies. Any drain on the mitochondrial H$^+$ gradient is counteracted by ATP synthase running in the opposite direction from that shown in Figure 12.3. ATP is hydrolyzed and H$^+$ ions are pushed out of the mitochondrion. The ADP/ATP exchanger also reverses its direction. ATP is regenerated by anaerobic respiration in the cytosol and is used up by ATP synthase in the mitochondrial matrix.

Chloroplasts do not contain the enzymes of the Krebs cycle, so at night they have no internal source of energy and are dependent on their host cell. ATP from the cytosol (most of it created by mitochondria) enters the chloroplast by the ADP/ATP exchanger, which is carrying ADP and ATP in the opposite direction from that shown in Figure 12.3. ATP synthase hydrolyzes ATP and maintains the H$^+$ gradient.

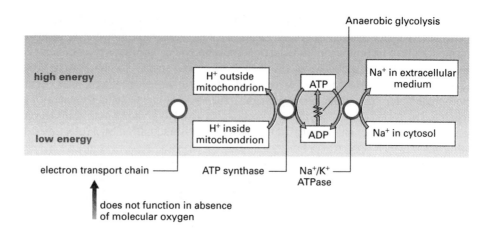

Figure 12.11. Energy flow between the currencies in an anaerobic cell.

Healthy cells maintain themselves in a steady state in which none of the energy currencies are allowed to run down. The direction in which energy moves between the four currencies depends on the primary source of energy for that cell.

IN DEPTH 12.3 Can It Happen? The Concept of Free Energy

We know from experience that some chemical reactions give out heat (the burning of organic matter for instance) while others absorb heat (so heat is used to make the reaction happen). The study of how energy affects matter and particularly how it affects chemical reactions is called **thermodynamics.** A vast amount of research in the late nineteenth and early twentieth centuries led to the development of this complex field. To make a very long story short, a process results in a change in two parameters: heat and **entropy.** Heat we are familiar with and understand as the random motion of molecules, but entropy may be unfamiliar. Entropy is the degree of disorder in a system: Entropy always tends to increase. The melting of ice is favorable because it results in an increase in entropy as the ordered, crystalline ice becomes more random water.

A physical process can absorb or emit heat and/or result in an increase or decrease in entropy. If we wish to know whether a process can occur or not, we need to consider both. The concept of free energy was formulated by the American physical chemist J. Willard Gibbs in 1878. It is now called the Gibbs free energy and given the symbol G in his honour. The change in free energy in a process, ΔG (Δ is the Greek upper-case delta used to denote a change) is

$$\Delta G = \Delta H - T \Delta S$$

Where ΔH is the heat change (called **enthalpy**) in the process, T is the absolute temperature in degrees Kelvin, and ΔS is the change in entropy. If ΔG for a process is negative, it will proceed, although the rate may be very slow in the absence of a catalyst. If it is not negative, the process cannot occur.

Unfortunately, we can only measure changes and not absolute values for G, H, and S in chemical processes. This difficulty is avoided by defining **standard** states so we can make comparisons. A free-energy change in a process under standard conditions is denoted $\Delta G°$. Biochemists use a slightly different standard state, which is in water at pH 7, and this is shown as $\Delta G°'$.

The free-energy change is related to the equilibrium constant for a reaction. Consider the reaction

$$A + B \rightleftharpoons C + D$$

$$K_{eq} = ([C][D])/([A][B])$$

And

$$\Delta G = \Delta G°' + RT \ln\{([C][D])/([A][B])\}$$

where R is the gas constant and T the absolute temperature.

At equilibrium $\Delta G = 0$ so

$$\Delta G°' = -RT \ln K'_{eq}$$

where K'_{eq} is the equilibrium constant under standard conditions.

Conditions in the cell are not usually the standard ones, but if we know $\Delta G°'$ for a reaction and the relevant concentrations, we can calculate ΔG for the reaction. For this book we have calculated ΔG values under human cellular conditions by using the following reasonable concentration values:

[ATP] = 6 mmol liter^{-1}
[ADP] = 0.6 mmol liter^{-1}
[AMP] = 0.2 mmol liter^{-1}
[Pi] = 4.5 mmol liter^{-1}
Glucose = 250 μmol liter^{-1}
Glucose-6-phosphate = 10 mmol liter^{-1}
[NAD$^+$] / [NADH] = 4 in the mitochondrial matrix
[O$_2$] = 21 μ mol liter^{-1}
Temperature 37°C

For example, we have used the hexokinase reaction (page 258) to illustrate coupling a reaction with a negative free-energy change with one which has a positive free energy change.

Glucose + ATP → Glucose-6-phosphate + ADP + H$^+$

Under biochemical standard conditions this reaction has a $\Delta G°'$ of −36 kJmol^{-1}. However, inside cells the concentration of glucose-6-phosphate is much higher than the concentration of unphosphorylated glucose; when this and other concentration effects are taken into account, the calculated overall free-energy change is −11 kJmol^{-1}: The reaction will certainly still proceed but is not as strongly favored as the $\Delta G°'$ would suggest.

The ΔG values we have given for the four energy currencies are those that apply in an aerobic animal cell, as shown in Figure 12.3, where each of the conversion reactions has a negative ΔG, as it must if it is to proceed as shown. Under conditions where one or more of the conversion reactions is operating in the opposite directions, the concentrations, and therefore ΔG values, will of course be different.

SUMMARY

1. Reactions with a positive Gibbs free-energy change (ΔG) can be caused to happen in a cell by linking them with a second reaction with a larger, negative ΔG. The second reaction drives the first.

2. The majority of such reactions in the cell are driven by one of four energy currencies— NADH, ATP, the hydrogen ion gradient across the inner mitochondrial membrane, and the sodium gradient across the plasma membrane.

3. The electron transport chain in the mitochondrial inner membrane converts between energy as NADH and energy in the hydrogen ion gradient.

4. ATP synthase in the mitochondrial inner membrane converts between energy in the hydrogen ion gradient and energy as ATP.

5. The sodium/potassium ATPase in the plasma membrane converts between energy in ATP and energy in the sodium gradient.

6. In a healthy cell, none of the energy currencies are allowed to run down. The direction of energy exchange between the four currencies depends on the primary source of energy available to the cell.

FURTHER READING

Voet, D., and Voet, J. D. 2003. *Biochemistry,* 3rd ed. New York: Wiley

✳ REVIEW QUESTIONS

For each question, choose the ONE BEST answer or completion.

1. A reaction that has a positive ΔG will
   A. proceed, although the rate may be very slow in the absence of an enzyme.
   B. only proceed in the presence of an enzyme.
   C. generate water.
   D. not proceed, even if an enzyme is present.
   E. be coupled to a carrier.

2. Which of the following is **not** a cellular energy currency?
   A. The sodium gradient across the plasma membrane
   B. Glucose
   C. ATP
   D. NADH
   E. The H^+ gradient across the mitochondrial inner membrane

3. Which of these statements about the proteins responsible for converting between the four energy currencies is **not** correct?
   A. They are all reversible.
   B. They are all located on the mitochondrial inner membrane.

C. Drugs that block them are highly toxic.

D. They are all carriers.

E. None of the above are true.

4. Hydrogen ions that enter the mitochondrial matrix through ATP synthase release a lot of energy because

A. the mitochondrial matrix is electrically negative with respect to the cytosol.

B. H^+ is at a higher concentration in the mitochondrial matrix than in the cytosol.

C. the mitochondrial inner membrane contains thermogenin.

D. H^+ is at a higher concentration in the cytosol than in the mitochondrial matrix.

E. There is a large electrochemical gradient favoring H^+ entry.

5. The overall reaction catalyzed by the electron transport chain is

A. Glucose + ATP → Glucose-6-phosphate + ADP

B. ATP + GDP ⇌ ADP + GTP

C. $NADH + H^+ + \frac{1}{2}O_2 \rightarrow NAD^+ + H_2O$

D. $ATP + H_2O \rightarrow ADP + H_2PO_4^-$

E. $NADH + OH^- + ATP \rightarrow NAD + H^+ + ADP + HPO_4^{2-}$

6. Which of the following statements is incorrect?

A. Mitochondria may have been derived in evolution from a prokaryote.

B. Mitochondria have their own DNA.

C. Mitochondria have a large negative voltage across their inner membrane.

D. Mitochondria increase their generation of ATP in cells starved of oxygen.

E. Mitochondria generate NADH.

7. The electron transport chain in mitochondria

A. accepts electrons from NADH, thus oxidizing it to NAD^+.

B. passes electrons to oxygen, thus reducing it to water.

C. uses the hydrophobic molecule ubiquinone to carry electrons between integral membrane protein complexes.

D. uses the hydrophilic molecule cytochrome c to carry electrons between integral membrane protein complexes.

E. all of the above.

ANSWERS TO REVIEW QUESTIONS

1. **D.** Reactions with negative ΔG can proceed; those with positive ΔG cannot proceed. Enzymes are catalysts and can therefore speed up the rate of a reaction, but they cannot cause a reaction to proceed if the ΔG is positive.

2. **B.** Although glucose is the major *extracellular* fuel supplied by the blood to our cells, it is not used as an energy currency *inside* cells. Rather, on entering cells energy is supplied *to* glucose to convert it to glucose-6-phosphate.

3. **B.** The sodium/potassium ATPase is located on the plasma membrane.

4. **E.** Answers B and C are false: H^+ is at a lower concentration in the mitochondrial matrix than in the cytosol, and thermogenin (an alternative route for H^+ entry expressed in a few cell types, especially brown fat) will not act to increase the energy released by H^+ as it enters. Of answers A, D, and E the last is the *best* answer because it states the complete source of the energy. It is true that hydrogen ions that enter the mitochondrial matrix through ATP synthase release energy because the mitochondrial matrix is electrically negative with respect to the cytosol, but this is not the only source of energy—the concentration gradient also favors H^+ entry. Similarly it is true that hydrogen ions that enter the mitochondrial matrix through ATP synthase release energy because H^+ is at a higher concentration in the cytosol than in the mitochondrial matrix, but this is not the only source of energy—the electrical gradient also favors H^+ entry.

5. *C.* Answers A, B, and D are reactions that occur somewhere in the cell, but they are not the reaction catalyzed by the electron transport chain. Answer E is a nonsense equation; in particular, there is no compound NAD as written, and nicotine adenine dinucleotide is always found as NADH or as NAD^+. Note that answers A, B, D, and E involve ATP: ATP is *not* a substrate or product of the electron transport chain.

6. *D.* Under anaerobic conditions mitochondria cannot generate ATP and are dependent on the rest of the cell for an ATP supply. In an anaerobic animal or yeast cell, the ATP is generated by glycolysis.

7. *E.* All of answers A through D are true. In particular, while ubiquinone and cytochrome *c* both function as electron carriers, ubiquinone is hydrophobic and remains within the lipid bilayer, while cytochrome *c* is a hydrophilic molecule found in the aqueous intermembrane space.

13

METABOLISM

In Chapter 12 we described the energy currencies NADH and ATP. In this chapter we will describe the chemical pathways that regenerate these currencies when their levels are depleted. We will then consider some other important chemical pathways, some that operate in all cells, others that are found only in certain types of organism or in specialized biochemical centers like the liver. Figure 13.1 is an overview of the main metabolic pathways within a cell.

All the processes that occur within a living cell are ultimately driven by energy taken from the outside world. Green plants and some bacteria take energy directly from sunlight. Other organisms take compounds made using sunlight and break them down to release energy, a process called **catabolism.** The most common way of breaking down these food compounds is to oxidize them, that is, to burn them but in a controlled way. The energy trapped in energy currencies can then be used for the building, repair, and homeostatic processes termed **anabolism.** The collective term for all of the reactions going on inside a cell is **metabolism.** All metabolic reactions share some general features:

- They are catalyzed by enzymes.
- They are universal in that all organisms show remarkable similarity in the main pathways.
- They involve relatively few types of chemical reaction.

Cell Biology: A Short Course, Second Edition, by Stephen R. Bolsover, Jeremy S. Hyams, Elizabeth A. Shephard, Hugh A. White, Claudia G. Wiedemann
ISBN 0-471-26393-1 Copyright © 2004 by John Wiley & Sons, Inc.

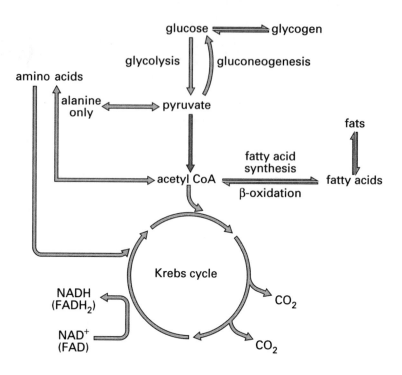

Figure 13.1. Overview of metabolism.

- They are controlled, often by modulation of key regulatory enzymes.
- They are compartmentalized within cells. In eukaryotes different sets of metabolic reactions are carried out in different organelles. In animals and plants this compartmentalization is carried further, so that in some cases different reactions take place in different body organs. Prokaryotes too show compartmentalization of a simpler sort—some processes are associated with regions on the inner face of the cell membrane.
- They usually involve coenzymes, molecules that are second substrates in a number of different reactions.
- The pathways that break particular molecules down are different from those used to synthesize them. This allows them to be controlled separately.

Molecules that are common second substrates for a number of reactions are called coenzymes. These are either energy currencies themselves (such as NADH and ATP) or carriers of chemical groups such as coenzyme A. ATP is used to drive other reactions by transfer of a phosphate group. The resultant ADP can be rephosphorylated back to ATP. The energy to do this can be derived from NADH passing its electrons to the electron transport chain, which converts the NADH back to its oxidized form NAD^+. The total amount of ATP + ADP in a cell is relatively constant, but the proportion that is ATP can vary. Similarly the amount of NADH + NAD^+ is constant, but the proportion present as NADH can vary.

❀ THE KREBS CYCLE: THE CENTRAL SWITCHING YARD OF METABOLISM

The fuels we take in in the diet are mainly fats, proteins, and carbohydrates. At the very center of metabolism is a cycle of reactions that takes place in the mitochondrial matrix. The cycle is named after its discoverer, Hans Krebs, and is also known as the tricarboxylic acid (TCA) cycle or the citric acid cycle. The foods we eat are converted to the two-carbon unit acetate, CH_3COO^-. The acetate is not free but is carried by a coenzyme called coenzyme A. Acetate bound to coenzyme A—acetyl-CoA for short—is then fed into the Krebs cycle and may be completely oxidized to carbon dioxide and water. In the process the energy currency NADH is produced. The Krebs cycle is central to carbohydrate, fat, and amino acid metabolism.

We shall first describe the Krebs cycle and then look at the other pathways that interact with it. The reactions are shown in Figure 13.2.

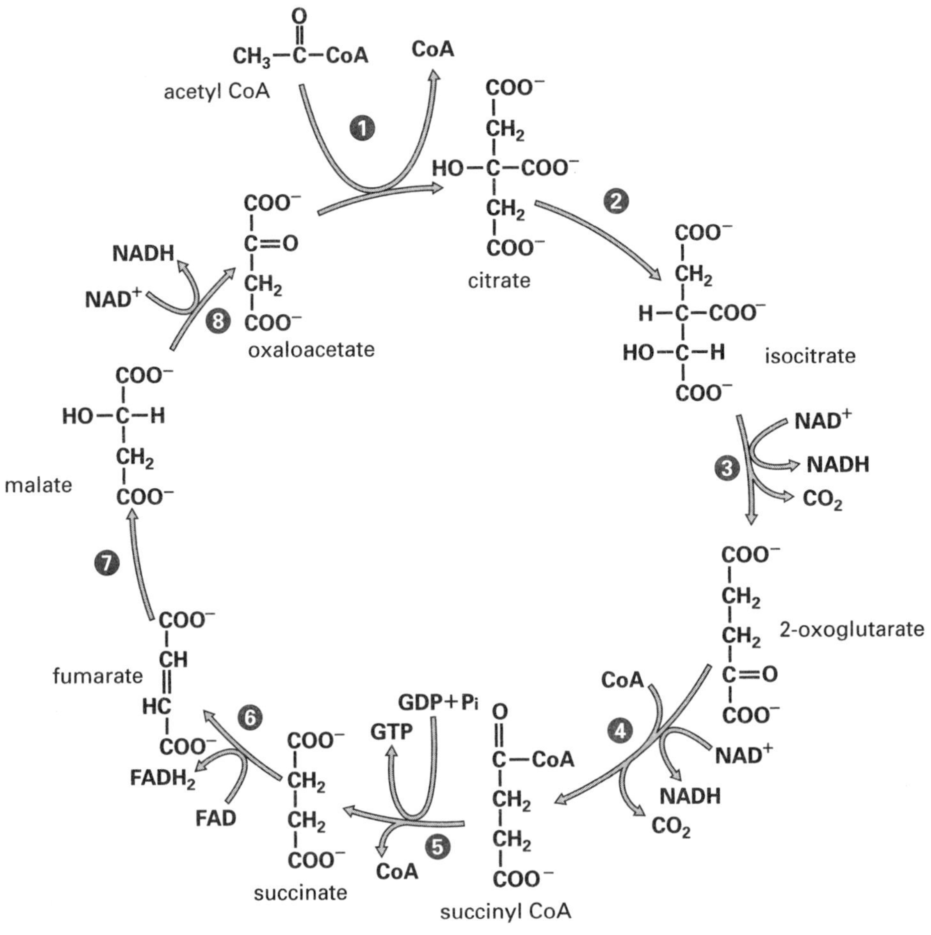

Figure 13.2. The Krebs cycle.

1. A molecule of acetyl-CoA enters the cycle, and the acetate (two carbons) is combined with the four-carbon molecule oxaloacetate, making citrate (as carboxyl groups are ionized at the pH in the cell, we normally speak of them as ions; so we say "citrate" rather than citric acid).

2. Citrate is rearranged to isocitrate.

3. In the first oxidation step isocitrate is oxidized to 2-oxoglutarate (sometimes called α-ketoglutarate). A carbon is lost as CO_2 and NAD^+ is reduced to NADH.

4. A second oxidation converts 2-oxoglutarate to succinyl-CoA. A second carbon leaves as CO_2 and again NAD^+ is reduced to NADH. Note that the product is attached to coenzyme A. This reaction is catalyzed by oxoglutarate dehydrogenase.

5. The bond between succinate and CoA is now broken and the energy released used to drive the phosphorylation of a GDP to GTP. γ-phosphate groups can be swapped between nucleotides, so this GTP can be used to regenerate ATP from ADP.

6. Succinate is oxidized to fumarate. The oxidant in this case is flavin adenine dinucleotide (FAD) and not NAD^+ so an $FADH_2$ is produced. $FADH_2$, the reduced form of FAD, does not carry as much energy as NADH but like NADH is used to drive H^+ up its electrochemical gradient out of the mitochondrial matrix (page 265). The enzyme involved is succinate dehydrogenase, which is actually part of the electron transport chain (page 267).

7. Water is added to the double bond in fumarate making malate.

8. Malate is oxidized to oxaloacetate in a reaction catalyzed by malate dehydrogenase. One NAD^+ is reduced to NADH. The starting compound oxaloacetate has been regenerated and is ready to accept another acetyl group to start another turn of the cycle.

The reactions of the Krebs cycle can be summarized as

$$CH_3CO\text{-}CoA + (3\ NAD^+ + FAD) + GDP + Pi + 3H_2O \rightarrow$$
$$CoA\text{-}H + (3\ NADH + FADH_2) + GTP + 2CO_2 + 3H^+$$

where Pi represents an inorganic phosphate ion.

✸ FROM GLUCOSE TO PYRUVATE: GLYCOLYSIS

Glucose is an important fuel for most organisms, and glycolysis is the main pathway that enables it to be used. The word *glycolysis* simply means the breakdown of glucose. It is an ancient pathway that can function without oxygen and indeed is thought to have evolved before there was much oxygen in the atmosphere. The pathway is present in almost all cells and takes place in the cytosol. It is shown in Figure 13.3.

1. Free glucose is phosphorylated on carbon 6 to produce glucose-6-phosphate. As we discussed earlier (page 258), this reaction, catalyzed by hexokinase, is driven by the free energy available from ATP.

2. Glucose-6-phosphate is isomerized to fructose 6-phosphate.

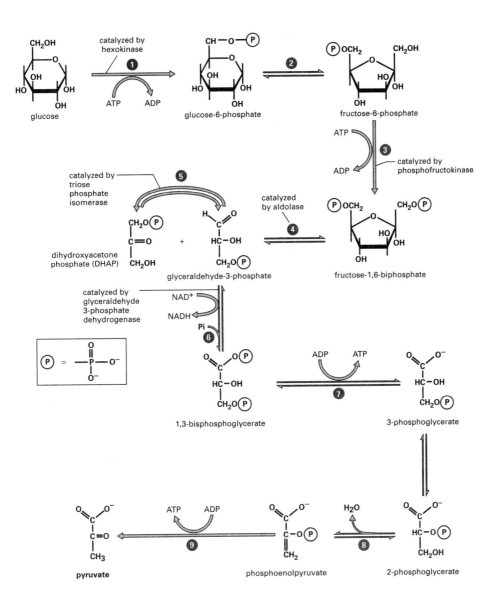

Figure 13.3. Glycolysis breaks glucose down into pyruvate.

3. Fructose-6-phosphate is phosphorylated to produce fructose-1,6-bisphosphate. Again ATP is used and the reaction is carried out by phosphofructokinase. This reaction commits the sugar to being broken down and used to provide energy rather than used for other purposes. Phosphofructokinase is regulated allosterically, being inhibited by ATP.

4. The fructose-1,6-bisphosphate is now split into two halves. Each of the halves has a phosphate attached. The two products are dihydroxyacetone phosphate and glyceraldehyde-3-phosphate. The enzyme is aldolase.

5. Triose phosphate isomerase interconverts dihydroxyacetone phosphate and glyceraldehyde-3-phosphate. This allows both halves of the original glucose to be used.

The glucose-6-phosphate has now been converted into two molecules of glyceraldehyde-3-phosphate. Each of the following reactions occurs twice for each molecule of glucose fed into the pathway.

6. Glyceraldehyde-3-phosphate is oxidized in a reaction that also attaches one inorganic phosphate ion to give 1,3-bisphosphoglycerate. NAD^+ is the oxidant and a NADH is produced. The enzyme is called glyceraldehyde-3-phosphate dehydrogenase.

7. One of the phosphate groups from bisphosphoglycerate is transferred to ADP. This is a process called substrate-level phosphorylation. ATP has been made from ADP by a single enzyme, without any involvement of ATP synthase. We are left with 3-phosphoglycerate.

8. After a rearrangment to 2-phosphoglycerate, water is removed leaving phosphoenolpyruvate.

9. In another substrate-level phosphorylation phosphoenolpyruvate transfers its phosphate group to ADP leaving pyruvate.

Figure 13.4 illustrates the various ways that the cell can use pyruvate. If it is to be used to make fatty acids or to be oxidized in the Krebs cycle, it is carried into the mitochondrial matrix. Here it is oxidized and decarboxylated by the complex enzyme pyruvate dehydrogenase (Fig. 13.4a). NAD^+ is used as the oxidant, producing NADH. Coenzyme A is also added. The product is therefore acetyl-CoA, which may enter the Krebs cycle or can be converted to fatty acids or some other molecules. Overall glycolysis has used up two ATP molecules but has produced four: a net gain of 2 ATP per glucose.

The Krebs cycle is limited by the availability of oxaloacetate. This may run low as some biosynthetic pathways use one or the other component of the Krebs cycle as their starting material. If this happens new oxaloacetate is made from pyruvate (Fig. 13.4b). Pyruvate may also be converted to the amino acid alanine in a transamination reaction (Fig. 13.4c; Fig. 11.8 on page 250).

Glycolysis Without Oxygen

The leg muscles of a sprinter cannot be supplied with oxygen rapidly enough to supply the mitochondria. Muscle cells need to make ATP by a method that does not require oxygen. As we have seen glycolysis itself produces two ATPs per glucose. Can the cell let the pyruvate pile up and use it when oxygen becomes available again? No, because glycolysis as far as pyruvate also converts one NAD^+ to NADH. If this was all that happened, the cell would quickly convert all its NAD^+ to NADH, and glycolysis would stop as the cell does not have oxygen available for the mitochondria to oxidize the NADH back to NAD^+. To solve this problem cells reduce pyruvate to lactate (Fig. 13.4d) and in doing so regenerate the NAD^+ needed to allow glycolysis to carry on. The buildup of lactic acid in poorly

in eukaryotes

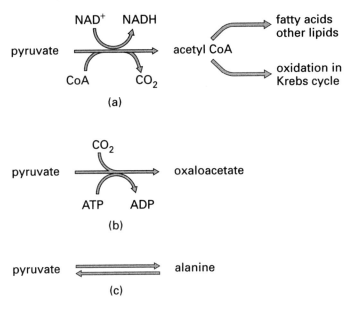

(a)

(b)

(c)

(d)

in microorganisms

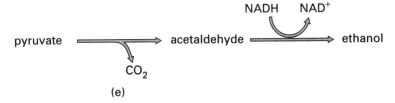

(e)

Figure 13.4. Pyruvate can be used in a number of ways.

oxygenated muscles is thought to cause the pain of cramp. The blood carries the lactate to the liver where it is reoxidized to pyruvate. When we stop using the muscle, the blood can supply more oxygen to the muscle and the need for lactate production abates. Red blood cells lack mitochondria and are entirely dependent on glycolysis for their energy needs.

> **Medical Relevance 13.1**
>
> **Sleeping Sickness**
>
> On page 259 we discussed chronic fatigue syndrome, a problem of unknown cause characteristic of Western countries. In contrast sleeping sickness, caused by a parasite called *Trypanosoma brucei*, is a serious disease in Africa. Trypanosomes are single-celled eukaryotes with a complex life cycle, part of which is spent in the human bloodstream. Here they reveal a prodigious appetite for glucose. Each tiny parasite can consume its own body weight in glucose every hour. The reason for the parasite's insatiable demand for glucose is twofold. First, trypanosomes dispense with mitochondrial ATP production and rely on anaerobic respiration to generate their ATP, so they only make 2 ATPs per glucose rather than the 30 or so that they could if their mitochondria were still working. Second, trypanosomes do not have any glycogen or fat stores and are therefore entirely dependent on a continuous supply of glucose in the host blood. The enormous consumption of glucose by the parasites leaves little for the host, who is overcome by extreme langor such that even keeping the eyes open is an unsurmountable effort.

Some microorganisms—particularly yeasts—regenerate their NAD^+ in a different way. Pyruvate is first decarboxylated to acetaldehyde (also called ethanal) and then reduced to ethanol by alcohol dehydrogenase, which uses NADH and regenerates NAD^+. One molecule of CO_2 is also produced (Fig. 13.4*e*).

Example 13.1 **Anaerobes Good and Bad**

The fact that yeasts regenerate their NAD^+ by making the gas CO_2 together with ethanol has been utilized in breadmaking, brewing, and winemaking since prehistoric times. Other microorganisms are like our muscles and generate lactate under anaerobic conditions. These too have their uses in the food industry: yogurts, many cheeses, sauerkraut, and dill pickles all rely on lactic acid released in anaerobic respiration. On the minus side, some food-spoiling bacteria can only function when there is no oxygen. These **obligate anaerobes** include the deadly *Clostridium botulinum* (page 232).

Glycogen Can Provide Glucose for Glycolysis

The polysaccharide glycogen (page 30) is used as a store of glucose particularly in liver and muscle cells. We saw in Chapter 2 how the glycosidic bond can be hydrolyzed with the broken ends of the bond being sealed with groups from a water molecule, so that a hydrogen atom is added to one side of the broken bond and a hydroxyl group is added to the other (page 44). The enzyme **glycogen phosphorylase** specifically breaks the $\alpha(1\rightarrow 4)$ glycosidic bond in glycogen but seals the broken ends with groups from inorganic phosphate, so that a hydrogen atom is added to one side of the broken bond and a phosphate group is added to the freed glucose monomer (Fig. 13.5). The resulting glucose-1-phosphate is readily converted to glucose-6-phosphate for glycolysis. Breaking up glycogen this way is more energy efficient than simply hydrolyzing it (as happens in the intestine) since the ATP that would otherwise be required to make glucose-6-phosphate from free glucose is saved. The occasional $\alpha(1\rightarrow 6)$ links that attach side arms to the glycogen chain are broken by other enzymes.

One of the many important roles of the liver is to maintain the level of glucose in the blood—it is the most important of the circulating fuels and is the primary fuel for red

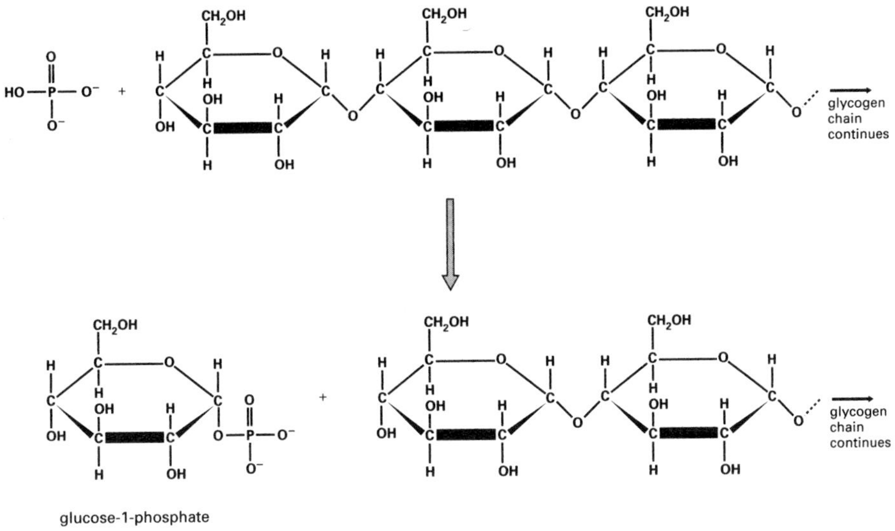

Figure 13.5. Glycogen phosphorylase cleaves a glucose monomer off glycogen and phosphory-lates it.

blood cells and the brain. Glycogen stores in the liver can provide glucose when none is available from the gut. The glucose carrier (page 238) allows glucose to enter and leave the liver cells but cannot transport phosphorylated sugars, so glucose-6-phosphate must be converted to free glucose for transport out into the extracellular medium and from there to the blood. To do this, liver has the enzyme glucose-6-phosphate phosphatase, which removes the phosphate group. Muscle also has stores of glycogen, but these are for its own use: it does not have glucose-6-phosphate phosphatase so cannot release glucose into the extracellular medium.

Glucose May Be Oxidized to Produce Pentose Sugars

Cells need the five-carbon sugar ribose to manufacture nucleotides. This is made from glucose in an oxidative pathway called the pentose phosphate pathway (Fig. 13.6). This pathway also provides reducing power for biosynthetic reactions in the form of NADPH. Glucose-6-phosphate undergoes two oxidations generating a molecule of a pentose phosphate, CO_2, and two molecules of NADPH. Like glycolysis, these reactions occur in the cytoplasm. The pentose phosphate pathway interacts with glycolysis, and this allows it to perform a different function. In a cell that does not need lots of ribose, but does need lots of NADPH for biosynthesis, the pentose is recycled by being combined with glycolytic intermediates. Going around the loop six times converts a glucose entirely to $6CO_2$, giving 12 NADPH for biosynthesis. Although the pentose phosphate pathway breaks up sugar to give CO_2 and the strong reducing agent NADPH, it is not used for energy production: only in the mitochondria, where NADH can be used to power the electron transport chain, can reducing agents be converted to other energy currencies.

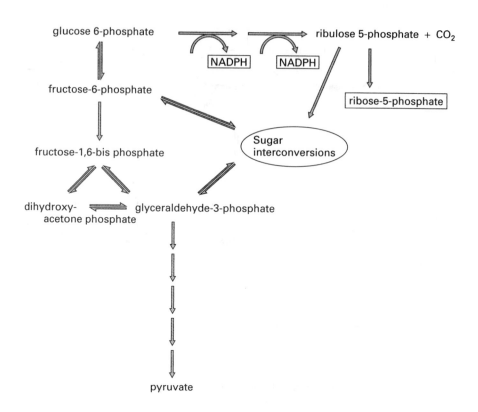

Figure 13.6. The reactions of the pentose phosphate pathway are shown in black with the reactions of glycolysis in green.

✺ FROM FATS TO ACETYL-CoA: β OXIDATION

Cells must be able to break down fatty acids (page 39) and use the energy released, whether the fatty acids come from the lipolysis of triacylglycerols in fat droplets or triacylglycerols and phospholipids in food. Before they can be broken down, the fatty acids are coupled to coenzyme A to give an acyl-CoA in a reaction driven by the conversion of ATP to AMP. A spiral of reactions in the mitochondrial matrix called β oxidation (Fig. 13.7) then oxidizes the fatty acyl-CoA. Each turn of the spiral shortens the fatty acyl chain by two carbons, releasing an acetyl-CoA and generating an NADH and an $FADH_2$ for each acetyl-CoA. The acetyl-CoA may be oxidized in the Krebs cycle or converted to other molecules.

During fasting, fat cells supply fuels for other parts of the body. Because fats are insoluble and fatty acids have only a limited solubility, fat cells convert fatty acids into soluble, circulating fuels called **ketone bodies** (Fig. 13.8) (the word **ketone** means any chemical containing a carbon atom with single bonds to two other carbons and a double bond to an oxygen). The fundamental ketone body is acetoacetate, which the liver synthesizes from acetyl-CoA. Acetoacetate is then reduced to 3-hydroxybutyrate. These two molecules are important circulating fuels in mammals. Heart muscle, for instance, prefers ketone bodies to glucose as a fuel source.

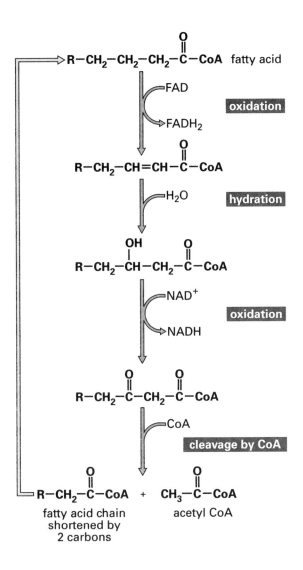

Figure 13.7. ß oxidation of fatty acids produces acetyl-CoA.

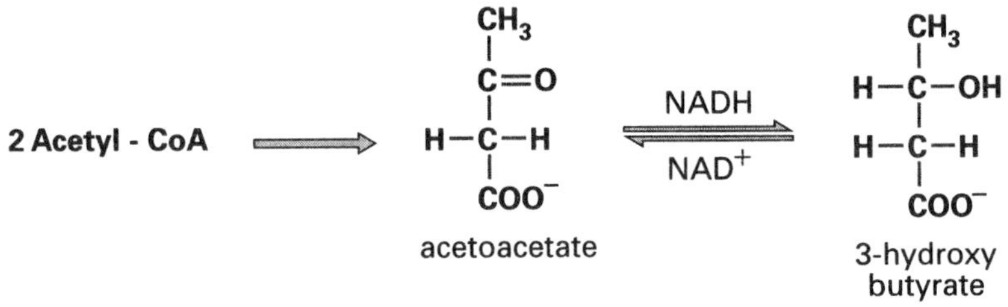

Figure 13.8. Formation of ketone bodies from acetyl-CoA.

❀ AMINO ACIDS AS ANOTHER SOURCE OF METABOLIC ENERGY

Protein forms a considerable part of the animal diet, even in vegetarians. It is broken down to free amino acids during digestion. These amino acids can be used for the biosynthesis of new proteins in the cell, but those in excess of this need can serve as metabolic fuels. To begin this process, the amino groups are removed in a process called transamination by aminotransferases such as alanine glyoxylate aminotransferase (page 249). The resulting carbon skeletons are then converted to intermediates in the Krebs cycle or to acetyl-CoA. Since there are 20 amino acids, there are many different steps, but the overall effect is that the amino groups are passed to oxaloacetate or 2-oxoglutarate to form aspartate and glutamate, respectively. The amino groups on glutamate and aspartate are then converted to urea (see below) for excretion.

●●●● IN DEPTH 13.1 The Urea Cycle—The First Metabolic Cycle Discovered

Dietary amino acids that are not required for protein synthesis have their amino groups passed to oxaloacetate or 2-oxoglutarate to form aspartate and glutamate, respectively. The amino groups on each of these amino acids are then used in the liver to make urea, which is excreted in the urine. The first step is the release of the amino group as a free ammonium ion.

(a)

This ammonium ion, plus the amino groups on aspartate, is then used to make urea in the urea cycle, which operates only in the liver. First, the ammonium ion is converted to carbamoyl phosphate. This is then combined with the α-amino acid ornithine to produce citrulline. In turn, citrulline is joined with aspartate, which carries a second nitrogen into the cycle, producing argininosuccinate. Argininosuccinate is now cleaved to release fumarate and arginine. Urea is removed from arginine to regenerate ornithine, which is now ready for another cycle. Urea can then be excreted in the urine, eliminating two nitrogen atoms per urea molecule.

The overall cycle converts two ATP to ADP and one ATP to AMP. This is energetically equivalent to a total of four ATP to ADP conversions, since a fourth ATP is required to turn the AMP into ADP in the reaction:

$$ATP + AMP \rightleftarrows 2ADP$$

The fumarate can enter the Krebs cycle and be converted to malate, which is oxidized to oxaloacetate, which can (among other things) be transaminated back to aspartate to carry in another nitrogen.

(b)

Ornithine plays the same role in the urea cycle as oxaloacetate in the Krebs cycle: it accepts the incoming molecule, undergoes a series of interconversions, and is regenerated allowing the cycle to begin again. The availability of ornithine determines the rate at which the cycle can operate. Arginine is a protein amino acid and is normally present in the diet as a source of ornithine. Vice versa the urea cycle can make arginine so this is not normally considered to be an essential amino acid for adults. Arginine is, however, essential in the diet of growing children as the net protein synthesis of growth needs more of it than can be supplied from the urea cycle without draining too much and so impeding the cycle.

Since the urea cycle only occurs in the liver, liver failure results in a buildup of ammonium ions in the blood. This is toxic to nerve cells, producing first mental confusion and finally coma and death.

It is interesting to note that of the three major energy sources in the diet—carbohydrate, fat, and protein—only protein is not used as an energy store in our bodies. We make specific proteins when we need them but never simply as a way of storing amino acids. Of course, if one eats a lot of protein, one puts on weight: once the amino acids have been converted to carbon skeletons or to acetyl-CoA, these can then be used to make glucose and hence glycogen (page 298) or to make fat (page 300).

Medical Relevance 13.2

Phenylketonuria

All the phenylalanine in the diet that is not required for protein synthesis is converted to tyrosine by the enzyme phenylalanine hydroxylase. If this enzyme is missing or defective through mutation, then there is a serious problem. The standard way the body deals with amino acids is for aminotransferases (page 249) to transfer the NH_3^+ group to aspartate or glutamate, which are then processed by the urea cycle (In Depth 13.1 on page 292). However, the product formed when phenylalanine has its NH_3^+ group transferred away, a phenylketone called phenylpyruvate, cannot be further metabolized. Both phenylalanine and phenylpyruvate therefore accumulate in the body. About 1 in 20,000 new-born babies have this defect, which is called phenylketonuria because the phenylpyruvate appears in the urine.

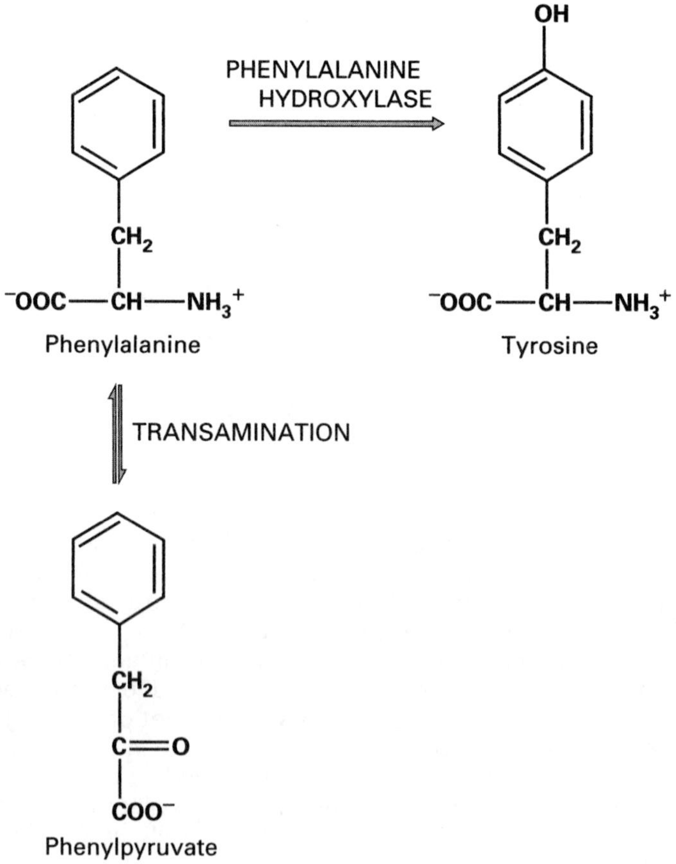

The disease is a devastating one, and, if left untreated, babies of only a few weeks old begin to suffer severe neurological damage. All newborn babies in the United States and in most other developed countries are therefore tested for phenylketonuria within the first week of life. This represents the largest genetic screening program carried out by the medical profession. The test, often called the heel-prick test or more correctly the Guthrie test after the scientist who developed it, is relatively simple. A drop of blood taken from a baby's heel is dried onto a small filter paper disk. Disks from hundreds of infants can be tested at the same time by placing their disks onto an agar plate containing bacteria that require phenylalanine for growth. If the bacteria grow, then the baby is at risk from phenylketonuria and blood from the baby will be retested a few days later to ensure the infant does indeed suffer from phenylketonuria. The Guthrie test is simple and cost-effective and can test for phenylketonuria irrespective of the mutation in DNA that has caused the problem.

If detected in the first weeks of life the prognosis for the patient is good. Affected infants are fed a strict diet that provides just enough phenylalanine for protein synthesis but no more. This treatment, carried on to maturity, is very effective and individuals develop normally.

✳ MAKING GLUCOSE: GLUCONEOGENESIS

Such is the importance of glucose that there is a pathway for its synthesis from other molecules. **Gluconeogenesis** enables animals to maintain their glucose levels even during starvation. It makes use of some of the enzymes of glycolysis but uses different enzymes to bypass the steps that are not freely reversible. Figure 13.9 shows the glycolytic pathway in green. The new reactions that allow the entire pathway to run in reverse are shown in black.

New step 1: Pyruvate is carboxylated to make oxaloacetate (a reaction that also serves to "top-up" levels of oxaloacetate for the Krebs cycle). The oxaloacetate is moved from the mitochondria into the cytosol where it is converted to phosphoenolpyruvate and CO_2. The source of the phosphate group is GTP, which is converted to GDP.

From phosphoenolpyruvate all of the reactions are reversible until fructose bisphosphate is reached. However, the reaction catalyzed by phosphofructokinase is not reversible and this step is avoided by:

New step 2: One of the phosphoester bonds on fructose bisphosphate is hydrolyzed by a phosphatase.

The interconversion of fructose-6-phosphate and glucose-6-phosphate is easily reversible. The final production of free glucose is accomplished by:

New step 3: The other phosphoester bond is hydrolyzed by glucose-6-phosphate phosphatase.

Gluconeogenesis is an expensive process: The conversion of 2 pyruvates to a glucose molecule uses 4 ATP, 2 GTP, and 2 NADH.

Different compounds can be fed into the gluconeogenesis pathway as appropriate. Glutamate is converted to 2-oxoglutarate (page 292), fed into the Krebs cycle, and tapped off as oxaloacetate to feed gluconeogenesis. Lactate and alanine are converted to pyruvate and hence oxaloacetate. Glycerol, released from lipids by hydrolysis, is phosphorylated and then oxidized to generate dihydroxyacetone phosphate. However, animals cannot make glucose from fats. Figures 13.1 and 13.2 (pages 282 and 283) might suggest that they could: use β oxidation to make acetyl CoA, feed the acetyl-CoA into the Krebs cycle, and tap off

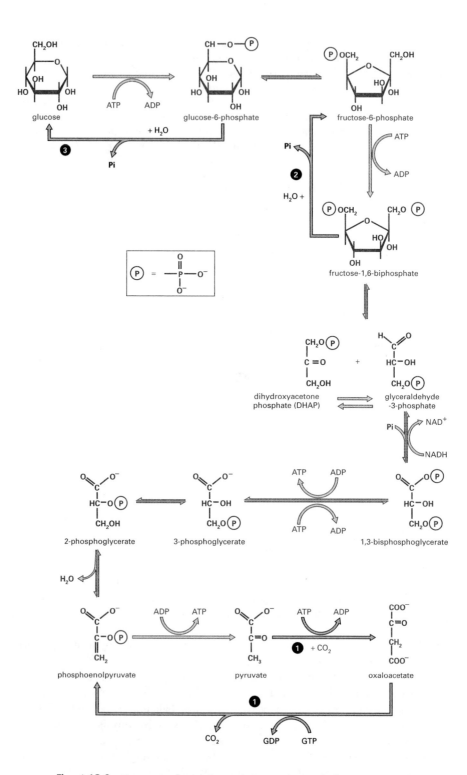

Figure 13.9. Gluconeogenesis allows glucose to be made from pyruvate.

oxaloacetate to feed gluconeogenesis. However, in passing around the Krebs cycle from citrate to oxaloacetate, the two carbons delivered by the acetyl-CoA have already been lost as CO_2.

IN DEPTH 13.2 The Glyoxylate Shunt

Many plants store triacylglycerol oils in their seeds. We have emphasized that mammals cannot make glucose from acetyl-CoA as two carbon dioxides are lost for every two carbons on acetyl-CoA that enter the Krebs cycle. Plants, however, can convert acetyl-CoA into glucose. Special organelles called glyoxysomes carry out a series of reactions called the glyoxylate pathway or glyoxylate shunt, which effectively short-circuits the Krebs cycle.

As in the Krebs cycle acetyl-CoA combines with oxaloacetate to give citrate, which is converted to isocitrate. However, rather than undergoing oxidation and loss of CO_2, the isocitrate is split to give succinate and the two carbon glyoxylate. Glyoxylate is then combined with another acetyl-CoA to yield malate, which is oxidized to oxaloacetate. In parallel the succinate is also converted to oxaloacetate using the standard Krebs cycle reactions. The overall effect of one complete cycle is therefore to generate one oxaloacetate that can enter gluconeogenesis (page 295).

Many bacteria can also carry out these reactions. Indeed, some bacteria can grow on acetate as the sole carbon source because they are able to combine free acetate with coenzyme A to form acetyl-CoA, which is then fed into the glyoxylate pathway as required.

✿ MAKING GLYCOGEN: GLYCOGENESIS

Glucose is stored as the polymer glycogen. Glucose polymerization, as a stand-alone reaction, has a positive ΔG and will not occur. The synthesis of glycogen is therefore driven by the hydrolysis of nucleoside triphosphates—not only ATP but also UTP. If phosphorylated glucose is not available, then it is made by hexokinase using ATP (page 284). Glucose-1-phosphate then reacts with UTP to make UDP-glucose (Fig. 13.10). Glycogen synthase then transfers glucose from the UDP to the growing glycogen chain (Fig. 13.11). Other enzymes insert the $\alpha(1 \rightarrow 6)$ branches at intervals.

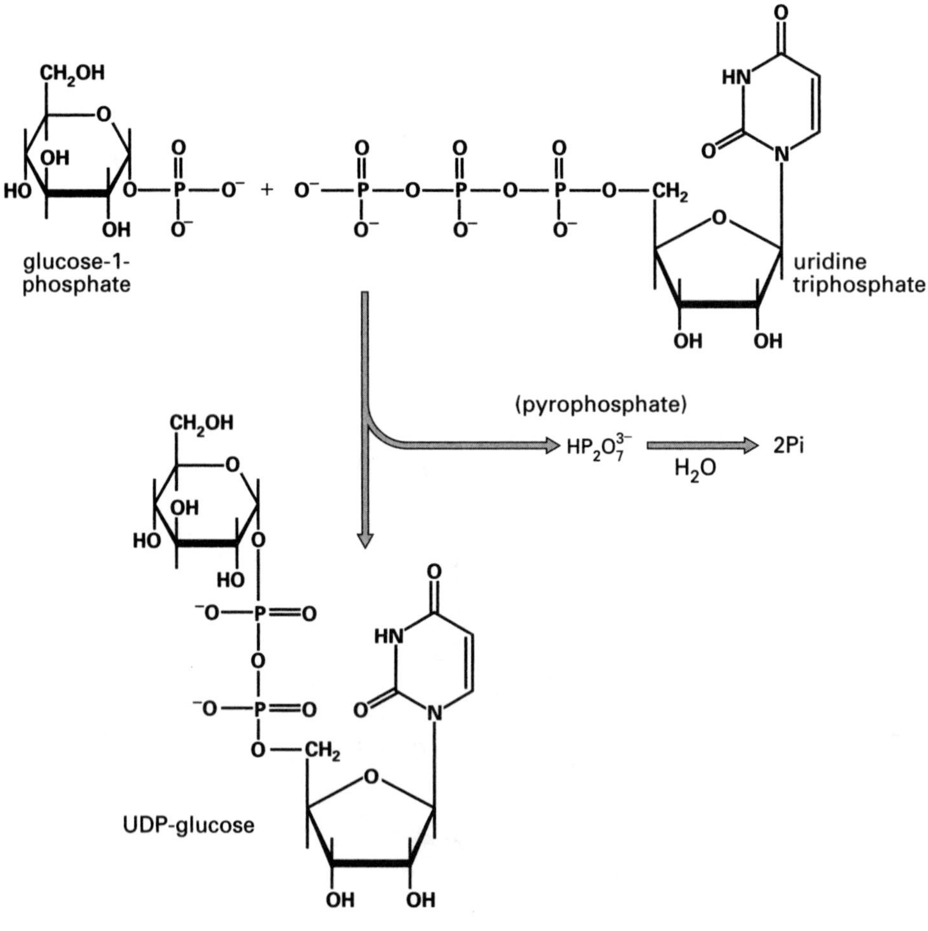

Figure 13.10. Uridine diphosphate glucose is synthesized from UTP and glucose-1-phosphate.

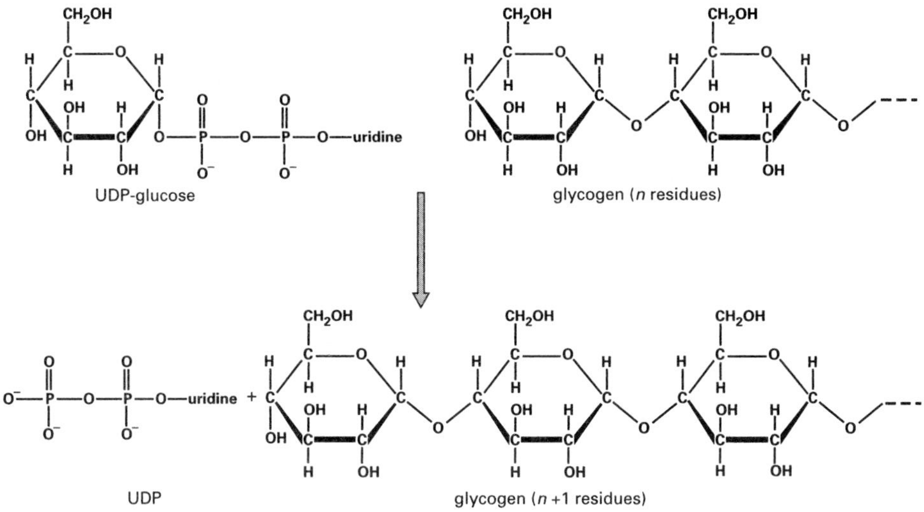

Glycogen synthesized from UDP-glucose monomers.

Figure 13.11. Glycogen is synthesized from UDP-glucose monomers.

Example 13.2 **Diabetes, Starvation, Ketone Bodies, and the Odor of Sanctity**

During starvation the only source of glucose is gluconeogenesis, which must use body protein amino acids as a source of precursors. It is clearly far better to use fat reserves than to degrade body protein, so those tissues that can shift to using ketone bodies as their main fuel source do so. Even the brain changes over to get three quarters of its energy needs from ketone bodies. Fat stores can keep a starving human being going for weeks.

Diabetes mellitus is characterized by a similar shift to using ketone bodies. Diabetes arises when either no insulin is produced (Type 1) or cells are unable to respond to this hormone (Type 2). In both cases the body switches to a starvation type metabolism and the ketone bodies 3-hydroxybutyrate and acetoacetate are made to excess. Acetoacetate is chemically unstable and slowly loses carbon dioxide to form acetone. When ketone bodies are at a high concentration, sufficient acetone is present to give the breath the fruity smell of acetone. This pathological condition is described as ketosis and is characteristic of untreated diabetes.

Medieval saints were given to mortification of the body by voluntary starvation. Perhaps the so-called odor of sanctity was simply acetone on the breath of starving and thus ketotic saints?

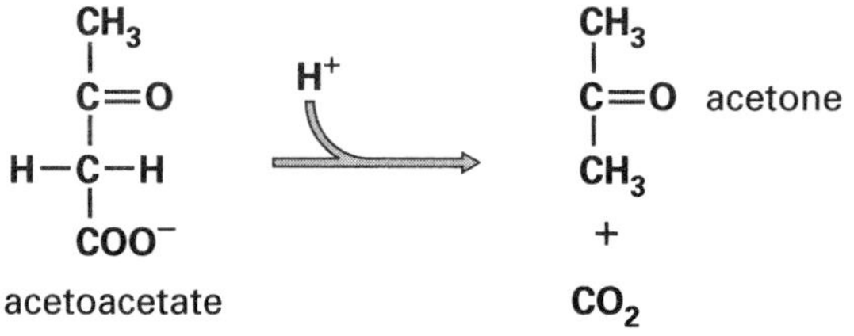

✳ MAKING FATTY ACIDS AND GLYCERIDES

All cells need fatty acids for membrane lipids. Fat cells make large amounts of fat (triacyl-glycerols) in times of plenty. The basic machinery is a multienzyme complex (in bacteria) or a multidomain protein (in eukaryotes) that uses the substrate acetyl-CoA. In both cases the growing fatty acid chain is not released: it swivels from enzyme to enzyme or domain to domain in the array, adding two carbons for each complete cycle until the limiting length of 16 carbons is reached: The product, palmitic acid, is then released. Although the reactions look similar (Fig. 13.12), the process is not a reversal of β oxidation (page 290). It uses entirely different enzymes, takes place in the cytosol rather than in the mitochondria, and is separately regulated. Like much of biosynthesis it is reductive, and the reducing power comes not from NADH but from the closely related dinucleotide NADPH.

Initially acetyl-CoA is carboxylated to malonyl-CoA. From here on, however, fatty acid synthesis does not use free coenzyme A to carry the growing chain but instead uses a protein called acyl carrier protein (ACP). The malonyl residue is transferred to ACP from malonyl-CoA. This condenses with a molecule of acetyl-ACP (made from an acetyl-CoA) to give a four-carbon molecule with the release of ACP and CO_2. The four-carbon acetoacetyl-ACP is next reduced to hydroxybutyryl-ACP. The next enzyme (molecule or domain) removes water, leaving a double bond, which is again reduced to give butyryl-CoA. Another malonyl-CoA is condensed with this and the cycle continues. Finally a chain 16 carbons long has been made (palmitic acid). At this point it is hydrolyzed from the ACP. Overall 14 NADPH molecules, 1 acetyl-CoA, and 7 malonyl-CoA molecules have been used to make palmitic acid. Palmitic acid is then used by enzymes on the endoplasmic reticulum that extend the chains and that can introduce double bonds. Mammals cannot, however, synthesize all the different kinds of fatty acids that they need for their membranes and must obtain essential fatty acids in food (page 42).

The main use of fatty acids is to make glycerides, both triacylglycerols for storage in fat globules and phospholipids for membranes. The process uses glycerol phosphate, which is usually generated by reduction of dihydroxyacetone phosphate. The fatty acid is then swopped in, replacing the phosphate group, which leaves as inorganic phosphate.

Example 13.3 Eating Well, Getting Fat

Glycerol phosphate, the building block from which lipids and phospholipids are made, is obtained in most cells by reducing dihydroxyacetone phosphate. This in turn is generated in the glycolytic pathway. This means that adipose cells can only make fats when glucose is abundant—when we are eating lots of sugar or carbohydrate. In contrast, liver cells have an enzyme called glycerol kinase that phosphorylates glycerol to make glycerol phosphate directly, so that some lipids and phospholipids can be made even during times of fasting.

✳ SYNTHESIS OF AMINO ACIDS

Nitrogen is an important constituent of proteins and nucleic acids and many other molecules important in cells. Although nitrogen gas is plentiful, making up 80% of the atmosphere, it is inert. It is chemically a very difficult task to break the triple bond and reduce nitrogen gas to ammonia that can be used for incorporation into biomolecules. Chemical fertilizers are made using the Haber process, which fixes nitrogen by the use of pressures of around

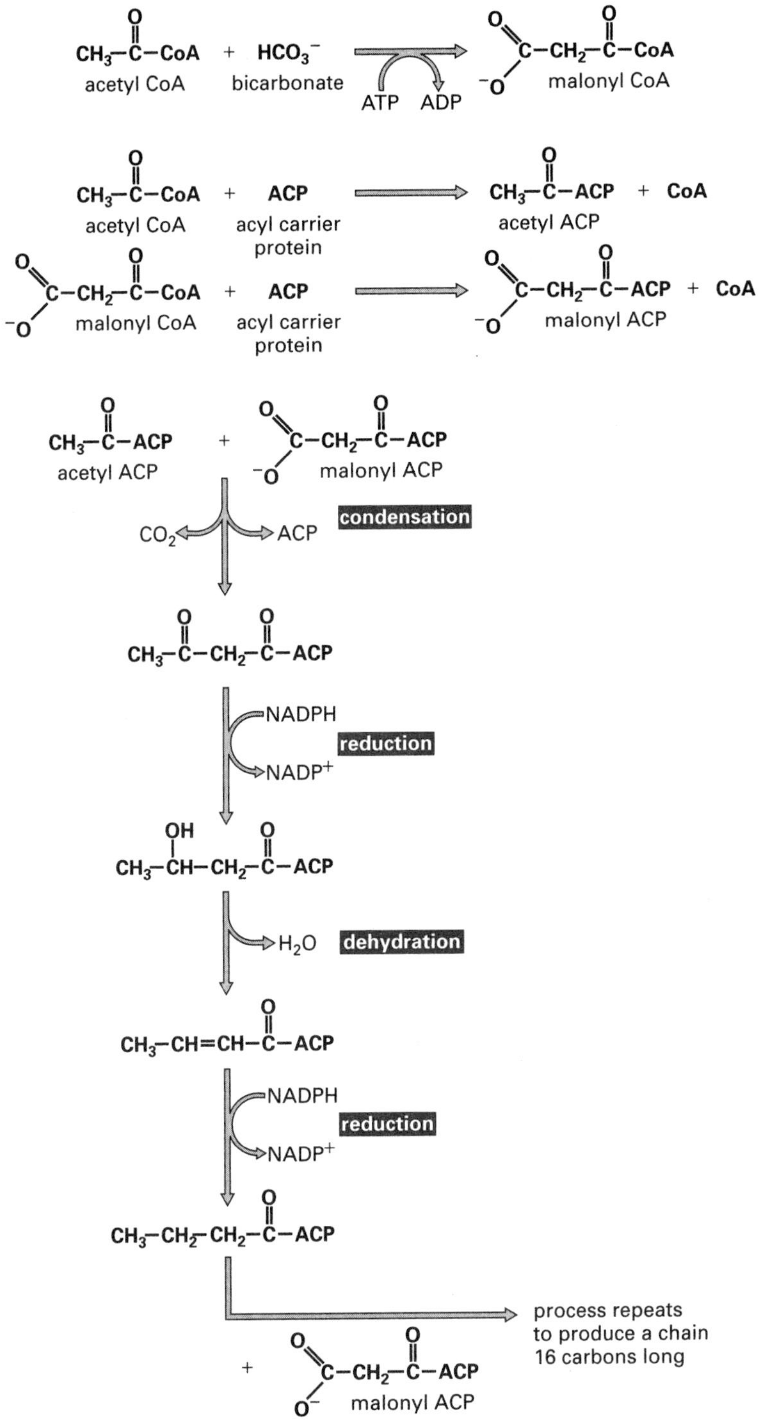

Figure 13.12. Synthesis of fatty acids.

300 atmospheres and temperatures of 500°C. Some nitrogen is fixed naturally by lightning, but most is fixed by various types of prokaryotes, which possess the complex enzyme called nitrogenase. Some of these nitrogen-fixing organisms are free-living while others form symbiotic relationships with plants: the root nodules of legumes are a good example of plants generating a special environment for their nitrogen-fixing symbionts.

Nitrogenase consists of two protein complexes: a reductase and an iron–molybdenum protein. Both proteins have iron–sulfur clusters. The reductase accepts electrons from donors such as ferredoxin (page 275), which can get its electrons through photosynthetic electron transport or from other reactions. The reductase passes electrons to the iron–molybdenum protein (also called dinitrogenase). Nitrogen gas is bound to the iron–molybdenum cofactor where it is reduced to ammonia. Ammonia in water forms ammonium ions $NH_4{}^+$, which can be incorporated into the amino acid glutamine and from there into other amino acids and other molecules.

The overall reaction is

$$N_2 + 8e^- + 8\,H^+ + 16\,\text{ATP} \rightarrow 2NH_3 + H_2 + 16\,\text{ADP} + 16\,\text{Pi}$$

where e^- represents an electron.

Nitrogenase is inhibited by oxygen. Plants with nitrogen-fixing symbiotic bacteria have evolved methods of restricting the oxygen concentration in the vicinity of the bacteria. For example, legumes (the pea family) surround the bacteria with cells that produce a molecule called leghaemoglobin, which is very similar to myoglobin (page 239). The leghaemoglobin binds oxygen, preventing it from reaching the nitrogenase.

Given ammonium ions, plants and bacteria can synthesise all 20 amino acids. Animals are more limited and must obtain some amino acids from their diet. Amino transferases (page 249) allow animals to move amino groups from an amino acid to an oxo-acid to generate a new amino acid, but there are some carbon skeletons that cannot be synthesized. Adult humans require histidine, isoleucine, leucine, lysine, methionine, phenylalanine, threonine, tryptophan, and valine in their diet—these are the **essential amino acids.** Adult humans are in nitrogen balance—we excrete the same amount of nitrogen as we take in. Growing infants, however, have a net uptake of nitrogen: they take in more nitrogen than they excrete. Growth clearly demands more amino acids, and in this case the limited ability we have to synthesize arginine is insufficient and so it must be present in the diet as well.

Other molecules, such as the nucleic acid bases, are synthesized from amino acid starting materials.

CARBON FIXATION IN PLANTS

We have already described how chloroplasts use the energy of light to oxidize H_2O to give O_2 and to generate ATP and NADPH (page 271). A series of so-called **dark reactions** in the chloroplast stroma uses ATP and NADPH in a process that grabs carbon dioxide from the air and builds it into larger molecules, a process called carbon fixation. These reactions are often called the Calvin cycle in honor of their discoverer (Fig. 13.13). The initial step is carried out by the enzyme ribulose bisphosphate carboxylase, said to be the most abundant protein on earth. CO_2 is combined with the five-carbon sugar ribulose bisphosphate to give a six-carbon intermediate that immediately splits into two molecules of 3-phosphoglycerate. Further reactions use NADPH and ATP to convert these three-carbon

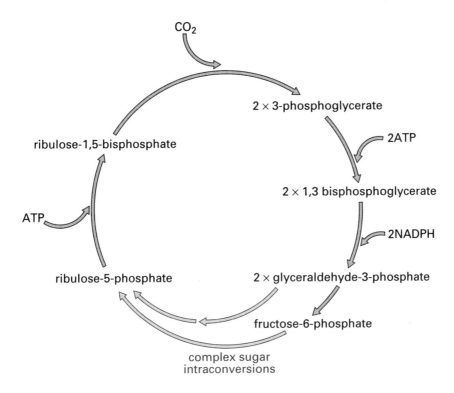

CO$_2$

2 × 3-phosphoglycerate

2ATP

ribulose-1,5-bisphosphate

2 × 1,3 bisphosphoglycerate

2NADPH

ATP

ribulose-5-phosphate 2 × glyceraldehyde-3-phosphate

fructose-6-phosphate

complex sugar
intraconversions

Figure 13.13. The dark reactions of photosynthesis capture carbon dioxide from the air.

units to fructose-6-phosphate. It is interesting that many of these reactions are similar to those of the pentose phosphate pathway (page 289).

✱ CONTROL OF ENERGY PRODUCTION

Feedback and Feedforward

We have seen how an energy currency that runs low is topped up by conversion from another currency. This is not enough to ensure a constant energy supply, however. Therefore the cell has more mechanisms that ensure that the supply of cellular energy is accelerated or slowed as appropriate. These mechanisms are of two types called feedforward and feedback. We will introduce the terms by analogy with real money. Consider a bank teller. During the day people deposit checks but draw out cash to spend. As time passes the stock of banknotes and change in the till gets low. The teller signals to the supervisor, who opens the bank vault, takes out more cash, and refills the teller's till. This is an example of **negative feedback.** In general negative feedback is said to occur when a change in some parameter activates a mechanism that reverses the change in that parameter. We have already met an analogous negative feedback system in the control of tryptophan biosynthesis (page 116). A downward change in the concentration of tryptophan in the bacterial cell activates the mechanism that causes the cell to make more tryptophan. **Positive feedback** is less common in both biology

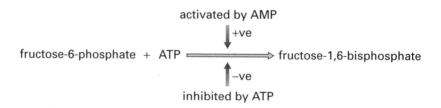

Figure 13.14. Phosphofructokinase is regulated by the binding of ATP or AMP at a regulatory site that is separate from the active site. Binding of ATP inhibits while binding of AMP activates.

and banking. It is said to occur when a change in some parameter activates a mechanism that accelerates the change. In banking this occurs when a rumor starts that a bank is about to fail. The lower a bank's reserves of money get, the more its depositors rush to take their money out before it is too late. Biological examples of positive feedback are unusual. Quorum sensing in luminescent bacteria (page 116) and the action potential (page 325) are two examples of positive feedback that we describe in this book.

What about feedforward? The bank is especially busy at lunchtime, with lots of cash withdrawn between 12:30 and 2:00. During this time, everyone is rushed off their feet. The supervisors do not wait until tellers signal that they are short of cash but instead open the vaults at 12 noon and bring out enough cash to see the tellers through the lunchtime rush. They are preparing for a future drain on cash by stocking up the tills before the drain occurs—this is feedforward. Feedforward happens in biological systems too, as we will see in this section.

Negative Feedback Control of Glycolysis

Phosphofructokinase catalyzes the first irreversible step in glycolysis after the paths from glucose and glycogen converge (Fig. 13.3). The enzyme is allosterically regulated by ATP (Fig. 13.14). When ATP concentrations are high, ATP binds to regulatory sites on phosphofructokinase and locks it into an inactive (low affinity for fructose phosphate) conformation. When the concentration of ATP is reduced, levels of AMP increase (because ADP is converted to ATP and AMP, page 259). AMP will compete with ATP so an increasing number of phosphofructokinase molecules come to have AMP in the regulatory sites, which causes the enzyme to switch to the active, high-affinity conformation. Fructose-1,6-bisphosphate is produced, feeding the glycolytic pathway that in turn feeds the mitochondria with pyruvate for the production of ATP. This process is negative feedback because changes in the concentration of ATP act, through its allosteric action on phosphofructokinase, to reverse the change in ATP concentration.

Feedforward Control in Muscle Cells

When a signal goes out from our brains to the muscles in our legs to tell them to start working, it causes the endoplasmic reticulum to release calcium ions into the cytosol. Calcium is acting as an intracellular messenger, a topic we will cover in more detail in Chapter 16. The increase of calcium activates several processes. It causes the muscle cell to contract, using the energy released by ATP hydrolysis to do mechanical work (page 393). At

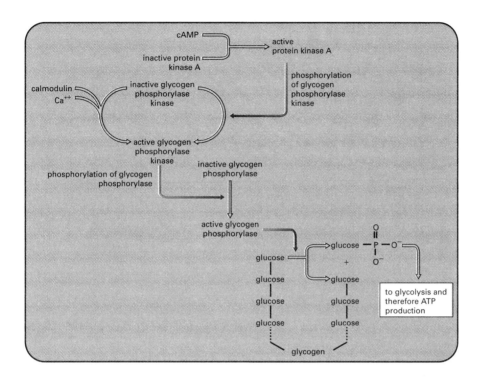

Figure 13.15. Calcium and cyclic-AMP both activate glycogen breakdown in muscle and liver.

the same time, other calcium ions pass through a channel into the mitochondrial matrix, attracted by the large negative voltage of the mitochondrion interior. Once there, calcium activates three key enzymes: pyruvate dehydrogenase, oxoglutarate dehydrogenase, and malate dehydrogenase (pages 286, 284). The cell has not waited until the ATP concentration starts to decrease before activating the Krebs cycle in the mitochondria, so this is feedforward control. Meanwhile, in the cytosol, calcium ions bind to the protein calmodulin (page 201), which in turn binds to and activates an enzyme called **glycogen phosphorylase kinase** (Fig. 13.15). Glycogen phosphorylase kinase is only able to phosphorylate one target, glycogen phosphorylase. Glycogen phosphorylase is activated by phosphorylation and proceeds to break down glycogen to release glucose-1-phosphate, which is fed into the glycolytic pathway. The cell has not waited until glucose concentration falls before activating glycogen breakdown, so this is feedforward control.

In fact, muscles can begin to break down glycogen even before the message goes out from the brain to tell them to contract. When the brain realizes that we are in a dangerous situation and might be going to have to run, it causes the release of the hormone adrenaline from the adrenal glands above the kidneys. Adrenaline binds to an integral membrane protein of the skeletal muscle cells called the β-adrenergic receptor. This causes the production of the intracellular messenger cyclic AMP (cAMP) within the cytosol of the muscle cell. This topic is dealt with in more detail in Chapter 16. cAMP then activates **cAMP-dependent protein kinase,** which is given the short name of **protein kinase A** (Fig. 13.15). Glycogen phosphorylase kinase, the enzyme that phosphorylates glycogen phosphorylase, is itself phosphorylated by protein kinase A, and is then active even when cytosolic calcium is low.

Thus even before we know for sure that we need to run, the muscles are breaking down glycogen and making the glucose they will use if running becomes necessary.

SUMMARY

1. Metabolism is the collective term for all of the reactions going on inside a cell. These reactions are divided into catabolic—those that break down chemical compounds to provide energy—and anabolic—those that build up complex molecules from simpler ones. Catabolic reactions are oxidative and anabolic reactions are reductive.

2. The Krebs cycle is at the center of the cell's metabolism. It can act to oxidize two carbon units derived from carbohydrates, fats, or amino acids and acts as a central switching yard for the molecules in metabolism.

3. Glycolysis converts glucose to pyruvate. If the pyruvate is reduced to lactate, glycolysis can continue in the absence of oxygen as this reduction regenerates the NAD^+ needed for glycolysis.

4. Glycogen can act as a reserve of glucose for glycolysis.

5. Glucose-6-phosphate can be converted to a pentose sugar for nucleotide manufacture with production of NADPH for biosynthesis.

6. Fats (triacylglycerols) are concentrated fuel stores. Their fatty acid components are oxidized to two-carbon units by β oxidation.

7. Amino groups must be removed before excess dietary amino acids can be used as fuels. This is done by transferring them to make aspartate or glutamate, and thence to urea for excretion.

8. Gluconeogenesis allows the synthesis of glucose from noncarbohydrate precursors (but mammals cannot make glucose from fatty acids).

9. Biosynthetic pathways for molecules follow different routes from the catabolic pathways. Good examples are fatty acid synthesis and breakdown (β oxidation) and glycogen synthesis and breakdown.

10. Plants use light energy to generate ATP and reducing power as NADPH and hence to fix carbon dioxide into sugars.

11. Metabolic reactions are controlled by feedforward and feedback mechanisms, which make use of allosteric control and covalent modification of key enzymes.

FURTHER READING

Bender, D. A. 2002. *An Introduction to Nutrition and Metabolism,* 3rd ed. London: Taylor and Francis.

Buchanan, B. B., Gruissem, W., and Jones, R. L. 2000. *Biochemistry and Physiology of Plants.* Rockville, Maryland: American Society of Plant Physiologists.

Devlin, T. M. 2002. *Textbook of Biochemistry with Clinical Correlations,* 5th ed. New York: Wiley.

Voet, D., and Voet, J. D. 2003. *Biochemistry,* 3rd ed. New York: Wiley.

❀ REVIEW QUESTIONS

For each question, choose the ONE BEST answer or completion.

1. Metabolism is
   A. the means by which DNA is stored.
   B. the sum of all the anabolic and catabolic reactions in a cell.
   C. the reactions that break fuel molecules down to gain energy.
   D. the reactions that synthesise molecules in the cell.
   E. the sum of all the reactions taking place in the mitochondrial matrix.

2. The Krebs cycle
   A. breaks down pyruvate.
   B. oxidizes two-carbon units to produce carbon dioxide and reduced coenzymes.
   C. oxidizes glycerol.
   D. drives H^+ out of the mitochondrial matrix.
   E. uses GTP rather than ATP to supply phosphate groups.

3. Which of the following statements about the glycolysis pathway in the cytosol is incorrect?
   A. It makes ATP.
   B. It makes acetyl-CoA.
   C. It interacts with the pentose phosphate pathway.
   D. All but three steps are reversible and therefore also operate in gluconeogenesis.
   E. It is oxidative under aerobic conditions.

4. β oxidation
   A. breaks the β-glycosidic bonds in cellulose.
   B. uses acetyl-CoA, NADH, and $FADH_2$ to produce long-chain fatty acids.
   C. generates pentose sugars.
   D. oxidizes fatty acids to give acetyl-CoA, NADH, and $FADH_2$.
   E. Is part of triacylglycerol biosynthesis.

5. Gluconeogenesis is not capable of making glucose from
   A. alanine.
   B. lactate.
   C. glycerol.
   D. palmitate.
   E. 2-oxoglutarate.

6. Which of the following statements is incorrect concerning photosynthesis in green plants?
   A. It generates ATP, NADPH, and oxygen.
   B. It adds carbon dioxide to a five-carbon sugar that is then cleaved to form two molecules of phosphoglycerate.
   C. Its enzymes are found in the mitochondria.
   D. It uses light energy to drive the splitting of water.
   E. Its enzymes are found in the chloroplasts.

7. Positive feedback is operating
   A. when adenosine monophosphate activates phosphofructokinase.
   B. when cAMP activates transcription of the *lac* operon.

C. when tryptophan inhibits transcription of the *trp* operon.

D. when *N*-acyl-HSL promotes transcription of the *lux* operon.

E. in all the examples above.

8. ATP

A. is converted to ADP as H^+ enters the mitochondrial matrix.

B. is a coenzyme.

C. passes through the outer mitochondrial membrane by simple diffusion.

D. cannot be made in the absence of oxygen.

E. concentrations fall to a low level in resting cells.

ANSWERS TO REVIEW QUESTIONS

1. *B.* It is the sum of all of the catabolic and anabolic reactions in a cell.

2. *B.* The Krebs cycle oxidizes acetyl-CoA and two carbon dioxides are lost for each turn of the cycle. The cycle does not use pyruvate, rather, pyruvate is first converted to acetyl-CoA by pyruvate dehydrogenase, and it is acetyl-CoA that is fed into the cycle. Concerning answer D: the electron transport chain drives H^+ out of the mitochondrial matrix. ATP synthase can also (by running in reverse) perform this function, but the enzymes of the Krebs cycle certainly do not. Concerning answer E: the Krebs cycle does not use either GTP or ATP to supply phosphate groups, though it does generate one GTP from GDP (step 5 in Fig. 13.2).

3. *B.* Glycolysis generates pyruvate. Pyruvate can then be converted to acetyl-CoA, but this takes place in mitochondria, not in the cytosol, and is not regarded as being part of the glycolysis pathway.

4. *D.* β oxidation produces acetyl-CoA from fatty acids and also produces NADH and $FADH_2$.

5. *D.* Glucose cannot be produced from any fatty acid, including palmitate.

6. *C.* Photosynthesis takes place in chloroplasts.

7. *D.* One of the genes within the *lux* operon is *luxI*, which codes for the enzyme that makes *N*-acyl-HSL. This is therefore an example of positive feedback. Concerning the other examples (A and C): these are examples of negative feedback. Activation of phosphofructokinase will cause more fuel to be sent to the mitochondria and will therefore help the mitochondria convert AMP and ADP to ATP. Inhibiting transcription of the *trp* operon will reduce the rate at which the cell synthesizes tryptophan. (B) This is a sensible control mechanism, causing the cell to use glucose when it is available in preference to lactose. However, there is no feedback involved: the products of the *lac* operon are concerned with lactose metabolism, while cAMP levels in bacteria are affected by the glucose concentration.

8. *B.* A coenzyme is a molecule that acts as a second substrate for many enzymes; ATP is a substrate for a wide range of enzymes; some we have already met are polynucleotide kinase (page 139), hexokinase (page 284), phosphofructokinase (page 285), glycogen phosphorylase kinase (page 305), and cAMP-dependent protein kinase (page 305). Answer A is false: As H^+ enters the mitochondria ADP is converted to ATP, not the other way around. Answer C is false: ATP is a highly charged ion and will certainly not pass through membranes by simple diffusion; it crosses the outer mitochondrial membrane by passing through porin. Answer D is false: in the absence of oxygen ATP can be made by anaerobic glycolysis. Answer E is false: in a healthy cell none of the energy currencies are allowed to run down.

14

IONS AND VOLTAGES

We described in Chapter 2 how membranes are composed of phospholipids arranged so that their hydrophobic tails are directed toward the center of the membrane, while the polar hydrophilic head groups face out. Membranes are a barrier to the movement of many solutes. In particular, small hydrophilic solutes such as ions and sugars cannot pass through membranes easily because, to do so, they would have to lose the cloud of water molecules that forms their hydration shell (page 21). Two consequences follow from the fact that membranes are barriers. First, the composition of the liquid on one side of a membrane can be different from the composition of the liquid on the other side. Indeed, by allowing cells to retain proteins, sugars, ATP, and many other solutes, the barrier property of the cell membrane makes life possible. Table 14.1 shows how five important ions have different concentrations in cytosol and extracellular medium. Second, the cell must make proteins called channels and carriers whose job it is to help hydrophilic solutes across the membrane. This chapter describes the properties of membranes, with particular emphasis on their role in energy storage.

❋ THE POTASSIUM GRADIENT AND THE RESTING VOLTAGE

Ions are electrically charged. This fact has two consequences for membranes. First, the movement of ions across a membrane will tend to change the voltage across that membrane. If positive ions leave the cytosol, they will leave the cytosol with a negative voltage, and vice versa. Second, a voltage across a membrane will exert a force on all the ions present. If the cytosol has a negative voltage, then positive ions such as sodium and potassium will

Cell Biology: A Short Course, Second Edition, by Stephen R. Bolsover, Jeremy S. Hyams,
Elizabeth A. Shephard, Hugh A. White, Claudia G. Wiedemann
ISBN 0-471-26393-1 Copyright © 2004 by John Wiley & Sons, Inc.

Table 14.1. Typical Concentrations for Five Important Ions in Mammalian Cytosol and Extracellular Medium[a]

Ion	Cytosol	Extracellular Medium
Sodium Na$^+$	10 mmol liter^{-1}	150 mmol liter^{-1}
Potassium K$^+$	140 mmol liter^{-1}	5 mmol liter^{-1}
Calcium Ca^{2+}	100 nmol liter^{-1}	1 mmol liter^{-1}
Chloride Cl$^-$	5 mmol liter^{-1}	100 mmol liter^{-1}
Hydrogen ion H$^+$ (really H$_3$O$^+$)	63 nmol liter^{-1} or pH 7.2	40 nmol liter^{-1} or pH 7.4

[a]Note the units n for nano is one million times smaller than m for milli.

be attracted in from the extracellular medium. In this chapter, we will begin to address the question of how ions and voltages interact by considering the effect of potassium movements on the voltage across the plasma membrane.

Potassium Channels Make the Plasma Membrane Permeable to Potassium Ions

The potassium channel (Fig. 14.1) is a protein found in the plasma membrane of almost all cells. It is a tube that links the cytosol with the extracellular fluid. Potassium ions, which cannot pass through the lipid bilayer of the plasma membrane, pass through the potassium

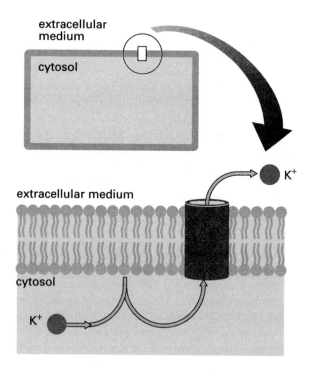

Figure 14.1. The positively charged potassium ion cannot cross the lipid bilayer but passes easily through a water-filled tube in the potassium channel.

channel easily. Other ions cannot go through. The precise shape of the tube, and the position of charged amino acid side chains within the tube, blocks their movement. The channel is selective for potassium.

We saw earlier that the Na^+/K^+ ATPase (page 270) uses the energy of ATP hydrolysis to drive sodium ions out of the cell and, at the same time, brings potassium ions into the cell. This ensures that potassium is much more concentrated in the cytosol than outside—typically 140 mmol liter^{-1} in the cytosol but only 5 mmol liter^{-1} in the extracellular medium. There is an apparent paradox here. If potassium can pass through the potassium channel, why is this ion much more concentrated inside the cell than outside? Why doesn't all the potassium rush out? To explain why, we must think about the effects of ion movement on transmembrane voltage. First a word on nomenclature. All cells have a voltage across their membrane when they are not being stimulated. This **resting voltage** is about −80 mV in a relaxed skeletal muscle cell. As soon as the muscle is stimulated to contract, there is a sudden sharp change in the transmembrane voltage called the action potential (a process described in more detail in Chapter 15). Many cells never change their transmembrane resting voltage.

Concentration Gradients and Electrical Voltage Can Balance

A few potassium ions do escape from the cell through the potassium channel. As they do so, they carry out their positive charge and leave the cytosol with a negative voltage that attracts positively charged ions like potassium. There is still a tendency for potassium ions to leave the cell down the concentration gradient, but there is now an electrical force pulling the positively charged potassium ions back inside. We have met a similar situation before, where we described how a concentration gradient and an electrical force combine to form an electrochemical gradient down which H^+ ions will rush into bacteria or mitochondria (page 261). However, in the case of potassium ions at the plasma membrane, the concentration gradient and the electrical force act in opposite directions. As potassium ions continue to leave the cell, carrying out their positive charge and leaving the cytosol at a more and more negative voltage, the electrical force pulling them back in gets increasingly strong. Soon, the opposing electrical and concentration gradients are equal, and the overall electrochemical gradient for potassium is zero. Potassium ions then stop leaving the cell, even though they are much more concentrated inside than outside.

For every ion present on both sides of a membrane, it is possible to calculate the transmembrane voltage that will exactly balance the concentration gradient. This voltage is called the **equilibrium voltage** for that ion at that membrane. For potassium at the plasma membrane of a normal animal cell, the equilibrium voltage is about −90 mV.

The departure of potassium ions through the potassium channels, leaving negative charge behind, produces the resting voltage across the plasma membrane. Because the potassium channels are the major pathway by which ions can cross the plasma membrane of an unstimulated cell, the resting voltage has a value close to the potassium equilibrium voltage. In some cells, such as white blood cells, potassium channels are the only channels in the plasma membrane, and potassium ions move out until the transmembrane voltage tending to pull them back in exactly balances their tendency to move out down their concentration gradient. The resting voltage of white blood cells therefore has a value equal to the potassium equilibrium voltage, about −90 mV. In other cells the situation is more complicated. In muscle cells, for instance, the resting voltage is −80 mV; in nerve cells it is −70 mV. Even in these cells, though, the major influence on the resting voltage is potassium movement through its channels, so the resting voltage does not deviate very far from the potassium equilibrium voltage.

The resting voltage set up by potassium movement turns the action of the Na^+/K^+ ATPase into an energetically asymmetrical one. Consider one conversion cycle: one molecule of ATP is hydrolyzed, three sodium ions are pushed out of the cell, and two potassium ions move in. Very little of the energy of ATP hydrolysis is used up in moving the two potassium ions into the cell because the electrochemical gradient for this ion is close to zero. Although potassium is being moved up a concentration gradient, it is also being pulled in by the negative voltage of the cytosol, and the two forces cancel. In contrast, pushing the three sodium ions out of the cell requires more energy than would be required if the cytosol were at the same voltage as the extracellular fluid. The sodium ions are positively charged, so they are attracted by the negative voltage inside the cell, which combines with the concentration gradient to form a large inward electrochemical gradient. All the energy released by ATP hydrolysis is needed to push the three sodium ions up this large electrochemical gradient. The presence of the potassium channels, and the resting voltage that they set up, means that almost all the energy of ATP hydrolysis by the Na^+/K^+ ATPase is stored in the sodium gradient, while potassium ions are close to equilibrium.

●●● IN DEPTH 14.1 The Nernst Equation

An ion that can pass across a membrane is acted on by two forces. The first derives from the concentration gradient. The ion tends to diffuse from a region where it is at high concentration to one where it is at low concentration. The second force derives from the transmembrane voltage. In the case of positively charged ions such as Na^+ and K^+, the ions tend to move toward a negative voltage. Negatively charged ions such as Cl^- tend to move toward a positive voltage. For each ion there is a value of the transmembrane voltage for which these forces balance, and the ion will not move. The ion is said to be at equilibrium, and this value of the transmembrane voltage is called the equilibrium voltage for that ion at that membrane.

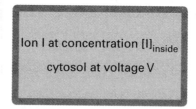

Ion I at concentration $[I]_{outside}$

Ion I at concentration $[I]_{inside}$

cytosol at voltage V

When the forces balance, then ions that move in will neither gain nor lose energy. This way of describing equilibrium is useful because it allows us to set equivalent the effects of the two very different gradients, concentration and voltage. For concentration, the free energy possessed by a mole of ions I by virtue of its concentration is

$$G = G_0 + RT \log_e[I] \quad \text{joules}$$

where G_0 is the standard free energy, R is the gas constant (8.3 J mol^{-1} $degree^{-1}$), and T is the absolute temperature.

A mole of I passing in therefore moves from a region where it had a free energy of

$$G_{outside} = G_0 + RT \log_e[I_{outside}] \quad \text{joules}$$

to one where its free energy is

$$G_{inside} = G_0 + RT \log_e[I_{inside}] \quad \text{joules}$$

One mole of ions I moving inward therefore gains by virtue of the concentration gradient free energy equal to

$$RT \log_e[I_{inside}] - RT \log_e[I_{outside}] \quad \text{joules}$$

Now consider the electrical force. The definition of a volt means that one coulomb of charge moving across a membrane with a transmembrane voltage of V volts gains V joules of free energy. However, we are working in moles, not coulombs. One mole of ions has a charge of zF coulombs, where z is the charge on the ion in elementary units. For Na^+ and K^+ z is 1; for Ca^{2+} z is 2; and for Cl^- z is -1. The term F is a number that relates the coulomb to the mole. It has the value 96,500. One mole of ions I moving inward gains by virtue of the transmembrane voltage free energy equal to

$$zF V \quad \text{joules}$$

This does not mean that an ion always gains energy from the transmembrane voltage when it moves inward: the term $zF V$ can just as easily be negative as positive.

When the effects of concentration and voltage just balance, a mole of ions moving inward neither gains nor loses free energy. Hence, at equilibrium

$$RT \log_e[I_{inside}] - RT \log_e[I_{outside}] + zF V_{eq} = 0$$

This can be simplified to

$$V_{eq} = \frac{RT}{zF} \log_e \left(\frac{[I_{outside}]}{[I_{inside}]} \right) \quad \text{volts}$$

This is the Nernst equation. At a typical mammalian body temperature of 37°C, the Nernst equation can be written in the more convenient form:

$$V_{eq} = \frac{62}{z} \log_{10} \left(\frac{[I_{outside}]}{[I_{inside}]} \right) \quad \text{mvolts}$$

Note the change in the type of logarithm and the units.

"In" and "out" can refer to any two solutions separated by a membrane. At the plasma membrane *in* is the cytosol and *out* is the extracellular medium, but when considering equilibria across the inner mitochondrial membrane *in* is the mitochondrial matrix and *out* is the intermembrane space.

❈ THE CHLORIDE GRADIENT

Chloride ions (Fig. 14.2) are at a lower concentration in the cytosol than in the extracellular medium. Typically, their concentration in the cytosol is 5 mmol liter^{-1} compared with 100 mmol liter^{-1} in the extracellular fluid. This is because of the resting voltage set up by the potassium channels. Chloride ions are repelled by the negative voltage of the cytosol. They leave until their tendency to enter the cell down their concentration gradient exactly matches their tendency to be repelled by the negative voltage of the cytosol.

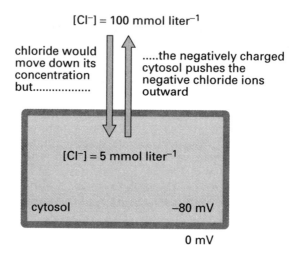

Figure 14.2. Chloride is close to equilibrium across most plasma membranes.

❈ GENERAL PROPERTIES OF CHANNELS

Channels are integral membrane proteins that form water-filled tubes through the membrane. We have already introduced three: the gap junction channel (page 55), porin (page 262), and the potassium channel. Channels that, like the potassium channel, are selective for particular ions can set up transmembrane voltages. The gap junction channel (page 55) is much less selective than the potassium channel. It forms a tube, 1.5 nm in diameter, through which any solute of $M_r \leq 1000$ can pass. The gap junction channel is not always open. It opens only when it connects with a second gap junction channel on another cell, forming a tube through which solutes can pass from the cytosol of one cell to the cytosol of the other. Channels that are sometimes open and sometimes shut are said to be **gated**. When a gap junction channel contacts another on another cell, its gate opens and solute can pass through; at other times the gate is shut. The usefulness of gating is obvious: if the gap junction channels not contacting others were open, many solutes, including ATP and sodium, would leak out into the extracellular fluid exhausting the cell's energy currencies.

Porin in the outer mitochondrial membrane plays an important role in energy conversion. It forms a very large diameter tube that allows all solutes of $M_r \leq 10,000$ to pass and seems to spend a large fraction of time open under most circumstances. This is why the outer mitochondrial membrane is permeable to most solutes and ions. Appendix 1, at the end of this book, lists all the different types of channels described in this book. It represents only a small fraction of the total number known.

Example 14.1 **Cytochrome *c*—Vital But Deadly**

We have described how the electron carrier cytochrome *c* resides in the intermembrane space between the outer and inner mitochondrial membranes and helps the electron transport chain to convert energy as NADH to energy as the hydrogen ion electrochemical gradient across the mitochondrial inner membrane (page 266). Although cytochrome *c* is a soluble protein of relative molecular mass 12,270, it cannot escape from the intermembrane space into the cytosol because porin, the channel of the outer mitochondrial membrane, only allows solutes of $M_r \leq 10,000$ to pass. Although cytochrome *c* is essential for mitochondrial function, it has another, deadly role. If cytochrome *c* comes into contact with a class of cytosolic enzymes called caspases, it activates them, turning on the process of cell suicide called apoptosis (page 417). Under certain conditions, porin can associate with other proteins to form a channel of larger diameter; when this happens, cytochrome *c* can leak out and the cell dies by apoptosis. This process seems to occur in hearts during heart attacks, and in the brain during a stroke: there is therefore a considerable research effort aimed at preventing this from occurring.

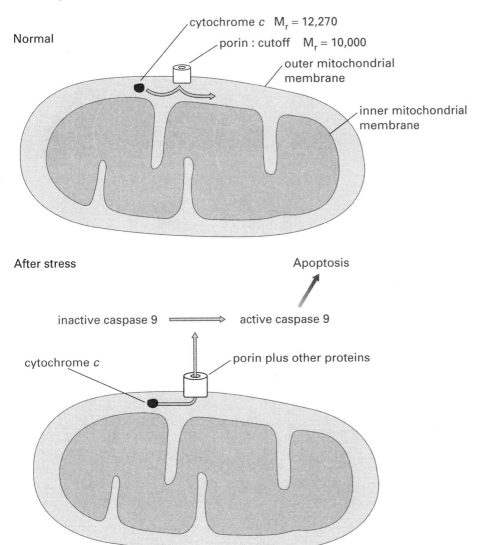

✽ GENERAL PROPERTIES OF CARRIERS

We have already met the three carriers that interconvert the four energy currencies of the cell (page 263). Carriers are like channels in that they are integral membrane proteins that allow solute to cross the membrane and, like channels, they form a tube across the membrane. However, there is a critical difference. In carriers the tube is never open all the way through; it is always closed at one or other end. Solutes can move into the tube through the open end. When the carrier changes shape, so that the end that was closed is open, the solute can move into the solution on the other side of the membrane.

The Glucose Carrier

One of the simplest carriers is the glucose carrier (Fig. 14.3). It switches freely between a form that is open to the cytosol and a form that is open to the extracellular medium. Inside the tube is a site to which a glucose molecule can bind. On the left, a glucose molecule is entering

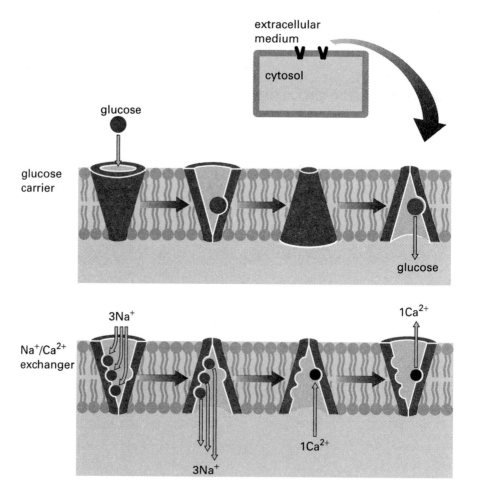

Figure 14.3. The glucose carrier and the sodium/calcium exchanger are sometimes open to the cytosol and sometimes open to the extracellular medium.

the tube and binding to the site. Sometimes glucose leaves the binding site before the carrier switches shape. In Figure 14.3 we see the other possibility: the carrier switches shape before the glucose has left. The binding site is now open to the cytosol, and the glucose can escape into the cytosol. It has been carried across the plasma membrane. Unlike channels, carriers never form open tubes all the way across the membrane. Instead, they bind one or more molecules or ions, then change shape to carry the molecules or ions across the membrane.

Example 14.2 The Glucose Carrier Is Essential

The cells of our bodies are bathed in a glucose-rich solution. However, glucose cannot cross the plasma membrane by simple diffusion because it is strongly hydrophilic: It can only get in via the glucose carrier. Some cells have glucose carriers in their membranes all the time, but others such as muscle and fat cells translocate the glucose carrier to the plasma membrane only in the presence of insulin. Insulin-dependent diabetics cannot produce their own insulin, and unless they inject synthetic insulin, their muscles and fat cells cannot take up glucose and therefore run out of energy, even though the concentration of glucose in the blood becomes very high. This is why muscular weakness is one symptom of diabetes.

The glucose carrier is very simple, whereas other carriers are more complex. We will next consider two carriers, the **sodium/calcium exchanger** and the calcium ATPase, which do much the same job—they push calcium ions up their concentration gradient out of the cell—but which take their energy from different currencies. The sodium/calcium exchanger uses the sodium concentration gradient while the calcium ATPase uses ATP.

The Sodium–Calcium Exchanger

Figure 14.3 shows that like the glucose carrier, the sodium/calcium exchanger can exist in two shapes, one open to the extracellular medium and one open to the cytosol. Inside the tube are three sites that can bind sodium ions and one site that can bind a calcium ion. The sodium/calcium exchanger is not free to switch between its two shapes at any time. Instead, it switches only if either all the sodium sites are filled and the calcium site is empty or if the calcium site is filled and all the sodium sites are empty.

On the left, the carrier is open to the extracellular medium. It can switch its shape so that it is open to the cytosol if one of two things happen. It could, as shown, bind three sodium ions (keeping the calcium site empty) or it could bind one calcium and keep the sodium sites vacant. This second option does not often happen, since sodium is at high concentration in the extracellular fluid, and one or more of the sodium sites is usually occupied. Therefore nearly all the switches from open-to-outside to open-to-inside are of the type shown. Once the tube has opened to the low sodium environment of the cytosol, the sodium ions tend to leave.

Once the carrier is open to the cytosol, it can switch back to the open-to-outside form by binding either one calcium or three sodium ions. Since sodium is scarce in the cytosol, the latter event is unlikely. More frequently a calcium ion will bind and will be carried out. The carrier is now ready to bind sodium again and perform another cycle.

The overall effect of one cycle is to carry three sodium ions into the cell down their electrochemical gradient, and one calcium ion out of the cell up its electrochemical gradient. A simple rule about when the carrier can switch shape has produced a machine that uses the energy currency of the sodium gradient to do work in pushing calcium ions out of the cell. Figure 14.4 shows a simple way of representing what is happening. The circle represents one cycle of operation, from open to extracellular medium back to open to extracellular medium again.

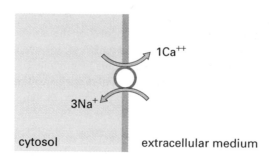

Figure 14.4. Action of the sodium/calcium exchanger.

Carriers with an Enzymatic Action: The Calcium ATPase

The electron transport chain, ATP synthase, and the Na^+/K^+ ATPase are carriers with an additional level of complexity in that they carry out an enzymatic action as well as their carrier function. Figure 14.5 illustrates another carrier with a linked enzymatic function: the Ca^{2+} ATPase. Like the sodium/calcium exchanger this is found in the plasma membrane. The transmembrane part of this carrier forms a tube that can be open to the cytosol or to the extracellular medium. Two $-COO^-$ groups in the tube can form a binding site for a calcum ion. Other domains of the protein lie in the cytosol and can hydrolyze ATP. Changes in the shape of the cytosolic region are transmitted to the transmembrane region and force it between the open-to-cytosol and open-to-outside shapes. Figure 14.5 shows our present understanding of how this might happen. (1) is the relaxed shape of the protein. ATP is held by noncovalent interactions in a crevice in one of the cytosolic domains. One of the COO^- groups in the tube is protonated at this time. (2) A calcium ion moving in from the cytosol pushes the tube open and binds to both binding sites inside the tube. The positively charged calcium ion displaces an H^+ ion and is then able to interact with both COO^- groups. The distortion in protein shape caused by the calcium ion pushing the tube open is transmitted to the cytosolic region and brings the γ phosphate of ATP close to an aspartate residue on a neighboring domain. (3) An intrinsic kinase activity of the two domains transfers the γ-phosphate group from ATP to the aspartate residue. As the insert shows, this means that a phosphate group, with two negative charges, is sitting very close to ADP, with three negative charges. (4) The repulsion between the phosphorylated aspartate and the ADP forces the two domains apart. The shape change is transmitted to the transmembrane region, which is forced into a wide open-to-outside shape. The calcium ion, which can no longer bind to the COO^- groups on both sides of the tube, is held only weakly and tends to escape into the extracellular medium. An H^+ ion moves in to protonate one of the COO^- groups. (5) A third cytosolic domain, which has phosphatase catalytic ability, swings in and dephosphorylates the aspartate residue. Meanwhile, ATP moves in and displaces ADP at the binding crevice. Once the aspartate has been dephosphorylated, the protein is no longer held in the open-to-outside shape by electrical repulsion and relaxes back to the resting shape shown in (1). It is now ready to accept another calcium ion from the cytosol and repeat the process.

Figure 14.6 summarizes what is happening. The circle represents one cycle of operation. One molecule of ATP is hydrolyzed to ADP and inorganic phosphate, one calcium ion moves out of the cell, and one H^+ moves in. The energy released by ATP hydrolysis has been used to drive one calcium ion up its electrochemical gradient out of the cell.

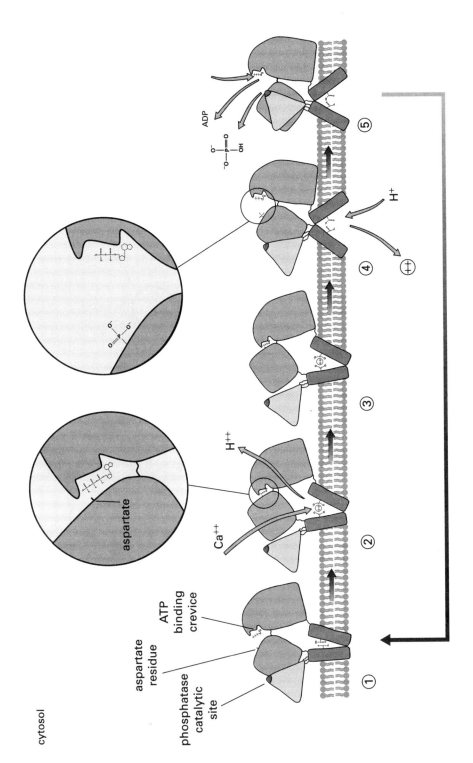

cytosol

aspartate
residue

ATP
binding
crevice

phosphatase
catalytic
site

aspartate

① ② ③ ④ ⑤

Ca++

H++

H+

⊕

ADP

Figure 14.5. Calcium ATPase undergoes a cycle of phosphorylation and dephosphorylation. These drive shape changes that in turn push calcium ions out of the cell.

Medical Relevance 14.1

Poisoned Hearts Are Stronger

Digitalis is used to treat heart failure. Digitalis inhibits the sodium/potassium ATPase and is extremely toxic. Nevertheless, a small dose, which inhibits the sodium/potassium pump just a little, causes the heart muscle to beat more strongly. The reason is that inhibiting the sodium/potassium pump just a little causes a small increase of cytosolic sodium concentration. Because the sodium/calcium exchanger has three binding sites for sodium, its activity is extremely sensitive to sodium concentration, and even a small increase of cytosolic sodium reduces its activity significantly. The calcium concentration in the cytosol therefore rises. The mechanical motor that drives heart contraction (Chapter 18) is controlled by calcium, so that a small increase of cytosolic calcium makes the heart beat more strongly.

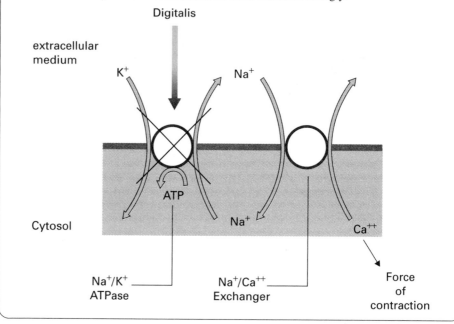

All the cells of our bodies have one of these two calcium pumps—the sodium/calcium exchanger or the calcium ATPase—and many have both. Because of the action of these carriers, the calcium concentration in the cytosol is much less than the concentration in the extracellular medium: usually about 100 nmol liter^{-1} compared with 1 mmol liter^{-1}. Because the resting voltage is attracting the positively charged calcium ions inward, the overall result is a large electrochemical gradient favoring calcium entry into cells.

●●●● IN DEPTH 14.2 Measuring the Transmembrane Voltage

In 1949 Gilbert Ling and Ralph Gerard discovered that when a fine glass micropipette filled with an electrically conducting solution impaled a cell, the plasma membrane sealed to the glass, so that the transmembrane voltage was not discharged. The voltage difference between a wire inserted into the micropipette and an electrode in the extracellular fluid could then be measured. By passing current through the micropipette, the transmembrane voltage could be altered.

Twenty-five years later Erwin Neher and Bert Sakmann showed that the micropipette did not have to impale the cell. If it just touched the cell, a slight suction caused the plasma membrane to seal to the glass. The technique, called

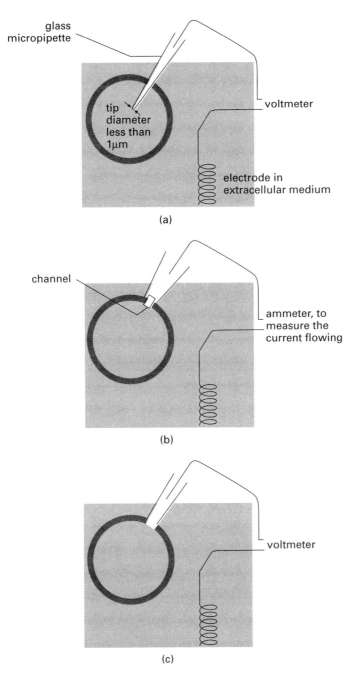

glass
micropipette

tip
diameter
less than
1μm

voltmeter

electrode in
extracellular medium

(a)

channel

ammeter, to
measure the
current flowing

(b)

voltmeter

(c)

cell-attached patch clamping, can measure currents through the few channels present in the tiny patch of membrane within the pipette.

Stronger suction bursts the membrane within the pipette. The transmembrane voltage can now be measured.

Alternatively, current can be passed through the micropipette to change the transmembrane voltage—this is the whole cell patch clamp technique. In 1991, Neher and Sakmann received the Nobel prize for medicine.

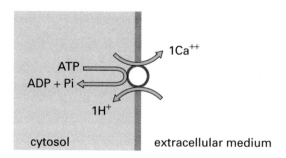

Figure 14.6. Action of the calcium ATPase.

SUMMARY

1. Channels are membrane proteins with a central water-filled hole through which hydrophilic solutes, including ions, can pass from one side of the membrane to the other. Changes in protein structure may act to gate the channel but are not required for movement of the solute from one side to the other.

2. The presence of potassium channels in the plasma membrane, and the resting voltage that they set up, means that almost all the energy of ATP hydrolysis by the Na^+/K^+ ATPase is stored in the sodium gradient, while potassium ions are close to equilibrium.

3. The resting voltage repels chloride ions from the cell interior.

4. Carriers, like channels, form a tube across the membrane, but the tube is always closed at one end. Solutes can move into the tube through the open end. When the carrier changes shape, so that the end that was closed is open, the solute can leave to the solution on the other side of the membrane.

5. The glucose carrier is present in the plasma membrane of all human cells.

6. The sodium/calcium exchanger is a carrier that uses the energy of the sodium gradient to push calcium ions out of the cell.

7. The electron transport chain, ATP synthase, the sodium/potassium ATPase, and calcium ATPase are both carriers and enzymes, the two actions being tightly linked.

FURTHER READING

Ashcroft, F. M. 2000. *Ion Channels and Disease*. San Diego: Academic Press.

Levitan, I. B., and Kaczmarek, L. K. 2002. *The Neuron*. New York: Oxford University Press.

✳ REVIEW QUESTIONS

For each question, choose the ONE BEST answer or completion.

1. Which of the following statements about potassium ions is false?

   A. Potassium ions are more concentrated in the extracellular medium than in cytosol.

B. Potassium ions are moved into cells by the Na^+/K^+ ATPase.

C. Most plasma membranes contain potassium-selective channels.

D. Potassium would flow out of most cells rapidly but for the effect of the negative resting voltage

E. Answers A though D are false.

2. Which of the following statements is false?

A. The potassium channel allows potassium ions to pass from one side of the plasma membrane to the other.

B. Porin allows chloride ions to pass from one side of the inner mitochondrial membrane to the other.

C. The glucose carrier allows glucose to pass from one side of the plasma membrane to the other.

D. The gap junction channel allows potassium ions to pass from the cytosol of one cell to the cytosol of its neighbor.

E. Thermogenin allows H^+ to pass from one side of the inner mitochondrial membrane to the other.

3. The resting voltage

A. is between -70 and -90 mV in most cells.

B. attracts chloride ions in from the extracellular medium.

C. is mainly determined by the chloride concentration gradient across the plasma membrane.

D. helps retain ATP and glucose within cells.

E. helps keep the sodium concentration low in the cytosol.

4. Solute movement through which channels is mainly responsible for the resting voltage?

A. Porin

B. Gap junction channels

C. Glucose

D. Potassium

E. All of the above contribute equally to the resting voltage.

5. Which of the following statements is false?

A. Channels form water-filled tubes through the membrane.

B. Channels are often gated.

C. Channels are only found in the plasma membrane.

D. Channels cannot transport solute up an electrochemical gradient.

E. Channels are often selective for a particular solute or type of solute.

6. Carrier proteins

A. never form a tube open all the way through a membrane.

B. can transport solute up an electrochemical gradient.

C. can often carry out an enzymatic function in addition to their carrier function.

D. are found in the plasma membrane and in intracellular membranes.

E. all of the above.

7. An ion X^+ is at equilibrium across the plasma membrane of a cell. The cell is voltage clamped, that is, the value of the transmembrane voltage is set by the experimenter, not by the properties of the membrane. Which of the following would cause X^+ to no longer be at equilibrium?

A. A change of transmembrane voltage

B. A change in the concentration of X^+ in the extracellular medium

C. Either answer A or B

D. Closure of the channels that allow X^+ to cross the membrane

E. A change in the selectivity of the channel that allows X^+ to cross the membrane so that Y^+ can cross as well

F. Either answer D or E

8. Which of the following is not a carrier?

A. The electron transport chain

B. Hexokinase

C. ATP synthase

D. The sodium/potassium ATPase

E. Calcium ATPase

F. The sodium/calcium exchanger

G. The ADP/ATP exchanger

ANSWERS TO REVIEW QUESTIONS

1. *A.* Potassium ions are more concentrated in the cytosol due to the action of the Na^+/K^+ ATPase.

2. *B.* Although porin will certainly allow chloride ions to pass, it is present in the outer mitochondrial membrane, not the inner. All the other statements are true, although, regarding answer D, note that the gap junction channel will allow many different solutes to pass; it is not in any way selective for potassium.

3. *A.* All the others are false. In particular, (B) chloride ions are repelled from the negative cytosol; (C) the resting voltage is mainly determined by the potassium concentration gradient across the plasma membrane; (D) ATP is negatively charged and is therefore not attracted by a negatively charged cytosol, it is retained by the cytosol because it is not able to cross the plasma membrane. Glucose is uncharged and is therefore unaffected by the transmembrane voltage; it is retained within cells because it is rapidly phosphorylated to glucose-6-phosphate, which is not able to cross the plasma membrane. (E) Sodium is attracted *into* cells by the negatively charged cytosol; the reason it has a low concentration in the cytosol is because it is pushed out of cells by the Na^+/K^+ ATPase.

4. *D.* Because most cells express potassium channels in their plasma membranes, potassium is the ion that can most easily cross the plasma membrane and therefore is mainly responsible for the resting voltage. Concerning answer C, note that there is no "glucose channel," although large channels, including porin and the gap junction channel, can carry glucose along with many other solutes.

5. *C.* Channels are found in all intracellular membranes. We have already discussed porin, a channel of the outer mitochondrial membrane. All the other statements are true. In particular, channels cannot transport solute up an electrochemical gradient; they simply make it easier for solute to pass down its electrochemical gradient.

6. *E.* Answers A through D are true. In particular, carrier proteins can transport solute up an electrochemical gradient by linking the movement with another process with negative ΔG. In most cases this is either the hydrolysis of ATP to ADP and inorganic phosphate or the movement of another solute down its electrochemical gradient. The electron transport chain links the movement of H^+ up its electrochemical gradient with the reaction

$$\text{NADH} + H^+ + \frac{1}{2}O_2 = \text{NAD}^+ + H_2O \quad \Delta G = -206 \text{ kJ mol}^{-1}$$

7. *C.* If X^+ is at equilibrium across the plasma membrane, its tendency to move one way down its concentration gradient is matched by its tendency to move the other way down the voltage gradient. Changing the amplitude of either gradient will cause these two effects to no longer balance. Neither answer D nor E is true. If X^+ is at equilibrium, then it will remain at equilibrium irrespective of how easily it or other ions can cross the membrane.

8. *B.* Hexokinase is a soluble protein of the cytosol; it catalyzes the vital metabolic step of phosphorylating glucose to glucose-6-phosphate but has no carrier action. All the others are carriers.

15

THE ACTION POTENTIAL

An action potential is an explosive change in the voltage across the plasma membrane. The most sophisticated electrically excitable cells—cells that can produce action potentials—are the nerve cells that allow our brains to carry out the complex electrical data processing called thought. However, we will begin with a much simpler electrically excitable cell: the sea urchin egg.

THE CALCIUM ACTION POTENTIAL IN SEA URCHIN EGGS

In most animals, eggs that are fertilized by more than one sperm fail to develop. A number of mechanisms exist to prevent such "polyspermy," one of which is based on action potentials.

Effect of Egg Transmembrane Voltage on Sperm Fusion

An experiment first carried out in 1976 by Laurinda Jaffe is illustrated in Figure 15.1*a*. She impaled a sea urchin egg with a micropipette (page 320). Instead of just measuring the natural resting transmembrane voltage (around −70 mV), she **depolarized** the plasma membrane by passing positive current through the pipette. We use the word *depolarization* to mean any positive shift in the transmembrane voltage, whatever its size or cause. In this case an electronic feedback circuit was used to alter the value of the current passed though the electrode so that the transmembrane voltage was depolarized to +10 mV and held at

Cell Biology: A Short Course, Second Edition, by Stephen R. Bolsover, Jeremy S. Hyams,
Elizabeth A. Shephard, Hugh A. White, Claudia G. Wiedemann
ISBN 0-471-26393-1 Copyright © 2004 by John Wiley & Sons, Inc.

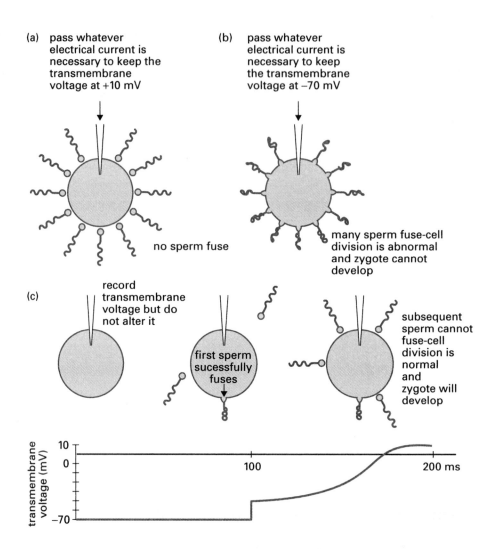

(a) pass whatever electrical current is necessary to keep the transmembrane voltage at +10 mV

(b) pass whatever electrical current is necessary to keep the transmembrane voltage at −70 mV

no sperm fuse

many sperm fuse-cell division is abnormal and zygote cannot develop

(c) record transmembrane voltage but do not alter it

first sperm sucessfully fuses

subsequent sperm cannot fuse-cell division is normal and zygote will develop

Figure 15.1. In sea urchin eggs transmembrane voltage controls sperm fusion.

that value. This technique, in which the experimenter, not the cell, determines the value of the transmembrane voltage, is called **voltage clamp.**

Jaffe observed that sperm would not fuse with eggs in which the transmembrane voltage was clamped at +10 mV, and the eggs remained unfertilized (Fig. 15.1a). However, an egg clamped at −70 mV (Fig. 15.1b) was fertilized by multiple sperm, leading to a defective embryo. The transmembrane voltage is able to control the fusion of sperm and egg.

When no current is passed through the micropipette, the transmembrane voltage is set by the channels in the egg plasma membrane (Fig. 15.1c). The resting voltage of the unfertilized egg is set by potassium movement through potassium channels. The first sperm that arrives can fuse because the transmembrane voltage is −70 mV. But within 100 ms, the transmembrane voltage depolarizes to +10 mV and further sperm fusion is blocked. Thus transmembrane voltage changes to prevent multiple sperm fusion. The sea urchin egg, which when unfertilized has a transmembrane voltage of -70 mV and is receptive to fertilization, depolarizes its membrane following successful fertilization to protect itself from subsequent fertilization. The result is shown on the right of Figure 15.1c. Only one sperm has fused, so the fertilized egg has the correct number of chromosomes and can go on to develop into an embryo and then into a mature sea urchin.

The Voltage-Gated Calcium Channel

The rapid depolarization that occurs upon fertilization of the sea urchin egg is an example of an action potential. Like many cells, urchin eggs can generate action potentials because their plasma membranes contain, together with potassium channels, an ion-conducting channel called the voltage-gated calcium channel.

Figure 15.2 shows how the voltage-gated calcium channel works. On the left is the shape of the protein when the transmembrane voltage is -70 mV, as in an unfertilized egg. The protein forms a tube, but it is not open all the way through the membrane. If the membrane depolarizes so that the cytosol is less negative or even becomes positive, positively charged amino acid residues on the protein are repelled, popping the tube into an open state. Calcium ions can now pass through the channel. If the membrane is then repolarized, the channel quickly recloses because the positive charges are pulled back toward the inside of the cell.

A protein domain in the cytosol called the inactivation plug is constantly jiggling about at the end of a flexible link and can bind to the inside of the open channel. The resulting blockage is called inactivation, and it occurs after about 100 ms. As long as the plasma membrane remains depolarized, the voltage-gated calcium channel will remain inactivated. When the plasma membrane is repolarized, the positive charges are attracted back toward the inside of the cell, squeezing the plug out. In summary:

1. When the transmembrane voltage is -70 mV, the voltage-gated calcium channel is gated shut.
2. When the plasma membrane is depolarized, the channel opens rapidly and after about 100 ms inactivates.

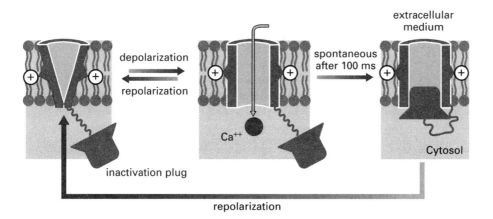

Figure 15.2. The voltage-gated calcium channel.

3. After the channel has gone through this cycle, it must spend at least 100 ms with the transmembrane voltage at the resting voltage before it can be opened by a second depolarization.

The voltage-gated calcium channel goes through this process whether or not calcium ions are present. To understand the action potential, we must now think about the effect that movements of calcium ions through this channel have on the transmembrane voltage.

The Calcium Action Potential

At point A in Figure 15.3 the egg is unfertilized. The potassium channels are open and set the transmembrane voltage to −70 mV. This negative voltage closes the voltage-gated calcium channels.

At point B we show what happens if an experimenter passes enough current down a micropipette to depolarize the plasma membrane by 10 mV, to −60 mV, and then stops passing current and once again allows the ion channels in the plasma membrane to determine the transmembrane voltage. This small depolarization opens only a very few voltage-gated calcium channels, so what happens next depends almost entirely on the movement of ions through the potassium channels. The slight decrease in the negative voltage of the cytosol reduces its attraction to potassium ions, and they move out of the cell down their concentration gradient. In so doing, they carry positive charge away from the cytosol, and the transmembrane voltage returns to its resting value of −70 mV.

At point C, the experiment is repeated, but this time the membrane is fleetingly depolarized by 20 to −50 mV. Again, most of the voltage-gated calcium channels stay shut, but some open, and calcium ions rush into the cell down their electrochemical gradient. The flow of positive charge inward on the calcium ions outweighs the outward flow of positive charge on potassium ions, so the cytosol is gaining positive charge and its voltage is moving in the positive direction. As a result more voltage-gated calcium channels open, and this allows more calcium ions to flood in. This is an example of positive feedback (page 303) because every calcium ion that moves in makes the cytosol more positive, causing more calcium channels to open, and therefore making it easier for the next calcium ion to move in. At point D, all the calcium channels are open, the inward calcium current is much greater than the outward potassium current, and the plasma membrane is rapidly depolarizing.

However, after 100 ms, the voltage-gated calcium channels begin to inactivate and the inward flow of calcium ions stops. The plasma membrane stops depolarizing and returns to −70 mV as potassium leaves through its channels.

This experiment shows the critical feature of an action potential: it is **all or nothing.** The depolarization at point B was too small to generate an action potential, and the transmembrane voltage simply fell back to its resting state. At point C the depolarization was big enough to start the process, and the action potential took off in an explosive, self-amplifying way until all the voltage-gated calcium channels were open and the plasma membrane had greatly depolarized. In all excitable cells there is a **threshold** for initiating an action potential; in the sea urchin egg it lies between −60 and −50 mV. Depolarizations to below the threshold elicit nothing; depolarizations to voltages more positive than the threshold elicit the complete action potential. At the heart of the action potential is a positive feedback loop. Depolarization causes voltage-gated calcium channels to open, and open calcium channels cause depolarization.

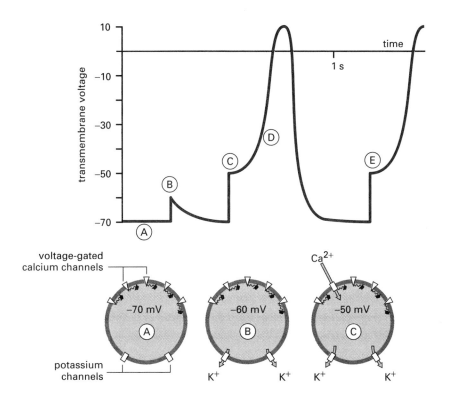

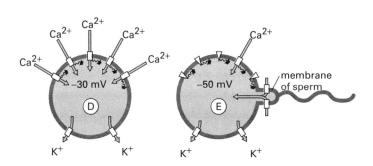

Figure 15.3. The calcium action potential in a sea urchin egg.

In nature the initial depolarization that sets off an action potential is produced by the sperm. This is illustrated at point E. The membrane of the sperm contains an unusual type of calcium channel that is open even at the resting voltage. When the membrane of the sperm is added to that of the egg, the inward current through the sperm's calcium channels depolarizes the plasma membrane of the egg by 20 mV to −50 mV, enough to cause the critical number of the egg's own voltage-gated calcium channels to open. The mechanism of the action potential takes over and the plasma membrane depolarizes rapidly to +10 mV. The graph has been cut off at the right-hand edge (where the voltage-gated calcium channels inactivate) where other mechanisms, not described here, take over and maintain a positive transmembrane voltage. The action potential has served its purpose in preventing second and subsequent sperm fusions in the 100 ms following the first fertilization.

❋ THE VOLTAGE-GATED SODIUM CHANNEL IN NERVE CELLS

The voltage-gated calcium channel appeared early in evolution and is found in all eukaryotes and in a wide variety of cell types. Later on in evolution, when multicellular organisms arose, there was a need for a system that could send electrical signals rapidly over long distances within the body. This system uses sodium rather than calcium ions.

The Voltage-Gated Sodium Channel

The cells in multicellular animals that are specialized for rapid conduction have a second, voltage-gated channel that is selective for sodium instead of calcium ions. It operates like the voltage-gated calcium channel, but both opening and inactivation are faster (Fig. 15.4). In summary:

1. When the transmembrane voltage is -70 mV, the voltage-gated sodium channel is gated shut.
2. When the plasma membrane is depolarized, the channel opens rapidly and then, after about 1 ms, inactivates.
3. After the channel has gone through this cycle, it must spend at least 1 ms with the transmembrane voltage at the resting voltage before it can be opened by a second depolarization.

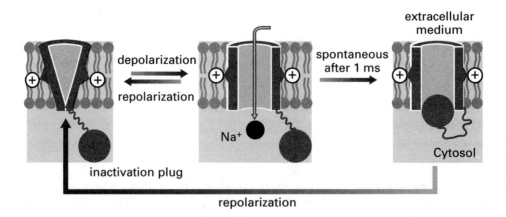

Figure 15.4. The voltage-gated sodium channel.

Example 15.1 Chewing off the Inactivation Plug

Proteases are enzymes that hydrolyze the peptide bonds within proteins. In 1976 Emilio Rojas and Bernardo Rudy investigated the effect of introducing protease into squid axons. The membrane of squid axons contains voltage-gated sodium channels that, like ours, normally inactivate about 1 ms after they are opened by a depolarization. However, after introduction of the protease, depolarization caused the voltage-gated sodium channels to open and remain open indefinitely, although the channels would close if the membrane was repolarized. Back in 1976 Rojas and Rudy had no idea why this should be so. Now we can understand the result—the protease cuts the linker between the main part of the voltage-gated sodium channel and the inactivation plug. The plug then floats off and is not available to block the open channel.

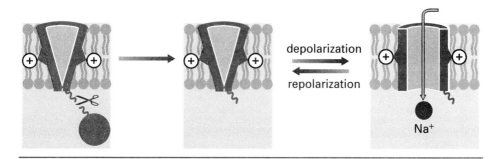

Like calcium, sodium is at a much higher concentration outside the cell than inside. In a mammal, the sodium concentration in the blood is 150 mmol liter^{-1}, whereas in the cytosol it is 10 mmol liter^{-1}. When voltage-gated sodium channels open, sodium ions rush into the cell carrying positive charge and depolarizing the plasma membrane. This then favors the opening of more voltage-gated sodium channels, and so on. This positive feedback produces a depolarization to +30 mV called the sodium action potential. Because the voltage-gated sodium channels inactivate so quickly, the sodium-based action potential lasts only for 1 ms.

Medical Relevance 15.1

A Sodium "Channelopathy"

Epilepsy, in which nerve cells in the brain fire action potentials in an uncontrolled fashion, is not uncommon in young children but often clears up as the child gets older. However, one form of childhood-onset epilepsy, termed GEFS$^+$ for "generalized epilepsy with febrile seizures that extend beyond six years of age," does not improve with age. Robyn Wallace and co-workers in Australia showed that this condition could be caused by a mutation that slows the rate at which the inactivation plug moves to block the voltage-gated sodium channel. The slow rate of inactivation makes it easier for stimuli to trigger action potentials in the nerve cells in the brain.

Although enough sodium ions move into the cell to dramatically change the transmembrane voltage, the concentration of sodium ions inside the cell is increased only very slightly. The amount depends on the electrical capacitance of the cell membrane and the cell volume, but, for example, it can be calculated that a single sodium-based action potential increases the sodium concentration in the nerve cell axon of Figure 15.7 by only 250 nmol liter^{-1}.

Electrical Transmission down a Nerve Cell Axon

Figures 15.5 and 15.6 show how one nerve cell, a pain receptor, uses sodium-based action potentials to signal to the central nervous system. The cell body is close to the spinal cord and extends an **axon** that branches to the spinal cord and out to the body. The particular pain receptor illustrated in Figure 15.5 sends its axon for almost a meter to the tip of a finger, an extraordinary distance for a single cell. Potentially damaging events are detected at the finger, and the message is passed on to another nerve cell (a pain relay cell) in the spinal cord. We will now explain how this function is performed.

Figure 15.5 shows the transmembrane voltage of the pain receptor at the two axon terminals, one near the tip of the finger (green line) and one at the opposite end of the

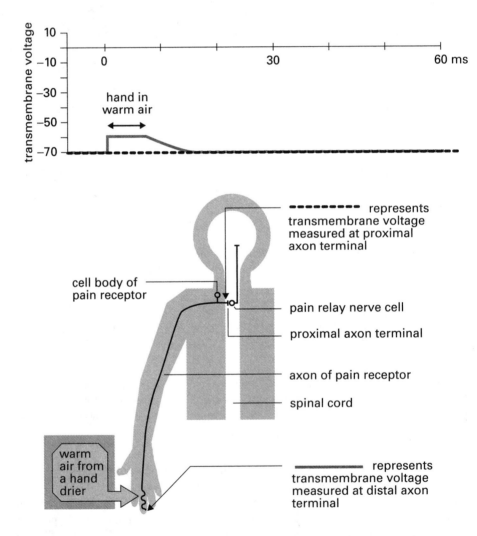

Figure 15.5. Mild heat stimulus is not felt as pain because the pain receptor is not depolarized to threshold.

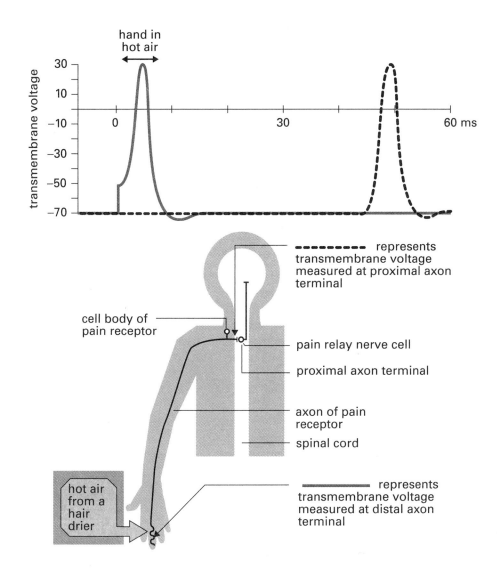

Figure 15.6. Stronger heat stimulus evokes a propagating action potential.

cell, in the spinal cord (dashed black line). The axon terminal in the skin is the **distal,** or far away, terminal and that in the spinal cord is the **proximal** one. When the finger is passed through the warm air from a hand dryer, the heat causes the plasma membrane of the distal terminal to depolarize by 10 mV (continuous green line). This depolarization is not enough to open more than a very few voltage-gated sodium channels. Even in the warm air, the transmembrane voltage of the distal axon terminal is below threshold, and as soon as the finger leaves the warm air jet it sags back to −70 mV under the influence of the potassium channels. The axon is so long that the small voltage change at the distal terminal does not affect the voltage at the proximal terminal at all (dashed black line). No signal passes to the pain relay cell, and the subject feels no sensation of pain.

Figure 15.6 shows what happens if we pass our finger through hotter air. The depolarization caused by the heat is greater and this causes more voltage-gated sodium channels to open (continuous green line). Just as in the sea urchin egg at fertilization, the plasma membrane then depolarizes rapidly (in this case as more and more sodium enters) until the transmembrane voltage is $+30$ mV. After only 1 ms, the voltage-gated sodium channels inactivate, and the transmembrane voltage returns to -70 mV, due to the action of the potassium channels. At the proximal terminal (dashed black line), a meter away, there is also an action potential after a short delay. The signal then passes on to the brain and the person notices the hot air in time to pull the hand away before too much damage is caused. The pain receptor is one example of how we are quickly made aware of changes in our environment by sodium action potentials traveling rapidly along nerve cell axons.

Myelination and Rapid Action Potential Transmission

The axon transmits the action potential so rapidly because for most of its length it is insulated by a fatty sheath called myelin made by glial cells (the glial cells outside the brain and spinal cord are often called **Schwann cells**). Figure 15.7 shows part of the axon of the pain receptor and its associated glial cells, each of which wraps around the axon to form an electrically insulating sheath. Only at short gaps called **nodes** is the membrane of the nerve cell exposed to extracellular fluid. Note that the vertical and horizontal scales of this diagram are completely different. The axon together with its myelin sheath is only 3 μm across, but the section of axon between nodes is 1 mm long, a large distance by normal cell standards. Between nodes the axon is an insulated electrical cable: just as voltage changes at one end of an insulated metal wire have an almost instantaneous effect on the voltage at the other end, so a change in transmembrane voltage at one node has an almost instantaneous effect at the next node along. The plasma membrane at the nodes has potassium and voltage-gated sodium channels and can therefore generate action potentials.

> **Example 15.2** **Local Anaesthetic, Overall Well-being**
>
> Nerve conduction is exploited when a dentist uses local anaesthetic. A patient feels pain from the drill because pain receptors in the tooth are depolarized and transmit action potentials toward the brain. The site at which they are being depolarized is inside the tooth and therefore inaccessible to drugs until the drill has made a hole. However, the axons of the pain receptors run through the gum. Local anaesthetics injected into the gum close to the axon bind to the nerve cell membrane at the node and prevent the opening of voltage-gated sodium channels. Drilling into the tooth still depolarizes the pain receptor membrane, and action potentials begin their journey toward the brain, but they cannot pass the injection site because the nodes there cannot generate an action potential. The patient therefore feels no pain.

Figure 15.7 represents a snapshot of one brief instant in the operation of the nerve cell. The node on the left is generating an action potential. Voltage-gated sodium channels are open, sodium ions are flooding in, and the transmembrane voltage is $+30$ mV. Some of the charge flowing in is carried out again by potassium ions flowing through the potassium channels. However, some positive charge flows axially up and down the cytosol of the nerve

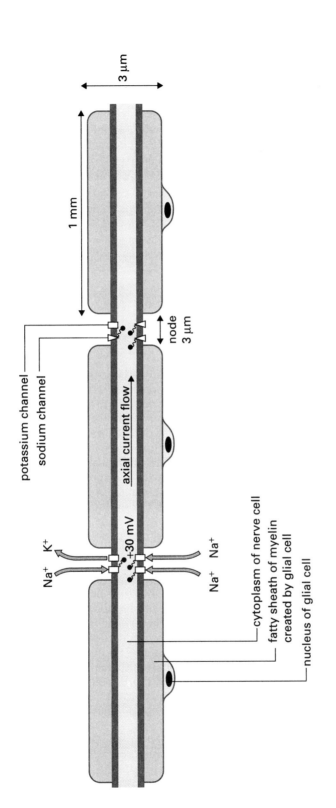

Figure 15.7. The myelin sheath and the node.

cell. Because myelin is an electrical insulator, this charge cannot leave the cytosol until the next node. At the node on the right the incoming charge depolarizes the plasma membrane to threshold, and voltage-gated sodium channels at this node open as well. Sodium rushes in, and this node generates an action potential. The action potential has jumped from node to node in the extremely short time of 50 μs. Now sodium ions will rush into the node on the right, current will pass axially up the interior of the axon to the next node up, and in a further 50 μs the action potential will have jumped to the next node, and so on, all the way up to the proximal terminal at an overall speed of 20 m s^{-1}. This process, in which the action potential jumps from node to node, is named after the Latin for "to jump," *saltere,* so it is **saltatory conduction.**

The combination of axon plus myelin sheath is called a nerve fiber. The nerve fiber shown in Figure 15.7 has an overall diameter of 3 μm. The human body contains myelinated nerve fibers with diameters ranging from 1 μm, conducting at 6 m s^{-1}, to 10 μm, conducting at 60 m s^{-1}. A number of axons are not myelinated and generally conduct more slowly than myelinated fibers. The axons of scent-sensitive nerve cells (page 350) are an example. Getting scent information from the nose to the brain as fast as possible is apparently not a significant advantage, given that a scent can take many seconds to waft from its source to our nose.

IN DEPTH 15.1 Frequency Coding in the Nervous System

At the distal axon terminal of the pain receptor, the intensity of the stimulus is represented by the amount of depolarization produced. The hotter the air blowing on the skin, the more positive the transmembrane voltage of the terminal becomes. Other sensory cells behave in the same way. For instance, in Chapter 16 we will meet the scent-sensitive nerve cells in the nose that detect smell chemicals. The higher the concentration of the smell chemical, the more these cells depolarize.

However, the signal that passes toward the brain is an all-or-none action potential. Either the stimulus produces a depolarization that is below threshold, in which case no signal passes to the brain, or the membrane depolarizes beyond threshold, in which case an action potential is generated. How then is the brain told about the intensity of the stimulus?

Action potentials continue to be produced and travel up toward the brain as long as the stimulus is maintained. The diagrams represent what happens when the hand is being held in a constant airstream. If the stimulus is strong, the membrane of the pain receptor depolarizes rapidly, so that after each action potential the threshold voltage is soon reached once more. Thus time between successive action potentials is short. The stronger the stimulus, the higher the frequency of action potentials generated.

Amplitude modulation (AM) and frequency modulation (FM) are familiar terms from radio. Here we are seeing exactly the same two coding strategies in the nervous system. AM is used in the distal terminals of the pain receptor. The strength of stimulus is coded for by the amplitude of the depolarization. FM is used in almost all axons, including those of the pain receptor. The strength of stimulus is coded for by the frequency of action potentials, each of which has the same amplitude.

Man-made systems use exactly the same strategy. On the control unit of a games console, for instance, the force with which the player presses on a button is coded in terms of the amplitude of a voltage change. This is then recoded in all-or-none digital pulses for transmission to the main console.

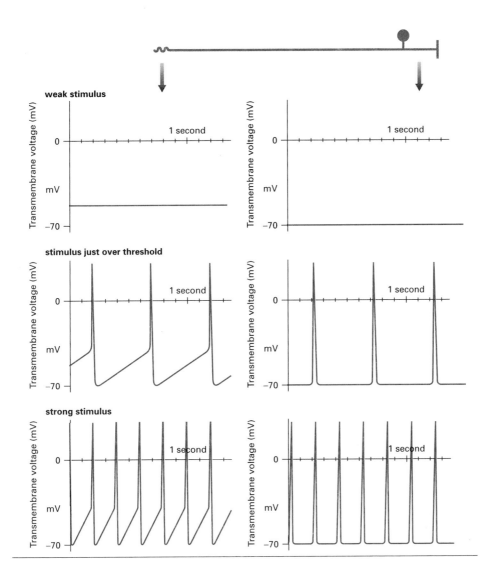

SUMMARY

1. The voltage-gated calcium channel is shut at the resting voltage but opens upon depolarization. After about 100 ms the channel inactivates.

2. While the calcium channel is open, calcium ions pour into the cell down their electrochemical gradient. The mutual effect of current through the channel on transmembrane voltage, and of transmembrane voltage on current through the channel, constitutes a positive feedback system. If the membrane is initially depolarized to threshold, the positive feedback of the calcium current system ensures that depolarization continues in an all-or-nothing fashion. Repolarization occurs when the

calcium channel inactivates. The entire cycle of depolarization and repolarization is called an action potential.

3. The same process occurs in cells expressing voltage-gated sodium channels. But, since these channels inactivate within about 1 ms, the action potential lasts only this long. Long nerve cell processes called axons transmit action potentials at speeds up to 60 m per second. They can transmit the signal at such high rates because myelin, a fatty sheath, insulates the 1 mm distances between nodes.

4. A number of axons are not myelinated; these generally conduct action potentials more slowly than do myelinated axons and are found where high conducation speed is not required.

FURTHER READING

Ashcroft, F. M. 2000. *Ion Channels and Disease*. San Diego: Academic Press.

Levitan, I. B., and Kaczmarek, L. K. 2002. *The Neuron*. New York: Oxford University Press.

❀ REVIEW QUESTIONS

For each question, choose the ONE BEST answer or completion.

1. A sea urchin egg is used in a voltage clamp experiment; that is, the value of the transmembrane voltage is set by the experimenter, not by the properties of the membrane. The transmembrane voltage is held at $+10$ mV for one minute. At the end of that time
   A. the voltage-gated calcium channels are all open.
   B. the voltage-gated calcium channels are all inactivated.
   C. the voltage-gated calcium channels have recovered from inactivation.
   D. the voltage-gated calcium channels are going through cycles of opening and inactivation as the cell generates repetitive action potentials.
   E. any of answer A though D above could occur, depending on the density of potassium channels.

2. In a calcium action potential
   A. the influx of calcium ions depolarizes the plasma membrane.
   B. depolarization of the plasma membrane opens voltage-gated calcium channels.
   C. open voltage-gated calcium channels carry an inward current of calcium ions.
   D. an inward current of calcium ions adds positive charge to the cytosol.
   E. all of the above occur.

3. Action potentials in nerve cell axons last a much shorter time than action potentials in sea urchin eggs because
   A. sodium has only a single positive charge while calcium has two.
   B. voltage-gated sodium channels are found only at the nodes, while voltage-gated calcium channels are found over the entire surface of the egg cell.
   C. sodium ions are removed rapidly from the cytosol by the Na^+/K^+ ATPase.
   D. voltage-gated sodium channels inactivate more rapidly than do voltage-gated calcium channels.
   E. calcium has a higher atomic weight than does sodium and therefore diffuses more slowly.

4. An action potential is being generated at the node of a myelinated axon. Which of the following statements is false?

A. Sodium is entering the axon interior through voltage-gated sodium channels.

B. Potassium is leaving the axon interior through potassium channels.

C. Calcium is entering the axon interior through voltage-gated calcium channels.

D. A net inward current is flowing across the plasma membrane of the node.

E. Current is leaving the region of the node axially, through the cytosol of the nerve cell.

5. The nodes of a myelinated axon are spaced a distance apart of approximately

A. 1 nm.

B. 1 μm.

C. 1 mm.

D. 1 cm.

E. 1 m.

6. A sperm can trigger an action potential in a sea urchin egg because

A. its cytosol contains a lot of calcium ions that, upon fertilization, rush into the egg.

B. its membrane contains ion channels that, upon fertilization, pass an inward current that depolarizes the egg plasma membrane.

C. its membrane contains ion channels that, upon fertilization, pass an outward current that repolarizes the egg plasma membrane.

D. its cytosol contains a chemical that tends to pop the voltage-gated calcium channels into an open state.

E. its cytosol contains a protease that modifies the behavior of voltage-gated calcium channels in the plasma membrane of the egg.

7. When a nerve cell axon is depolarized to threshold, the flow of potassium through potassium channels in the plasma membrane is

A. outward, and of a greater absolute amplitude than at the resting voltage.

B. inward, and of a greater absolute amplitude than at the resting voltage.

C. outward, but of a smaller absolute amplitude than at the resting voltage.

D. inward, but of a smaller absolute amplitude than at the resting voltage.

E. zero.

ANSWERS TO REVIEW QUESTIONS

1. *B.* They are all inactivated. Upon depolarization they opened and inactivated, and since the transmembrane voltage was never returned to the resting value, the channels have not recovered from inactivation. Regarding answer D, note that the cell cannot generate repetitive action potentials because it is voltage clamped: The value of the transmembrane voltage is set by the experimenter, not by the properties of the membrane. For this reason the density of potassium channels (answer E) is irrelevant—the density of potassium channels will affect the current through those channels, but since the cell is voltage clamped, the current through potassium channels will have no effect on transmembrane voltage and therefore no effect on the behavior of the voltage-gated calcium channels.

2. *E.* Answers A through D are true. The influx of calcium acts, through the voltage change it causes, to increase the influx of calcium—this is why the generation of an action potential is an example of positive feedback.

3. *D.* Statements A, B, C, and E are all true in themselves but are not reasons why action potentials in nerve cell axons are shorter than those in sea urchin eggs. Concerning answer C, the sodium ions do not have to be removed from the cytosol in order for the action potential to terminate. The very small amounts of sodium that do enter (see page 331) are removed over a timescale of seconds by the Na^+/K^+ ATPase. If the Na^+/K^+ ATPase is blocked by the toxin digitalis (page 271), a nerve cell can continue to generate action potentials as normal. Only after many thousands of action potentials have occurred in a digitalis-poisoned nerve cell will the cytosolic sodium concentration rise to a concentration sufficient to significantly affect the threshold for action potential generation.

4. *C.* There are no voltage-gated calcium channels at the node. All the other statements are true: although some charge is being carried out of the axon at the node by potassium, the current carried by the inward movement of sodium is larger, so there is a net inward current. The excess charge is carried out of the node axially.

5. *C.*

6. *B.* Statements A, D, and E are untrue. Statement C is true in itself in that, like most cells, the sperm plasma membrane will contain some potassium channels that will pass an outward current, but as the answer states, the action of these will be to repolarize the egg membrane, not depolarize it. The depolarizing effect of the calcium influx through the sperm calcium channels, which are open even at the resting voltage, is greater.

7. *A.* Potassium ions are usually attracted to the negative interior of the cell, and this force opposes the tendency of potassium ions to leave down their concentration gradient. Depolarization reduces the inward electrical force, so potassium ions leave at a greater rate than at rest. At the same time, voltage-gated sodium channels are carrying an inward current, but that movement is not relevant to the question.

16

INTRACELLULAR SIGNALING

The behavior of cells is not constant. Cells need to be able to alter their behavior in response to internal changes or to external events, and the internal signaling mechanisms that allow them to do so are varied and complex. We will begin by continuing the story begun in Chapter 15 and explain how calcium ions carry a signal from the plasma membrane of nerve cells to vesicles deep within the cytosol. The rest of the chapter will describe some other very different methods of intracellular signaling.

✳ CALCIUM

Calcium ions are present at a very low concentration (about 100 nmol liter^{-1}) in the cytosol of a resting cell. An enormous number of processes in many types of cells are activated when the concentration of calcium rises. Calcium can move into the cytosol from two sources: the extracellular medium or the endoplasmic reticulum.

Calcium Can Enter from the Extracellular Medium

In Chapter 15, we saw how heating the hand causes action potentials to travel from the hand to the spinal cord along the axons of pain receptors. When the action potential reaches the proximal axon terminal, the amino acid glutamate is released onto the surface of another nerve cell, the pain relay cell. Glutamate is a transmitter that stimulates the pain relay nerve cell, so that the message that the finger is being damaged is passed toward the brain.

Cell Biology: A Short Course, Second Edition, by Stephen R. Bolsover, Jeremy S. Hyams,
Elizabeth A. Shephard, Hugh A. White, Claudia G. Wiedemann
ISBN 0-471-26393-1 Copyright © 2004 by John Wiley & Sons, Inc.

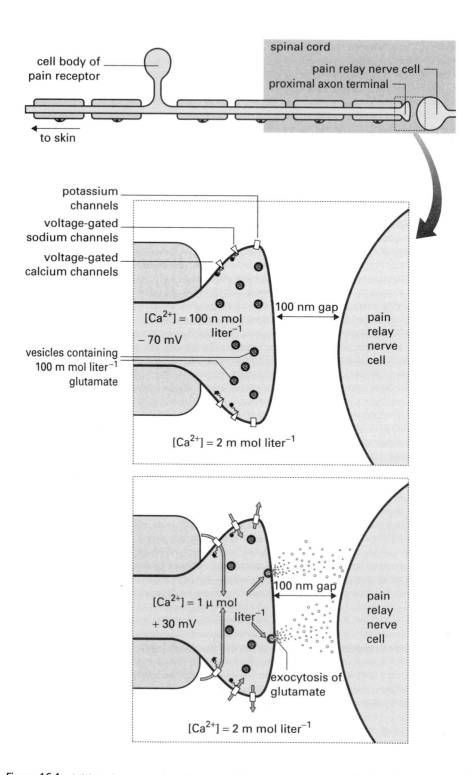

Figure 16.1. Calcium ions entering the cytosol from the extracellular fluid activate regulated exocytosis in the proximal axon terminal of the pain receptor nerve cell.

The proximal axon terminal of the nerve cell is unmyelinated, so that the plasma membrane is exposed to the extracellular medium (Fig. 16.1). The membrane contains not only potassium channels and voltage-gated sodium channels, but also voltage-gated calcium channels. These are closed in a resting cell, but when an action potential travels in from the skin and depolarizes the plasma membrane of the proximal axon terminal, the voltage-gated calcium channels open. Calcium ions pour in, increasing their concentration in the cytosol by 10 times, from the normal concentration of 100 nmol liter^{-1} to 1 μmol liter^{-1}. At the proximal axon terminal the cytosol of the nerve cell contains regulated exocytotic vesicles (page 229). In the case of the pain receptor, these vesicles are filled with sodium glutamate. In response to the increased concentration of cytosolic calcium the regulated exocytotic vesicles move to the plasma membrane and fuse with it, releasing their contents into the extracellular medium. The glutamate then diffuses across the gap to the pain relay cell, stimulating it (we will describe how on page 372). As the action potential in the axon terminal of the pain receptor cell is over in 1 ms, the voltage-gated calcium channels do not have time to inactivate. They simply return to the ready-to-open state and can be reopened immediately by the next action potential that arrives.

Example 16.1 **Visualizing Calcium Signals**

(a)

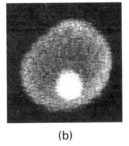

(b)

Calcium signals in cells can be visualized by using dyes that fluoresce brightly when they bind calcium. The images in the figure show a cell body of a pain receptor cell that has been filled with a calcium indicator dye. In the first image, at rest, the cell is dim. The second image was acquired after 100 ms of depolarization. In this the edges of the cell are bright because calcium ions have entered through voltage-gated calcium channels. The bright object in the lower part of the cell is the nucleus.

In the process that we have described, calcium ions act as a link between the depolarization of the plasma membrane and the regulated exocytotic vesicles within the cytosol.

Calcium ions are **intracellular messengers.** Regulated exocytotic vesicles that are triggered by an increase of cytosolic calcium concentration were first discovered in nerve cells but are now known to be a feature of almost all cells.

Before we continue with the topic of intracellular messengers, we will review two general points raised by our description of the pain receptor proximal axon terminal. The first point is that different nerve cells release different transmitters. Glutamate is in the exocytotic vesicles of many nerve cells, but other nerve cells release other transmitters. The second point concerns nomenclature. Many nerve cells, like the pain receptor, have their axon terminals close to a second cell and release their transmitter onto it. In such cases the complete unit of axon terminal, gap, and the part of the second cell that receives the transmitter is called a **synapse.** The part of the axon terminal that releases transmitter is called a **presynaptic terminal,** and the cell upon which the transmitter is released is called the **postsynaptic cell.** Many nerve cells do not come close enough to a specific second cell to form a synapse but simply release transmitter from their axon terminals into the extracellular medium.

Calcium Can Be Released from the Endoplasmic Reticulum

Cells may show an increase of cytosolic calcium not because of an action potential but because of the appearance of a transmitter or other chemical in the extracellular medium. The presence of the chemical is detected by integral membrane proteins, each one a receptor that recognizes a particular chemical with high affinity. These receptors then participate in a more general mechanism, the end result of which is the release of calcium ions from the smooth endoplasmic reticulum into the cytosol. This mechanism will be illustrated with the particular example of blood platelets. We will then discuss how much of the mechanism is general to a wider range of cells.

Platelets are common in the blood. They are small fragments of cells and contain no nucleus, but they do have a plasma membrane and some endoplasmic reticulum. Blood platelets use the release of calcium from the endoplasmic reticulum as one step in the mechanism of blood clotting (Figs. 16.2 and 16.3). Two new mechanisms are involved in calcium release from the endoplasmic reticulum. The inositol trisphosphate-gated calcium channel (Fig. 16.3), like the voltage-gated calcium channel, allows only calcium ions to pass. Most of the time its gate is shut, and no ions flow. It is not opened by a change in the voltage across the endoplasmic reticulum membrane. Instead, when the intracellular solute inositol trisphosphate (IP_3 for short) binds to the cytosolic face of the channel, the channel changes to an open shape.

The second new mechanism that we must describe is the one that makes IP_3 (Fig. 16.2). If a blood vessel is cut, cytosol from damaged cells at the edge of the cut can leak into the blood. The appearance of solutes that normally are found only inside cells is a sure sign that damage has occurred. Adenosine diphosphate (ADP) is one such solute, and it acts to stimulate platelets, causing them to begin a blood clot to help plug the damaged vessel. The plasma membrane of the platelet contains a protein receptor that binds ADP, that is, ADP is its ligand. When the ADP has bound, the receptor, which is free to move in the plasma membrane, becomes a guanine nucleotide exchange factor (page 218) for a **trimeric G protein** called G_q. Like the GTPases Ran, ARF, and Rab that we have met earlier, trimeric G proteins are GTPases that activate target proteins when they have GTP bound, but turn themselves off by hydrolyzing the GTP to GDP. Trimeric G proteins have a slight additional complication in that, as the name indicates,

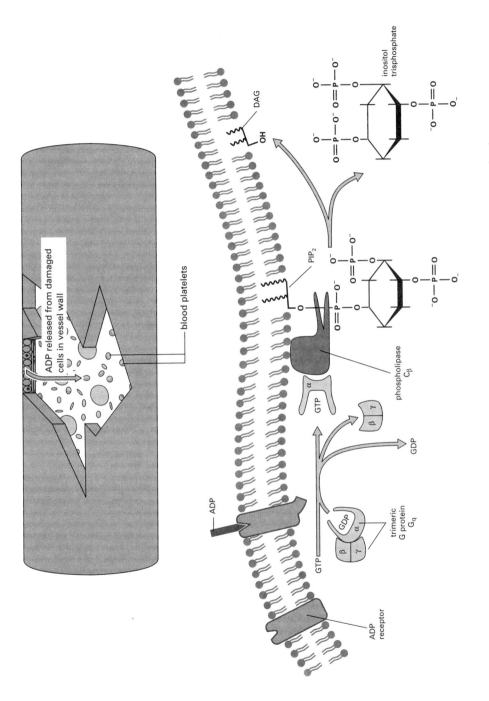

Figure 16.2. ADP from damaged cells activates G_q and hence phospholipase $C\beta$ in platelets.

345

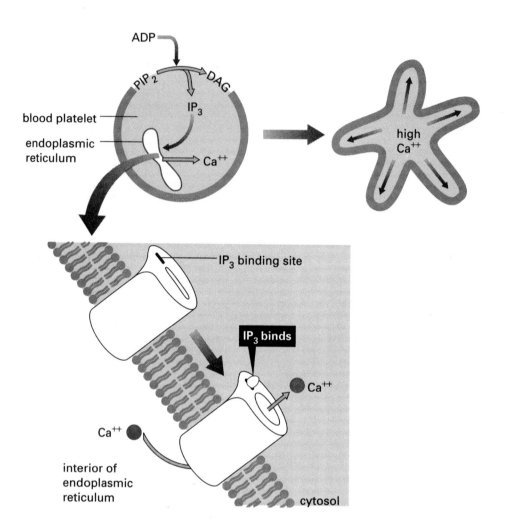

Figure 16.3. The inositol trisphosphate-gated calcium channel receptor is a calcium-selective channel in the membrane of the endoplasmic reticulum. An increase of cytosolic calcium concentration in platelets makes them sticky, initiating blood clotting.

they are composed of three subunits. The α subunit is homologous to the GTPases we have met before, while the β and γ subunits dissociate from the α subunit when it has GTP bound and only reassociate when the GTP has been hydrolyzed. GTP-loaded G_q activates the β isoform of an enzyme called phosphoinositide phospholipase C, which we will call **phospholipase C** or just **PLC.** (The capital C refers to the bond hydrolyzed by the enzyme; A, B, and D phospholipases also exist, but we will not meet them in this book). PLC specifically hydrolyzes **phosphatidylinositol bisphosphate** (**PIP$_2$** for short), a phospholipid in the plasma membrane that has phosphorylated inositol as its polar head group. Hydrolysis releases the IP$_3$ to diffuse freely in the cytosol, leaving behind the glycerol backbone with its attached fatty acids: this residual lipid is called diacylglycerol, or DAG. As it diffuses through the cytosol, IP$_3$ reaches the endoplasmic reticulum and binds to the inositol trisphosphate-gated calcium channels. The inositol trisphosphate-gated calcium

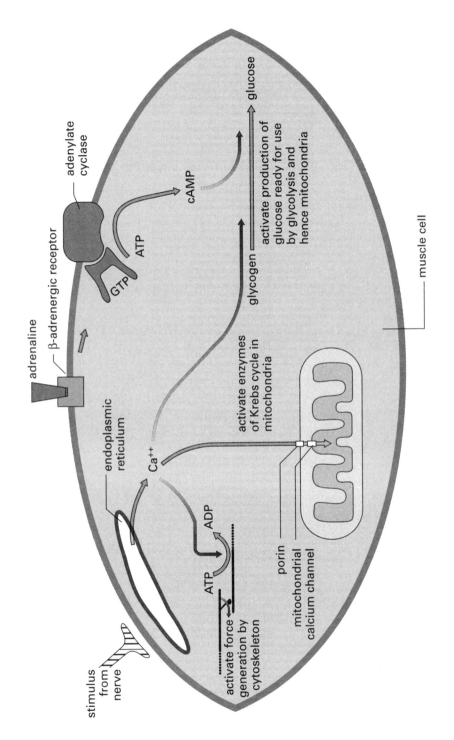

Figure 16.4. Calcium and cyclic-AMP activate distinct but overlapping sets of target processes in skeletal muscle cells.

347

channels open, and calcium ions pour out of the endoplasmic reticulum into the cytosol. The increase of calcium concentration causes the platelet to change shape and to become very sticky, so platelets begin to clump together in a clot.

Phosphatidylinositol bisphosphate, G_q, PLCβ, and the inositol trisphosphate-gated calcium channel are found in almost all eukaryotic cells, but the distribution of the ADP receptor is more restricted. Only cells that express the ADP receptor will show an increase of cytosolic calcium concentration in response to ADP. Other cells respond to other chemicals. Each produces a receptor specific for that chemical, which then activates G_q. Over 100 such receptors are known. We will meet two more in the next chapter.

Processes Activated by Cytosolic Calcium Are Extremely Diverse

The targets activated by an increase of cytosolic calcium concentration differ between different cells. An increase of calcium is a crude signal that says "do it" but contains no information about what the cell should do. This depends on what the cell was designed to do. Cells and cell regions designed for regulated exocytosis (e.g., salivary gland cells and axon terminals) exocytose when cytosolic calcium increases. Cells designed to contract (e.g., muscle cells) contract when calcium increases, and so on. In each case, the calcium ions bind to a calcium-binding protein; the calcium ion–calcium-binding protein complex activates the target process.

In skeletal muscle cells (page 14) a transmitter released from nerve axon terminals leads to the escape of calcium from the endoplasmic reticulum (Fig. 16.4). This activates several processes. First, calcium ions bind to a protein called troponin that is attached to the cytoskeleton. This causes the cytoskeleton to contract, using the energy released by ATP hydrolysis to do mechanical work (page 393). Second, the calcium binds to the protein calmodulin, which in turn activates glycogen phosphorylase kinase and hence glycogen breakdown as part of the feedforward control of energy metabolism (page 305).

Mitochondria are also affected by calcium, which passes down its electrochemical gradient from the cytosol into the mitochondrial matrix through a calcium channel. Once there, calcium stimulates the mitochondria to increase the production of NADH and ATP (page 305).

●●●● IN DEPTH 16.1 Ryanodine Receptors

Many cells contain a second type of calcium channel in the membrane of the endoplasmic reticulum. This channel was initially distinguished from the IP$_3$ receptor by the fact that it bound the plant alkaloid ryanodine, so it was called the **ryanodine receptor**. Ryanodine receptors were first identified in skeletal muscle. In these cells they are physically linked to voltage-gated calcium channels in the plasma membrane (left-hand side of diagram). When the voltage-gated calcium channels switch to their open configuration, they induce the ryanodine receptors below them to switch into an open configuration, allowing calcium to flow out of the smooth endoplasmic reticulum into the cytosol.

No such direct linkage between ryanodine receptors and plasma membrane channels is found in other cell types. Instead, ryanodine receptors are opened when the concentration of calcium ions in the cytosol rises above a critical level. For example, in heart muscle (right-hand side of diagram) depolarization causes voltage-gated calcium channels in the plasma membrane to open. The calcium ions that enter through this channel bind to the cytosolic aspect of the ryanodine receptor, causing it to open too, allowing calcium to flow out of the smooth endoplasmic reticulum into the cytosol.

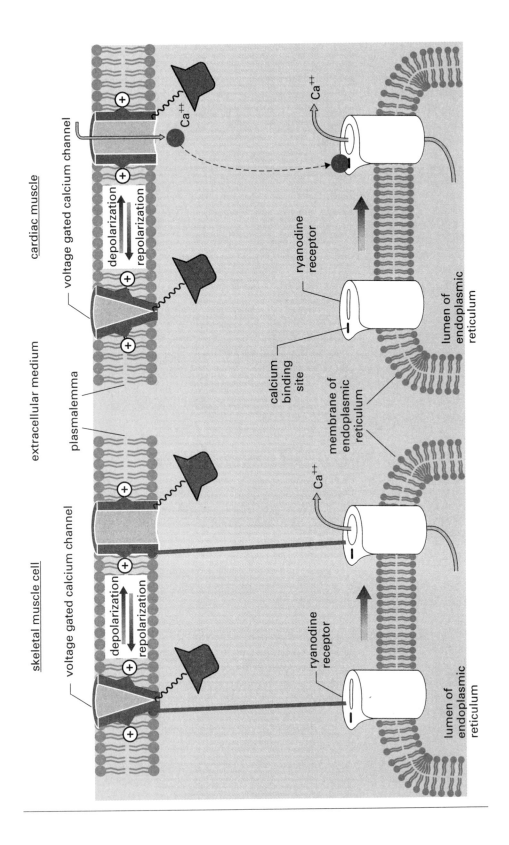

The same simple intracellular messenger, calcium, has many different actions inside the skeletal muscle cell. Under its influence the cytoskeleton begins to use ATP, glycogen phosphorylase releases more glucose, and the mitochondria produce more ATP. The skeletal muscle cell is an example of how diverse mechanisms inside a cell can be integrated by the action of intracellular messengers.

Example 16.2 **Training for Anaerobic Respiration**

Muscle cells that are contracted for long periods, cutting off the blood supply—such as those in the arms of rock climbers—have a much enhanced capacity for anaerobic respiration compared to those that are used briefly and then relaxed (such as those in the arms of baseball pitchers). This occurs because the calcium signals that activate muscle contraction (page 393) also act to increase the transcription of the genes coding for glycolysis (via calcineurin and NFAT, page 222).

Return of Calcium to Resting Levels

As soon as a stimulus, be it depolarization or extracellular chemical, disappears, cytosolic calcium concentration falls again. Calcium ions are pumped up their electrochemical gradients from the cytosol into the extracellular medium or into the endoplasmic reticulum by two carriers. The Ca^{2+} ATPase (page 318) uses the energy released by ATP hydrolysis to move calcium ions up their electrochemical gradient out of the cytosol, while the Na^+/Ca^{2+} exchanger (page 317) uses the energy released by sodium ion movement into the cytosol to do the same job.

✳ CYCLIC ADENOSINE MONOPHOSPHATE

We have already met the nucleotide cyclic adenosine monophosphate, or cAMP (Fig. 16.5), in the context of the regulation of the *lac* operon (page 114). (It is worth mentioning that cAMP is always pronounced "cy-clic A-M-P," with five syllables.) In eukaryotes, cAMP is also important and acts as an intracellular messenger in a great many cells, including the scent-sensitive nerve cells in our nose (Fig. 16.6). These cells have their cell bodies in the skin of the air passages in the nose. Each cell sends an axon into the brain, and shorter processes, called dendrites, into the mucus lining the air passages. Scent-sensitive nerve cells are stimulated by scents in the air. Particular chemicals in the air stimulate these cells because the cells have protein receptors that specifically bind the scent (Fig. 16.7). When the scent binds, the receptor becomes a guanine nucleotide exchange factor for a trimeric G protein called G_s. GTP-loaded G_s activates an enzyme called adenylate cyclase that converts ATP to cAMP. The next stage in the detection of scents is a channel in the plasma membrane called the cAMP-gated channel. Like the inositol trisphosphate-gated calcium channel, the cAMP-gated channel is opened by a cytosolic solute, in this case cAMP. When the channel is open, it allows sodium and potassium ions to pass through. The electrochemical gradient pushing sodium ions into the cell is much greater than the electrochemical gradient pushing potassium ions out of the cell. Thus, when the cAMP-gated channel opens, sodium ions pour in, carrying their positive charge and depolarizing the plasma membrane. The plasma membrane also contains voltage-gated sodium channels. Thus, when enough cAMP-gated channels open, the transmembrane voltage reaches threshold for generation of sodium action potentials. These then propagate along the axon to the brain, and the person becomes aware of the particular scent for which that scent-sensitive nerve cell was specific.

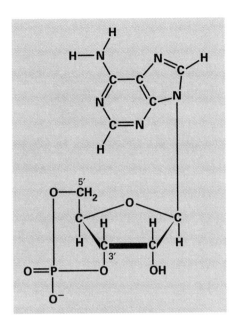

Figure 16.5. Cyclic adenosine monophosphate, also called cyclic-AMP or just cAMP.

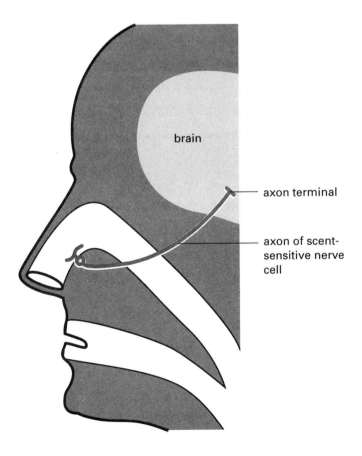

Figure 16.6. Scent-sensitive nerve cells send axons to the brain.

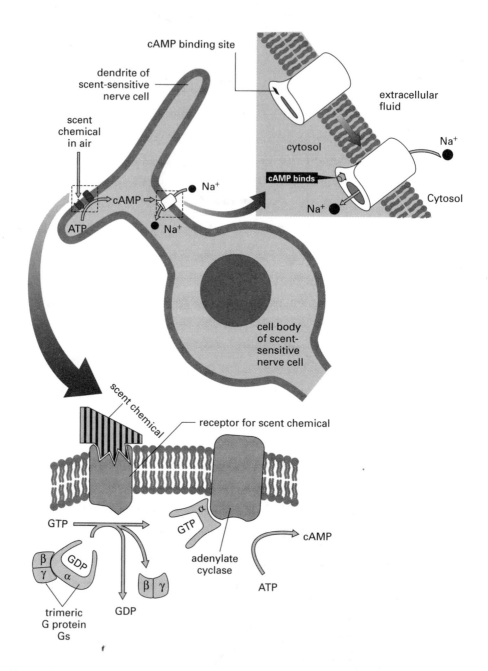

Figure 16.7. Scent chemicals activate G_s and hence adenylate cyclase in scent-sensitive nerve cells. cAMP then opens a nonselective cation channel in the plasma membrane.

G_s and adenylate cyclase are found in many cells, but only scent-sensitive nerve cells are sensitive to scents, because only they have specific scent receptors in their plasma membranes. Other cells that use cAMP as an intracellular messenger are sensitive to other specific chemicals because each makes a receptor that binds the chemical, whatever it may be, and then activates G_s. Once the stimulating chemical has gone, the cAMP concentration returns to resting levels. The enzyme **cAMP phosphodiesterase** hydrolyzes cAMP to AMP, which is inactive at cAMP-gated channels and other cAMP-binding proteins.

Most of the symptoms of the deadly disease cholera are caused by a toxin released by the gut bacterium *Vibrio cholera*. The toxin is an enzyme that enters the cytosol of the cells lining the gut and attaches an ADP ribosyl group to the catalytic domain of G_s, preventing it from hydrolyzing GTP. G_s is therefore locked in the active state, and activates adenylate cyclase nonstop. The cAMP concentration in the cytosol then shoots up. Ion channels in the plasma membrane are opened by the increase of cAMP, allowing ions to leak from the cells into the gut contents. If untreated this loss of ions, and of the water that accompanies them, leads to death from dehydration.

✺ CYCLIC GUANOSINE MONOPHOSPHATE

Another nucleotide that acts as a intracellular messenger is cGMP. In light-sensitive nerve cells called photoreceptors, cGMP plays a role like that of cAMP in scent-sensitive nerve cells. In the dark, the enzyme guanylate cyclase makes cGMP from GTP. cGMP binds to and opens a channel that allows sodium and potassium ions to pass, so that in the dark the photoreceptor is depolarized to about -40 mV because of the constant influx of sodium ions through the cGMP-gated channel. In the light the concentration of cGMP in the cytosol falls, and the transmembrane voltage changes to the more typical resting voltage of -70 mV. The changing transmembrane voltage is transmitted to other nerve cells, making us aware of the pattern of light and dark.

✺ MULTIPLE MESSENGERS

Many cells use more than one intracellular messenger at once. Skeletal muscle cells are a good example (Fig. 16.4). In the excitement before a race, the runner's adrenal glands release adrenaline into the blood. This binds to a receptor on skeletal muscle cells called the β-adrenergic receptor. The complex of adrenaline plus β-adrenergic receptor can now activate G_s and hence adenylate cyclase. This in turn generates cAMP, which activates **cAMP-dependent protein kinase.** This enzyme, which is given the short name of **protein kinase A,** phosphorylates glycogen phosphorylase kinase and turns on the latter enzyme even when cytosolic calcium is low (see Fig. 13.15 on page 305). The end result is that even before the runner begins to run, the muscles break down glycogen and make the glucose-6-phosphate (page 288) they will need once the race begins.

✺ BIOCHEMICAL SIGNALING

Receptor Tyrosine Kinases and the MAP Kinase Cascade

When platelets are stimulated, they not only help the blood to clot, but also trigger the damaged blood vessel to repair itself. They do this by regulated exocytosis of a protein called **platelet-derived growth factor** (**PDGF**) (Fig. 16.8). The plasma membranes of the cells of the blood vessel contain receptors for this protein. PDGF receptors (Fig. 9.15 on page 198) comprise an extracellular domain, which binds PDGF, a single polypeptide chain that crosses the plasma membrane, and a cytosolic domain with tyrosine kinase activity. The PDGF receptor is therefore able to phosphorylate tyrosine residues on other proteins.

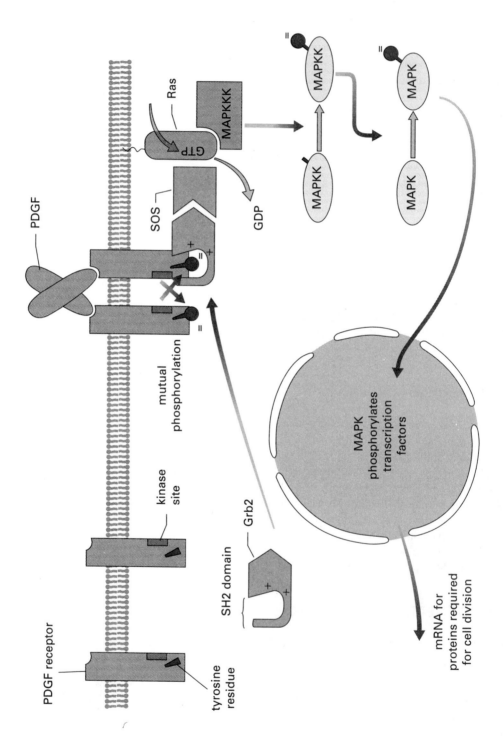

Figure 16.8. The PDGF receptor, like other growth factor receptors, activates the GTPase Ras and therefore the MAP kinase cascade.

In the absence of PDGF, the PDGF receptor does not come in contact with proteins to phosphorylate. PDGF can bind to two receptor molecules, drawing them close to each other and allowing them to phosphorylate each other on tyrosine residues.

A number of cytoplasmic proteins have a domain called **SH2** that is just the right shape to stick to phosphorylated tyrosine. At the bottom of a deep pocket in the protein surface is a positively charged arginine. Although proteins can be phosphorylated on amino acids other than tyrosine, only tyrosine is long enough to insert down the pocket so that the negative phosphate can stick to the positive arginine. Proteins with SH2 domains therefore stick to dimerized PDGF receptors. In contrast, SH2 domains do not stick to solitary PDGF receptors whose tyrosines do not carry the negatively charged phosphate group.

One protein that has an SH2 domain is called **growth factor receptor binding protein number 2 (Grb2)**. Grb2 has no catalytic function but recruits a second protein called SOS, and SOS is a guanine nucleotide exchange factor for a GTPase called **Ras,** allowing it to discard GDP and bind GTP. Active Ras turns on a chain of protein kinases, each of which phosphorylates and hence activates the next, culminating in a kinase called **mitogen-associated protein kinase (MAP kinase** or **MAPK)**. The kinase that phosphorylates MAP kinase is called **MAP kinase kinase (MAPKK)** while the one that phosphorylates MAPKK is called **MAP kinase kinase kinase (MAPKKK)**. MAPKKK is activated by Ras.

The word *mitogen* means a chemical that tends to cause mitosis, reflecting the fact that MAP kinase is activated by transmitters such as PDGF that turn on cell division. When MAP kinase is phosphorylated, it moves to the nucleus and phosophorylates transcription factors that in turn stimulate the transcription of the genes for **cyclin D** (page 413) and other proteins that are required for DNA synthesis and cell division.

Platelet-derived growth factor is one of many growth factors. All work in much the same way: Their receptors are tyrosine kinases that are triggered to dimerize and phosphorylate their partners upon tyrosine when the growth factor binds. The general term for this type of receptor is **receptor tyrosine kinase.** Phosphorylated tyrosine then recruits proteins with SH2 domains including Grb2, which in turn allows activation of Ras and the MAP kinase pathway, leading to DNA synthesis and cell division.

Medical Relevance 16.1

Blocking Growth Factor Receptors

Active growth factor receptors cause cell division, by activating the MAP kinase pathway, and keep cells alive, by activating PKB (page 417). Turning off growth factor receptors therefore tends both to stop cells dividing and to kill them. The drug trastuzumab is effective in slowing down the progression of breast cancer because it prevents a growth factor, called epidermal growth factor (EGF), from binding to its receptor tyrosine kinase. This slows cell division and promotes cell death in the cancer cells.

Medical Relevance 16.2

R115777 Blocks Ras

Like other GTPases, Ras turns itself off by hydrolyzing its bound GTP. Mutant forms of Ras without GTPase activity are therefore **constitutively active:** they are always in the on state and will therefore be activating the pathway that terminates in MAP kinase and cell division at all times, even in the absence of growth factor. Such mutant forms of Ras are found in about 20% of all human cancers. The drug R115777 prevents Ras from activating MAPKKK and should therefore help stop the uncontrolled cell division that is characteristic of cancer cells.

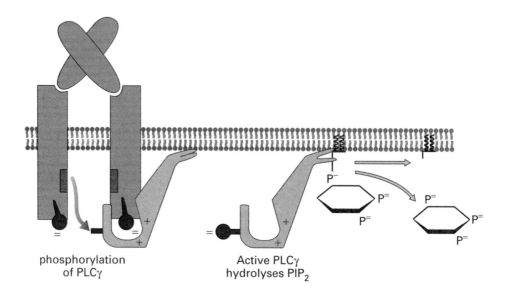

<u>Figure 16.9.</u> The PDGF receptor, like other growth factor receptors, phosphorylates and hence activates phospholipase Cγ.

Growth Factors Can Trigger a Calcium Signal

A second protein that contains an SH2 domain and that is therefore recruited to phosphorylated receptor tyrosine kinases is another isoform of phosphoinositide phospholipase C, **PLCγ.** Binding to the phosphorylated tyrosine holds PLCγ at the growth factor receptor long enough for it itself to be phosphorylated (Fig. 16.9), and this activates its enzymatic action. Active PLCγ hydrolyzes PIP$_2$ into diacylglycerol and inositol trisphosphate, and in turn inositol trisphosphate triggers the release of calcium from the smooth endoplasmic reticulum.

Protein Kinase B and the Glucose Transporter: How Insulin Works

Figure 16.10 shows the **insulin receptor.** Like other receptor tyrosine kinases, the insulin receptor has an extracellular domain that can bind the transmitter, in this case the protein insulin, a single polypeptide chain that crosses the plasma membrane, and a cytosolic domain with tyrosine kinase activity. Unlike growth factor receptors, the insulin receptor exists as a dimer even in the absence of its ligand, insulin. When insulin binds, the shape and orientation of the individual insulin receptors change a little, and this allows each receptor to phosphorylate its partner upon tyrosine. An associated protein called the **insulin receptor substrate number 1 (IRS-1)** is also phosphorylated on tyrosine. Proteins with SH2 domains are therefore recruited, either to the phosphorylated tyrosines on the insulin receptors themselves or to the phosphorylated tyrosines on IRS-1. The most important protein to be recruited is **phosphoinositide 3-kinase (PI 3-kinase).** Binding to the phosphorylated tyrosine holds PI 3-kinase close to the insulin receptor long enough for it to

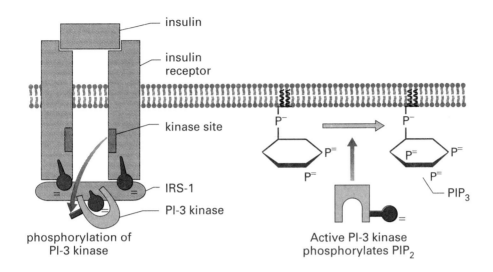

Figure 16.10. The insulin receptor phosphorylates and hence activates PI 3-kinase.

become phosphorylated, and this activates its enzymatic action. PI 3-kinase operates on the same substrate, PIP_2, that PLC does, but instead of hydrolyzing it, PI 3-kinase adds another phosphate group to the inositol head group, generating the intensely charged lipid **phosphatidylinositol trisphosphate (PIP_3)**.

Just as phosphotyrosine is bound by a wide variety of proteins that contain SH2 domains, so highly phosphorylated inositols, such as those in PIP_3, are bound by a domain called the **PH domain** found on many proteins. Most important among these is a protein kinase called **protein kinase B (PKB)**. PKB is itself activated by phosphorylation, but the kinase that does this is located at the plasma membrane and therefore only gets a chance to phosphorylate PKB when PKB is held at the plasma membrane through its binding to PIP_3 (Fig. 16.11).

In many cell types, particularly fat cells and muscle cells, the final stage in trafficking (page 226) of the glucose carrier (page 316) from the Golgi apparatus to the plasma membrane requires active PKB. At the same time, glucose carriers are being endocytosed and are only returned to the plasma membrane if PKB is active. When we eat a large meal, insulin concentrations in the blood increase. Activation of the insulin receptor therefore causes an increase in the activation of PKB and hence a translocation of glucose carriers to the plasma membrane. This allows muscle and fat cells to take up large amounts of glucose from the extracellular medium. Muscle cells convert the glucose to glycogen (page 298); fat cells convert the glucose to fat (page 300). The action of PKB on a protein called **BAD** that is shown in Figure 16.11 will be described in Chapter 19.

Crosstalk—Signaling Pathways or Signaling Webs?

Cell biologists often talk of signaling pathways in which a transmitter activates a chain of events culminating in an effect. The vertical arrows in Figure 16.12 show four signaling pathways that we have discussed in this chapter. We have already seen one example of

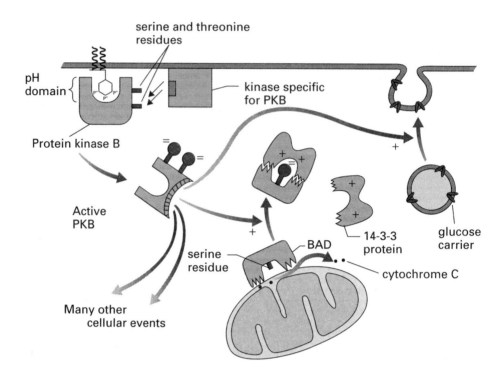

Figure 16.11. PIP₃ recruits protein kinase B (PKB) to the plasma membrane, where it is activated. Active PKB has many effects including exocytosis of vesicles containing glucose carriers and inactivation of BAD.

how these pathways can interact: growth factor receptors can trigger a calcium signal by activating phospholipase Cγ. In fact, the pathways can interact in many ways and at many levels. The green arrows in Figure 16.12 show some of the most important. PI 3-kinase can bind to phosphotyrosine on growth factor receptors via its SH2 domain and be activated by phosphorylation, thus growth factors, as well as insulin, will activate protein kinase B (arrow X in Fig. 16.12). In Chapter 19 we will see that this process is literally vital for our cells—if it stops, they die (page 417).

Many of the actions of cAMP occur through the phosphorylation of other proteins by protein kinase A. In particular, cAMP can activate transcription of many eukaryotic genes (arrow Y in Fig. 16.12), but it does so by causing the phosphorylation of transcription factors, and hence activating them. This is in sharp contrast to the cAMP-CAP system in *Escherichia coli* (page 114).

Calcium ions can activate a number of protein kinases, notably **protein kinase C** and **calcium-calmodulin–activated protein kinase.** These kinases phosphorylate many of the same targets as protein kinase A and hence can activate the same downstream events (arrows Z in Fig. 16.12). Paradoxically, calcium-calmodulin also activates the phosphatase calcineurin, which dephosphorylates and hence activates the transcription factor NFAT (page 222). The more we know about intracellular signaling systems, the more they come to resemble a web of interactions rather than a set of discrete pathways.

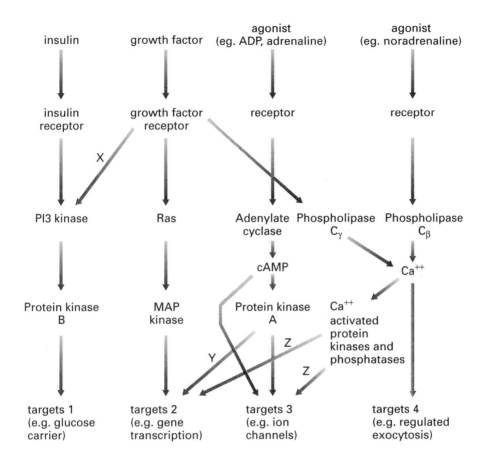

Figure 16.12. Interactions of signaling pathways.

SUMMARY

1. An intracellular messenger is an intracellular solute whose concentration changes in response to cell stimulation; it activates or modulates a variety of cellular processes. The most important intracellular messengers are calcium ions, cyclic adenosine monophosphate (cAMP), and cyclic guanosine monophosphate (cGMP).

2. The increased cytosolic calcium may be derived from two sources: the extracellular fluid and the smooth endoplasmic reticulum.

3. A complex cascade of reactions causes release of calcium from the smooth endoplasmic reticulum in response to stimulation of the cell. Binding of extracellular chemical to a cell surface receptor activates the trimeric G protein G_q, which in turn activates the enzyme phospholipase $C\beta$, which cleaves the hydrophilic head group inositol trisphosphate (IP_3) from the lipid phosphatidylinositol bisphosphate. Inositol trisphosphate then binds to and opens a calcium channel in the membrane of the endoplasmic reticulum.

4. A different set of receptors activates G_s and hence adenylate cyclase, which makes cyclic AMP from ATP. Many of the actions of cAMP are mediated by cAMP-dependent protein kinase (protein kinase A).

5. The majority of transmitters are released from cells by exocytosis induced by an increase of cytosolic calcium concentration.

6. Many nerve cells have their axon terminals close to a second cell, and they release their transmitter onto it. The complete unit of axon terminal, gap, and the part of the cell that receives the transmitter is called a synapse.

7. Receptor tyrosine kinases are caused to dimerize when their ligand binds. This allows each partner to phosphorylate the other upon tyrosine; this in turn recruits proteins containing SH2 domains.

8. Grb2 is an SH2-domain-containing protein that serves to bring together SOS and Ras, allowing activation of Ras and therefore activation of the MAP kinase pathway. This culminates in the transcription of genes necessary for DNA synthesis and cell division.

9. Phospholipase Cγ and phosphoinositide 3-kinase, two enzymes that act on the membrane lipid phosphatidylinositol bisphosphate (PIP_2), both have SH2 domains and are both recruited to, and phosphorylated by, receptor tyrosine kinases: this activates them.

10. Protein kinase B is recruited to the plasma membrane by PIP_3; this allows it to be phosphorylated, which activates it.

11. In many cells including fat and skeletal muscle, glucose transporters only appear at the plasma membrane when PKB is kept active through the activation of the insulin receptor.

FURTHER READING

Downward, J. 1998. Mechanisms and consequences of activation of protein kinase B/Akt. *Curr. Opin. Cell. Biol.* 10: 262–267.

Marshall, C. J. 1996. Ras effectors. *Curr. Opin. Cell Biol.* 8: 197–204.

Toescu E. C., and Verkhratsky, A. 1998. *Integrative Aspects of Calcium Signalling.* New York: Plenum Press.

❋ REVIEW QUESTIONS

For each question, choose the ONE BEST answer or completion.

1. Which of the following does not exist?
   A. Protein kinase A
   B. Protein kinase B
   C. Phospholipase Cα

D. Phospholipase Cβ

E. Phospholipase Cγ

2. Which of the following is an enzyme, one of whose substrates is a nucleotide triphosphate?

A. The PDGF receptor

B. The insulin receptor

C. Ras

D. Protein kinase A

E. All of the above

3. Arrival of a scent chemical at a scent-sensitive nerve cell causes that cell to

A. activate adenylate cyclase.

B. open cAMP-gated channels in the plasma membrane.

C. open voltage-gated sodium channels in the plasma membrane.

D. generate action potentials.

E. all of the above.

4. Which of the following statements is false?

A. An increase of cytosolic cAMP concentration can activate a protein kinase.

B. An increase of cytosolic calcium concentration can activate a protein kinase.

C. Dimerization of growth factor receptors can activate a protein kinase.

D. Appearance of PIP$_3$ at the plasma membrane can activate a protein kinase.

E. Hydrolysis of GTP by Ras can activate a protein kinase.

5. Depolarization of axon terminals causes calcium ions to flow into the cytosol through

A. voltage-gated calcium channels.

B. inositol trisphosphate-gated calcium channels.

C. ryanodine receptors.

D. calcium ATPases.

E. cAMP-gated calcium channels.

6. Inositol trisphosphate-gated calcium channels are located in

A. the plasma membrane.

B. the smooth endoplasmic reticulum.

C. the inner mitochondrial membrane.

D. regulated exocytotic vesicles.

E. all of the above.

7. Phosphatidylinositol bisphosphate is used in two different signaling pathways. These convert it, respectively, into

A. IP$_3$ and PIP$_3$.

B. IP$_3$ and cAMP.

C. PIP$_2$ and PIP$_3$.

D. PIP$_2$ and cAMP.

E. cAMP and cGMP.

ANSWERS TO REVIEW QUESTIONS

1. *C.* While protein kinase A, protein kinase B, phospholipase Cβ, and phospholipase Cγ are enzymes with important roles in signaling, there is no enzyme with the name phospholipase Cα.

2. *E.* Protein kinases transfer the γ-phosphate group of ATP to a protein. Thus all protein kinases (including the PDGF receptor and the insulin receptor, both of which are protein kinases) are

enzymes that operate on ATP, a nucleotide triphosphate. Ras, in contrast, hydrolyzes the nucleotide triphosphate GTP.

3. **E.** These statements constitute a summary of signaling in scent-sensitive nerve cells.

4. **E.** Hydrolysis of GTP by Ras turns Ras off, so that it is no longer able to activate MAPKK kinase. The other statements are true. In particular (A) cAMP activates protein kinase A, (B) calcium activates a number of protein kinases, (C) dimerization activates growth factor receptors, and (D) PIP_3 recruits PKB to the plasma membrane, where it is phosphorylated and hence activated.

5. **A.** Statements B, D, and E are completely false, in particular (B) inositol trisphosphate-gated calcium channels are opened by inositol trisphosphate, which is not generated by depolarization. (D) Calcium ATPases push calcium ions out of the cytosol—they do not allow it in. (E) cAMP-gated calcium channels are opened by cAMP, which is not generated by depolarization. Some scientists think that (C) may operate in a restricted set of axon terminals, that is, depolarization of the axon terminal allows calcium to flow in through voltage-gated calcium channels, and this calcium then opens ryanodine receptors (page 348). However, this is certainly not a general phenomenon.

6. **B.** Answers C through E are completely false. Some rare cells (the best studied is a sensory cell in lobsters) do express inositol trisphosphate-gated calcium channels in their plasma membranes (answer A), but this is certainly not typical.

7. **A.** Phospholipase C hydrolyzes phosphatidylinositol bisphosphate (PIP_2), releasing the head group as IP_3, while phosphatidylinositol 3-kinase phosphorylates the inositol head group of PIP_2, generating PIP_3. Each of these products has an important role in cell signaling. IP_3 acts to cause release of calcium from the endoplasmic reticulum, while PIP_3 causes activation of protein kinase B.

17

INTERCELLULAR COMMUNICATION

The millions of cells that make up a multicellular organism can work together only because they continually exchange the chemical messages called transmitters. Here we describe how systems of cells use transmitters to cooperate for the good of the organism. Most of the many transmitters known are found in all animals and probably evolved with ancestral multicellular organisms more than a billion years ago. We will introduce the general principles of intercellular communication and then illustrate how transmitters operate in a single tissue, the **gastrocnemius muscle**.

✳ CLASSIFYING TRANSMITTERS AND RECEPTORS

Transmitter mechanisms can be classified in two ways. The first depends on their lifetime in the extracellular fluid. A transmitter that is rapidly broken down or taken up into cells acts only near its release site. One that is broken down slowly can diffuse a long way and may act on cells a long way away. The shortest lived transmitters of all are those released at synapses (page 344), where the distance from release site to receptor is only 100 nm.

At the other extreme are transmitters that last many minutes or, sometimes, even longer. **Hormones** are long-lived transmitters that are released into the blood and travel around the body before being broken down. Most are released by specialized groups of secretory cells that form a structure called an endocrine gland. **Paracrine transmitters** also last many

Cell Biology: A Short Course, Second Edition, by Stephen R. Bolsover, Jeremy S. Hyams,
Elizabeth A. Shephard, Hugh A. White, Claudia G. Wiedemann
ISBN 0-471-26393-1 Copyright © 2004 by John Wiley & Sons, Inc.

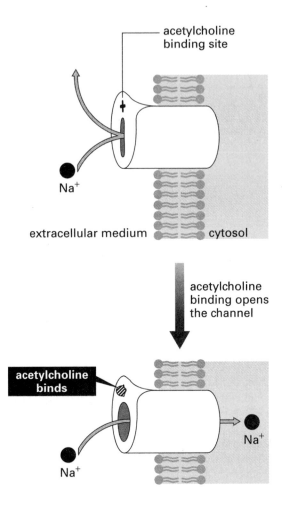

Figure 17.1. The nicotinic acetylcholine receptor is an ionotropic cell surface receptor that opens when acetylcholine in the extracellular medium binds.

minutes before being broken down, but they are released into specific tissues rather than into the blood, and only diffuse within the tissue before they are destroyed.

The second way of classifying transmitter mechanisms depends on the location and action of the receptor on the target cell. There are three types of receptors: **ionotropic cell surface receptors, metabotropic cell surface receptors,** and **intracellular receptors.**

Ionotropic Cell Surface Receptors

Ionotropic cell surface receptors are channels that open when a specific chemical binds to the extracellular face of the channel protein. The **nicotinic acetylcholine receptor** (Fig. 17.1), so named because the drug nicotine binds to it, is one example. In the absence of a transmitter called acetylcholine, the channel is closed. When acetylcholine binds, the channel opens and allows sodium and potassium ions to pass through. The electrochemical gradient pushing sodium ions into the cell is much greater than that pushing potassium ions out of the cell, so when the channel opens, sodium ions pour in carrying positive charge and depolarizing the plasma membrane. This mechanism is similar to that of two channels we met in Chapter 16, the inositol trisphosphate-gated calcium channel and the cAMP-gated channel. There is, however, a major difference: Those two channels were opened by cytosolic solutes, while ionotropic cell surface receptors are opened by extracellular solutes.

Metabotropic Cell Surface Receptors

Metabotropic cell surface receptors are linked to enzymes. We have already met a number of metabotropic cell surface receptors in Chapter 16. When the ADP receptor (page 344) binds extracellular ADP, it activates G_q and hence phospholipase $C\beta$. The receptor for smell chemicals (page 350) and the β-adrenergic receptor (page 353) are linked to G_s and hence adenylate cyclase so ligand binding increases cytosolic cAMP. Receptor tyrosine kinases are themselves protein kinases activated when their ligand binds.

The α-**adrenergic receptor** (Fig. 17.2) is another receptor that causes cytosolic calcium concentration to increase. The α- and β-adrenergic receptors are distinct proteins that bind the same transmitters, adrenaline and noradrenaline. To simplify the issue somewhat, we can say that noradrenaline acts mainly on α receptors and adrenaline acts mainly on β receptors. Because the α and β receptors are distinct proteins, it is possible to design drugs (α and β blockers) that interfere with one or the other.

Intracellular Receptors

Intracellular receptors lie within the cell (in the cytosol or in the nucleus) and bind transmitters that diffuse through the plasma membrane. They always exert their effects by activating enzymes. The receptors for nitric oxide and steroid hormones are two examples.

Nitric oxide, or NO, is a transmitter in many tissues. It is not stored ready to be released but is made from arginine at the time it is needed. NO diffuses easily through the plasma membrane and binds to various cytosolic proteins that are NO receptors. One particularly important NO receptor is the enzyme **guanylate cyclase,** which in the presence of NO converts the nucleotide GTP to the intracellular messenger cyclic guanosine monophosphate, or cGMP.

Steroid hormones have intracellular receptors such as the glucocorticoid receptor (page 121). In the absence of hormone this receptor remains in the cytosol and is inactive because it is bound to an inhibitor protein. However, when the glucocorticoid hormone binds to its receptor, the inhibitor protein is displaced. The complex of the glucocorticoid receptor with its attached hormone now moves into the nucleus. Here two molecules of the complex bind to a 15-bp sequence known as the hormone response element (HRE), which lies upstream of the TATA box (page 122). The HRE is a transcriptional enhancer sequence.The binding of the glucocorticoid hormone receptor to the HRE stimulates transcription.

✤ INTERCELLULAR COMMUNICATION IN ACTION: THE GASTROCNEMIUS MUSCLE

The gastrocnemius muscle illustrates how transmitters and receptors operate in a single tissue. This is the calf muscle at the back of the lower leg. When it contracts, it pulls on the Achilles tendon so that the toes push down on the ground. Most of the bulk of the muscle is made up of one type of cell, skeletal muscle cells.

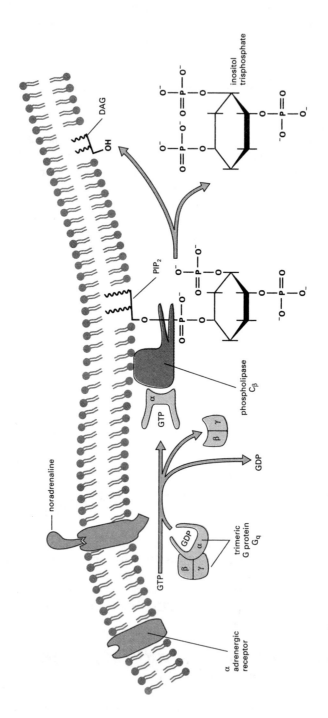

Figure 17.2. Noradrenaline activates G_q and hence phospholipase C_β in many cells including smooth muscle.

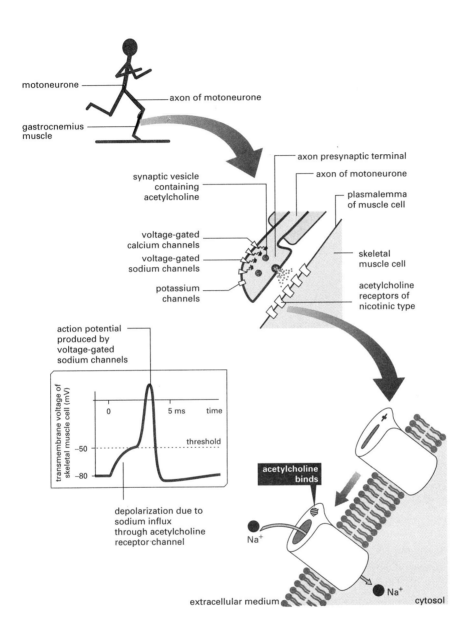

Figure 17.3. Motoneurones release the transmitter acetylcholine that binds to nicotinic receptors on the muscle cells. The plasma membrane of the muscle cell is depolarized to threshold and fires an action potential.

Telling the Muscle to Contract: The Action of Motoneurones

All skeletal muscle cells are relaxed until they receive a command to contract from nerve cells called **motoneurones.** Motoneurone cell bodies are in the spinal cord while the myelinated axons run to the muscle. The axon terminal of the motoneurone that controls the gastrocnemius muscle contains vesicles of the transmitter acetylcholine (Fig. 17.3). To press the foot down, action potentials travel from the spinal cord down the motoneurone axon to its terminal, opening voltage-gated calcium channels in the plasma membrane. Calcium ions pour in, raising their cytosolic concentration from 100 nmol liter^{-1} to 1 μmol liter^{-1}. The calcium torrent causes the synaptic vesicles containing acetylcholine to fuse with the plasma membrane, releasing the transmitter into the extracellular fluid.

Acetylcholine survives only 0.2 ms in the extracellular fluid since it is quickly hydrolyzed into choline and acetate by the enzyme **acetylcholinesterase.** However, its target— the skeletal muscle cell—is only 100 nm away. It hits its mark before it is broken down and binds to the nicotinic acetylcholine receptors present in the plasma membrane of the skeletal muscle cell. These open, allowing both potassium and sodium ions to pass. There is no large electrochemical gradient for potassium, but sodium is at a high concentration in the extracellular fluid and rushes in, depolarizing the plasma membrane of the skeletal muscle cell. The plasma membrane of the skeletal muscle cell also contains voltage-gated sodium channels. When the flow of sodium ions through the nicotinic acetylcholine receptors depolarizes the plasma membrane to threshold, the skeletal muscle cell fires an action potential, which in turn causes release of calcium from the endoplasmic reticulum (page 348). The resulting increase of calcium concentration in the cytosol of the muscle cell causes it to contract (page 393).

Controlling the Blood Supply: Paracrine Transmitters

Figure 17.4 is a cut-away drawing of a blood vessel in the muscle. Lining the tube is a thin layer of endothelial cells. Wrapped around these are muscle cells of a different type, smooth muscle cells. Both endothelial cells and smooth muscle cells are much smaller than skeletal muscle cells. A small blood vessel may be only as large as a single skeletal muscle cell. Smooth muscle cells have no input from motoneurones. Instead, like many other internal organs, they are supplied by a distinct class of nerve cells that are called **autonomic.** Autonomic nerve cells usually use one of two transmitters: acetylcholine or noradrenaline. Two other transmitters, adrenaline and nitric oxide, also help to adjust the blood flow on a timescale of seconds to minutes. The nerve cells that release noradrenaline onto the smooth muscle of blood vessels are called **vasoconstrictors** because they cause blood vessels to contract. The smooth muscle cells have α-adrenergic receptors in their plasma membranes. Binding of noradrenaline to α-adrenergic receptors activates PLCβ, which generates IP$_3$, which in turn releases calcium from the endoplasmic reticulum into the cytosol. The increase of cytosolic calcium concentration causes the smooth muscle cells to contract, constricting the blood vessel and reducing the flow. Vasoconstrictor nerves are used to restrict the flow of blood to muscles and other organs that are not in heavy use.

The hormone adrenaline is chemically related to noradrenaline but is more stable, lasting a minute or so in the extracellular fluid before being broken down. It is released from an endocrine gland (the adrenal gland) during times of stress and spreads around the body in the blood. Adrenaline that diffuses to the skeletal muscle cells stimulates them to begin breaking down glycogen to make glucose-6-phosphate (page 305). The smooth muscle cells of blood vessels within skeletal muscles also have β-adrenergic receptors connected to adenylate cyclase. However, they do not contain glycogen, and cAMP has another effect in these cells. When cAMP rises, it activates cAMP-dependent protein kinase. This in turn phosphorylates proteins that relax the smooth muscle cell. The action of adrenaline is

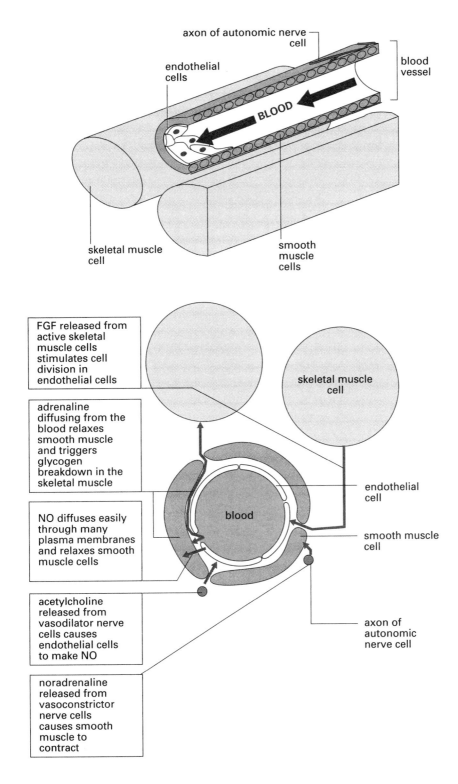

Figure 17.4. Transmitters regulate the blood supply to muscles.

therefore to increase the blood supply to all the muscles of the body in preparation for flight or fight. If we are very frightened and too much adrenaline is released, so much blood is diverted from the brain to the muscles that we faint.

Just as autonomic vasoconstrictor nerve cells are used to shut down blood flow to particular regions, autonomic vasodilator nerves are used to dilate blood vessels in muscles that are about to be used. They release acetylcholine, but neither smooth muscle cells nor endothelial cells have nicotinic acetylcholine receptors. Endothelial cells have, instead, a different receptor called the **muscarinic acetylcholine receptor.** This is named for a chemical that binds it: muscarine, from the poisonous fly agaric mushroom *Amanita muscaria* (page 376). Like the α-adrenergic receptor, the muscarinic acetylcholine receptor is linked to G_q and therefore to phospholipase $C\beta$, so that binding of acetylcholine causes an increase of cytosolic calcium concentration. Endothelial cells do not contract in response to a calcium increase. However, they have an enzyme that makes nitric oxide, and this enzyme is activated by calcium. When endothelial cells are stimulated with acetylcholine, they make nitric oxide. This intercellular transmitter easily passes through the plasma membranes of both endothelial and smooth muscle cells and reaches its receptor within the smooth muscle cells. Here it activates guanylate cyclase, causing an increase of cGMP concentration.

Just as cAMP exerts many of its effects through cAMP-dependent protein kinase, called protein kinase A for short, so cGMP exerts many of its effects through cGMP-dependent protein kinase, also called protein kinase G. One of the targets of protein kinase G is calcium ATPase (page 318). When this is phosphorylated by protein kinase G it works harder, reducing the concentration of calcium ions in the cytosol. This has the effect of relaxing the smooth muscle cells. Nitric oxide lasts for only about 4 s before being broken down. It is therefore a paracrine transmitter, able to diffuse through and relax all the smooth muscle cells surrounding the endothelial cells, but without lasting long enough to pass into more remote tissues.

Example 17.1 Nitroglycerine Relieves Angina

The discovery in 1987 that nitric oxide was a transmitter explained why nitroglycerine (more familiar as an explosive than as a medicine) relieved angina pectoris. Angina is a pain felt in an overworked heart. Nitroglycerine spreads throughout the body via the bloodstream and slowly breaks down, releasing nitric oxide that then dilates blood vessels. The heart no longer has to work so hard to drive the blood around the body.

Example 17.2 Viagra

Just as cAMP phosphodiesterase hydrolyzes cAMP to AMP and hence terminates its action as an intracellular messenger, so cGMP is inactivated by cGMP phosphodiesterase. There are a number of isoforms of this enzyme in different human tissues. The drug sildenafil, sold as Viagra, inhibits the form of the enzyme found in the penis. If cGMP is not being made, this has little effect on blood flow to the region. However, when cGMP is made in response to a local production of NO, its concentration in blood vessel smooth muscle increases much more than would otherwise occur because its hydrolysis to GMP is blocked. This in turn causes a greater activation of protein kinase G, a greater activation of the calcium ATPase, a lower cytosolic calcium concentration, a greater relaxation of blood vessel smooth muscle, and therefore greater blood flow.

New Blood Vessels in Growing Muscle

All the phenomena we have discussed so far occur on a timescale of seconds. However, when a muscle is repeatedly exercised over many days, it becomes stronger: the individual skeletal muscle cells enlarge. This is because high cytosolic calcium acts via NFAT (page 222) to stimulate the transcription of genes coding for structural proteins. Furthermore, new blood vessels sprout and grow into the enlarging muscle. A growth factor called FGF is released by stimulated muscle. The receptor tyrosine kinase for FGF, like that for PDGF (page 353) is found on endothelial and smooth muscle cells and like the PDGF receptor signals via Ras and MAP kinase (page 355) to trigger cell division and hence the growth of new blood vessels. (FGF stands for fibroblast growth factor, but FGF is effective on a vast range of cell types including both endothelial and smooth muscle cells).

In our study of the gastrocnemius muscle and its blood supply, we have seen examples of all types of transmitter mechanism. Acetylcholine acts as a synaptic transmitter at the axon terminal of the motoneurone. Adrenaline is a hormone. The other transmitters are paracrine. The nicotinic acetylcholine receptor is an ionotropic cell surface receptor. The nitric oxide receptor is intracellular. The other receptors are metabotropic cell surface receptors. There is a wide variety of timescales of action. The acetylcholine released from the axon terminal of the motoneurone causes a contraction of the skeletal muscle cell within 5 ms, by which time the acetylcholine has already been destroyed. Adrenaline lasts 1 min and dilates blood vessels for all this time. FGF lasts 10 min, but its effects are much longer lasting. FGF triggers the synthesis of proteins, which then act to cause cell proliferation that lasts for

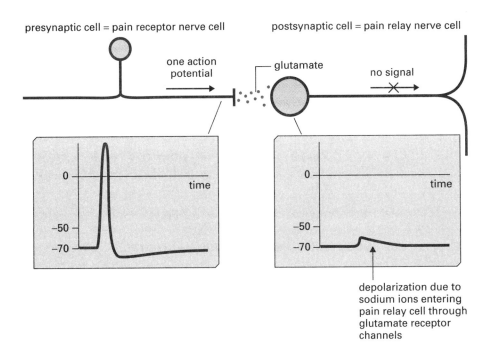

Figure 17.5. At most synapses, one presynaptic action potential is insufficient to depolarize the postsynaptic nerve cell to threshold.

days. A similar pattern of intercellular communication, using some of the same transmitters and receptors but also many others, is found in every tissue of the body.

❊ SYNAPSES BETWEEN NEURONS

The synapse between a motoneurone and a skeletal muscle cell is unusual in that a single action potential in the presynaptic cell releases enough transmitter to depolarize the postsynaptic cell to threshold and hence produce an action potential in the postsynaptic cell. The synapse between a pain receptor and a pain relay nerve cell (page 341) is more typical (Fig. 17.5). An action potential in the axon terminal of the pain receptor raises cytosolic calcium and releases the transmitter glutamate. This diffuses to the pain relay cell where it binds to glutamate receptors, which are ionotropic and allow both potassium and sodium ions to pass. The postsynaptic nerve cell depolarizes as sodium ions move in through the glutamate receptor channels. However, the depolarization is not enough to take the transmembrane voltage to threshold. As soon as glutamate is removed from the extracellular medium, the transmembrane voltage returns to the resting level. In the axon of the relay cell, some distance from the synapse, the transmembrane voltage does not change, and no message passes on to the brain. The subject does not feel pain.

Example 17.3 **A Toxic Glutamate Analogue**

The grasspea is a protein-rich crop that has been cultivated since ancient times and is still an important source of calories and protein in India, Africa, and China. The peas contain the amino acid β-N-oxalyl-L-α-β-diaminopropionic acid. If untreated peas are eaten, the toxin binds to and opens the glutamate receptor on nerve cells. The resulting long-lasting depolarization damages and finally kills the nerve cells. Boiling the peas during cooking destroys the toxin; but in times of famine, when fuel is scarce, many people are poisoned, and the resulting brain damage is irreversible.

$$
\begin{array}{cc}
 & COO^- \\
 & | \\
 & C{=}O \\
COO^- & | \\
| & NH \\
CH_2 & | \\
| & CH_2 \\
CH_2 & | \\
| & \\
{}^-OOC{-}CH{-}NH_3{}^+ & {}^-OOC{-}CH{-}NH_3{}^+ \\
\end{array}
$$

glutamate β-N-oxalyl-L-α-β-diaminopropionic acid

A subject *does* feel pain when they move their finger into the hot air from a hair dryer (Fig. 17.6). This is because the hot air heats a large area and causes many pain receptors to fire action potentials. Many of the pain receptors synapse onto one relay cell, which

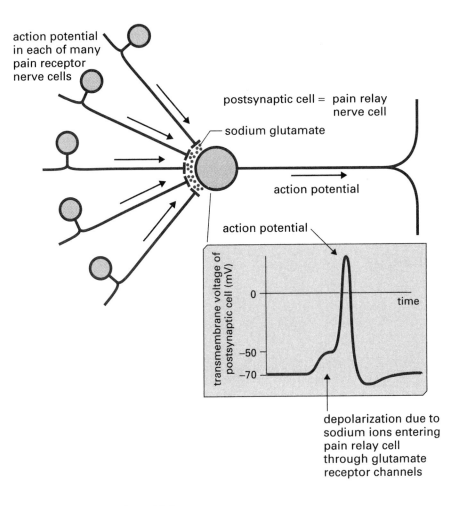

Figure 17.6. Spatial summation at a synapse.

therefore receives many doses of glutamate. Because of this **spatial summation,** enough glutamate receptor channels open to depolarize the relay cell to threshold, and an action potential travels along the axon of the relay cell to the brain. The subject feels pain.

An intense stimulus to a small area is also painful. When the subject is jabbed with a needle, only one pain receptor is activated, but that receptor is intensely stimulated and fires a rapid barrage of action potentials, each of which causes the release of glutamate and an extra depolarization of the pain relay cell (Fig. 17.7). Soon the transmembrane voltage of the relay cell reaches threshold, and an action potential travels along its axon toward the brain and the subject feels pain. Such **temporal summation** only occurs if the presynaptic action potentials are frequent enough to ensure that the depolarizations produced in the postsynaptic cell add.

Another amino acid, γ-amino butyric acid (GABA) (page 37), is also a synaptic transmitter. Its receptor is an ionotropic cell surface receptor that forms a channel selective

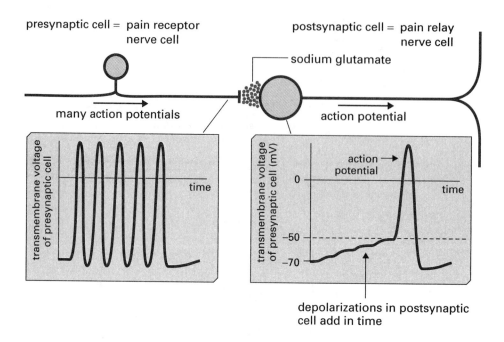

Figure 17.7. Temporal summation at a synapse.

for chloride ions. Figure 17.8 illustrates a single nerve cell bearing both glutamate and γ-amino butyric acid receptors.

At Ⓐ an action potential in the GABA-secreting axon releases GABA onto the surface of the postsynaptic cell, causing the GABA receptor channels to open. Although chloride ions could now move into or out of the cell, they do not, because their tendency to travel into the cell down their concentration gradient is balanced by the repulsive effect of the negative voltage of the cytosol: chloride ions are at equilibrium (page 314). The opening of GABA channels therefore causes no ion movements and therefore does not alter the transmembrane voltage.

At Ⓑ, action potentials occur simultaneously in six glutamate-secreting axons. In this example, this activity provides enough glutamate to depolarize the postsynaptic cell to threshold, and the postsynaptic nerve cell fires an action potential.

At Ⓒ, action potentials occur simultaneously in six glutamate-secreting axons and also in the GABA-secreting axon. The same number of glutamate receptor channels open as before, and the same number of sodium ions flow into the postsynaptic nerve cell, tending to depolarize it. However, as soon as the transmembrane voltage of the postsynaptic cell deviates from the resting value, chloride ions start to enter through the GABA receptor channels because the cytosol is no longer negative enough to prevent them from entering the cell down their concentration gradient. The inward movement of negatively charged chloride ions neutralizes some of the positive charge carried in by sodium ions moving in through the glutamate receptor channels. The postsynaptic nerve cell therefore does not

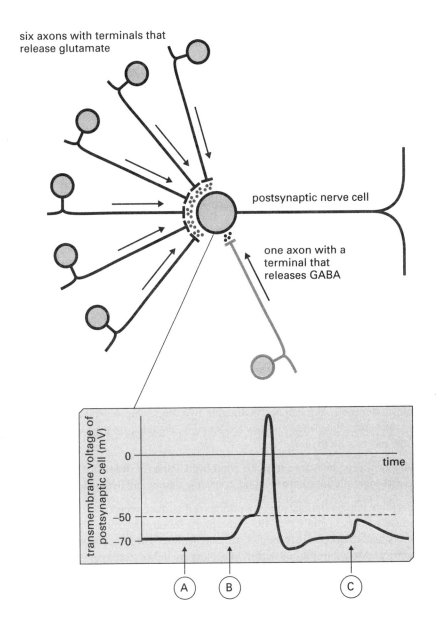

six axons with terminals that
release glutamate

postsynaptic nerve cell

one axon with a
terminal that
releases GABA

Figure 17.8. Inhibition by a GABAergic synapse.

depolarize as much and does not reach threshold. No action potential is generated in the
postsynaptic nerve cell.

Antianxiety drugs such as Valium act on the GABA receptor and increase the chance of
its channel opening. Nerve cells exposed to the drug are less likely to depolarize to threshold.
Valium therefore reduces action potential activity in the brain, calming the patient.

Medical Relevance 17.1

Cigarettes, Mushrooms, and Insecticides

Nicotine is one of the most addictive substances known. The reason for this is still unclear. The muscle weakness that new smokers experience is due to nicotine binding to and blocking receptors on the muscle cells, but we are still not sure how nicotine acts in the brain to cause a pleasurable experience.

The fly agaric mushroom *Amanita muscaria* is seldom taken recreationally. Its popular use is more prosaic, in killing flies, hence its English and Latin names. However, Robert Graves has argued that it was used by the classical Greeks and by the Vikings to produce a state of berserk strength with hallucinations. Any such use would be highly dangerous: muscarine acts by stimulating the muscarinic acetylcholine receptor, and the difference between a dose of muscarine large enough to cause pharmacological effects and a lethal one is very small.

Amanita muscaria has an even more dangerous relative, *A. phalloides,* the death cap mushroom. The toxin it produces, phalloidin, does not affect acetylcholine signaling. Instead, it interferes with the actin cytoskeleton (page 390).

Farmers and others who suffer insecticide toxicity show similar symptoms to those of muscarine poisoning. This is because these insecticides work by blocking the enzyme acetylcholinesterase, which breaks down acetylcholine. With the enzyme blocked, acetylcholine hangs around in the extracellular medium, stimulating the receptor for longer than it should. Nerve gases developed to kill people (like Sarin, released in the Tokyo subway by members of the Aum cult in 1995) work in the same way. Paradoxically, sufferers are treated with another toxin, atropine, from the deadly nightshade plant. Atropine turns off the muscarinic acetylcholine receptor.

SUMMARY

1. Transmitter molecules can be classified in two ways. One depends on their lifetime and the extracellular medium they are found in, the other on the location and action of the receptor on the target cell.

2. Synaptic transmitters are extremely short lived. Paracrine transmitters and hormones have a longer lifetime and are found in specific tissues and in the blood, respectively.

3. Receptors can be divided into ionotropic cell surface receptors, metabotropic cell surface receptors, and intracellular receptors. Ionotropic cell surface receptors are ion channels that open in response to ligand binding. The effect upon the target cell is electrical. Metabotropic cell surface receptors are linked to enzymes. The effect upon the target cell is mediated through a biochemical process. Intracellular receptors lie within the target cell and bind transmitters that are able to cross the cell membrane by simple diffusion.

4. The gastrocnemius muscle provides examples of all types of intercellular signaling operating in concert to fit the operation of the tissue to the requirements of the organism.

5. Synapses between nerve cells do not generally mediate a 1-to-1 transmission of the action potential. The presynaptic signal must show summation in time or space to elicit a postsynaptic action potential.

6. Stimulation of ionotropic cell surface receptors that pass chloride ions makes it more difficult for a nerve cell to be depolarized to threshold.

FURTHER READING

Seeley, R. R., Stephens, T. D., and Tate, P. 1995. *Anatomy and Physiology*. St. Louis: Mosby.

Siegel, G. J., Agranoff, B. W., Albers, R. W., Fisher, S. K., and Uhler, M. D. 1999. *Basic Neurochemistry*. Philadelphia: Lippincott-Raven.

❊ REVIEW QUESTIONS

For each question, choose the ONE BEST answer or completion.

1. Ionotropic cell surface receptors
   A. are ion channels permeable to both sodium and potassium.
   B. allow the transmembrane voltage of the cell possessing them to change in response to transmitter.
   C. are coupled to, or are themselves, enzymes.
   D. are only found on nerve cells.
   E. open when the concentration of acetylcholine in the extracellular medium increases.

2. Metabotropic cell surface receptors
   A. are ion channels.
   B. exert their effects by activating enzymes.
   C. are tyrosine kinases.
   D. cause the concentration of cAMP in the cytosol to increase.
   E. make it more difficult for a postsynaptic cell to be depolarized to threshold.

3. Application of adrenaline to a tissue tends to increase blood flow through it, while noradrenaline has the opposite effect. This is because
   A. noradrenaline, but not adrenaline, is able to diffuse through plasma membranes.
   B. the receptor at which noradrenaline is most effective is a cAMP-gated channel while the receptor at which adrenaline is most effective is a calcium channel.
   C. adrenaline is a hormone, while noradrenaline is released close to the target tissue.
   D. the electrochemical gradient for noradrenaline is inward, so it tends to make the cytosol more positive, while the electrochemical gradient for adrenaline is outward.
   E. the receptor at which noradrenaline is most effective is coupled to phospholipase C while the receptor at which adrenaline is most effective is coupled to adenylate cyclase.

4. Nitric oxide
   A. is generated in endothelial cells when the cytosolic calcium concentration increases.
   B. easily diffuses across plasma membranes.
   C. acts at intracellular receptors.
   D. causes an increase of cGMP in the target cell.
   E. All of the above.

5. When seven action potentials arrive in close succession at a particular axon terminal in the brain, the postsynaptic cell is depolarized to just negative of threshold. If, one minute later, the same train of seven action potentials arrives again, the postsynaptic cell is more likely to fire an action potential if
   A. simultaneously, another nerve cell fires an action potential and releases glutamate onto the same postsynaptic cell.

B. simultaneously, another nerve cell fires an action potential and releases GABA onto the same postsynaptic cell.

C. simultaneously, another nerve cell fires an action potential and releases noradrenaline onto the same postsynaptic cell.

D. the same seven action potentials arrive but spread over a longer period of time, that is, their frequency is less.

E. PDGF is present in the extracellular medium.

6. Which of the following statements about the gastrocnemius muscle and its blood supply is untrue?

A. Acetylcholine released from motoneurons causes skeletal muscle cells to contract.

B. Acetylcholine released from sympathetic neurons causes smooth muscle cells to contract.

C. NO released from endothelial cells causes smooth muscle cells to relax.

D. PDGF released from platelets causes endothelial cells to enter the cell cycle.

E. FGF released from skeletal muscle cells causes endothelial cells to enter the cell cycle.

7. The cells in the adrenal gland that secrete adrenaline have, in their plasma membranes, nicotinic acetylcholine receptors, muscarinic acetylcholine receptors, and voltage-gated calcium channels. Application of acetylcholine will therefore cause

A. an increase of cytosolic calcium concentration, the vast majority of which is due to influx of calcium from the extracellular medium.

B. an increase of cytosolic calcium concentration, the vast majority of which is due to release of calcium from the smooth endoplasmic reticulum.

C. an increase of cytosolic calcium concentration, due to a combination of influx of calcium from the extracellular medium and release of calcium from the smooth endoplasmic reticulum.

D. a simultaneous increase of cytosolic calcium concentration and a shift in the transmembrane voltage to a more negative value.

E. a reduction in the cytosolic calcium concentration.

ANSWERS TO REVIEW QUESTIONS

1. *B.* Ionotropic cell surface receptors pass current when they open and therefore change the transmembrane voltage. Considering the other answers: (A) Some ionotropic cell surface receptors are ion channels permeable to both sodium and potassium, but others (such as the GABA receptor) are not. (C) Metabotropic, not ionotropic, receptors are coupled to, or are themselves, enzymes. (D) Many other cells have ionotropic cell surface receptors. Skeletal muscle cells are one example. (E) One class of ionotropic cell surface receptors (nicotinic acetylcholine receptors) respond to acetylcholine, but others respond to other transmitters.

2. *B.* This is the definition of a metabotropic cell surface receptor. Of the other answers: (A) Metabotropic cell surface receptors are not ion channels. (C) Some metabotropic cell surface receptors are tyrosine kinases, but many are not, including those that signal to G_q or G_s. (D) The subset of metabotropic cell surface receptors that signal to G_s will increase the concentration of cAMP, but many others will not. (E) Metabotropic receptors affect many aspects of cell function, so some will indeed make it more difficult for a postsynaptic cell to be depolarized to threshold, but this is not a general rule.

3. *E.* Both receptors are on smooth muscle cells. The receptor at which noradrenaline is most effective is coupled to phospholipase C and therefore causes an increase of cytosolic calcium concentration, leading to contraction and hence a closing of the blood vessel. The receptor at which adrenaline is most effective is coupled to adenylate cyclase and therefore causes an increase of cAMP concentration, leading to phosphorylation and therefore relaxation of the contractile machinery and

18

MECHANICAL MOLECULES

✳ THE CYTOSKELETON IS BOTH STRONG AND MOTILE

Eukaryotic cells are supported by a network of struts and cables called the cytoskeleton. In animal cells, which lack a rigid cell wall, it is the cytoskeleton that determines cell shape. In addition the cytoskeleton is responsible for cell locomotion, for moving components from place to place within the cell, and for cell division. The cytoskeleton is composed of three cytoplasmic filament networks: **microtubules, microfilaments,** and **intermediate filaments** (Fig. 18.1). Although all three are physically strong, microtubules and microfilaments have the additional role of organizing movement, both of entire cells and of structures within them. The term cytoskeleton implies a rigid set of "bones" within the cell. Nothing could be further from the truth, and all three filament systems are highly dynamic, altering their organization in response to the needs of the cell. The molecules that make up the three filament systems have been highly conserved throughout evolution; the cytoskeletal proteins present in the cells of complex organisms such as humans are much the same as those in a simple organism such as a yeast. Although the individual molecules making up the cytoskeleton are below the limit of resolution of the light microscope, the cytoskeleton itself can be readily observed within the cell by using fluorescence microscopy (e.g., page 6).

✳ MICROTUBULES

Microtubules possess a combination of physical properties that allows them to participate in multiple cellular functions. They can form bundles of rigid fibers that make excellent

Cell Biology: A Short Course, Second Edition, by Stephen R. Bolsover, Jeremy S. Hyams,
Elizabeth A. Shephard, Hugh A. White, Claudia G. Wiedemann
ISBN 0-471-26393-1 Copyright © 2004 by John Wiley & Sons, Inc.

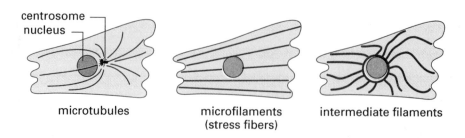

Figure 18.1. Typical spatial organization of microtubules, stress fibers (one form of microfilaments), and intermediate filaments.

structural scaffolds and hence serve an important role in the determination of cell shape. They have an inherent structural polarity that defines a polarity (front vs. back) to the cell. They provide a system of intracellular highways that supports a two-way traffic of organelles and small vesicles powered by enzymes that interact with the microtubule surface. They can be rapidly formed and broken down, a property that allows the cell to respond to subtle environmental changes. Finally, they play a role in one of the most exquisite and precise of all movements within the cell, the segregation of chromosomes at mitosis and meiosis (Chapter 19).

Animal cells contain a network of several thousand microtubules, each 25 nm in diameter. All the cell's microtubules can be traced back to a single structure called the **centrosome,** which is tightly attached to the surface of the nucleus at the cell centre (Fig. 18.1). The centrosome is the microtubule organizing center of the cell and consists of amorphous material enclosing a pair of **centrioles** (Fig. 18.2). Centrioles have a characteristic nine-way pattern that we will meet again in cilia and flagella (page 386). Microtubules in plant cells have a quite different organization, lying immediately beneath the cell membrane, oriented at right angles to the direction of cell expansion. The role of microtubules in plants is to direct the deposition of cellulose fibers on the outside of the cell membrane. **Cellulose synthase,** a multiprotein enzyme complex that spans the cell membrane, is thought to move along microtubules forming new parallel tracks of cellulose microfibrils on the cell surface (Fig. 18.3). As a consequence of the position of the microtubules beneath the membrane, cellulose is laid down in hoops that encase the plant cell in a rigid corset and allow it to expand only in one direction. Plant cells do not contain centrioles, and it remains unclear whether their microtubules arise from a defined organizing center.

Microtubules are composed of a protein called **tubulin** that consists of two dissimilar subunits designated α and β. In the human genome there are about five α-tubulin genes and roughly the same number for β tubulin. There is a third member of the tubulin superfamily, γ tubulin, which does not itself contribute to microtubule structure but which is found at the centrosome and plays a role in initiating microtubule assembly. α-Tubulin/β-tubulin dimers assemble into chains called protofilaments, 13 of which make up the microtubule wall (Fig. 18.4). Within each protofilament the tubulin dimers are arranged in a "head-to-tail" manner, α β α β and so on. This gives the microtubule an built-in molecular polarity that is reflected in the way it grows. Tubulin subunits are added to, and lost from, one end much more rapidly than the other. By convention, the fast growing end is referred to as the $(+)$ end and the slow growing end as the $(-)$ end.

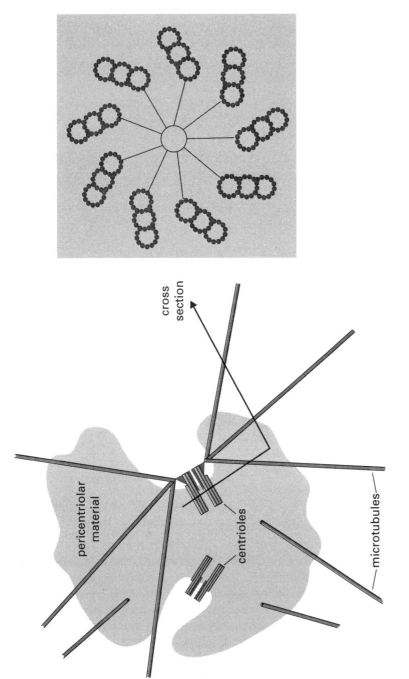

Figure 18.2. The microtubule organizing center or centrosome consists of amorphous material enclosing a pair of centrioles.

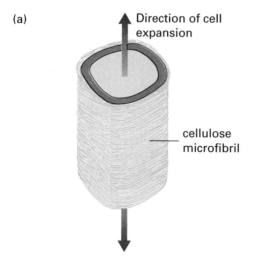

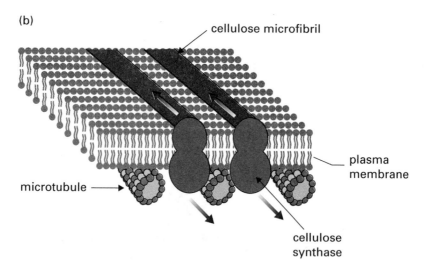

Figure 18.3. Cellulose synthase runs on microtubule tracks.

In cells, the minus end of each microtubule is embedded in the centrosome so that only the plus ends are free to grow or shrink. This process is surprisingly complex. Individual microtubules undergo periods of slow growth followed by rapid shrinkage, sometimes disappearing completely. This phenomenon is referred to as dynamic instability. By chance, the growing ends of certain microtubules may be captured by sites at the cell membrane and stabilized, so that they are protected from shrinkage. Their further growth influences the shape of the cell (Fig. 18.5). Groups of microtubules having a common orientation make an excellent structural framework. Because microtubules are dynamic, the framework can be continually remodeled as the needs of the cell change.

One of the most important tools in establishing microtubule function in cells has been the plant alkaloid colchicine. Extracted from the corms of the autumn crocus, *Colchicum*

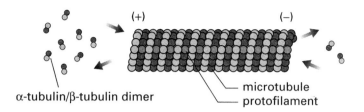

Figure 18.4. Microtubule structure.

autumnale, colchicine has been used since Roman times as a treatment for gout. Cells exposed to colchicine lose their shape, and the movement of organelles within the cytoplasm ceases. When the drug is washed away, microtubules reassemble from the centrosome and normal functions are resumed (Fig. 18.6). Another drug, taxol, obtained from the bark of the Pacific yew, *Taxus brevifolia,* has the opposite effect, causing large numbers of very stable microtubules to form in the cell, an effect that is difficult to reverse.

Unlike animal cells, plant cells do not use their cytoskeletons to define their cell shape, but instead use their extracellular cell walls. The direction in which a growing plant cell extends is determined in the first instance by the direction in which the cellulose micro-filaments run. However, these are laid down by cellulose synthase following microtubule tracks so in fact the cytoskeleton has defined the direction of cell growth, but indirectly rather than by itself acting as a load-bearing component.

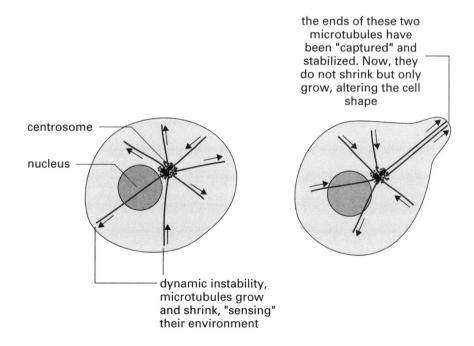

Figure 18.5. Microtubules show dynamic instability.

Figure 18.6. Effects of taxol and colchicine on microtubules.

✿ MICROTUBULE-BASED MOTILITY

Cells move for a variety of reasons. Human spermatozoa in their millions swim frantically toward an ovum; the soil amoeba, *Acanthamoeba* (said to be the most abundant organism on Earth), crawls over and between soil particles, engulfing bacteria and small organic particles as it does so. Cells in the early human embryo show similar crawling movement as they reorganize to form tissues and organs; the invasive properties of some cancer cells is due to their reverting to this highly motile embryonic state. Of course, not all cells show these obvious forms of motility, but careful observation of even the most sedentary eukaryotic cells often reveals a remarkable repertoire of intracellular movements. Both microtubules and actin filaments play important roles in cell motility. We will describe microtubule-based motility first; actin-based motility will be covered following the section on microfilaments.

Cilia and Flagella

Cilia and flagella appeared very early in the evolution of eukaryotic cells and have remained essentially unchanged to the present day. The terms **cilium** (meaning an eyelash) and **flagellum** (meaning a whip) are often used arbitrarily. Generally, cilia are shorter than flagella (<10 μm compared to >40 μm) and are present on the surface of the cell in much greater numbers (ciliated cells often have hundreds of cilia but flagellated cells usually have a single flagellum). The real difference, however, lies in the nature of their movement (Fig. 18.7). Cilia row like oars. The movement is biphasic, consisting of an **effective stroke** in which the cilium is held rigid and bends only at its base and a **recovery stroke** in which the bend formed at the base passes out to the tip. Flagella wriggle like eels. They generate waves that pass along their length, usually from base to tip at constant amplitude. Thus the movement of water by a flagellum is parallel to its axis while a cilium moves water perpendicular to its axis and, hence, perpendicular to the surface of the cell.

Cilia are such a conspicuous feature of some protists that they are called ciliates. The swimming of a paramecium (Fig. 18.8), for example, is generated by the coordinated motion of several thousand cilia on the cell surface. Cilia also play a number of important roles in the human body. The respiratory tract, for example, is lined with about 0.5 m^2 of ciliated epithelium, bearing something like 10^{12} cilia. The beating of these cilia moves a belt of mucus containing inhaled particles and microorganisms away from the lungs. This activity is paralyzed by cigarette smoke, causing mucus to accumulate in the smoker's lung and causing the typical cough.

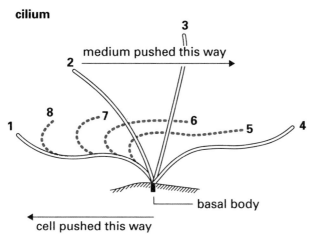

cilium

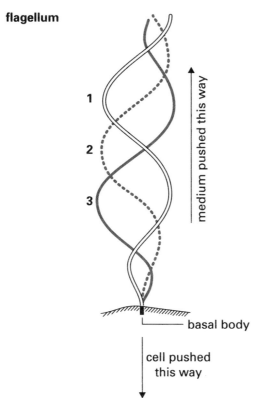

flagellum

Figure 18.7. How cilia and flagella bend.

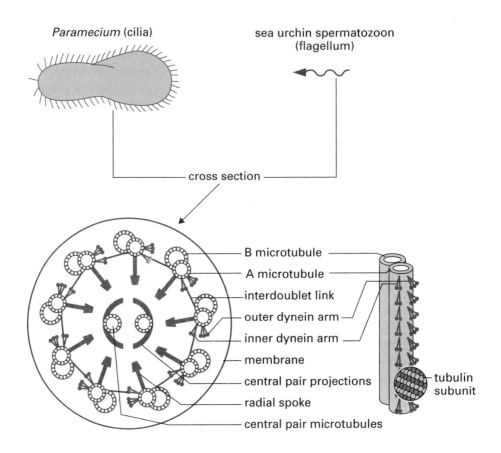

Figure 18.8. Cilia and flagella have identical structure.

Despite their different pattern of beating, cilia and flagella are indistinguishable structurally (Fig. 18.8). Both consist of a regular arrangement of microtubules and associated proteins with the nine-way pattern we have also seen in centrioles. Unlike centrioles, cilia and flagella have a central pair of microtubules, so that the overall structure is called the **9 + 2 axoneme.** The axoneme is enclosed by an extension of the cell membrane. Attached to the nine outer doublet microtubules are projections, or arms, composed of a motor protein called **dynein.** Dynein is an ATPase that converts the energy released by ATP hydrolysis into the mechanical work of ciliary and flagellar beating. Using ATP produced by mitochondria near the base of the cilium or flagellum as fuel, the dynein arms push on the adjacent outer doublets, forcing a sliding movement to occur between adjacent outer doublets. Because the arms are activated in a strict sequence both around and along the axoneme and because the amount of sliding is restricted by the radial spokes and interdoublet links, sliding is converted into bending.

Bacterial flagella (page 264) use a fundamentally different mechanism. Like the propellor of a boat, the motion of the bacterial flagellum is entirely driven by the rotary motor at its base. The bacterial flagellum itself is a specialized piece of extracellular cell wall, made of one protein (flagellin) that has no similarity to tubulin or dynein. Cilia and flagella are full of cytosol all the way to their tips, and use the ATP in that cytosol to generate force all the way along their length.

IN DEPTH 18.1 Calcium and *Paramecium* Swimming

Paramecium, although it is a single-celled organism, is a fast-swimming predator. Its surface is covered with hundreds of cilia that are arranged in rows that lie obliquely to the cell axis. Cilia beat with their effective stroke oriented toward the rear, propelling the cell forward in a spiral motion. The beating of the individual cilia is coordinated to produce "metachronal waves" that pass from the front of the animal to the rear, rather like the wind passing across a corn field. *Paramecium* needs millisecond-to-millisecond control of its cilia as it swims through pondweed in search of prey. One reflex, the "avoiding reaction," was first described almost a century ago. When a *Paramecium* is swimming forward, the internal calcium concentration is kept low, at about 100 nmol liter^{-1}. When the front of the animal bumps into an object, the plasma membrane depolarizes, opening voltage-gated calcium channels in the cilia all the way down the animal as a calcium action potential is generated. The influx of calcium ions raises the intracellular calcium concentration, and this in turn causes the cilia to swing through 180° so that the effective stroke is now directed toward the anterior. This causes the *Paramecium* to reverse a few body lengths, taking it out of contact from the object with which it collided. Calcium is now pumped out of the cell and the internal calcium concentration falls back to 100 nmol liter^{-1}. The cilia now swing back to their original position and forward swimming is resumed.

Medical Relevance 18.1

A Sinister Complaint

A small number of men seeking the help of fertility clinics turn out to have not only nonswimming sperm but a failure of all the cilia and flagella in their bodies. On questioning, they are often found to suffer from chronic respiratory disease because the cilia lining their airways are also immobile. Most surprising is the fact that half of these individuals also exhibit Kartegener's syndrome, a condition in which the internal organs have the reversed orientation from normal, that is, the heart is on the right, the appendix on the left, and so on. It turns out that cilia set up a flow from right to left across the early embryo. Transmitters carried in this flow tell the left-hand side to develop its characteristic organs, and vice versa. If a mutation renders the cilia immobile, the left–right gradient of transmitter does not get set up, and the body organs are positioned by chance and not design.

Intracellular Transport

The beating of cilia and flagella is an obvious manifestation of movement generated by a microtubule-based structure having a defined and geometric shape. However, motility is a general property of microtubules within cells. It is seen particularly clearly in some specialized cell types such as the pigment cells called **chromatophores** in the skin of fish and amphibia. The inward and outward movement of pigment granules along radial arrays of microtubules underlies the remarkable color changes such animals are able to display (Fig. 18.9). But this is just an exaggerated example of a process that occurs less

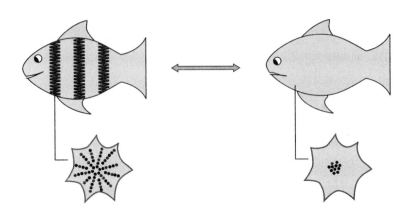

Figure 18.9. Pigment migration in fish chromatophores.

spectacularly in all cells. For instance, nerve cell axons extend up to 1 m from the cell body. Organelles, small vesicles, and even mRNA are transported in both directions in a phenomenon referred to as **axonal transport** (Fig. 18.10). This is subdivided into outward or **anterograde** transport and inward or **retrograde** transport. Both are dependent upon microtubules that are abundant in nerve cells.

The two forms of axonal transport are dependent upon different molecular motors. Dynein, sometimes referred to as **cytoplasmic dynein** to distinguish it from its relative in cilia and flagella, moves vesicles and organelles in the retrograde direction while another protein, **kinesin,** is the motor for movement along microtubules in the anterograde direction. Their directionality is specified by the polarity of the microtubules. Thus, dynein moves along a microtubule in a plus to minus direction while kinesin works in the opposite direction. Both proteins consist of a tail that binds to the cargo to be transported and two (kinesin) or three (dynein) globular heads that interact with the surface of the microtubule, generating movement. Specificity is imparted to this process by having multiple dyneins and kinesins in a single cell, each responsible for transporting a specific type of cargo. This multiplicity of dyneins and kinesins is created by alternative splicing of a smaller number of genes (page 118).

✿ MICROFILAMENTS

Microfilaments are fine fibers, about 7 nm in diameter, that are made up of subunits of the protein actin. Because it is a globular protein, the actin monomer is designated G-actin while the filament that forms from it is referred to as F-actin. Each actin filament is composed of two chains of actin monomers twisted around one another like two strands of beads (Fig. 18.11). In animal cells, actin is particularly associated with the cell periphery. When they are grown in plastic cell culture dishes, nonmotile cells have two main types of microfilaments: bundles of actin filaments called **stress fibers** run across the cell and help to anchor it to the dish (Fig. 18.1) while under the cell membrane can be seen loose meshwork of filaments that give the edges of the cell structural strength. In actively moving cells,

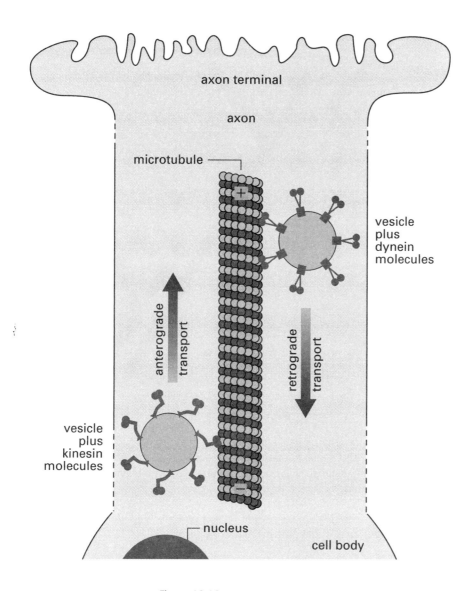

Figure 18.10. Axonal transport.

the stress fibers disappear and actin becomes concentrated at the leading edge. Projections from the cell surface such as microvilli (page 13) are maintained by rigid bundles of actin filaments.

The equilibrium between G- and F-actin is affected by many factors including **actin-binding proteins** (Fig. 18.11). Filament growth is prevented by **profilin,** which binds G-actin monomers and prevents their polymerization, and by capping proteins that bind to F-actin filament ends. In contrast, filament nucleation proteins such as **actin-related protein (Arp)** act as a base on which new filaments can form. Crosslinking proteins stick to two existing filaments, forming a mechanically strong lattice. Of these **villin,** found in

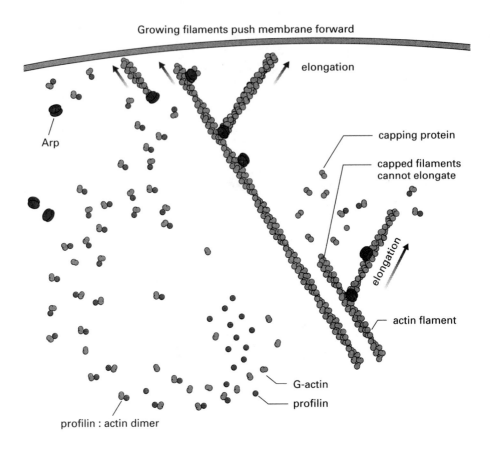

Figure 18.11. Actin polymerization is regulated by actin-binding proteins.

microvilli, generates parallel bundles, while related proteins bind criss-crossing filaments together to form a viscous, three-dimensional cytoplasmic gel.

Example 18.1 **Spearing Eggs**

Upon contacting an egg the sperm of the marine organism *Thyone* suddenly extends a needlelike process called an **acrosome** from its front end. This 90-μm-long needle helps the sperm to penetrate the egg. Prior to activation, a vesicle at the sperm front end contains G-actin at concentrations way above that which would normally result in spontaneous polymerization into F-actin filaments. The reason it does not do so is that all the actin is bound to profilin. Upon contact with the egg, salts from seawater flow into the vacuole and cause dissociation of the profilin–actin dimers. The actin then immediately polymerizes to form the long microfilaments of the acrosome.

Cells are anchored to the extracellular matrix through transmembrane proteins such as **integrins.** These are dimeric proteins that have an extracellular domain that binds to collagen and other extracellular matrix proteins and an intracellular domain that attaches to actin microfilaments on the inside (Fig. 18.12).

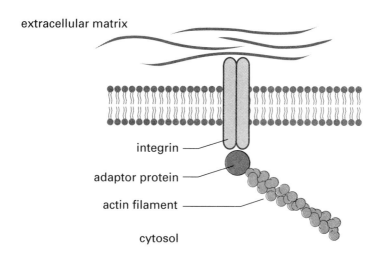

extracellular matrix

integrin

adaptor protein

actin filament

cytosol

<u>Figure 18.12.</u> Integrins anchor the actin cytoskeleton to the extracellular matrix.

Medical Relevance 18.2

Some Bacteria Highjack the Cytoskeleton for Their Own Purposes

We are all familiar with the ways that bacteria and viruses spread from person to person by various forms of social or antisocial contact, but how does a bacterium spread through the cells of a human body? A number of bacteria including the important pathogens *Listeria* (responsible for sepsis and meningitis in immunocompromised patients and for infections of the fetus during pregnancy) and *Shigella* (which causes dysentery), avoid contact with the human body's antibodies and white blood cells by remaining hidden within the cytoplasm of our cells as they spread through our tissues. The bacteria use actin to power their journey from cell to cell. *Listeria* has a protein called **ActA** at one end of its rod-shaped body. Like Arp, ActA promotes the polymerization of actin filaments. This generates enough force to drive the bacterium through the cytoplasm in the opposite direction. Movement is random, but occasionally the bacterium will bump into the cell membrane causing the formation of a fingerlike projection from the cell surface. The membrane of the projection fuses with the membrane of the neighboring cell, transferring the bacterium in a membrane sac from which it quickly escapes to repeat the procedure. The really clever thing (from the bacterium's point of view) about such a transfer mechanism is that the bacterium never leaves the safety of the cytoplasm.

Muscle Contraction

Striated muscle, the kind of muscle found attached to bones or in our hearts, gains that appearance because passing along the cell one encounters in turn regions of parallel microfilaments (which are relatively transparent) then regions of **thick filaments** of a second protein called myosin (Fig. 18.13). The complete repeating unit is called a sarcomere and is delineated by the **Z disc,** which holds the microfilaments in a regular pattern, so that striated muscle has an almost crystalline appearance in transverse section. The myosin molecule has a distinctive structure, consisting of a tail and two globular heads. The thick filament is formed from a large number of myosin molecules arranged in a tail-to-tail manner. This design means that the thick filament is a bipolar structure with myosin heads at both ends. In a resting striated muscle, the myosin heads are unable to operate on the actin microfilaments.

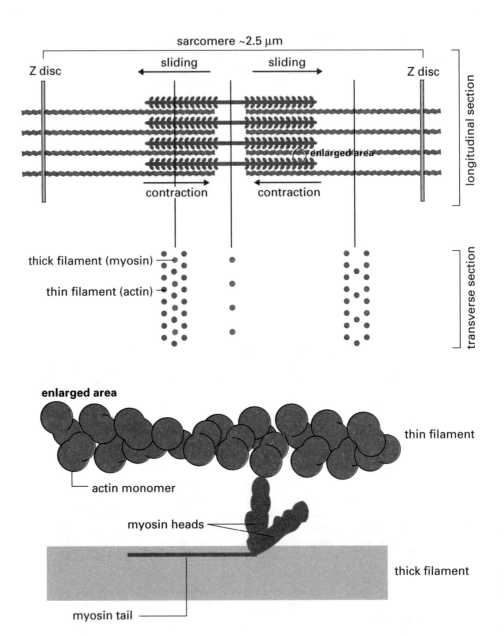

Figure 18.13. Muscle contraction.

When the cytosolic calcium concentration increases, calcium binds to a protein called tro-ponin that changes shape to allow myosin access to the actin. The myosin heads now crawl along the actin (thin) filaments and generate sliding that is powered by the hydrolysis of ATP. Because of the geometry of the system, the two Z discs are pulled toward each other and the cell shortens (Fig. 18.13).

Several types of myosin are found in nonmuscle cells. One of them, called **myosin II,** is very similar to muscle myosin but does not assemble into filaments to the same extent,

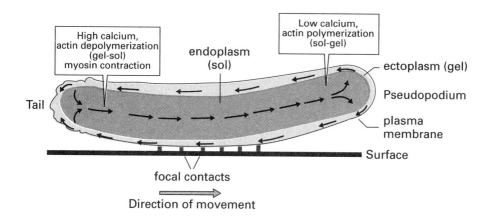

Figure 18.14. Amoeboid movement resembles the progression of a military tank.

probably because the levels of force required within nonmuscle cells is relatively small. The primary role of myosin II is in cell division (page 404). **Myosin V** is also two-headed and is responsible for carrying cargo (vesicles or organelles) along actin filaments. Unlike the microtubule-associated motors dynein and kinesin, which can make long journeys, myosin V can only make short excursions along an actin filament before falling off.

Cell Locomotion

For a cell such as an amoeba to crawl across a surface it must generate traction. Less than 1% of the ventral surface of a crawling cell is in contact with the surface over which it is moving, so that the cell does not slither along on its belly but rather "tiptoes" delicately across the surface. The transient points of attachment with the surface are called **focal contacts.** At the leading edge actin polymerization (Fig. 18.11) causes the cell to extend projections or **pseudopodia** in the direction of progress. The overall increase in actin polymerization and crosslinking in this region causes the cytoplasm to form a viscous **ectoplasm** (Fig. 18.14). In contrast, raised cytosolic calcium concentrations in the tail activate **actin-severing proteins** such as **gelsolin.** Calcium also activates myosin II, the resulting contraction driving the now fluid endoplasm forward and displacing the existing actin cytoskeleton backward. Since actin is connected via the focal contacts to the surface, the backward movement of the ectoplasm drives the cell forward, operating rather like the tracks of a military tank.

Cytoplasmic Streaming

Even though the cells themselves do not move, the cytoplasm of many plant cells also shows active cytoplasmic streaming. In the epidermal cells of the leaves of the water plant, *Elodea,* a belt of moving cytosol carries the chloroplasts and other organelles in a continuous, unidirectional stream around the central vacuole. In the giant cells of the alga *Nitella,* the chloroplasts are embedded in a layer of viscous ectoplasm while other organelles are carried along in the flowing endoplasm. Cytoplasmic streaming is driven by the interaction of myosin molecules attached to the moving organelles with cables of actin filaments in the ectoplasm.

✸ INTERMEDIATE FILAMENTS

Intermediate filaments were so named originally because their diameter, 10 nm, lay between that of the thin and thick muscle filaments. They are the most stable of the cytoplasmic filament systems. Although intermediate filaments from different mammalian tissues look much the same in the electron microscope, they are in fact composed of different protein monomers (and can therefore be distinguished by immunofluorescence, page 6). Nerve cells contain **neurofilaments,** muscle cells contain **desmin** filaments, fibroblasts contain **vimentin** filaments, and epithelial cells contain filaments composed of **keratin,** the protein that gives our skin its protective coating, forms our hair and finger nails, and makes the horns and hooves of our domestic animals and pets. These different proteins can generate a common structure because all share the same basic design. This consists of a central, α-helical rod and a nonhelical head and tail. Most of the variation between intermediate filament proteins is in the head and tail, and these regions probably confer subtly different properties on different intermediate filament classes. The basic building block of intermediate filaments is a tetramer of pairs of subunit proteins joined by their central region to form coiled coils.

In the cell, intermediate filaments tend to form wavy bundles that extend from the nucleus to the cell surface (Fig. 18.1). It appears that the nucleus is suspended by intermediate filaments stretching to the plasma membrane, rather like a sailor in a hammock. Indeed intermediate filaments share a high degree of amino acid sequence homology with the nuclear **lamins** that reinforce the nuclear envelope on the inside (page 57).

Anchoring Cell Junctions

The cells that form tissues in multicellular organisms are attached together with anchoring junctions (Fig. 18.15). Integral membrane proteins called **cell adhesion molecules,** of which **cadherin** is an example, extend out from each cell and bind tightly together, while their cytosolic domains attach to the cytoskeleton. There are two basic types of anchoring junction. In **adherens junctions** the cell adhesion molecules are linked to actin microfilaments by linking proteins such as catenin. In **desmosomes** the cell adhesion molecules are linked to intermediate filaments. Tissues that need to be mechanically strong, such as the epithelial cells of the gut (page 14) and cardiac muscle, have many anchoring junctions linking the cytoskeletons of the individual cells. Anchoring junctions are one of the three types of cell junctions, the others being tight junctions (page 55) and gap junctions (page 55).

Example 18.2 **Protected by the Dead**

The epidermis of the skin is made up of a layer of living cells called keratinocytes covered by a protective layer of their dead bodies. Dead keratinocytes form a good protective layer because while alive they generate a dense internal cytoskeleton of the intermediate filament keratin, with adjacent cells being linked by desmosomes. When the cells die, the keratin fibers remain because intermediate filaments are stable. Since the intermediate filaments were joined by desmosomes, the resulting protective fibers do not stop at the edge of the now dead cell, but are strongly connected with the fibers in the next cell, and the next, and so on, forming an extremely strong network of fibers.

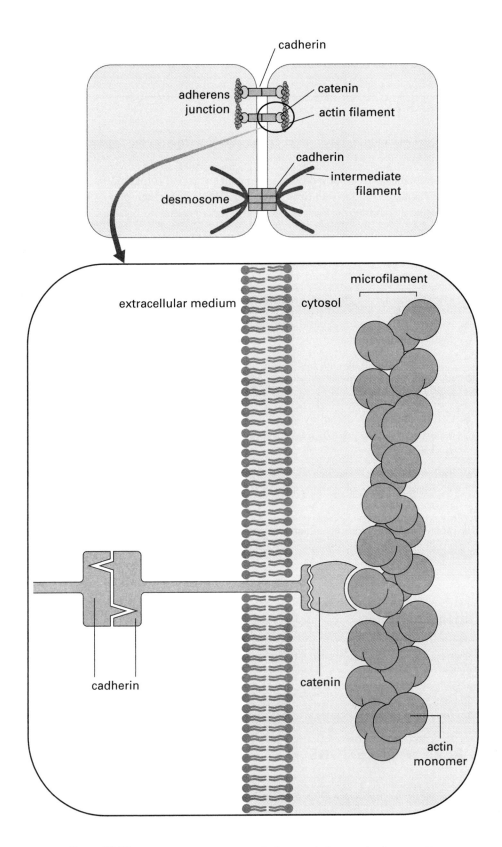

Figure 18.15. Anchoring junctions attach the cytoskeletons of adjacent cells.

SUMMARY

1. The cytoskeleton is made up of microtubules, microfilaments, and intermediate filaments.

2. Microtubules are composed of equal amounts of α and β tubulin. In animal cells their minus $(-)$ end is stabilized at the centrosome while their plus $(+)$ end shows dynamic instability.

3. Cilia and flagella contain a 9+2 axoneme of microtubules plus the motor protein dynein.

4. Dynein and kinesin transport cargo along microtubules.

5. Microfilaments are composed of actin.

6. Actin-binding proteins control actin polymerization and determine the organization of the actin lattice.

7. The motor protein myosin operates in all cell types but is particularly prominent in muscle cells. It operates on actin microfilaments when cytosolic calcium concentration rises.

8. Cell locomotion is driven by spatially distinct zones of actin polymerization and depolymerization, aided by myosin. A similar cycle underlies cytoplasmic streaming in plants.

9. The proteins that form intermediate filaments differ in different cells. Intermediate filaments are more stable than microtubules and microfilaments and serve a structural role only.

10. Anchoring cell junctions attach the cytoskeletons of adjacent cells together. They are divided into adherens junctions, which link to actin, and desmosomes, which link to intermediate filaments.

FURTHER READING

Hirokawa N. 1998. Kinesin and dynein superfamily proteins and the mechanism of organelle transport. *Science* 279: 519–526.

Preston, T. M., King, C. A., and Hyams, J. S. 1990. *The Cytoskeleton and Cell Motility.* Glasgow: Blackie.

✳ REVIEW QUESTIONS

For each question, choose the ONE BEST answer or completion.

1. Which of the following are *not* components of the cytoskeleton?
   A. Actin filaments

B. Collagen fibers

C. Microtubules

D. Microfilaments

E. Intermediate filaments

2. Which of the following statements are *incorrect*?

A. Dynein mediates retrograde axonal transport.

B. Myosin molecules move along actin filaments.

C. Muscle thin filaments are composed of actin.

D. Bacterial flagella have a 9+2 structure.

E. Cilia are abundant in the human respiratory tract.

3. Arp, profilin, and villin are all

A. cell adhesion molecules.

B. molecular motors.

C. actin-binding proteins.

D. intermediate filament proteins.

E. nonmuscle myosins.

4. Plant cells lack

A. microtubules.

B. centrioles.

C. actin.

D. gelsolin.

E. cellulose synthase.

5. In cross section cilia and flagella have nine equally spaced sets of microtubule structures. This organization—nine equally spaced units—is also shown by

A. nuclear lamins.

B. the nuclear pore.

C. bacterial flagellin.

D. centrosomes.

E. connexins.

6. Motor proteins acting on actin microfilaments include

A. dynamin.

B. myosin.

C. kinesin.

D. dynein.

E. all of the above.

7. Desmosomes differ from adherens junctions in that

A. desmosomes are structurally strong, while adherens junctions block the movement of extracellular medium.

B. desmosomes block the movement of extracellular medium, while adherens junctions are structurally strong.

C. desmosomes connect to intermediate filaments while adherens junctions connect to microtubules.

D. desmosomes connect to intermediate filaments while adherens junctions connect to microfilaments.

E. desmosomes anchor the nucleus in position within the cell while adherens junctions connect cells together.

ANSWERS TO REVIEW QUESTIONS

1. *B.* Collagen is a component of the extracellular matrix.

2. *D.* The bacterial flagellum is made of the protein called flagellin; it does not have a 9 + 2 structure. Tubulin and dynein, the components of the 9 + 2 axoneme, are only found in eukaryotes.

3. *C.*

4. *B.* Plant cells do not contain centrioles, and it remains unclear whether their microtubules arise from a defined organizing center.

5. *D.* Centrosomes are made of tubulin and in cross section have an organization similar to that of cilia and flagella. Bacterial flagellin has no obvious subdivision; nuclear lamins, the nuclear pore, and connexins are formed of subunits arranged in patterns of 4, 8, and 6, respectively.

6. *B.* Myosin acts on actin. In contrast kinesin and dynein act on microtubules, and dynamin causes the pinching off of clathrin-coated vesicles.

7. *D.* Both desmosomes and adherens junctions are types of anchoring cell junction, desmosomes connecting to intermediate filaments while adherens junctions connect to microfilaments.

19

CELL CYCLE AND CONTROL
OF CELL NUMBER

All cells arise by the division of an existing cell. The life of a cell from the time it is generated by the division of a progenitor cell to the time it in turn divides is called the **cell division cycle** or just the **cell cycle** for short. The duration of the cell cycle varies from 2 to 3 h in a single-celled organism like the budding yeast *Saccharomyces cerevisiae* to around 24 h in a human cell grown in a culture dish. During this period the cell doubles in mass, duplicates its genome and organelles, and partitions these between two new progeny cells. These events have to be carried out with great precision and in the correct order, and cells have established sophisticated control processes to ensure that the cell cycle proceeds with the required accuracy. In humans the cells of some tissues, such as the skin, the lining of the gut, and the bone marrow continue to divide throughout life. Others, such as the light-sensitive cells of the eye and skeletal muscle cells, show almost no replacement. The latter, laid down in infancy, must last a lifetime. Not only must cell division be precisely controlled in terms of which cells divide when, the whole process has to know when to stop. A human being is bigger than a mouse and smaller than an elephant. Allowing for some minor differences in basic design, this is because humans are made up of more cells than a mouse (and, hence, are the result of more cell cycles) and less than an elephant. How are these differences achieved? Why don't humans grow to be the size of elephants or whales? What has become increasingly clear over recent years is that cells contain what have become known as "cell cycle control genes." The proper functioning of such genes not only determines how big we are, it also prevents cell division becoming "out of control" and leading to cancer.

Cell Biology: A Short Course, Second Edition, by Stephen R. Bolsover, Jeremy S. Hyams,
Elizabeth A. Shephard, Hugh A. White, Claudia G. Wiedemann
ISBN 0-471-26393-1 Copyright © 2004 by John Wiley & Sons, Inc.

Under the microscope it is possible to distinguish two elements of the cell cycle. **Interphase** occupies about 90% of the cell cycle and is a period of synthesis and growth, during which the cell roughly doubles in mass but without displaying obvious morphological changes. Once interphase is complete, the cell enters **mitosis,** which is a brief period of profound structural changes. The focal point of mitosis is the behavior of the chromosomes, and it is this that we will deal with first.

🌸 STAGES OF MITOSIS

Mitosis is designed to produce two progeny cells each containing an identical set of the progenitor cell's chromosomes. To achieve this, the chromosomes execute a precisely chore-ographed sequence of movements that were first described well over a century ago. Classically, mitosis is divided into five stages, each of which is characterized by changes in the appearance of the chromosomes and their organization with respect to a cellular structure, called the **mitotic spindle,** that is responsible for their segregation. The stages of mitosis are shown diagramatically in Figure 19.1.

One copy of our genome is encoded in 23 chromosomes and is made up of 3×10^9 base pairs. Human cells contain 46 chromosomes, 23 inherited from each parent. This state, in which each cell has two complete sets of chromosomes, is called **diploid.** For simplicity

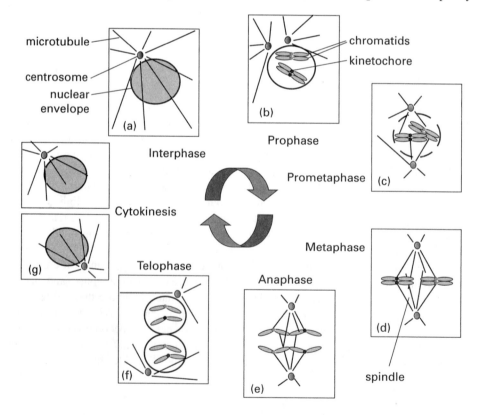

Figure 19.1. Stages of mitosis. One pair of chromosomes is shown, one chromosome originating from the father (light green kinetochores), and the matching one from the mother (dark green kinetochores).

Figure 19.1 shows just one pair of chromosomes, one that came from the father (shown with light green kinetochores), and the matching one from the mother (dark green kinetochores).

Prophase. The first evidence of mitosis in most cells is the compaction of the threads of chromatin that existed through interphase into compact chromosomes that are visible in the light microscope. As the chromosomes compact, each can be seen to be paired structures composed of two **chromatids.** This is the visible effect of the DNA molecules having been replicated (page 87) in interphase. Chromosome condensation reduces the chance of long DNA molecules becoming tangled and broken. Each chromosome has a constriction called the **kinetochore,** a structure that forms around a region rich in satellite DNA (page 100) called the **centromere.** The kinetochore is the point of attachment of the chromosome to the spindle. At the same time as the chromosomes are condensing within the nucleus, the centrosomes, which lie on the cytoplasmic side of the nuclear envelope, begin to separate to establish the mitotic spindle.

Prometaphase. At the breakdown of the nuclear envelope, the chromosomes become free to interact with the forming spindle. Microtubule assembly from the centrosomes is random and dynamic. The growing ends of individual microtubules make chance contact with and are captured by the kinetochores. Because of the random nature of these events, the kinetochores of chromatid pairs are initially associated with different numbers of microtubules, and the forces acting upon each chromosome are unbalanced. Initially, therefore, the spindle is highly unstable and chromosomes make frequent excursions toward and away from the poles. Gradually, a balance of forces is established and the chromosomes become aligned at the equator, with the kinetochores of each member of a chromatid pair oriented toward opposite poles.

Example 19.1 **Counting Chromosomes**

The DNA molecules of an interphase cell can only be observed using an electron microscope (page 8). However, at mitosis, when they condense, individual chromosomes can be made out easily using a standard light microscope (page 76). Mitotic cells can then be screened for chromosome abnormalities such as the presence of three copies of chromosome 21, a condition that causes Down syndrome, and the "Philadelphia chromosome," a rearrangement between chromosomes 9 and 21 (which results in a form of leukemia).

Metaphase. Metaphase is the most stable period of mitosis. The system can be regarded as being at steady state with the chromosomes lined up rather like athletes at the start of a race (albeit half facing in one direction and half the other). The metaphase spindle consists of two major groups of microtubules: those connecting the chromatids to the poles and a second group extending from each pole toward the other. The second group overlap at the spindle equator. If a single chromatid becomes detached from the spindle (Fig. 19.2), it sends a signal that tells the other chromosomes not to start anaphase without it. The other chromosomes dutifully wait until the missing chromatid reattaches, whereupon all of the chromosomes enter anaphase together.

Anaphase. The trigger for the separation of the paired chromatids and the start of their journey to the spindle poles is the degradation of the protein **cohesin,** which acts as the glue holding the pairs of chromatids together. In anaphase A the microtubules holding the chromosomes shorten, pulling the chromosomes to the spindle poles. The chromosomes move as a "V" with the kinetochores, at which the force for chromosome movement is

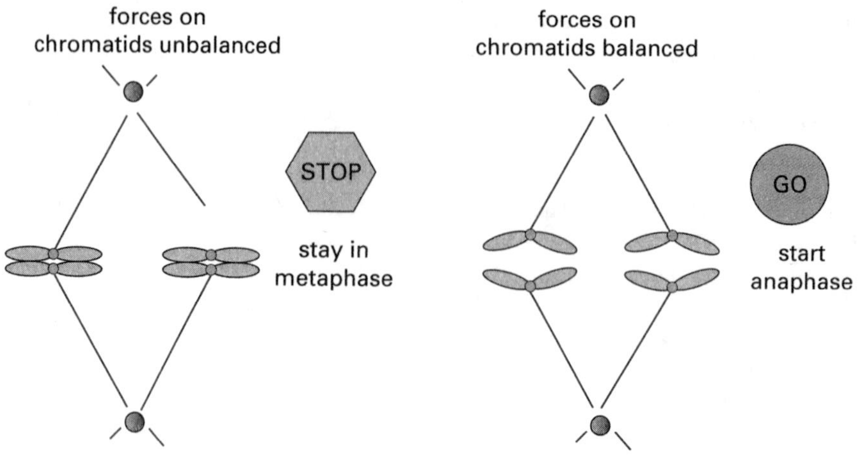

Figure 19.2. Anaphase only begins when all chromatids are correctly attached to spindle microtubules.

applied, leading the way. In contrast, in anaphase B the microtubules that overlap at the spindle equator lengthen, extending the distance between the poles. Compared to other forms of cell motility, the movement of chromosomes at anaphase is extremely slow, less than 1 μm per minute.

Example 19.2 Taxol Stops Mitosis

During cell division, cytoskeletal microtubules break down and the tubulin monomers reform as the mitotic spindle. Taxol (page 385) stabilizes existing microtubules, making it impossible for the cell to form spindles and therefore preventing cell division. For this reason taxol, usually referred to by its generic name palitaxel, is a valuable anti-cancer drug.

Telophase. This stage sees the reversal of many of the events of prophase; the chromosomes decondense, the spindle disassembles, the nuclear envelope reforms, Golgi apparatus and endoplasmic reticulum reform, and the nucleolus reappears. Each progeny nucleus now contains one complete copy of the genome from the father and one copy from the mother. Since in Figure 19.1 we show only one pair of chromosomes, our diagram shows each nucleus containing two chromosomes, one that originated from the father plus one that originated from the mother.

Cytokinesis. During the last stages of telophase, the cell itself divides. In animal cells, a **cleavage furrow** made of actin and its motor protein myosin II constricts the middle of the cell; in plants, a structure called the **phragmoplast** forms at the equator of the spindles where it directs the formation of a new cell wall.

MEIOSIS AND FERTILIZATION

In sexually reproducing organisms, the germ cells that give rise to the eggs and sperm arise by a different type of cell division from the **somatic cells** that make up most of the body. In regular cell division DNA synthesis is followed by mitosis so that the two progeny end

chromosome from the other parent with a functional phenylalanine hydroxylase gene, then they can make enough phenylalanine hydroxylase to convert phenylalanine to tyrosine and are perfectly normal. However, when a baby inherits a nonfunctional phenylalanine hydroxylase gene from each parent, then they have the potentially disastrous condition called phenylketonuria (page 294). People with one functional gene and one nonfunctional one are called **carriers.** If two carriers of the same defective gene have a child, that child has a one in four chance of getting a defective gene from both parents and exhibiting the recessive condition, be it nonbrown eyes or phenylketonuria.

Medical Relevance 19.1

Why Marrying Your Cousin Is Dangerous

Marriage between first cousins is illegal in 30 states of the United States. The scientific rationale behind these laws is that cousin marriage provides an easy mechanism for a single carrier of a recessive genetic disease to generate greatgrandchildren with the disease. The accompanying figure shows how this could happen in an imaginary family tree. Harry and Ingrid are first cousins: they share a grandmother, Alice, who was a carrier for phenylketonuria. We indicate this in the letters below Alice's name: P indicates the gene coding for a functional phenylalanine hydroxylase while p indicates the gene coding for a defective protein. Both Alice's children, Constance and David, inherited her defective gene and were therefore carriers. Of their four children, Harry, Ingrid, and Jill inherited the defective gene and are carriers. The children of Harry and Ingrid therefore each had a one-in-four chance of inheriting one defective gene from Harry plus one defective gene from Ingrid. One of their children, Maria, does so and has phenylketonuria.

In the case we have drawn we know that Constance, David, Harry, and Ingrid are carriers. If we did not know this, what is the probability that a child of first-cousin parents will have phenylketonuria, given that one shared grandparent is a carrier? This is easy to calculate: each child of a carrier has a 1-in-2 chance of inheriting the defective gene, while each child of two carriers has a 1-in-4 chance of having the disease. The overall probability is therefore $1/2 \times 1/2 \times 1/2 \times 1/2 \times 1/4$ or under 1.6% per child. The extent to which this risk is significant enough to require legislation, given the low frequency of cousin marriage in those countries and states where it is allowed, is debatable.

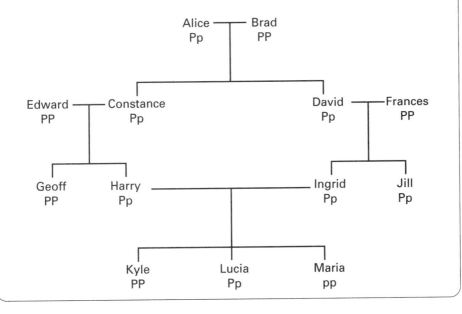

Dominant Genetic Disease

More unusually, a defective gene may cause a problem even if the individual has one good copy, that is, the defective gene is dominant over the normal one. One example is familial Creutzfeldt–Jacob disease. As we described earlier (page 207), the brain damage that occurs in Creutzfeldt–Jacob disease results when an alternatively folded form of PrP protein called PrP^{sc} causes all the PrP protein around it to fold up in the alternative way. Sufferers from familial Creutzfeldt–Jacob disease have a mutant form of the PrP gene that generates protein that spontaneously folds into the PrP^{sc} form. The PrP^{sc} then triggers the normal protein, generated by the second, normal copy of the gene, to fold into the PrP^{sc} shape. Thus one copy of the defective gene is sufficient to cause the disease.

Crossing Over and Linkage

Prophase I lasts such a long time because the chromosomes deliberately tie themselves in knots and then untangle (using topoisomerase II, page 74). Figure 19.4 shows what happens. At the top we see the paternal and maternal chromosomes, each composed of two chromatids, lined up side by side. As before we show the chromosome that originated from the father in light green and the maternal one in dark green. During prophase 1 the chromosomes are cut and resealed at points called **chiasmata** (singular **chiasma**) so that lengths of paternal chromosome are transferred to a maternal one and vice versa. The rest of meiosis I proceeds, followed by meiosis II, and the end result is that some gametes contain chromosomes that are neither completely paternal nor completely maternal but are a **recombination** of the two.

The biological advantage of sexual reproduction is that it allows organisms to possess a random selection of the genes from their ancestors. Those individuals with a complement of genes that makes them better suited to their environment tend to do better, allowing evolution by natural selection of the individuals posessing the better genes. Without crossing over this could not happen: Those genes that are located on the same chromosome would remain **linked** down the generations, greatly reducing the number of gene permutations possible at each generation. Crossing over allows a child to inherit, for example, his grandmother's green eyes without also inheriting her defective sodium channel gene (page 331), although both genes are on chromosome 19. Even with crossing over, genes on the same chromosome are inherited together more than they would be if they were on different chromosoomes. The closer the genes, the less likely it is that a chisma will form between them, and therefore the greater the probability that they will be inherited together. This phenomenon is used to help identify the genes responsible for specific diseases such as cystic fibrosis (page 426).

✸ CONTROL OF THE CELL DIVISION CYCLE

Although mitosis is the most dramatic period of the cell cycle, it is the culmination of a sequence of biochemical and structural events that occur during interphase. The most important of these is the duplication of the genome in **S phase** (the "S" stands for DNA synthesis). In this cell cycle shorthand, mitosis is referred to as **M phase.** S phase and M phase must occur (a) in the correct order and (b) only once per cell cycle. To make sure this is the case, S phase and M phase do not follow immediately after one another but are separated by "gaps" that allow the cell to check that everything is in order before going

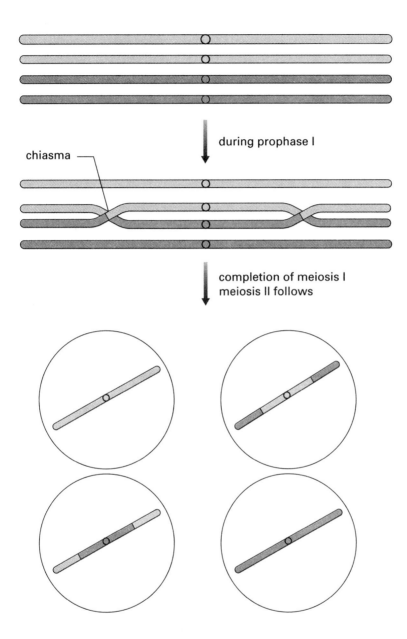

Figure 19.4. Chiasmata allow crossing over of genetic material during prophase I of meiosis.

on to the next stage. The gap between M phase and S phase is called **G1 (gap 1),** and the gap between S phase and M phase is called **G2 (gap 2).** These four phases, G1, S, G2, and M make up the classic cell cycle "clock" (Fig. 19.5). Nondividing or **quiescent** cells are said to be in the **G0 (gap 0)** phase of the cell cycle. Cells can only enter G0 from G1. G0 cells can remain viable for months or even years, and most of the cells in the human body are in fact in this nondividing state. Cells are said to be terminally differentiated if they are unable to return to the cell division cycle. Nerve cells are one such example. In contrast

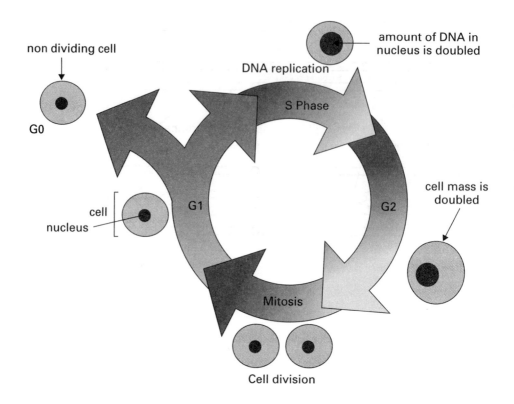

Figure 19.5. Cell division cycle.

other differentiated cells, such as glial cells (page 13), can return to the cycle if they receive the correct signals from their neighbors.

For dividing cells, two points in the cell cycle are particularly critical. The first is at the G1/S boundary at which point the cell is committed to DNA replication. The second is at the G2/M transition when it is committed to mitosis. These are the major control points of the cell cycle, and before crossing them the cell must be sure that conditions are such that S phase and M phase can be executed successfully. At G1/S, the cell must decide whether it is big enough and whether nutritional conditions are appropriate to begin the crucial process of replicating its genome. In G2 the primary concern is that its DNA is in perfect condition before entering mitosis. There are sensitive mechanisms for detecting the presence of unreplicated or damaged DNA, and cells will not commit themselves to mitosis until any defects have been attended to.

Molecular Regulation of the G2/M (Interphase/Mitosis) Cell Cycle Control Point

As cells enter mitosis, they undergo a remarkable sequence of structural changes. Not only do the chromosomes condense as we have seen above but, in addition, the nuclear envelope breaks down, the nucleolus disperses, the membranes of the Golgi apparatus and endoplasmic reticulum fragment, and the cytoskeleton undergoes remodeling to form a completely new cellular structure, the mitotic spindle. How are so many and diverse

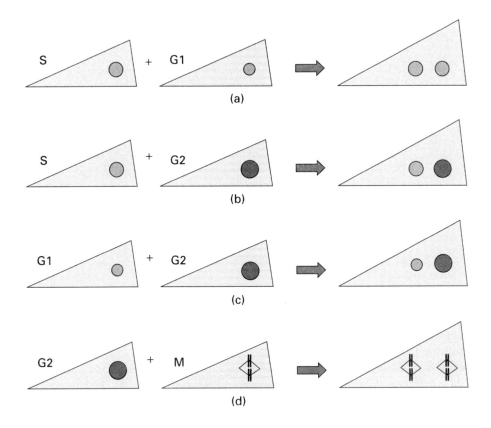

Figure 19.6. Cell fusion experiments reveal how constituents of the cytoplasm control the behavior of the nucleus. Mitotic cells are identified by the presence of the mitotic spindle. Nuclei in S, G1, and G2 phases are represented by different color shades.

structural changes coordinated in a narrow window of the cell cycle, often lasting only a few minutes? During the 1960s some remarkable experiments started the hunt that led to the identification of a cytoplasmic master switch that activated all these changes. These involved fusing together cultured mammalian cells at different stages of the cell cycle (Fig. 19.6). These experiments demonstrated that cell cycle progress was under the control of cytoplasmic factors and that cells had to complete certain cell cycle steps before entering subsequent ones. In particular, row (*d*) shows how fusion of an M-phase cell with a G2 cell causes the G2 nucleus to enter mitosis. Similar results were obtained when cytoplasm from unfertilized *Xenopus* oocytes (which are naturally arrested in meiotic M phase) was injected into interphase oocytes (Fig. 19.7). The implication of these experiments was that the cytoplasm of dividing cells contained a factor capable of controlling entry into mitosis. This activity was given the name **M-phase promoting factor,** or **MPF.** Defining MPF in biochemical terms took almost 20 years and culminated in the award of the 2001 Nobel prize in physiology or medicine to Paul Nurse, Tim Hunt, and Leland Hartwell for their pioneering work on the identification of the molecules that drive the cell division cycle.

Remarkably, given the variety of the changes it induces, MPF is a complex of only two proteins. These are an enzyme called **cyclin-dependent kinase 1 (Cdk1)** and its regulatory subunit called **cyclin B.** Cdk1 can act as a cell cycle switch because it is a protein kinase, an

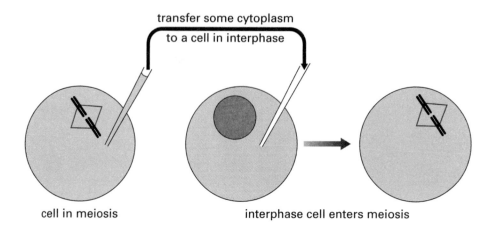

Figure 19.7. Injection of cytosol from a mitotic cell into an interphase one triggers mitosis in the interphase nucleus.

enzyme that can modify the function of other proteins by phosphorylation. The substrates of Cdk1 are proteins associated with the chromosomes, the nuclear envelope, nucleolus, centrosomes, and so on and phosphorylation of such proteins sets in train the structural changes that culminate in mitosis. However, this activity must be activated only at the G2/M boundary and then must be quickly inactivated if the cell is not to undergo multiple rounds of mitosis. Figure 19.8 shows how Cdk1 activity is regulated through the cell cycle. Without cyclin B Cdk1 is very slow at phosphorylating its target but, to ensure that it remains completely inactive through interphase, it is itself phosphorylated by another protein kinase called Wee1 (because mutants, with unphosphorylated Cdk1, divide too early, while they are still "wee"—Scottish for small!). Wee1 adds two phosphates to Cdk1, one on a threonine residue at amino acid number 14, the other on a tyrosine residue at amino acid number 15. These two amino acids lie within the ATP-binding site of Cdk1, and the presence of two phosphates prevents ATP binding, the first step of the phosphorylation reaction. Immediately following mitosis, the concentration of cyclin B in the cell is vanishingly small. Cyclin B is made steadily so its concentration gradually increases, and so more and more Cdk1 comes to have cyclin B attached (Fig. 19.8b). However, this heterodimer is inactive because it is phosphorylated on T14 and Y15. At the G2/M boundary the phosphates are removed by a second enzyme called **Cdc25**. MPF is now active and can phosphorylate its target proteins. Once mitosis is initiated, it is essential that Cdk1 be deactivated for the reasons outlined above. This is achieved by the destruction of cyclin B and the rephosphorylation of Cdk1 by Wee1.

What About the G1/S Control Point?

The decision to begin DNA synthesis is arguably an even more critical decision than entry into M phase because once a cell begins, it is committed to eventually dividing. (Cells that replicate their DNA but then fail to undergo mitosis in a reasonable time commit suicide by apoptosis (page 415) rather than hang around in a **polyploid** state). Cell fusion experiments again indicated that factors in the cytoplasm controlled advance to the next stage in the cell

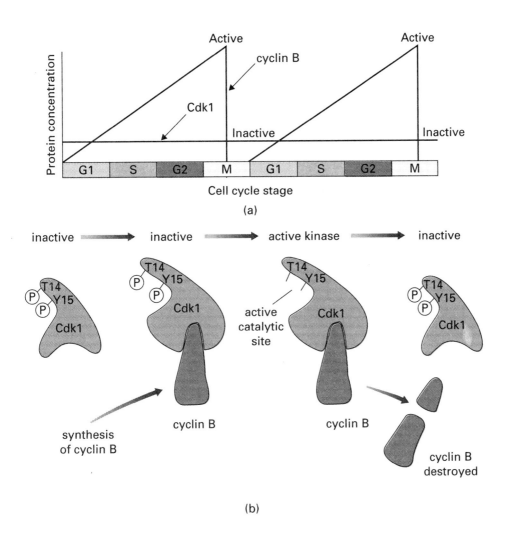

Figure 19.8. Control of Cdk1 by cyclin B and by phosphorylation.

cycle. In particular, fusion of an S-phase cell with a G1 cell caused the G1 nucleus to begin DNA synthesis (Fig. 19.6a). We now know that entry into S phase is regulated not by one but by three cyclin-dependent kinases called **Cdk2, Cdk4,** and **Cdk6,** which, like Cdk1, are active at high efficiency only when coupled to a cyclin partner, in this case **cyclin D** for Cdks 4 and 6 and **cyclin E** for Cdk2. The function of Cdk4 is most clearly understood and is illustrated in Figure 19.9. DNA replication requires a specific set of proteins not used in quiescent cells (page 87), and transcription of the genes coding for these proteins is activated by a transcription factor called **E2F-1.** In quiescent cells E2F-1 is prevented from activating transcription by being bound to **RB,** the product of a gene that is mutated in the cancer **retinoblastoma.** The main function of Cdk4 seems to be to phosphorylate RB, preventing its association with E2F-1, which is therefore freed to bind to enhancer sequences, activate transcription, and hence trigger DNA synthesis. Growth factors trigger cell division in part through activating transcription of the cyclin-D gene and hence allowing

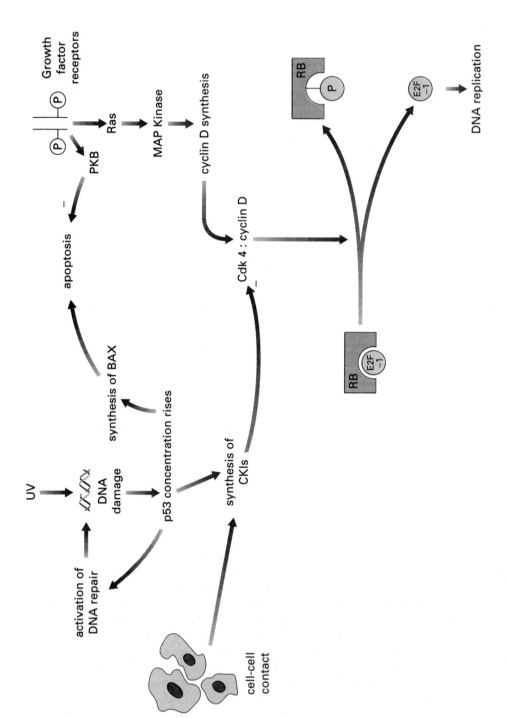

Figure 19.9. Retinoblastoma protein RB sequesters E2F-1, the critical transcription factor controlling entry into S phase.

Cdk4 to operate. Once S phase has begun, RB is dephosphorylated to prevent further rounds of DNA replication.

The G1 Cdks have an extra level of regulation above that described for Cdk1–cyclin B because they are inhibited by an additional class of proteins called **CKIs** for **cyclin-dependent kinase inhibitor** proteins. There are many CKIs mediating a number of pathways all with the same outcome: stopping cell division. One of these pathways mediates **contact inhibition.** Cells in a culture dish, or at the edge of a wound, divide until they touch each other. When they contact neighboring cells, they stop dividing because contact causes production of two CKIs called p16^{INK4a} and p27^{KIP1}. These inhibit the G1 Cdks and therefore prevent DNA synthesis. Loss of contact inhibition is one of the first changes seen in the **transformation** of normal cells into cancer cells.

Another CKI called p21^{CIP1} is produced in response to the transcription factor **p53.** p53 is produced all the time in cells but is broken down rapidly, so its concentration is usually low. However, if the DNA is damaged, for example, by ultraviolet light (page 94), the destruction of p53 is halted. As p53 concentrations increase, a number of processes are activated. The systems that repair DNA are activated (page 94). *p21^{CIP1}* is produced, inhibiting the G1 CDKs and therefore preventing replication of the defective DNA. Lastly, if the DNA cannot be repaired within a set time (usually a few hours), the cell is triggered to self-destruct in the process called apoptosis.

✳ APOPTOSIS

An adult human is made up of about 30 trillion (3×10^{13}) cells, all of which originate from a single fertilized egg. If this first cell divides into 2, the 2 progeny cells into 4, and so on, it would take only about 45 rounds of division to produce the number of cells required to make an adult human. In fact, cell division occurs constantly through our lifetimes, such that we generate a new complete set of 3×10^{13} cells every 2 weeks. The reason that multicellular organisms do not become infinitely large is because the proliferation of cells is balanced by cell death. Cells die for two quite different reasons. One is accidental, the result of mechanical trauma or exposure to some kind of toxic agent, and often referred to as **necrosis.** This is the only type of death seen in unicellular organisms. The other type of death is deliberate, the result of an built-in suicide mechanism known as **apoptosis** or **programmed cell death.** The two types of cell death are quite distinct. In cells that are injured, ATP concentrations fall so low that the Na$^+$/K$^+$ ATPase can no longer operate, and therefore ion concentrations are no longer controled. This causes the cells to swell and then burst. The cell contents then leak out, causing the surrounding tissues to become inflamed. Cells that die by suicide on the other hand shrink, and their cell contents are packaged into small membrane-bound packets called blebs. The nuclear DNA becomes chopped up into small fragments, each of which becomes enclosed in a portion of the nuclear envelope. The dying cell modifies its plasma membrane, signaling to macrophages (page 13), which respond by engulfing the blebs and the remaining cell fragments and by secreting cytokines that inhibit inflammation. The changes that occur during apoptosis are the result of hydrolysis of cellular proteins by a family of proteases called caspases (short for cysteine-containing, cleaving at aspartate). All the cells of our body contain caspases, but they are normally locked in an inactive form by an integral inhibitory domain of the protein. Proteolysis cleaves the inhibitory domain off, releasing the active caspases. The advantage

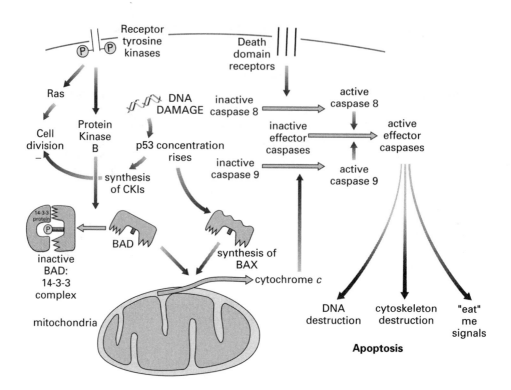

Figure 19.10. Pathways controlling apoptosis.

to the cell of this strategy is that no protein sythesis is required to activate the apoptotic pathway—all the components are already present. Thus, for example, if a virus infects a cell and takes over all protein synthesis, the cell can still commit suicide and hence prevent viral replication. Figure 19.10 summarizes the complex control systems that regulate the decision to die or survive. Cells activate apoptosis in response to three types of event.

Instructed Death: Death Domain Receptors

Cells are instructed to die when a ligand binds to one of a family of **death domain** receptors. This occurs, for example, in the case described above: If a cell is infected by a virus, white blood cells recognize viral proteins on the cell surface and activate **Fas,** a death domain receptor on the surface of the unlucky cell. On binding its ligand, a death domain receptor causes caspase 8 to activate. In turn, caspase 8 can hydrolyze and hence activate the **effector caspases** that begin the processes of cell destruction.

Default Death: Absence of Growth Factors

Cells that are not required by the organism die. To make sure that this occurs, death is the default option for the cells of a multicellular organism: Only if a cell receives growth factors from other cells will it survive. We have already described the first part of this pathway (page 356). Active receptor tyrosine kinases recruit and in turn activate phosphatidyli-nositol 3-kinase, which generates the highly charged membrane lipid phosphoinositide trisphosphate. This causes protein kinase B to visit the membrane, where it is itself phos-phorylated and hence activated. Protein kinase B phosphorylates one of a family of **bcl-2 family proteins** called bcl-2 associated death promoter, or **BAD.** Phosphorylated BAD is inactive and is sequestered by binding to a protein called 14-3-3. However, if PKB is ever allowed to stop working, BAD loses its phosphates and is then active. Active BAD makes holes in the outer mitochondrial membrane (how it does this is unclear) allowing cy-tochrome c to escape into the cytosol. Although cytochrome c plays a vital role in allowing mitochondria to generate the H^+ gradient energy currency (page 266), once in the cytosol it is deadly. It activates another caspase, number 9, which then hydrolyzes and hence activates the effector caspases.

Example 19.3 Neurotrophin Trafficking

During fetal development motoneurons die unless they succeed in growing their axons all the way to an appropriate muscle. Those that do find the target are bathed in **neurotrophin 3,** a growth factor released by the muscle cells. Neurotrophin 3 binds to its receptor tyrosine kinase, called Trk C, on the nerve cell surface. However, the activated kinase then has to activate protein kinase B in the cell body in order to prevent apoptosis, and the cell body may be many millimeters away. To achieve this, the neurotrophin–receptor complex is endocytosed using the clathrin mechanism (page 229). The endocytotic vesicles are then transported back to the cell body by dynein (page 390). Once in the cell body, protein kinase B is activated and apoptosis prevented.

The nerve that contains motoneurons also contains pain receptor axons. Instead of Trk C these express a related receptor, Trk A, which requires a different neurotrophin (number 1): Trk A cannot be activated by neurotrophin 3. However, both motoneurons and pain receptors express a second, unrelated receptor called the p75 neurotrophin receptor, which binds both neurotrophin 1 and neurotrophin 3. p75 is a death domain receptor. The overall result is that when an axon arrives at a target, it will automatically receive a signal to die. However, if it has arrived at the correct location, it will receive a countermanding signal to survive.

The Sick Are Left to Die: Stress-Activated Apoptosis

If a unicellular organism is damaged, it will try to repair itself since the alternative is the death of the organism. However, if a cell of a multicellular organism is stressed or damaged, it may be more efficient to allow the cell to undergo a quick and polite suicide and replace it by cell division of a healthy neighbor. There are therefore a number of mechanisms that trigger apoptosis in response to cell stress. One mechanism operates directly on mitochondria. When mitochondria are stressed, they can spontaneously release cytochrome c: This appears to happen when porin, a channel of the outer mitochondrial membrane, associates with other proteins to form a channel of large enough diameter to allow cytochrome c to leak out. This is a major medical problem because it can occur in tissues that are unable to rebuild by cell division: in hearts during heart attacks and in brain nerve cells during a stroke.

A second mechanism operates through the transcription factor p53, which we have already met in the context of cell cycle control (page 415). p53 concentrations increase when DNA is damaged. DNA repair mechanisms are activated, but so is transcription of the gene for another bcl-2 family protein called BAX. Like BAD, BAX allows cytochrome *c* to escape from mitochondria, so if the DNA is not repaired in time, BAX concentrations increase, cytochrome *c* escapes, and apoptosis results. Without this mechanism, cancers, which result from changes in the DNA of our body cells, would be far more abundant.

Example 19.4 **Sunburn, Skin Cancer, and Cell Death**

Without knowing it, we have all observed apoptosis in action. All of us at some time or other have been out in the sun without proper protection. The ultraviolet (UV) light causes damage to the DNA of the skin cells and activates p53. If the DNA damage is minor, p53 simply arrests the cell cycle until the DNA repair machinery has had time to repair the lesions. However, if the damage was more severe, the skin cells activate the apoptoptic pathway—and the sunburned skin dies and sloughs off.

A third mechanism (not shown in Fig. 19.10) operates through a protein kinase called **p38.** Although p38 is related to MAP kinase, it is activated not by growth factors but rather by cell stresses such as swelling, shrinkage, or radiation. The effects of p38 and another **stress-activated protein kinase,** called JNK, are complex, and we will consider only two. p38 phosphorylates p53, which protects it from the rapid breakdown that normally occurs. An increase of p53 concentration then triggers the usual consequences, including synthesis of BAX. p38 also has a direct effect on the founder member of the bcl-2 family, bcl-2 itself. Unlike BAD and BAX, bcl-2 cannot trigger cytochrome *c* release from mitochondria. Indeed, bcl-2 is antiapoptotic because it can bind to BAD and BAX and prevent them from releasing cytochrome *c*. p38 phosphorylates bcl-2, preventing it from associating with BAD and BAX. When BAD and BAX are freed, they can trigger cytochrome *c* release.

Since caspases can self-activate by mutual proteolysis, there is a steady slow rate of activation even in a healthy cell. To prevent this triggering apoptosis, cells produce **apoptosis inhibitor proteins,** which block caspase action. If the extent of caspase activation exceeds the capacity of the inhibitory proteins, death results.

•••● IN DEPTH 19.1 A Worm's Eye View of Cell Death

Our understanding of the mechanisms and importance of cell death has been greatly influenced by work on the nematode worm, *Caenorhabditis elegans.* This tiny creature (about 1 mm long) is a true animal with nervous, digestive, and muscular systems but is small enough to be grown in Petri dishes (where it is fed a diet of *Escherichia coli*). Crucially, every *C. elegans* worm is made up of a precise number of cells. In the newly hatched worm this number is 558, rising to 959 somatic cells and a variable number of germ cells in the adult. Because the worm is transparent, it has been possible to trace the lineage of every somatic cell back to the original fertilized egg. This pattern is identical in every worm examined. Strikingly, exactly 131 cells die during the developmental process as the result of programmed cell death.

Workers trying to understand this process isolated *ced* (= cell death defective) mutants in which various organs contained abnormally high numbers or abnormally low numbers of cells because the cell death program was abnormal. When these genes were cloned, it became clear that similar genes were present in other animals including humans. For example, *ced-3* was found to code for a proteolytic enzyme. Screening the human genome for similar enzymes revealed the family of apoptotic proteases we now know as caspases.

The contribution of the simple nematode to biology was recognized by the award of the 2002 Nobel prize in physiology or medicine to the pioneers of nematode research, Sydney Brenner, Robert Horvitz, and John Sulston.

SUMMARY

1. Haploid cells contain only one copy of each chromosome. Diploid cells contain two copies, one from the organism's father, one from the mother.

2. The cell cycle comprises S, G1, M, and G2 phases. G2, S, and G1 phases together constitute interphase. DNA is replicated in S phase, so that each chromosome becomes a pair of identical chromatids. The cell physically divides in mitosis, or M phase.

3. Mitosis comprises prophase, prometaphase, metaphase, anaphase, telophase, and cytokinesis. The end result of mitosis is two diploid cells whose chromosome complement is the same as that of the original cell before it underwent S phase.

4. Meiosis generates haploid germ cells: in vertebrates, eggs and sperm. Like mitosis, it follows an S phase but comprises two cycles of cell division, so that the end result is four cells whose chromosome complement is only half that of the original cell before it underwent S phase.

5. During meiosis I homologous chromosomes undergo recombination, a physical resplicing of homologous chromosomes that allows information on chromosomes originating from father and mother to be mixed.

6. Recessive genes are usually those that fail to make functional protein. Examples are the gene for blue eyes and the gene for phenylketonuria. The functional gene is called dominant because an individual need inherit only one copy to be able to make functional protein.

7. In a few cases a defective gene, for example, the gene for familial Creutzfeldt–Jacob disease, is dominant over a normal one.

8. Cells enter mitosis when cyclin-dependent kinase 1 is active. This in turn requires that cyclin B be present in high enough concentration and that cyclin-dependent kinase 1 be dephosphorylated by Cdc25. Once mitosis is initiated, cyclin-dependent kinase 1 is rapidly turned off through phosphorylation by WEE1 in parallel with the proteolytic destruction of cyclin B.

9. The entry of cells into S phase is a more complex decision involving cyclin-dependent kinases 2, 4, and 6. The main effect of active cyclin-dependent kinase 4 is to phosphorylate RB, causing it to release the transcription factor E2F-1 and hence allowing the synthesis of proteins required for DNA synthesis. Important components of the decision are a raised concentration of cyclin D as a result of MAP kinase activity and a low concentration of CKIs. Cell–cell contact upregulates CKIs so that when an organ has filled the space available it stops growing.

10. p53 is continually produced in cells but is as quickly destroyed. An increase in the concentration of p53 follows DNA damage or other cell stress and has three main effects: (a) activation of DNA repair mechanisms, (b) synthesis of CKIs, preventing cell division, and (c) activation of apoptosis.

11. In contrast to necrosis, which causes inflammation, apoptosis is a regulated mechanism of cell suicide that has little effect on the surrounding tissue. The final effectors of apoptosis are a family of proteases called caspases.

12. Apoptosis can be triggered in three ways: (a) binding of ligand to death domain receptors, (b) denial of growth factors, and (c) cell stress.

FURTHER READING

Hyams, J. S. 2002. The cell cycle and mitosis. *Biol. Sci. Rev.* 14: 37–41.

Jacobson, M. D., Weil, M., and Raff, M. C. 1997. Programmed cell death in animal development. Cell 88: 347–354.

Michael D., Jacobson, M. D., and McCarthy, N. (eds.) 2002. *Apoptosis: The Molecular Biology of Programmed Cell Death.* Oxford: Oxford University Press.

Mitchison T. J., and Salmon E. D. 2001. Mitosis: A history of division. *Nature Cell Biol.* 3: E17–E21.

Nicklas R. B. 1997. How cells get the right chromosomes. *Science* 275: 632–637.

Sharp, D. J., Rogers, G. C., and Scholey, J. M. 2000. Microtubule motors in mitosis. *Nature* 407: 41–47.

Vaux D. L., and Korsmeyer, S. J. 1999. Cell death in development. *Cell* Jan 96: 245–254.

REVIEW QUESTIONS

For each question, choose the ONE BEST answer or completion.

1. Mitosis proceeds in the order
   A. cytokinesis, prophase, prometaphase, telophase, metaphase, and anaphase.
   B. telophase, anaphase, prophase, prometaphase, metaphase, and cytokinesis.
   C. prophase, prometaphase, metaphase, anaphase, telophase, and cytokinesis.
   D. prophase, prometaphase, metaphase, cytokinesis, anaphase, and telophase.
   E. prophase, prometaphase, cytokinesis, metaphase, anaphase, and telophase.

2. Paired chromatids separate and begin to move toward the spindle poles in mitotic
   A. prophase.
   B. prometaphase.
   C. metaphase.
   D. anaphase.
   E. telophase.

3. Meiosis differs from mitosis in that
   A. the centrioles do not separate during the first meiotic division.
   B. meiotic divisions are always asymmetric.

C. meiosis generates cells with half as much DNA as the progeny of mitotic division have.

D. movement of chromosomes toward the spindles is slower.

E. All of the above.

4. Cdk1 can only be fully active when

A. it is phosphorylated on threonine 14.

B. it is phosphorylated on tyrosine 15.

C. it is bound to cyclin A.

D. it is dephosphorylated by Cdc25.

E. All of the above.

5. The main target of cdk4 is

A. RB.

B. E2F-1.

C. P53.

D. nuclear lamins.

E. Wee1.

6. Caspases are activated by

A. phosphorylation.

B. dimerization.

C. trimerization.

D. proteolysis.

E. binding to bcl-2 family members.

7. Phenylketonuria is said to be a recessive inherited disease because

A. the fraction of babies with the disease falls in successive generations.

B. the disease is not observed in individuals with one mutant gene and one normal one.

C. phenylketonuria accumulates at the back of the foot.

D. myelin recedes from axons, causing failure of action potential transmission.

E. the condition only appears following meiotic recombination.

ANSWERS TO REVIEW QUESTIONS

1. *C.* Mitosis proceeds in the order prophase, prometaphase, metaphase, anaphase, telophase, and cytokinesis.

2. *D.* Anaphase begins when the protein cohesin, the glue holding the pairs of chromatids together, is degraded, allowing the chromosomes to begin traveling toward the spindle poles.

3. *C.* Like mitosis, meiosis is preceded by one round of DNA synthesis but generates four progeny cells rather than two. Concerning the other answers: (A) The centrioles always separate in cell division, becoming the spindle poles on either side of the nucleus. If you answered answer A you may have been thinking of the fact that in the first meiotic division the centromeres of paired chromatids do not separate, rather, homologous pairs of chromosomes separate. (B) Although meiosis in females is asymmetric, producing one large egg cell and three small polar bodies, in males meiosis is symmetrical, producing four equally sized spermatids. (D) Meiosis often takes much longer than mitosis, but this is because of a lengthy prophase I rather than any change in the duration of anaphase.

4. *D.* For Cdk1 to be fully active, it must be dephosphorylated (not phosphorylated) on threonine 14 and tyrosine 15 by Cdc25 and must be bound to cyclin B (not cyclin A, which does play a role in the cell division cycle but which is not described in this book).

5. *A.* Cdk4 phosphorylates RB, which is then unable to sequester E2F-1. None of the others are targets of Cdk4, although nuclear lamin is certainly a target of Cdk1.

6. *D.* As generated at the ribosome, caspases are locked into an inactive form by an integral inhibitory domain. Proteolysis removes this inhibitory domain.

7. *B.* Recessive diseases are so called because the defective gene is overruled by the *dominant* normal gene if the latter is present.

20

CASE STUDY: CYSTIC FIBROSIS

❊ INTRODUCTION

In the final Chapter of the book we describe the biochemical basis of the disease cystic fibrosis (CF), the search for the *CF* gene, the prospects for CF gene therapy, and prenatal diagnosis. We have selected this disease as a case study to show how the combined efforts of biochemistry, genetics, molecular and cell biology, and physiology were needed to find the cause of CF and the function of the protein encoded by the *CF* gene. The basic principles of many of the techniques used in achieving this knowledge are described in Chapters 1–19.

❊ CYSTIC FIBROSIS IS A SEVERE GENETIC DISEASE

Among Caucasians (white, non-Jewish) about 1 child in 2500 is born with CF. Inheritance is simple: when both parents are carriers, there is a 1 in 4 chance that a pregnancy will result in a child with the disease. The disease is a distressing one. Most of its symptoms arise from faults in the way the body moves liquids, leading in particular to a build up of inadequately hydrated, sticky mucus in various parts of the body. In the lungs this leads to difficulty in breathing, a persistent cough, and a greatly increased risk of infection. A bacterium known as *Pseudomonas aerigunosa* thrives in the lung mucus and is particularly recalcitrant to antibiotic treatment. The pancreas—which provides a digestive secretion that flows to the intestine—is also affected and may be badly damaged (which explains

Cell Biology: A Short Course, Second Edition, by Stephen R. Bolsover, Jeremy S. Hyams,
Elizabeth A. Shephard, Hugh A. White, Claudia G. Wiedemann
ISBN 0-471-26393-1 Copyright © 2004 by John Wiley & Sons, Inc.

the disease's full name, cystic fibrosis of the pancreas). Often, there are digestive problems because the damaged pancreas is not producing enzymes. The reproductive system is also harmed and most males who survive to adolescence are infertile. Even though there are so many varied symptoms, the simple Mendelian pattern of inheritance led researchers to believe that the cause was an abnormality or absence of a single protein.

Until recently babies with CF did not survive to their first birthday. Today, the life expectancy is 22 years. This remarkable improvement is due to intensive therapy designed to reverse individual symptoms. Digestive enzymes are taken by mouth to replace those proteins the pancreas fails to produce. Physiotherapy—helping patients to cough up the mucus in their lungs by slapping their backs—reduces the severity of the lung disease. Yet, however remarkable the increase in survival, each life that extends only to early adulthood is still a tragedy.

✣ THE FUNDAMENTAL LESION IN CYSTIC FIBROSIS LIES IN CHLORIDE TRANSPORT

According to the *Almanac of Children's Songs and Games from Switzerland,* *"the child will soon die whose brow tastes salty when kissed."* Cystic fibrosis has long been recognized as a fatal disease of children. The link between a salty brow and a disease known for its effects on the pancreas and lungs was not, however, made until 1951. During a heat wave in New York that summer, Paul Di Sant' Agnese and co-workers noticed that babies with cystic fibrosis were more likely than others to suffer from heat prostration. This prompted them to test their sweat, and they found that it contained much more salt than the sweat of unaffected babies.

Sweat glands have two regions that perform different jobs (Fig. 20.1). The secretory region deep in the skin produces a fluid that has an ionic composition similar to that of extracellular medium; that is, it is rich in sodium and chloride. If the sweat glands were simply to pour this liquid onto the surface of the skin, they would do a good job of cooling, but the body would lose large amounts of sodium and chloride. A reabsorptive region closer to the surface removes ions from the sweat, leaving the water (plus a small amount of sodium chloride) to flow out of the end of the gland. CF patients produce normal volumes

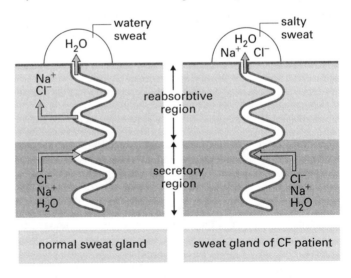

Figure 20.1. Sodium, chloride, and water transport in the sweat gland.

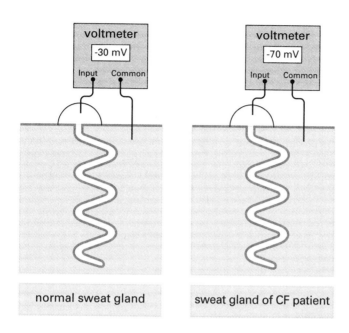

Figure 20.2. Measuring the transepithelial voltage of human sweat glands in situ. The voltmeter measures the voltage of the input terminal with respect to the common terminal. In this case a reading of −30 mV indicates that the droplet of sweat has a voltage 30 mV negative to the extracellular medium.

of sweat, but this contains lots of sodium and chloride, implying that the secretory region is working fine but that the reabsorptive region has failed. The pathways by which sodium and chloride ions are removed are distinct. Which one has failed in CF? A simple electrical test gave the answer. A normal sweat gland has a small voltage difference across the gland epithelium (Fig. 20.2). In CF patients the inside of the gland is much more negative than usual. This result tells us at once that it is the chloride transport (and not sodium transport) that has failed. In CF sweat glands the reabsorptive region has the transport systems to move sodium, but the chloride ions remain behind in the sweat, giving it a negative voltage. The sodium transport system cannot continue to move sodium ions out of the sweat in the face of this larger electrical force pulling them in, so sodium movement stops too, and sodium chloride is lost in the sweat, which then tastes salty. Every one of the symptoms suffered by CF patients is caused by a failure of chloride transport.

✺ HOMING IN ON THE *CF* GENE

The first family studies were carried out more than 50 years ago, and family pedigrees showing CF as a classic case of recessive inheritance were published in 1946. However, there was no obvious way to identify the millions of people who carry a single copy of the defective *CF* gene. There were many attempts to find a test for CF. For a time it seemed that a simple stain might do the job; but this was abandoned. Other tests were more eccentric. For example, there was a claim that an extract from the blood of CF patients, and perhaps even from that of their parents, who must be carriers, slowed the beating of the cilia of oysters.

Most other claims that particular gene products were peculiar to CF carriers also failed to stand up or proved to be symptoms rather than cause. For instance, trypsinogen, a precursor of the digestive enzyme trypsin, is found in fluid bathing fetuses homozygous for CF, and not in others—but this proves to be because the pancreas is already dying at this time. In the 1980s there was hope that this result could be used to diagnose affected pregnancies, but the approach was quickly overtaken by developments in studying the gene itself. As for most inborn diseases, the big problem was that nobody had any real idea what the faulty gene did; looking for a defect in an unknown protein is a task that requires hard work and good ideas. After much time-consuming work, in 1985 the research group led by Lap-Chee Tsui published data indicating that the gene for CF lay on chromosome 7. Researchers then used linkage measurements (page 408) to try to pinpoint the region of chromosome 7 containing the *CF* gene. An international scientific collaboration analyzed DNA from 200 families. The results indicated that the *CF* gene must lie between two marker regions on chromosome 7, named *met* and D7S8. The distance between the markers was 2 million base pairs. It was now possible to think about isolating the *CF* gene.

✹ CLONING THE GENE FOR CF

Sheer hard work won the day when it came to cloning the *CF* gene. The scientists knew that the gene was on chromosome 7, somewhere between the *met* and D7S8 markers. Their strategy was to "walk" along the chromosome (page 152), away from each marker, and hopefully approach the *CF* gene from either side. In the 1980s, the only genomic vectors available were cosmids (insert size 40,000 bp) and those based on the bacteriophage lambda (insert size 20,000 bp) (page 140). So each genomic clone covered only a tiny part of chromosome 7. Clever tricks had to be devised to speed up the process, and finally a series of cosmid clones, thought to possibly contain regions of the *CF* gene, were isolated. But how could they tell which clones were the right ones?

Since CF affects so many organs, the gene is likely to be a critical one conserved through mammalian evolution. A zoo blot (page 146) was therefore used in the hope of identifying the gene. In turn, each of the human cosmid clones was radiolabeled and used to probe a zoo blot. Researchers found three clones that hybridized to the DNA of all the test mammals. The next step was to check whether these clones contained sequences that code for mRNAs expressed in normal tissues that are affected by CF. Each clone was used to probe a northern blot (page 148) of sweat gland mRNA. One sequence passed the test—with luck, it would contain part of the *CF* gene. With hindsight we now know that the scientists' luck had almost run out: The one positive sequence did contain part of the gene, but only a tiny part—113 bp, less than 1% of the total gene.

Things moved swiftly from then on. The 113-bp section was used to screen cDNA libraries (page 134) made from sweat gland mRNA. This gave scientists cDNA for the entire *CF* mRNA, which could then be sequenced. Meanwhile, the cDNA was used to screen a genomic library to find the rest of the *CF* gene. The gene is 220,000 bp long and has 24 exons.

✹ THE *CFTR* GENE CODES FOR A CHLORIDE ION CHANNEL

Once *CF* cDNA had been sequenced, it could be read, revealing that *CF* codes for a protein of 1480 amino acids. Although we knew that CF was caused by a defect in chloride

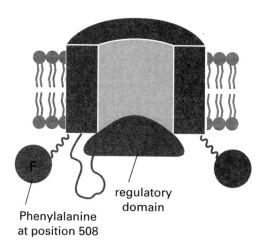

Figure 20.3. The CFTR protein forms a chloride ion channel in the plasma membrane of epithelial cells.

ion transport, it was not immediately clear that the product of the *CF* gene was itself a chloride channel. Although hydropathy plots (page 204) indicated that the CF protein was a transmembrane protein, the investigators had to consider the possibility that it was, for example, a cell surface receptor whose activation triggered expression of the proteins mediating chloride transport. The protein was therefore given the catch-all name "cystic fibrosis transmembrane regulator" (CFTR) and its gene was renamed as the *CFTR* gene. However, for once things were simple: CFTR is indeed itself the chloride ion channel in the plasma membrane (Fig. 20.3). We know this because if purified CFTR is introduced into lipid bilayers, chloride currents can be measured (see In Depth 20.1). In resting cells the channel mouth is blocked by a plug called the regulatory domain. When the regulatory domain is phosphorylated by cAMP-dependent protein kinase (page 353), the plug leaves the channel, allowing chloride ions to flow through.

Comparing the sequence of the normal *CFTR* gene to the one in people with CF revealed the mutations that caused the disorder. In 70% of CF patients, three nucleotides are missing, generating a protein that is missing the amino acid phenylalanine at position 508 (Fig. 20.4). The protein is a functional chloride channel, but the mutation affects its ability to fold properly (page 192), and so the mutant protein fails to be inserted into the plasma membrane. In the remaining 30% of patients more than 300 *CFTR* gene mutations have been documented. These include missense, nonsense, and frameshift mutations (page 80). In some people mutations that cause incorrect exon–intron splicing (page 118) occur. The severity of CF symptoms is dependent on the particular mutation. Some cause a complete failure to express the protein and produce severe symptoms. Other mutations produce a protein that is correctly trafficked to the plasma membrane and that works, but not as well as the normal protein.

✹ GENE THERAPY FOR CF

It is an exciting prospect to think that we might one day treat disease by introducing a normal gene into the cells of patients and correct a defect. However, **gene therapy** is a difficult procedure, still in its infancy, and one that is not without risk to the patient. At present, permission has only been granted to carry out gene therapy on somatic cells. In this method the defect is corrected in specific cells or tissues rather than in the germ line cells that will give rise to eggs and sperm. Before trying out gene therapy on CF patients, an animal study was conducted. A knock-out mouse (page 157), lacking the *CFTR* gene, was created (Fig. 20.5). The mouse showed the symptoms of CF. A plasmid containing the CFTR cDNA was used to prepare an aerosol. When the mice inhaled the aerosol droplets, the results were spectacular. The lung symptoms disappeared for many days. The CFTR cDNA had been expressed into protein and the protein targeted to the plasma membrane. Could this approach work in humans?

●●●● IN DEPTH 20.1 Lipid Bilayer Voltage Clamp

In 1992 Christine Bear and co-workers published a definitive demonstration that the CFTR is a chloride ion channel. They used a technique called the lipid bilayer voltage clamp, which is illustrated in the accompanying figure. A lipid bilayer is constructed from phospholipid and used to plug a hole between two chambers filled with salt solution. Since a lipid bilayer is a barrier to ion movement, no current is recorded when a voltage is imposed across the bilayer. However, when artificial vesicles containing integral membrane proteins are added to one bath, the vesicles fuse with the artificial membrane so that the proteins now span the barrier bilayer. When Bear and her co-workers did this with purified CFTR, they recorded currents with the direction and amplitude expected of a chloride-selective channel. In particular, no current flowed when the voltage difference across the membrane was equal to the calculated chloride equilibrium voltage (page 312). No other integral membrane proteins were present in this experiment, so the CFTR must itself be a chloride channel.

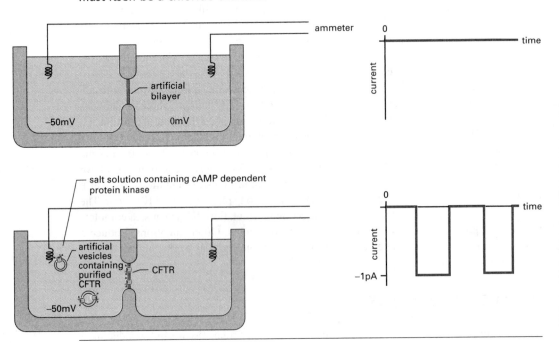

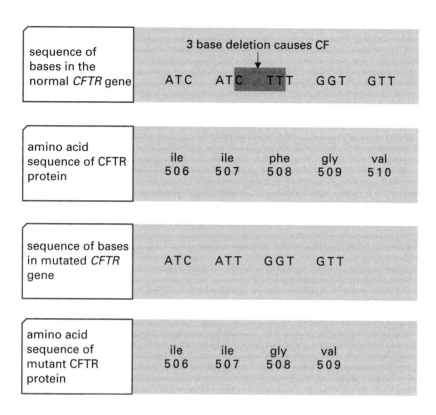

sequence of bases in the normal *CFTR* gene	3 base deletion causes CF

ATC ATC TTT GGT GTT

| amino acid sequence of CFTR protein | ile 506 | ile 507 | phe 508 | gly 509 | val 510 |

| sequence of bases in mutated *CFTR* gene | ATC ATT GGT GTT |

| amino acid sequence of mutant CFTR protein | ile 506 | ile 507 | gly 508 | val 509 |

Figure 20.4. The mutation seen in 70% of CF patients: a deletion of three nucleotides from the *CFTR* gene has removed a phenlyalanine at position 508.

In 1992 the National Institutes of Health in the United States granted permission for a *CFTR* gene therapy trial. Because viruses are so good at infecting our cells, they are useful tools for transferring DNA into somatic cells. Viral vectors for gene therapy have been engineered to remove their harmful genes. The gene to be transferred into the patient is inserted into the modified viral genome. A modified adenovirus vector containing the CFTR cDNA was introduced into the lung cells of four patients. Each of the patients made CFTR protein, but only for a short time. Some patients showed adverse reactions to the therapy, probably due to the adenovirus itself. Subsequent trials have delivered an adenovirus, containing the CFTR cDNA, to the nasal epithelium of CF patients. However, again adverse reactions to the treatment developed.

On a more positive note, a method called liposomal-mediated gene transfer shows some promise. This technique has the advantage that it avoids the use of viruses. The plasmid containing the CFTR cDNA is coated with a synthetic lipid bilayer to form a liposome (page 229). The liposome can fuse with the epithelial cell membrane and deliver its DNA contents into the cell. In the case of CF this approach seems to work when the liposomes are delivered to the lungs in the form of an aerosol spray. The CFTR cDNA is expressed and

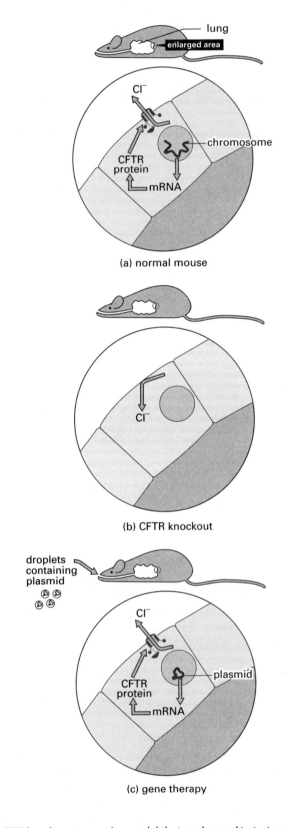

(a) normal mouse

(b) CFTR knockout

(c) gene therapy

Figure 20.5. The *CFTR* knock-out mouse is a model that can be used to test gene therapy stategies.

some chloride ion channels form in the plasma membrane. The defect is not permanently corrected and continual use of the aerosol will be necessary to alleviate the lung symptoms of CF. Other organs, such as the pancreas, are affected in CF. Gene therapy on pancreatic cells will prove much more difficult. These cells are not as easily accessible as the epithelial cells of the lung.

❄ DIAGNOSTIC TESTS FOR CF

Because 70% of CF patients carry the same mutation, the deletion of phe at postion 508, a diagnostic test was developed. The test, based on **PCR** (page 150), requires a few cells from the inside of the mouth and two DNA primers that will amplify a stretch of DNA that will include the region encoding amino acid 508 (Fig. 20.6). The PCR products are dotted onto two membrane filters. One membrane is incubated with an oligonucleotide containing the sequence encoding phe508 and the other is incubated with an oligonucleotide lacking the codon for phe508. By comparing the patterns of hybridization, carriers of CF can be detected. Other mutations that cause CF can also be tested for. However, because there are so many *CFTR* mutations, many families may require a new predictive test to be designed to match their version of the defective gene.

PRENATAL IMPLANTATION DIAGNOSIS FOR CF. In 1992 the first preimplantation diagnosis for CF was carried out. Three sets of parents took part in the trial. Each parent was a carrier for the phe508 mutation. Oocytes were removed from the women

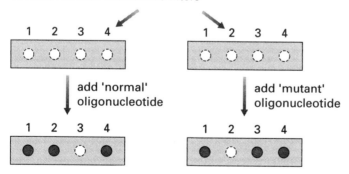

1. PCR amplify section of *CFTR* gene
   spanning codon for phe 508

2. Dot DNA onto membrane filters

add 'normal' oligonucleotide

add 'mutant' oligonucleotide

Individual 1: CF carrier
Individual 2: both copies of gene are normal
Individual 3: both copies of gene are mutant: CF patient
Individual 4: CF carrier

Figure 20.6. A diagnostic test for the most common *CFTR* gene mutation.

and fertilized with sperm from their partners. Success was reported for only one couple. Six of their oocytes developed into fertilized embryos. A single cell was tested from each embryo, using PCR, for the presence of the defective *CFTR* gene. Five embryos were characterized. Two embryos each had two copies of the normal *CFTR* gene, one on each of their chromosome 7's. Two embryos each had two copies of the phe508 deletion. One embryo tested positive for both the normal gene and the phe508 deletion and was therefore a carrier. One normal/normal embryo and the carrier embryo were implanted into the mother. A baby girl was born. Her DNA showed that both copies of her *CFTR* gene were normal.

❉ THE FUTURE

Gene therapy strategies for CF have progressed, but much needs to be improved before this type of treatment can offer real hope for patients. The best gene therapy can offer is an alleviation of some of the symptoms. A cure for CF does not seem a likely prospect. Detection of CF carriers and prenatal diagnosis together with genetic counseling informs parents of the risks and gives them the choice of terminating an affected pregnancy. Preimplantation diagnosis raises ethical questions about selecting a specific embryo. In the future it may be possible to carry out *germ line therapy* in the egg or sperm of carriers. Most countries regard such interference as unethical.

Medical Relevance 20.1

Success and a Setback with SCID

Children with severe combined immunodeficiency disease (SCID) show an almost complete failure to fight infection and die soon after birth if not kept in sterile conditions. SCID is more widely known as "bubble boy disease" because confinement to a germ-free plastic bubble was the way in which patients were kept alive for a number of years. One rare form of SCID, caused by a failure to make an enzyme called adenosine deaminase, can be treated by regular injections of recombinant protein (page 149). In 1990 the first trial of a genetic therapy began at the National Institutes of Health in Washington. White blood cells from patients were transfected with a plasmid encoding adenosine deaminase, which inserted into random positions in the genome and began to be transcribed. Although the therapy does seem to have helped the patients, they still need to be periodically injected with recombinant adenosine deaminase.

The commonest form of the disease, X-SCID, affects only boys and occurs in about 1 in every 150,000 births. X-SCID is caused by a defective receptor for interleukin 2 (see page 222 for more on interleukin 2). In March 1999 a team at INSERM in Paris initiated a trial in which white blood cells from X-SCID patients were transfected with a viral vector containing the normal, interleukin-2 receptor gene, which inserted randomly into the genome. So far 10 patients have been treated in Paris and one at University College London. Of these, 9 are now essentially cured; they are at home living a normal life and are unlikely to need any further treatment. Tragically, however, one of the children has developed leukemia—one of the transfected cells began dividing out of control. Apparently, the plasmid inserted into the gene for a transcription factor called LMO-2 and the viral promoter caused the transfected cells to make lots of LMO-2 as well as lots of the desired interleukin-2 receptor. Expression of LMO-2 causes white blood cells to divide uncontrollably.

At the time of writing, it is uncertain whether this method of treatment for X-SCID will continue. In the long term, this tragic accident will increase the drive toward treating genetic disorders by homologous recombination (page 157) so that only the defective gene is changed.

SUMMARY

1. Cystic fibrosis is the most common serious single-gene inherited disease in the Western world. Many organs are affected. Sticky mucus builds up in the reproductive tract and in the lungs. The pancreas is always affected and usually fails completely.

2. Electrical measurements show that the basic problem is in chloride transport.

3. CF is a classic recessive disorder. The gene is on chromosome 7.

4. A combination of hard work and novel techniques helped isolate a small part of the *CFTR* gene. This sequence helped to isolate the normal CFTR cDNA from a sweat gland clone library.

5. From the cDNA it was possible to identify the gene and infer its amino acid sequence in both normal and mutated forms. The gene codes for a chloride channel protein called CFTR. Hundreds of different *CFTR* mutations have now been found.

6. For the time being gene therapy only holds out hope of alleviating the lung problems, leaving other less accessible organs untreated. Tests have been devised for prenatal diagnosis and to detect carriers of CF.

FURTHER READING

Bragonzi, A., and Conese, M. 2002. Non-viral approach toward gene therapy of cystic fibrosis lung disease. Curr. Gene Therapy 2: 295–305.

Handyside, A. H., Lesko, J. G., Tarin, J. J., Winston, R. M., and Hughes, M. R. 1992. Birth of a normal girl after in vitro fertilization and preimplantation diagnostic testing for cystic fibrosis. N. Engl. J Med. 327: 905–909.

Kerem, B., Rommens, J. M., Buchanan, J. A., Markiewicz, D., Cox, T. K., Chakravarti, A., Buchwald, M., and Tsui, L. C. 1989. Identification of the cystic fibrosis gene: genetic analysis. *Science* 245: 1073–1080.

Rich, D. P., Anderson, M. P., Gregory, R. J., Cheng, S. H., Paul, S., Jefferson, D. M., McCann, J. D., Klinger, K. W., Smith, A. E., and Welsh, M. J. 1990. Expression of cystic fibrosis transmembrane conductance regulator corrects defective chloride channel regulation in cystic fibrosis airway epithelial cells. *Nature* 347: 358–363.

Riordan, J. R., Rommens, J. M., Kerem, B., Alon, N., Rozmahel, R., Grzelczak, Z., Zielenski, J., Lok, S., Plavsic, N., Chou, J. L., Drumm, M. L., Iannuzzi, M. C., Collins, F. S., and Tsui, L-C. 1989. Identification of the cystic fibrosis gene: Cloning and characterization of complementary DNA. *Science* 245: 1066–1073.

Welsh, M. J., and Smith, A. E. 1995. Cystic fibrosis. *Sci. Am.* 273: 52–59.

Bear, C. E., Li, C. H., Kartner, N., Bridges, R. J., Jensen, T. J., Ramjeesingh, M. and Riordan, J. R. 1992. Purification and functional reconstitution of the cystic fibrosis transmembrane conductance regulator (CFTR). *Cell* 68: 809–818.

Benjamini, E., Coico, R. and Sunshine, G. 2000. *Immunology, a Short Course,* 4th ed. New York: Wiley-Liss, pp. 348–353.

Check, E. 2002. A tragic setback. *Nature* 420: 116–118.

✵ REVIEW QUESTIONS

For each question, choose the ONE BEST answer or completion.

1. A mother and a father, both CF carriers, have two children that do not suffer from CF. The chance of a third pregnancy producing a child with the disease is
   A. zero.
   B. 1 in 4.
   C. 1 in 3.
   D. 1 in 2.
   E. very high—the child will almost certainly have CF.

2. Cystic fibrosis results from a defect in a protein that forms
   A. a plasma membrane sodium ion channel.
   B. a plasma membrane potassium ion channel.
   C. a plasma membrane calcium ion channel.   *c*
   D. a plasma membrane chloride ion channel.
   E. a plasma membrane phenylalanine carrier.

3. A zoo blot helps to detect DNA sequences that
   A. are mutating at a fast rate.
   B. are conserved between species.
   C. show no similarity between species.
   D. are lost due to species extinction.
   E. are processed pseudogenes.

4. Somatic gene therapy attempts to correct a gene defect by introducing the normal gene into
   A. the egg.
   B. the sperm.
   C. the fertilized egg.
   D. cultured embryonic stem (ES) cells.
   E. cells in the patient's body other than eggs or sperm.

5. The most common mutation that causes CF
   A. is a nonsense mutation at codon 508 that causes premature chain termination of the CFTR protein.
   B. is a frameshift mutation such that all amino acids following phe at 508 are different from the normal protein.
   C. is a deletion of three nucleotides that causes phe508 to be absent in the mutant protein.
   D. is a nonsense mutation causing failure to express a receptor tyrosine kinase whose activation triggers expression of the proteins mediating chloride transport.
   E. only causes disease in boys.

6. Cystic fibrosis is said to be "linked" to another gene if
   A. *CFTR* is part of an operon that includes the other gene.
   B. *CFTR* and the other gene are located near each other on the same chromosome.
   C. CFTR is translated as a chimera with the other protein.
   D. both CFTR and the protein product of the other gene are activated when phosphorylated by cAMP-dependent protein kinase.
   E. the function of the other gene requires chloride transport.

7. CFTR and the voltage-gated calcium channel are similar in that
   A. both pass calcium ions.
   B. both are opened by depolarization.
   C. both are caused to open by an increase in the concentration of an intracellular messenger.
   D. both are closed when a cytosolic plug enters the channel lumen and blocks it.
   E. both play a critical role in brain cells.

ANSWERS TO REVIEW QUESTIONS

1. **B.** Each pregnancy has a 1-in-4 chance of producing a child with CF. The presence or absence of the disease in earlier children is irrelevant.

2. **D.** CF is characterized by a defect in chloride ion transport.

3. **B.** Zoo blots detect protein-coding sequences that have been conserved through evolution and are therefore present in many species.

4. **E.** In somatic gene therapy the normal gene is introduced into a somatic cell where expression of the new gene will alleviate or cure the disease. This is any cell other than the gametes, the egg, and sperm.

5. **C.** Seventy percent of individuals with CF lack three nucleotides in their *CFTR* gene. This causes the phe at 508 in the normal protein to be absent from the mutant protein.

6. **B.** Concerning the other answers: (A) This would certainly produce genetic linkage, but operons are not found in eukaryotes. (C) Chimeras, which are characterized by the physical linkage on the same polypeptide chain of two proteins that are normally independent, can be produced by naturally ocurring mutations, but this is not what we mean by linkage. Answers D and E are not relevant to the concept of linkage.

7. **D.** The voltage-gated calcium channel is inactivated when a cytosolic plug enters the channel lumen and blocks it, while the CFTR is opened when phosphorylation of its cytosolic plug causes the plug to vacate the channel lumen. Of the other answers: (A) CFTR does not pass calcium ions. (B) CFTR is not opened by depolarization. (C) This answer is not completely wrong because while the CFTR opens when phosphorylated by cAMP-dependent protein kinase, phosphorylation by cAMP-dependent protein kinase does also make it easier for depolarization to open the voltage-gated calcium channel. However, phosphorylation alone cannot open the voltage-gated calcium channel, so this answer is not as good as answer D. (E) As far as we know, CFTR plays no role in the brain. CF patients are cognitively normal.

APPENDIX:
CHANNELS AND CARRIERS

All the channels and carriers described in the book are listed here for reference.

Name	Page for Details	Location	Selective for	Opened by	Comments
CHANNELS					
Large Channels					
Gap junction channel	55	Plasma membrane of many cells	Any solute of $M_r < 1000$	Contact with second gap junction channel	Allows solutes to pass from cytosol to cytosol
Porin	262	Outer mito-chondrial membrane	Any solute of $M_r < 10,000$	Probably spends a large fraction of time open under most circumstances	Stress can cause switch to a larger form that allows cytochrome c to pass
Potassium channels					
Potassium channel	310	Plasma membrane of all cells	Potassium ions	Some isoforms are opened by depolarization; some are open all the time	Responsible for the resting voltage

(*Continued*)

Cell Biology: A Short Course, Second Edition, by Stephen R. Bolsover, Jeremy S. Hyams,
Elizabeth A. Shephard, Hugh A. White, Claudia G. Wiedemann
ISBN 0-471-26393-1 Copyright © 2004 by John Wiley & Sons, Inc.

Name	Page for Details	Location	Selective for	Opened by	Comments
Calcium channels					
Voltage-gated calcium channel	327	Plasma membrane of sea urchin egg, axon terminal, muscle, and other cells	Calcium ions	Depolarization of plasma membrane	The inward movement of calcium ions changes the transmembrane voltage and raises the calcium concentration in the cytosol
Inositol trisphosphate-gated calcium channel	344	Membrane of endoplasmic reticulum	Calcium ions	Binding of cytosolic inositol trisphosphate	Part of the system that allows an extracellular solute to raise cytosolic calcium
Ryanodine receptor; skeletal muscle isoform.	348	Membrane of endoplasmic reticulum	Calcium ions	Direct physical link to voltage-gated calcium channels in the plasma membrane	Allows rapid increase of cytosolic calcium concentration in response to depolarization of plasma membrane
Ryanodine receptor; standard isoforms	348	Membrane of endoplasmic reticulum	Calcium ions	Binding of cytosolic calcium	Allows a small increase of cytosolic calcium concentration to trigger a larger increase
Mitochondrial calcium channel	305	Inner mito-chondrial membrane	Calcium ions	Not easily studied, probably always open	Allows an increase of cytosolic calcium concentration to activate mitochondria.
Sodium and unselective cation channels					
Voltage-gated sodium channel	330	Plasma membrane of nerve and muscle cells	Sodium ions	Depolarization of plasma membrane	Produces brief action potentials

(Continued)

Name	Page for Details	Location	Selective for	Opened by	Comments
Nicotinic acetylcholine receptor channel	364	Plasma membrane of skeletal muscle cells and some nerve cells	Sodium and potassium	Extracellular acetylcholine	An ionotropic cell surface receptor that causes the plasma membrane to depolarize
Glutamate receptor channel	372	Plasma membrane of some nerve cells	Sodium and potassium, some isoforms also pass calcium	Extracellular glutamate	An ionotropic cell surface receptor that causes the plasma membrane to depolarize
cAMP-gated channel	350	Plasma membranes of scent-sensitive nerve cells	Sodium and potassium	cAMP in the cytosol	Depolarizes the plasma membrane so that the cell produces action potentials
cGMP-gated channel	353	Plasma membranes of photore-ceptor cells	Sodium and potassium	cGMP in the cytosol	Depolarizes the plasma membrane
Chloride channels					
GABA receptor	373	Plasma membrane of some nerve cells	Chloride ions	Extracellular GABA	When this channel is open, it is more difficult to depolarize the cell to threshold
CFTR	426	Plasma membrane of sweat gland cells, airway epithelial cells, and of many other cells	Chloride ions	Phosphorylation by protein kinase A	Required for several transport processes in glands

Name	Page for Details	Location	Mode of Action	Comments
CARRIERS **Carriers with no enzymatic action**				
Glucose carrier	316	Plasma membrane of all cells	Glucose	Required by all human cells
ADP/ATP exchanger	271	Inner mitochondrial membrane	ADP / ATP	Gets ATP and ADP across inner mitochondrial membrane
Na^+/Ca^{2+} exchanger	317	Plasma membrane of many cells	$3 Na^+$ / Ca^{2+}	Pushes calcium ions out of the cytosol
β-Galactoside permease	111	Bacterial plasma membrane	H^+ Lactose	Brings lactose into the cell. One of the products of the *lac* operon
Carriers with an enzymatic action as well as a carrier action				
ATP synthase	272	Inner mitochondrial membrane	Inter-membrane space $10H^+$ 3ADP + 3P$_i$ 3ATP Matrix	Converts between energy in the H^+ gradient and energy as ATP
Ca^{2+} ATPase	318	Plasma membrane of many cells	Extra-cellular medium H^+ ATP ADP + P$_i$ Ca^{2+}	Pushes calcium ions out of the cytosol. All cells have either this or the Na^+/Ca^{2+} exchanger; many have both
Na^+/K^+ ATPase	270	Plasma membrane of all cells	Extra-cellular medium $2K^+$ ATP ADP + P$_i$ $3Na^+$	Converts between energy as ATP and energy in the Na^+ gradient
Electron transport chain	265	Inner mitochondrial membrane	Inter-membrane space NADH + H^+ + $\frac{1}{2}O_2$ NAD^+ + H_2O $12H^+$ Matrix	Converts between energy as NADH and energy in the H^+ gradient

GLOSSARY

14–3–3: protein that binds phosphorylated BAD, hence preventing it from triggering apoptosis.

7-methyl guanosine cap: modified guanosine found at the 5′ terminus of eukaryotic mRNA. A guanosine is attached to the mRNA by a 5′- 5′-phosphodiester link and is subsequently methylated on atom number 7 of the guanine.

9 + 2 axoneme: structure of cilium or flagellum; describes the arrangement of nine peripheral microtubules surrounding two central microtubules.

A (adenine): one of the bases present in DNA and RNA; adenine is a purine.

α-amino acid: amino acid in which the carboxyl and amino groups are attached to the same carbon.

α adrenergic receptor: a receptor for the related chemicals adrenaline and noradrenaline. The α adrenergic receptor activates G_q and therefore generates a calcium signal. To a first approximation, α receptors respond to noradrenaline, while adrenaline acts mainly on β adrenergic receptors.

α helix: a common secondary structure in proteins, in which the polypeptide chain is coiled, each turn of the helix taking 3.6 amino acid residues. The nitrogen atom in each peptide bond forms a hydrogen bond with the oxygen four residues ahead of it in the polypeptide chain.

Cell Biology: A Short Course, Second Edition, by Stephen R. Bolsover, Jeremy S. Hyams,
Elizabeth A. Shephard, Hugh A. White, Claudia G. Wiedemann
ISBN 0-471-26393-1 Copyright © 2004 by John Wiley & Sons, Inc.

A site: (aminoacyl site) site on a ribosome occupied by an incoming tRNA and its linked amino acid.

acceptor: in a hydrogen bond, the atom (oxygen, nitrogen, or sulfur) that is not covalently bonded to the hydrogen but that nevertheless accepts a small share of the electrons.

acetone: CH_3—CO—CH_3. A chemical with a fruity smell produced from acetoacetate during the ketosis that can occur in diabetes.

acetylcholine:

$$CH_3-\underset{\underset{O}{\|}}{C}-O-CH_2-CH_2-\underset{\underset{CH_3}{|}}{\overset{\overset{CH_3}{|}}{N^+}}-CH_3$$

A transmitter released by various nerve cells including motoneurones and autonomic vasodilator neurones.

acetylcholine receptor: integral membrane protein that binds acetylcholine. There are two types: the nicotinic acetylcholine receptor is ionotropic while the muscarinic acetylcholine receptor is metabotropic, linked via G_q to phospholipase $C\beta$.

acetylcholinesterase: enzyme that hydrolyzes acetylcholine, terminating its action at synapses.

acid: molecule that readily gives H^+ to water. Most organic acids are compounds containing the group —COOH, although —SH is also weakly acidic.

acrosome: actin-rich structure that helps the sperm to penetrate the egg at fertilization.

ActA: protein on the surface of the bacterium *Listeria* from which actin filaments polymerize.

actin-binding proteins: proteins that bind to and modulate the function of G-actin or F-actin.

actin-related protein (Arp): an actin nucleation protein. New actin filaments can grow out from an Arp base.

actin-severing protein: one of a group of enzymes that cuts actin microfilaments. Gelsolin is an example.

action potential: explosive depolarization of the plasma membrane.

activation energy: height of the energy barrier between the reactants and the products of a chemical reaction.

acyl group: group having the general formula

$$C_nH_m-\underset{\underset{O}{\|}}{C}-$$

Acyl groups are formed when fatty acids are attached to other compounds by ester bonds.

adenine (A): one of the bases present in DNA and RNA; adenine is a purine.

adenosine: adenine linked to the sugar ribose. Adenosine is a nucleoside.

adenosine diphosphate (ADP): adenosine with two phosphates attached to the 5′ carbon of ribose.

adenosine monophosphate (AMP): adenosine with one phosphate attached to the 5′ carbon of ribose.

adenosine triphosphate (ATP): adenosine with three phosphates attached to the 5′ carbon of ribose. ATP is a coenzyme and one of the cell's energy currencies.

adenylate cyclase: enzyme that converts ATP to the intracellular messenger cyclic AMP (cAMP). Also called adenyl cyclase.

adenyl cyclase: enzyme that converts ATP to the intracellular messenger cyclic AMP (cAMP). Also called adenylate cyclase.

adherens junctions: type of anchoring junction in which the cell adhesion molecules are linked to actin microfilaments.

adipocytes: cells that store fats (triacylglycerols).

adipose tissue: type of fatty connective tissue.

ADP: adenosine diphosphate; adenosine with two phosphates attached to the 5′ carbon of ribose.

ADP/ATP exchanger: carrier in the inner mitochondrial membrane. ADP is moved in one direction and ATP in the other.

adrenaline: hormone released into the blood when an individual is under stress. Adrenaline acts at β-adrenergic receptors to activate G_s and hence adenylate cyclase.

adrenergic receptor: receptor for the related chemicals adrenaline and noradrenaline. There are two isoforms, α and β. To a first approximation, noradrenaline acts mainly on α receptors linked to G_q and therefore generates a calcium signal, while adrenaline acts mainly on β receptors linked to G_s and therefore generates a cAMP signal.

α helix: common secondary structure in proteins in which the polypeptide chain is coiled, each turn of the helix taking 3.6 amino acid residues. The nitrogen atom in each peptide bond forms a hydrogen bond with the oxygen four residues ahead of it in the polypeptide chain.

Akt: earlier name for protein kinase B, a protein kinase that is activated when it is itself phosphorylated; this in turn only occurs when Akt is recruited to the plasma membrane by phosphatidylinositol trisphosphate (PIP_3). Akt phosphorylates proteins on serine and threonine residues (e.g., the bcl-2 family protein BAD).

alkali: strong base (that will take H^+ from water) such as sodium hydroxide or potassium hydroxide.

alkaline: of a solution: one with a low concentration of H^+ (really H_3O^+) so that the pH is greater than 7.0.

allolactose: disaccharide sugar. Lactose is converted to allolactose by the enzyme β-galactosidase. Allolactose is an inducer of *lac* operon transcription.

all or nothing: phenomenon that, once initiated, proceeds to completion even if the stimulus is removed. A process that involves positive feedback, such as an action potential, will tend to be all or nothing.

allosteric; allostery: when the binding of a ligand to one site on a protein affects the binding at another site. These interactions can be between sites for the same ligand or different ligands. Molecules that are allosteric almost always have a quaternary structure.

alternative splicing: phenomenon in which a single eukaryotic primary mRNA transcript can be processed to yield a number of different processed mRNAs and can therefore generate a number of different proteins.

α-amino acid: amino acid in which the carboxyl and amino groups are attached to the same carbon.

amino acid: chemical that has both a —COOH group and an —NH$_2$ group. In an α-amino acid, both the —COOH and —NH$_2$ groups are attached to the same carbon atom. All proteins are generated using the genetically encoded palette of 19 α-amino acids plus proline.

aminoacyl site (A site): site on a ribosome occupied by an incoming tRNA and its linked amino acid.

aminoacyl tRNA: tRNA attached to an amino acid via an ester bond.

aminoacyl tRNA synthases: family of enzymes, each of which attaches an amino acid to the appropriate tRNA.

γ-amino butyric acid: a γ-amino acid that acts to open chloride channels in the plasma membrane of sensitive nerve cells. Usually called γ-amino butyric acid rather than γ-amino butyrate, even though the latter ($^-$OOC—CH$_2$—CH$_2$—CH$_2$—NH$_3{}^+$) is the form in which it is found at neutral pH.

amino terminal: end of a peptide or polypeptide that has a free α-amino group.

amorphous: without form. Not a specifically scientific word.

AMP: adenosine monophosphate; adenosine with one phosphate attached to the 5$'$ carbon of ribose.

amphipathic: "hating both"—a molecule with a hydrophobic region and a hydrophilic region is said to be amphipathic.

anabolism: those metabolic reactions that build up molecules: biosynthesis.

anaerobic: without air. Obligate anaerobes are poisoned by oxygen and therefore can only function anaerobically. Other cells, such as yeast and skeletal muscle, can switch to using anaerobic respiration when they are denied oxygen.

anaerobic respiration: partial breakdown of sugars and other cellular fuels that can be accomplished in the absence of oxygen.

anaphase: period of mitosis or meiosis during which sister chromatids or homologous chromosome pairs separate; consists of anaphase A and anaphase B.

anaphase A: part of anaphase in which the chromosomes move to the spindle poles.

anaphase B: part of anaphase in which the spindle poles are separated.

anaphase I: anaphase of the first meiotic division (meiosis I).

anaphase II: anaphase of the second meiotic division (meiosis II).

anchoring junction: class of cell junction that attaches the cytoskeleton of one cell to the cytoskeleton of its neighbor, forming a physically strong connection. There are two types: desmosomes, which connect to intermediate filaments, and adherens junctions, which connect to actin microfilaments.

anion: negatively charged ion, e.g., chloride, Cl^-, or phosphate, HPO_4^{2-}.

anode: positively charged electrode, e.g., in a gel electrophoresis apparatus used for SDS–PAGE.

antenna chlorophyll: chlorophyl molecules that capture light and pass the energy to reaction center chlorophylls.

anterograde: forward movement; when applied to axonal transport it means away from the cell body.

anthrax: lethal bacterial disease. When food supplies are exhausted, anthrax bacteria produce spores that can survive for long periods in the soil or in (illegal) stocks of biological weapons.

antibiotic: chemical that is produced by one type of organism and kills others—often by inhibiting protein synthesis. The most useful antibiotics to humans are those that are selective for prokaryotes.

antibody: protein formed by the immune system that binds to and helps eliminate another chemical. Antibodies can be extremely selective for their ligand and are useful in many aspects of cell biology such as immunofluorescence microscopy and western blotting.

anticodon: three bases on a tRNA molecule that hydrogen-bond to the codon on an mRNA molecule.

antiparallel β sheet: β sheet in which alternate parallel polypeptide chains run in opposite directions.

AP endonuclease: DNA repair enzyme that cleaves the phosphodiester links on either side of a depurinated or depyrimidinated sugar residue. AP stands for apurinic/apyrimidinic.

apoptosis: process in which a cell actively promotes its own destruction, as distinct from necrosis. Apoptosis is important in vertebrate development, where tissues and organs are shaped by the death of certain cell lineages.

apoptosis inhibitor protein: proteins that block the action of caspases and hence help prevent apoptosis.

aporepressor: protein that binds to an operator region and represses transcription only when it is complexed with another molecule.

aqueous: watery.

ARF: GTPase that plays a critical role in the formation of coated vesicles.

Arp (actin-related protein): actin nucleation protein. New actin filaments can grow out from an Arp base.

A site (aminoacyl site): site on a ribosome occupied by an incoming tRNA and its linked amino acid.

assay: term for a chemical measurement, e.g., one in which the activity of an enzyme reaction is measured.

ATP: adenosine triphosphate; adenosine with three phosphates attached to the $5'$ carbon of ribose. ATP is a coenzyme and one of the cell's energy currencies.

ATP/ADP exchanger: carrier in the inner mitochondrial membrane. ADP is moved in one direction and ATP in the other.

ATP synthase: carrier of the inner mitochondrial membrane that is built around a rotary motor. Ten H^+ enter the mitochondrial matrix for every three ATP made.

ATP synthetase: another name for ATP synthase; a carrier of the inner mitochondrial membrane that is built around a rotary motor. Ten H^+ enter the mitochondrial matrix for every three ATP made.

autoinduction: process of self-induction that occurs when a product of a reaction stimulates the production of more of itself. An example is the activation of the *lux* operon by the small molecule *N*-acyl-HSL; transcription of the *lux* operon then causes the production of more *N*-acyl-HSL.

autonomic nerves: nerves that control processes of which we are usually not consciously aware, such as blood vessel size.

autoradiography: process that detects a radioactive molecule. For example, in a Southern blot experiment, the membrane that has been hybridized to a radioactive gene probe is placed in direct contact with a sheet of X-ray film. Radioactive decay activates the silver grains on the emulsion of the X-ray film. When the film is developed, areas that have been in contact with radioactivity will show as black.

axon: long process of a nerve cell, specialized for the rapid conduction of action potentials.

axonal transport: movement of material along microtubules within a nerve cell process; can be outward (anterograde) or inward (retrograde).

β adrenergic receptor: a receptor for the related chemicals adrenaline and noradrenaline. The β adrenergic receptor activates G_s and therefore generates a cAMP signal. To a first approximation, β receptors respond to adrenaline, while noradrenaline acts mainly on α adrenergic receptors.

β oxidation: process by which fatty acids are broken down into individual two-carbon units coupled to CoA to form acetyl-CoA. The process, which takes place in the mitochondrial matrix, generates both NADH and $FADH_2$.

β sheet: common secondary structure in proteins in which lengths of fully extended polypeptide run alongside each other, hydrogen bonds forming between the peptide bonds of the adjoining strands.

B-DNA: right-handed DNA double helix.

BAC (bacterial artificial chromosome): cloning vector used to propagate DNAs of about 300,000 bp in bacterial cells.

bacteriophage: (sometimes shortened to phage) a virus that infects bacterial cells.

BAD: a bcl-2 family protein that causes the release of cytochrome c from mitochondria, triggering apoptosis. BAD is phosphorylated and thereby inactivated by protein kinase B.

basal body: structure from which cilia and flagella arise; has the same structure as the centriole.

base: there are two meanings in cell biology: (1) A chemical that will accept an H^+. Many organic bases contain the group —NH_2, which accepts an H^+ to become —NH_3^+. (2) One of a group of ring-containing nitrogenous compounds that combine with a sugar to create a nucleoside; the members include adenine, guanine, hypoxanthine, cytosine, thymine, uracil, and nicotinamide.

base excision repair: process that repairs DNA double helices that have lost a purine (depurination) or in which a cytosine has been deaminated to uracil (U). The entire damaged monomer is removed and a correct deoxyribonucleotide inserted in its place.

basement membrane: thin layer of extracellular fibers that supports epithelial cells.

base pair: The Watson–Crick model of DNA showed that guanine in one DNA strand would fit nicely with cytosine in another strand while adenine would fit nicely with thymine. The two hydrogen-bonded bases are called a base pair. RNA can also participate in base pairing: instead of thymine it is uracil that now pairs with adenine. The rare base inosine, found in some tRNAs, can base pair with any of uracil, cytosine, or adenine.

BAX: bcl-2 family protein that causes the release of cytochrome c from mitochondria, triggering apoptosis.

bcl-2: antiapoptotic protein. By binding to BAD and BAX, it prevents them causing release of cytochrome c from mitochondria.

bcl-2 family: family of related proteins that regulate cytochrome c release from mitochondria.

B-DNA: right-handed DNA double helix.

β-galactosidase: enzyme that cleaves the disaccharide lactose to produce glucose and galactose and that also catalyzes the interconversion of lactose and allolactose. β Galactosidase is a product of the *lac* operon.

binding site: region of a protein that specifically binds a ligand. A property of the protein's tertiary structure.

bioluminesence: production of light by a living organism.

bisphosphate: of a compound, bearing two independent phosphate groups (as opposed to a diphosphate, which bears a chain of two phosphates in a line). Fructose-1,6-bisphosphate is an example.

blastocyst: early embryo.

Bloom's syndrome: disease resulting from a deficiency in helicase. Affected individuals cannot repair their DNA and are susceptible to developing skin cancer and other cancers.

blue-green algae: old name for the photosynthetic prokaryotes now known as cyanobacteria.

blunt ends: ends of a DNA molecule produced by an enzyme that cuts the two DNA strands at sites directly opposite one another.

β oxidation: process by which fatty acids are broken down into individual two-carbon units coupled to CoA to form acetyl-CoA. The process, which takes place in the mitochondrial matrix, generates both NADH and $FADH_2$.

bright-field microscopy: most basic form of light microscopy. The specimen appears against a bright background and appears darker than the background because of the light it has absorbed or scattered.

brown fat: tissue specialized for generating heat. Thermogenin in the inner mitochondrial membrane allows H^+ to enter down its electrochemical gradient so that the rate at which the electron transport chain works is not limited by the availability of ADP.

β sheet: common secondary structure in proteins in which lengths of fully extended polypeptide run alongside each other, hydrogen bonds forming between the peptide bonds of the adjoining strands.

bulky lesion: distortion of the DNA helix caused by a thymine dimer.

C (cytosine): one of the bases present in DNA and RNA; cytosine is a pyrimidine.

Ca^{2+} ATPase: carrier that uses the energy released by ATP hydrolysis to move calcium ions up their concentration gradient out of the cytosol. Different isoforms of calcium ATPase are located at the plasma membrane and in the membrane of the endoplasmic reticulum.

cadherin: cell adhesion molecule that helps form adherens junctions.

calcineurin: calcium-calmodulin–activated phosphatase, that is, an enzyme that removes phosphate groups from proteins, opposing the effects of kinases. Calcineurin is inhibited by the immunosuppressant drug cyclosporin.

calcium action potential: action potential driven by the opening of voltage-gated calcium channels and the resulting calcium influx.

calcium ATPase: carrier that uses the energy released by ATP hydrolysis to move calcium ions up their concentration gradient out of the cytosol. Different isoforms of calcium ATPase are located at the plasma membrane and in the membrane of the endoplasmic reticulum.

calcium-binding protein: any protein that binds calcium. Calmodulin, troponin and calreticulin are examples found in the cytosol, attached to actin filaments in striated muscle, and in the endoplasmic reticulum respectively.

calcium-calmodulin–activated protein kinase: important regulatory enzyme, activated when calcium-loaded calmodulin binds, which phosphorylates target proteins on serine and threonine residues.

calcium-induced calcium release: process in which a rise of calcium concentration in the cytoplasm triggers the release of more calcium from the endoplasmic reticulum. The best understood mechanism of calcium-induced calcium release is via ryanodine receptors.

calcium pump: carrier that moves calcium ions up their electrochemical gradient out of the cytosol into the extracellular medium or into the endoplasmic reticulum. There are two important calcium pumps: the sodium/calcium exchanger is found on the plasma membrane, while different isoforms of the calcium ATPase are found on the plasma membrane and on the membrane of the endoplasmic reticulum.

calmodulin: calcium-binding protein found in many cells. When calmodulin binds calcium, it can then activate other proteins such as the enzymes calcineurin and glycogen phosphorylase kinase.

Calvin cycle: series of reactions in the chloroplast that act to fix atmospheric carbon dioxide and build it into larger molecules.

CaM-kinase: another name for calcium-calmodulin–activated protein kinase; an important regulatory enzyme, activated when calcium-loaded calmodulin binds, that phosphorylates target proteins on serine and threonine residues.

cAMP (cyclic adenosine monophosphate): nucleotide produced from ATP by the action of the enzyme adenylate cyclase. cAMP is an intracellular messenger in many cells.

cAMP-dependent protein kinase (protein kinase A; PKA): protein kinase that is activated by the intracellular messenger cyclic-AMP. PKA phosphorylates proteins (e.g., glycogen phosphorylase kinase) on serine and threonine residues.

cAMP-gated channel: channel found in the plasma membrane of scent-sensitive nerve cells. The channel opens when cAMP binds to its cytoplasmic face and allows sodium and potassium ions to pass.

cAMP phosphodiesterase: enzyme that hydrolyzes cyclic AMP, producing AMP and hence turning off signaling through the cAMP system.

cap: methylated guanine added to the $5'$ end of a eukaryotic mRNA.

CAP (catabolite activator protein): protein found in prokaryotes that binds to cAMP. The CAP–cAMP complex then binds within the promoter region of some bacterial operons and helps RNA polymerase to bind to the promoter.

carbohydrates: monosaccharides and all compounds made from monosaccharide monomers.

carbon fixation: process of reducing atmospheric carbon dioxide and building it into larger molecules.

carboxylation: introduction of a carboxyl group (—COOH).

carboxyl group: —COOH group. Carboxyl groups give up hydrogen ions to form the deprotonated group —COO^-, so molecules that bear carboxyl groups are usually acids.

carboxyl terminal: end of a peptide or polypeptide that has a free α-carboxyl group. This end is made last at the ribosome.

cardiac muscle: form of striated muscle that is found in the heart.

carrier: there are two meanings used in this book: (1) An integral membrane protein that forms a tube through the membrane that is never open all the way through. Solutes can move into the tube through the open end. When the channel changes shape, so that the end that was closed is open, the solute can leave on the other side of the membrane. (2) A person who has one nonfunctional or mutant copy of a gene, but who shows no effects because the other copy produces sufficient functional protein.

caspase: cysteine-containing protease that cleaves at aspartate residues. Caspases are responsible for the degradative processes that occur during apoptosis.

catabolism: metabolic reactions that break down molecules to derive chemical energy.

catabolite activator protein (CAP): protein that binds to cAMP. The CAP–cAMP complex then binds within the promoter region of some bacterial operons and helps RNA polymerase to bind to the promoter.

catalyst: chemical or substance that reduces the activation energy of a reaction, allowing it to proceed more quickly. Many biological reactions would proceed at an infinitesimal rate without the aid of enzymes, which are protein catalysts.

catalytic rate constant (k_{cat}): proportionality constant that relates the maximal initial velocity (V_m) of an enzyme-catalyzed reaction to the enzyme concentration. $k_{cat} = V_m / [E]$. The units are reciprocal time. *See also Turnover number* and *Maximal velocity.*

cation: positively charged ion, for example, Na^+, K^+, and Ca^{2+}.

cdc: cell division cycle. The acronym is usually used to denote genes, especially of the yeasts *Saccharomyces cerevisiae* and *Schizosaccharomyces pombe*, which, when mutated, cause the cell division cycle to be abnormal.

Cdc25: protein phosphatase involved in the regulation of **CDK1.**

Cdk1 (cyclin-dependent kinase 1): protein kinase involved in the regulation of the G2/M transition of the cell cycle. Associates with cyclin B to form MPF.

Cdk2 (cyclin-dependent kinase 2): protein kinase involved in the regulation of the G1 phase of the cell cycle. Associates with cyclin E.

Cdk4 (cyclin-dependent kinase 2): protein kinase involved in the regulation of the G1 phase of the cell cycle. Associates with cyclin D.

Cdk6 (cyclin-dependent kinase 2): protein kinase involved in the regulation of the G1 phase of the cell cycle. Associates with cyclin D.

cDNA (or complementary DNA): DNA copy of an mRNA molecule.

cDNA library: collection of bacterial cells each of which contains a different foreign cDNA molecule.

cell: fundamental unit of life. A membrane-bound collection of protein, nucleic acid, and other components that is capable of self-replication using simpler building blocks.

cell adhesion molecule: integral membrane protein responsible for sticking cells together. The extracellular domain binds a cell adhesion molecule on another cell while the intracellular domain binds to the cytoskeleton, either directly or via a linker protein.

cell center: point immediately adjacent to the nucleus of eukaryotes where the centrosome and Golgi apparatus are located.

cell cycle (cell division cycle): ordered sequence of events that must occur for successful cell division; consists of G1, S, G2, and M phases.

cell division cycle: another name for the cell cycle: the ordered sequence of events that must occur for successful cell division; consists of G1, S, G2, and M phases.

cell division cycle (CDC) mutants: mutants, especially of the yeasts *Saccharomyces cerevisiae* and *Schizosaccharomyces pombe,* that are temperature-sensitive for cell division.

cell junctions: points of cell–cell interaction in tissues; includes tight junctions; anchoring junctions and gap junctions.

cell membrane: membrane that surrounds the cell; also known as the plasmalemma or plasma membrane.

cell surface membrane: another name for the plasma membrane.

cellulose: major structural polysaccharide of the plant cell wall; a $\beta 1 \rightarrow 4$ polymer of glucose.

cellulose synthase: multiprotein complex of the plant plasma membrane that makes cellulose.

cell wall: rigid case that encloses plant and fungal cells and many prokaryotes. The cell wall lies outside the plasma membrane. Plant cell walls are composed of cellulose plus other polysaccharide molecules such as hemicellulose and pectin.

central dogma (of molecular biology): "DNA makes RNA makes protein"—the concept that the sequence of bases on DNA defines the sequence of bases on RNA, and the sequence of bases on RNA then defines the sequence of amino acids on protein.

centriole: structure found at the centrosome (= microtubule organizing center) of animal cells; composed of microtubules.

centromere: region of the chromosome at which the kinetochore (where the microtubules of the mitotic or meiotic spindle attach) is formed.

centrosome (microtubule organizing center): structure from which cytoplasmic microtubules arise.

CF (cystic fibrosis): inherited disease characterized by failure of the pancreas and by thick sticky mucus in the lungs leading to fatal lung infection unless treated. Cystic fibrosis is caused by failure to make, or properly target, plasma membrane chloride channels.

cGMP-dependent protein kinase (PKG; protein kinase G): protein kinase that is activated by the intracellular messenger cyclic GMP. PKG phosphorylates proteins (e.g., the calcium ATPase) on serine and threonine residues.

chain release factors: proteins that occupy the A site in the ribosome when a stop codon is encountered and that act to trigger termination of polypeptide synthesis.

channel: integral membrane protein that forms a continuous water-filled hole through the membrane.

chaotropic reagents: reagents such as urea, which cause proteins to lose all their higher levels of structure and adopt random, changing conformations.

chaperone: protein that helps other proteins to remain unfolded for correct protein targeting, or to fold into their correct three-dimensional structure.

charge: (1) excess or deficit of electrons giving a negative or positive charge, respectively; (2) transfer RNA is said to be charged when it has an amino acid attached.

charged tRNA: tRNA attached to an amino acid.

chiasmata (singular chiasma): structures formed during crossing over between the chromatids of homologous chromosomes during meiosis; the physical manifestation of genetic recombination.

chimera: structure formed from two different parts. Chimeric proteins are generated by joining together all or part of the protein coding sections of two distinct genes. Chimeric organisms are formed by mixing two or more distinct clones of cells.

chimeric protein: proteins generated by joining together all or part of the protein coding sections of two distinct genes, for example, GFP and a protein of interest.

chiral: structure whose mirror image cannot be superimposed on it. Organic molecules will be chiral if a carbon atom has four different groups attached to it and is therefore asymmetric.

chlorophyll: major photosynthetic pigment of plants and algae.

chloroplast: photosynthetic organelle of plant cells.

chromatid: complete DNA double helix plus accessory proteins subsequent to DNA replication in eukaryotes. At mitosis, the chromosome is seen to be composed of two chromatids; these then separate to form the chromosomes of the two progeny cells.

chromatin: complex of DNA and certain DNA-binding proteins such as histones.

chromatophore: pigment cell in the skin of fish and amphibia.

chromosome: single, enormously long molecule of DNA, together with its accessory proteins. Chromosomes are the units of organization of the nuclear chromatin and carry many genes. In eukaryotes chromosomes are linear; in prokaryotes they are circular.

chromosome walking: investigating a chromosome bit by overlapping bit, each bit being used to clone the next.

cilium (plural cilia): locomotory appendage of some epithelial cells and protozoa.

CIP1: gene encoding a cyclin-dependent kinase inhibitor protein that prevents cells entering the S phase of the cell cycle. $p21^{CIP1}$ is upregulated by cell–cell contact.

cis: side to which material is added. Of the Golgi complex, the surface that receives vesicles from the endoplasmic reticulum.

cisternae: flattened membrane-bound sacs, e.g., those that make up the Golgi apparatus.

CKI (cyclin-dependent kinase inhibitor): type of cell cycle regulatory protein. Binds to and inactivates CDKs.

clathrin: protein that functions to cause vesicle budding in response to binding of specific ligand.

clathrin adaptor protein: protein that binds to specific transmembrane receptors and which in turn recruits clathrin to form a coated vesicle. The vesicle therefore contains the molecule for which the receptor is specific.

cleavage furrow: in animal cells, the structure that constricts the middle of the cell during cytokinesis.

clone: a number of genetically identical individuals.

clone library: collection of bacterial clones where each clone contains a different foreign DNA molecule.

cloning: strictly, the creation of a number of genetically identical organisms. In molecular genetics, the term is used to mean the multiplication of particular sequences of DNA by an asexual process such as bacterial cell division.

cloning vector: DNA molecule that carries genes of interest, can be inserted into cells, and which will then be replicated inside the cells. Cloning vectors range in size from plasmids to entire artificial chromosomes.

closed promoter complex: structure formed when RNA polymerase binds to a promoter sequence prior to the start of transcription.

coatamer: protein complex that encapsulates one class of coated vesicle. Formation of the coatamer coat on a previously flat membrane forces the membrane into a curved shape and therefore drives vesicle formation.

coated vesicle: cytoplasmic vesicle encapsulated by a protein coat. There are two types of coated vesicles, coated by coatamer and clathrin, respectively.

codon: sequence of three bases in an mRNA molecule that specifies a particular amino acid.

coenzyme: molecule that acts as a second substrate for a group of enzymes. ATP/ADP, NADH/NAD$^+$, and Acyl-coenzyme A/coenzyme A are all coenzymes.

cofactor: nonprotein molecule or an ion necessary for the activity of a protein. Cofactors are associated tightly with the protein but can be removed. Examples are pyridoxal phosphate in aminotransferases and zinc in zinc finger proteins.

cohesin: protein that holds two chromatids together. Degradation of cohesin allows the two chromatids to separate at the start of anaphase in mitosis and in meiosis II.

colchicine: plant alkaloid from the autumn crocus, *Colchicum autumnale;* binds to tubulin.

collagen: major structural protein of the extracellular matrix.

columnar: taller than it is broad. Used as a description of some types of epithelial cells.

competent: in molecular genetics this refers to a bacterial culture treated with a solution such as calcium chloride so that uptake of foreign DNA is enhanced.

complementary: two structures are said to be complementary when they fit into or associate with each other. The two strands of the DNA double helix are complementary, as are the anticodon and codon of transfer and messenger RNA.

complementary DNA (cDNA): DNA copy of an mRNA molecule.

complex (as a noun): association of molecules that is held together by noncovalent interactions and often can be readily dissociated.

complex I, complex II, complex III, complex IV: large multimolecular complexes that make up the electron transport chain in the inner mitochondrial membrane.

condenser lens: lens of light and electron microscopes that focuses light (or electrons) onto the specimen.

connective tissue: tissue that contains relatively few cells within a large volume of extracellular matrix.

connexon (gap junction channel): channel in the plasma membrane with a central hole about 1.5 nm in diameter. Gap junction channels only open when they contact a second channel on another cell, in this case they open and form a water-filled tube that runs all the way through the plasma membrane of the first cell, across the small gap between the cells, and through the plasma membrane of the second cell, so allowing passage of solute from the cytosol of one cell to the cytosol of the other.

constitutive: operating all the time without obvious regulation. Housekeeping genes, expressed all the time, are sometimes called constitutive genes. Proteins that are secreted all the time are said to use the constitutive route.

constitutively active: mutant or modified protein that is always in the "on" state. A constitutively active enzyme is active in the absence of its normal regulators. A constitutively active GTPase activates its downstream targets in the absence of GTP exchange factors. Paradoxically, the easiest way to generate a consitutively active GTPase is to eliminate its enzymatic activity, so that it does not hydrolyze GTP and therefore remains in the active GTP-bound state.

constitutive secretion: secretion that continues all the time, without the need for a signal such as an increase of cytosolic calcium concentration.

contact inhibition: inhibition of cell division by cell–cell contact. Contact inhibition allows cells to proliferate to fill a gap and then stop dividing.

contigs: overlapping lengths of DNA that can together be used to build up a map of the genome.

—COOH: carboxyl group. Carboxyl groups give up hydrogen ions to form the deprotonated group —COO^-, so molecules that bear carboxyl groups are usually acids.

corepressor: chemical species that binds to a repressor protein, the complex then binding to an operator region on DNA to prevent transcription. Tryptophan is a corepressor at the *trp* operon.

cortex: outer part of any organ or structure. For instance, the tissue outside the vascular tissue of plants, and the tissue that forms the outer region of the brain, are called cortex.

covalent bond: strong bond between two atoms in which electrons are shared.

cristae: name given to the folds of the inner membrane of mitochondria.

crossing-over: physical exchange of material that takes place between homologous chromosomes during recombination and is manifest in the formation of chiasmata.

crosstalk: two messenger systems show crosstalk when one messenger can produce some or all of the effects of the other.

C-terminal (carboxyl terminal): end of a peptide or polypeptide that has a free α-carboxyl group. This end is made last on the ribosome.

cyanobacteria: photosynthetic prokaryotes that were formerly known as blue-green algae.

cyclic adenosine monophosphate (cAMP): nucleotide produced from ATP by the action of the enzyme adenylate cyclase. cAMP is an intracellular messenger in many cells and exerts many of its actions by activating protein kinase A.

cyclic guanosine monophosphate (cGMP): nucleotide produced from GTP by the action of the enzyme guanylate cyclase. cGMP is an intracellular messenger in many cells and exerts many of its actions by activating protein kinase G.

cyclin: one of a family of proteins whose level oscillates (cycles) through the cell division cycle. Cyclins associate with and activate cyclin-dependent kinases and hence allow progression through cell cycle control points.

cyclin B: one of the two proteins that make up maturation promoting factor (MPF), the other being CDK1; one of a family of proteins whose level oscillates (cycles) through the cell division cycle.

cyclin D: protein that binds to CDK4 and 6 to form a complex that is active at the G1 control point of the cell division cycle; one of a family of proteins whose level oscillates (cycles) through the cell division cycle.

cyclin-dependent kinase (CDKs): one of the family of protein kinases that regulate the cell cycle. Cyclin-dependent kinases are only active when bound to one of the family of cyclin proteins. For example, CDK1 associates with cyclin B and regulates the G2/M transition while CDK2 associates with cyclin E and regulates the G1/S transition.

cyclin E: protein that binds to CDK2 to form a complex that is active at the G1 control point of the cell division cycle; one of a family of proteins whose level oscillates (cycles) through the cell division cycle.

cystic fibrosis (CF): inherited disease characterized by failure of the pancreas and by thick sticky mucus in the lungs leading to fatal lung infection unless treated. Cystic fibrosis is caused by failure to make, or properly target, plasma membrane chloride channels.

cystine: double amino acid formed by two cysteine molecules joined by a disulfide bond.

cytochemistry: use of chemical compounds to stain specific cell structures and organelles.

cytochrome: proteins with a heme prosthetic group that are able to transfer electrons. Cytochromes form a critical part of the electron transport chain of mitochondria and also form part of the cytochrome P450 detoxification system in the liver.

cytochrome *c*: soluble protein of the mitochondrial intermembrane space, often found loosely associated with the inner mitochondrial membrane. Cytochrome *c* transports electrons between components of the electron transport chain. If it is allowed to escape from mitochondria cytochrome *c* activates caspase 9 and hence triggers apoptosis.

cytokinesis: process by which a cell divides in two; part of the M phase of the cell division cycle.

cytology: study of cell structure by light microscopy.

cytoplasm: semiviscous ground substance of the cell. All the volume outside the nucleus and inside the plasma membrane is cytoplasm.

cytoplasmic dynein: motor protein that moves organelles along microtubules in a retrograde direction.

cytoplasmic streaming: movement of cytoplasm that is commonly seen in plant cells and in amoebae; generated by actin and myosin.

cytosine: one of the bases present in DNA and RNA; cytosine is a pyrimidine.

cytoskeleton: cytoplasmic filament system consisting of microtubules, microfilaments, and intermediate filaments.

cytosol: viscous, aqueous medium in which the organelles and the cytoskeleton are bathed.

DAG (diacylglycerol): two acyl groups (fatty acid chains) joined by ester bonds to a glycerol backbone. Diacylglycerol is produced by the action of phospholipase C on phospholipid and helps to activate protein kinase C.

dark reactions: reactions occurring in plants that use ATP and reducing power provided by NADPH to fix atmospheric carbon dioxide and use it to make sugar.

deamination: removal of an amino group. Deamination of cytosine to form uracil is a form of DNA damage.

death domain: domain found on proteins concerned with regulating apoptosis, such as Fas and the p75 neurotrophin receptor. When activated, death domain proteins turn on caspase 8 and hence initiate apoptosis.

denature: to cause to lose three-dimensional structure by breaking noncovalent intramolecular bonds.

dendrite: branching cell process. The term is commonly used to name those processes of nerve cells that are too short to be called axons.

deoxyribonucleic acid (DNA): polymer of deoxyribonucleotides. DNA specifies the inherited instructions of a cell.

deoxyribonucleotide: building block of DNA made up of a nitrogenous base and the sugar deoxyribose to which a phosphate group is attached.

deoxyribose: ribose that instead of —OH has only —H on carbon 2. Deoxyribose is the sugar used in the nucleotides that make up DNA.

depolarization: any positive shift in the transmembrane voltage, whatever its size or cause.

depth of focus: distance toward and away from an object over which components of the object remain in clear focus.

depurination: removal of either of the purine bases, adenine and guanine, from a DNA molecule. Depurination is a form of DNA damage.

desmin: protein that makes up the intermediate filaments in muscle cells.

desmosome: type of anchoring junction that joins the intermediate filaments of neighboring cells. Desmosomes are common in tissues such as skin.

diabetes: The word *diabetes* simply refers to conditions in which a patient produces lots of urine. The only form of diabetes we discuss in this book is diabetes mellitus in which the patient produces large volumes of urine containing sugar. Diabetes mellitus is caused by either a failure in the endocrine gland that produces insulin or a failure of the tissues to respond adequately to insulin.

diacylglycerol (DAG): two acyl groups (fatty acid chains) joined by ester bonds to a glycerol backbone. Diacylglycerol is produced by the action of phospholipase C on phospholipid and helps to activate protein kinase C.

dideoxyribonucleotide: man-made molecule similar to a deoxyribonucleotide but lacking a $3'$-hydroxyl group on its sugar residue. Used in DNA sequencing.

differentiation: process whereby a cell becomes specialized for a particular function.

diffusion: movement of a substance that results from the individual small random thermal movements of its molecules.

dimeric: formed of two parts. Of a molecule, formed of two parts that are not covalently linked.

diphosphate: of a compound, bearing a chain of two phosphates in a line (as opposed to a bisphosphate, which bears two independent phosphate groups). Adenosine diphosphate is an example.

diploid: containing two sets of chromosomes, in humans, this means two sets of 23, one from the father, one from the mother. Most of the cells of the body (the somatic cells) are diploid.

dipolar bond: unusual form of covalent bond found in trimethylamine N-oxide.

disaccharide: dimer of two monosaccharides. Examples are lactose (galactose $\beta 1 \rightarrow 4$ glucose) and sucrose (glucose $\alpha 1 \rightarrow \beta 2$ fructose).

distal: far from the center.

disulfide bond (bridge): covalent bond between two sulfur atoms. In proteins disulfide bonds form by oxidation of two thiol (—SH) groups of cysteine residues. Found chiefly in extracellular proteins.

DNA (deoxyribonucleic acid): polymer of deoxyribonucleotides; DNA specifies the inherited instructions of a cell.

DNAa: DNA binding protein that causes the two strands of the double helix to separate in the first stages of DNA replication.

DNAb: helicase that moves along a DNA strand, breaking hydrogen bonds, and in the process unwinding the helix.

DNAc: DNA-binding protein that serves to bring DNAb to the DNA strands.

DNA chip: tiny glass wafer to which cloned DNAs are attached. Also known as gene chips or microarrays.

DNA excision: process of cutting out damaged DNA prior to repair.

DNA fingerprint: individual pattern of DNA fragments determined by the number and position of specific repeated sequences.

DNA ligase: enzyme that joins two DNA molecules by catalyzing the formation of a phosphodiester bond.

DNA polymerase: enzyme that synthesizes DNA by catalyzing the formation of a phosphodiester bond. DNA is always synthesized in the $5'$ to $3'$ direction.

DNA repair enzymes: enzymes that detect and repair altered DNA.

DNA replication: process in which the two strands of the double helix unwind and each acts as a template for the synthesis of a new strand of DNA.

DNA sequencing: determining the order of bases on the DNA strand.

docking protein (signal recognition particle receptor): receptor on the endoplasmic reticulum to which the signal recognition particle binds during the process of polypeptide chain synthesis and import into the endoplasmic reticulum.

dogma: belief. The "central dogma" of molecular biology is that "DNA makes RNA makes protein"—the concept that the sequence of bases on DNA defines the sequence of bases on RNA, and the sequence of bases on RNA then defines the sequence of amino acids on protein.

domain: separately folded segment of the polypeptide chain of a protein.

dominant: gene that exerts its effect even when only one copy is present. Most dominant genes are dominant because they produce functional protein while the recessive gene does not. However, some mutant genes are dominant because even having 50% of one's protein as the mutant form is enough to cause an effect; the gene causing familial Creutzfeldt–Jacob disease is an example.

donor: that which gives. In a hydrogen bond, the donor is the atom (oxygen, nitrogen, or sulfur) to which the hydrogen is covalently bonded and that gives up some of its share of electrons to a second electron-grabbing atom.

double helix: structure formed when two filaments wind about each other, most commonly applied to DNA, but also applicable to, e.g., F-actin.

downstream: general term meaning the direction in which things move. When applied to the DNA within and adjacent to a gene, it means lying on the side of the transcription start site that is transcribed into RNA. When applied to signaling pathways it means in the direction in which the signal travels, thus MAP kinase is downstream of Ras.

duplication: doubling-up of a particular sequence of the genetic material.

dynamic instability: term that describes the behavior of microtubules in which they switch from phases of growth to phases of shortening.

dynamin: GTPase that plays a critical role in the formation of clathrin-coated vesicles.

dynein: motor protein. Cytoplasmic dynein moves vesicles along microtubules while dynein arms power ciliary and flagellar beating by generating sliding between adjacent outer doublet microtubules.

E2F-1: transcription factor required for DNA synthesis. In quiescent cells E2F-1 is prevented from activating transcription by being bound to RB, the product of the retinoblastoma gene. E2F-1 is released when RB is phosphorylated by CDK4.

ectoplasm: viscous, gel-like outer layer of cytoplasm.

effective stroke: part of the beat cycle of a cilium that pushes on the extracellular medium.

effector caspase: caspase that digests cellular components in the process of apoptosis. Effector caspases are differentiated from caspases 8 and 9, which do not themselves digest cellular components but which activate effector caspases by hydrolyzing particular peptide bonds in them.

electrically excitable: able to produce action potentials.

electrochemical gradient: free energy gradient for an ion in solution. The arithmetical sum of gradients due to concentration and voltage.

electron gun: source of electrons in electron microscopes.

electron microscope: microscope in which the image is formed by passage of electrons through, or scattering of electrons by, the object.

electron transport chain: series of electron acceptor/donator molecules found in the inner mitochondrial membrane which transport electrons from the reduced carriers NADH and $FADH_2$ (thus reoxidizing them) to oxygen. The entire complex forms a carrier that uses the energy of NADH and $FADH_2$ oxidation to transport hydrogen ions up their electrochemical gradient out of the mitochondrion.

electrophoresis: method of separating charged molecules by drawing them through a filtering gel material using an electrical field.

electrostatic bond: a strong attraction between ions or charged groups of opposite charge.

electrostatic interaction: attraction or repulsion between ions or charged groups.

elongation factors: proteins that speed up the process of protein synthesis at the ribosome. Elongation factor tu (EF-tu) is a GTPase.

embryonic stem cell: cells derived from an early embryo. Embryonic stem cells have the capability to divide indefinitely and to become any cell type in the body.

endocytosis: inward budding of plasma membrane to form vesicles. Endocytosis is the process by which cells retrieve plasma membrane and take up material from their surroundings.

endonuclease: enzyme that digests nucleic acids by cleaving internal phosphodiester bonds.

endoplasm: fluid, inner layer of cytoplasm that streams during cytoplasmic streaming.

endoplasmic reticulum (ER): network (reticulum) of membrane-delimited tubes and sacs that extends from the outer membrane of the nuclear envelope almost to the plasma membrane. There are two types of ER, rough endoplasmic reticulum (RER) with a surface coating of ribosomes and smooth endoplasmic reticulum (SER).

endosome: organelle to which newly formed endocytotic vesicles are translocated and with which they fuse.

endosymbiotic theory: proposal that some of the organelles of eukaryotic cells originated as free-living prokaryotes.

endothelial cells: cells that line blood vessels and other body cavities that do not open to the outside.

endothelium: layer of cells that lines blood vessels and other body cavities that do not open to the outside.

energy currency: source of energy that the cell can use to drive processes that would otherwise not occur because their ΔG is positive. Energy currencies can be coenzymes such as ATP and NADH, which give up energy on conversion to, respectively, ADP and NAD^+, or electrochemical ion gradients such as the hydrogen ion gradient across the mitochondrial membrane and the sodium gradient across the plasma membrane.

enhancer sequence: specific DNA sequence to which a protein binds to increase the rate of transcription of a gene.

enthalpy: thermodynamic description of heat.

entropy: degree of disorder in a system.

envelope: closed sheet enclosing a volume. The term is used to describe, among other things, the double membrane layer enclosing certain organelles, and the outer membrane of certain viruses.

enzyme: catalyst made mainly of protein. Like all catalysts, enzymes work by reducing the activation energy of the reaction.

epidermis: protective outer cell layer of an organism.

epithelial cells: cells that make up an epithelium.

epithelium: sheet of cells covering the surface of the human body and those internal cavities such as the lungs and gut that open to the outside. In contrast endothelium lines internal spaces such as blood vessels.

equilibrium: total balance of opposing forces. A process or object is in equilibium if the tendency to go in one direction is exactly equal to the tendency to go in the other direction. For an ion this condition is equivalent to saying that the electrochemical gradient for that ion across the membrane is zero. For a chemical reaction, equilibrium occurs when the rate of the forward reaction is equal to the rate of the reverse reaction, e.g., $2H_2O \rightleftarrows OH^- + H_3O^+$.

equilibrium voltage: transmembrane voltage that will exactly balance the concentration gradient of a particular ion.

ER (endoplasmic reticulum): network (reticulum) of membrane channels that extends from the outer membrane of the nuclear envelope almost to the plasma membrane; there are two types of ER, rough endoplasmic reticulum (RER) with a surface coating of ribosomes and smooth endoplasmic reticulum (SER).

ES (embryonic stem) cell: cell derived from a very early embryo. ES cells have not yet determined their developmental fate and can therefore, depending on the conditions, generate the entire range of tissue types.

essential amino acid: amino acid that the organism needs but cannot synthesize. Histidine, isoleucine, leucine, lysine, methionine, phenylalanine, threonine, tryptophan, and valine are essential for humans.

essential fatty acid: fatty acid that the organism needs but cannot synthesize. Linoleic and linolenic acids are essential for humans.

ester bond: bond formed between the hydrogen of an alcohol group and the hydroxyl of a carboxyl group by the elimination of water.

euchromatin: that portion of the nuclear chromatin that is not tightly packed. Euchromatin contains genes that code for proteins that are being actively transcribed.

eukaryotic: organism whose cells contain distinct nuclei and other organelles; includes all known organisms except prokaryotes (bacteria and cyanobacteria).

exocytosis: fusion of a vesicle with the plasma membrane. Exocytosis causes the soluble contents of the vesicle to be released to the extracellular medium, while integral membrane proteins of the vesicle become integral membrane proteins of the plama membrane.

exon: in a eukaryotic gene, exons are those parts that after RNA processing leave the nucleus. In contrast introns are spliced out before the RNA leaves the nucleus.

exonuclease: enzyme that digests nucleic acids by cleaving phosphodiester links successively from one end of the molecule.

expression (of a gene): appearance of the protein which the gene encodes.

expression vector: cloning vector containing a promoter sequence recognized by the host cell, thus enabling a foreign DNA insert to be transcribed into mRNA.

extracellular matrix: meshwork of filaments and fibers that surrounds and supports mammalian cells. The major protein of the extracellular matrix is collagen.

extracellular medium: aqueous medium outside cells. For a unicellular organism the extracellular medium is the outside world. For a multicellular organism such as a human being the extracellular medium is the fluid between the cells.

extragenic DNA: DNA that can neither be identified as coding for protein or RNA nor as being promoters or enhancers regulating transcription.

Fas: metabotropic cell surface receptor that activates caspase 8 and hence triggers apoptosis.

fat: triacylglycerol (triglyceride) that is solid at room temperature. In contrast, oils are liquid at room temperature.

fat cells (adipocytes): cells that store fats (triacylglycerols).

fatty acid: carboxyl group attached to a long chain of carbon atoms with attached hydrogens, that is, a chemical of the general form

$$C_nH_m - \underset{\underset{O}{\|}}{C} - OH$$

feedback: process in which the result of a process modifies the mechanisms carrying out that process to increase or decrease their rate. In negative feedback a change in some parameter activates a mechanism that reverses the change in that parameter; an example is the effect of tryptophan on expression of the *trp* operon. In positive feedback a change in some parameter activates a mechanism that accelerates the change; an example is the effect of depolarization on the opening of voltage-gated sodium channels.

feedforward: control process in which a change in a parameter is *predicted*, and mechanisms initiated that will act to reduce the change in that parameter.

ferredoxin: protein component of the electron transport system of chloroplasts.

fibroblast: cell found in connective tissue. Fibroblasts synthesize collagen and other components of the extracellular matrix.

fibroblast growth factor: paracrine transmitter that opposes apoptosis and promotes cell division in target cells. Although named for its effect on fibroblasts, FGF triggers proliferation of many tissues and plays critical roles in determining cell fate during development.

fission: breakage into two parts. The word is used for prokaryote replication. In addition it describes the division of a single membrane-bound organelle into two and the process whereby a vesicle breaks away from a membrane.

flagellin: protein that is the bacterial flagellum.

flagellum: swimming appendage. In eukaryotes flagella are extensions of the cell that use a dynein/microtubule motor system. In prokaryotes flagella are extracellular proteins rotated by a motor at their base.

fluid mosaic model: generally accepted hypothesis of how cell membranes are formed from a lipid bilayer plus protein.

fmet (formyl methionine): methionine modified by the attachment of a formyl group; fmet is the first amino acid in all newly made bacterial polypeptides.

focal contact: points at which a locomoting cell makes contact with its substrate.

formyl methionine (fmet): methionine modified by the attachment of a formyl group; fmet is the first amino acid in all newly made bacterial polypeptides.

fourteen-three-three (14-3-3): protein that binds phosphorylated BAD, hence preventing it from triggering apoptosis.

frameshift mutation: mutation that changes the mRNA reading frame, caused by the insertion or deletion of a nucleotide.

G (guanosine): guanine linked to the sugar ribose. Guanosine is a nucleoside.

G0: describes the quiescent state of cells that have left the cell division cycle.

G1 (gap 1): period of the cell division cycle that separates mitosis from the following S phase.

G2 (gap 2): period of the cell division cycle between the completion of S phase and the start of cell division or M phase.

GABA (γ-amino butyric acid): γ-amino acid that acts to open chloride channels in the plasma membrane of sensitive nerve cells. Usually called γ-amino butyric acid rather than γ-amino butyrate, even though the latter (^-OOC—CH_2—CH_2—CH_2—NH_3^+) is the form in which it is found at neutral pH.

G-actin: globular, subunit form of actin.

gamete: sperm or egg. Gametes are haploid; that is, they contain just one set of chromosomes (23 in humans).

gamma-amino butyric acid (γ-amino butyric acid, GABA): γ-amino acid that acts to open chloride channels in the plasma membrane of sensitive nerve cells. Usually called γ-amino butyric acid rather than γ-amino butyrate, even though the latter (^-OOC—CH_2—CH_2—CH_2—NH_3^+) is the form in which it is found at neutral pH.

GAP (GTPase activating protein): protein that speeds up the rate at which GTPases hydrolyze GTP and therefore switch from the active to the inactive state.

gap 0 (G0): describes the quiescent state of cells that have left the cell division cycle.

gap 1 (G1): period of the cell division cycle that separates mitosis from the following S phase.

gap 2 (G2): period of the cell division cycle between the completion of S phase and the start of cell division or M phase.

gap junction: type of cell junction that allows solute to pass from the cytosol of one cell to the cytosol of its neighbor without passing through the extracellular medium. Gap junctions consist of many paired gap junction channels or connexons.

gap junction channel (connexon): channel in the plasma membrane with a central hole about 1.5 nm in diameter. Gap junction channels only open when they contact a second channel on another cell, in this case they open and form a water-filled tube that runs all the way through the plasma membrane of the first cell, across the small gap between the cells, and through the plasma membrane of the second cell, so allowing passage of solute from the cytosol of one cell to the cytosol of the other.

gastrocnemius muscle: muscle at the back of the shin. When it contracts the toes move down.

gated, gating: a channel is gated if it can switch to a shape in which the tube through the membrane is closed.

gated transport: transport of fully folded proteins through intracellular pores that open to allow their passage.

GDP: guanosine diphosphate; guanosine with two phosphates attached to the 5′ carbon of ribose.

GEF (guanine nucleotide exchange factor): protein that accelerates the rate at which GDP leaves a GTPase to be replaced by GTP, thus switching the GTPase from its inactive state to its active state.

GEFS (generalized epilepsy with febrile seizures): relatively common form of childhood epilepsy, usually clearing up spontaneously. GEFS$^+$ is the less common condition in which the seizures still occur past the age of 6 years.

gelsolin: type of actin-binding protein that binds to and fragments actin filaments.

gene: fundamental unit of heredity. In many cases a gene contains the information needed to code for a single polypeptide.

gene chip: tiny glass wafer to which cloned DNAs are attached. Also known as microarrays or DNA chips.

gene family: group of genes that share sequence similarity and usually code for proteins with a similar function.

gene probe: cDNA or genomic DNA fragment used to detect a specific DNA sequence to which it is complementary in sequence. The probe is tagged in some way to make it easy to detect. The tag could be, for example, a radioactive isotope or a fluorescent dye.

gene therapy: correction or alleviation of a genetic disorder by the introduction of a normal gene copy into an affected individual.

genetically modified (GM): organism with a genome that has been modified by modern molecular techniques, usually by the addition of novel gene(s) or by swopping in new DNA to replace existing gene(s).

genetic code: relationship between the sequence of the four bases in DNA and the amino acid sequence of proteins.

genome: complete set of genes in an organism.

genomic DNA library: collection of bacterial clones each of which contains a different fragment of foreign genomic DNA.

germ cells: cells that give rise to the eggs and sperm.

Gibbs free energy: if the change of Gibbs free energy during a reaction is negative, the reaction is favored and will proceed if the activation energy is sufficiently low.

glial cells: electrically inexcitable cells found in the nervous system.

glucocorticoid: steroid hormone produced by the adrenal cortex that forms part of the system controlling blood sugar levels.

glucocorticoid receptor: intracellular receptor to which glucocorticoid hormone binds.

gluconeogenesis: synthesis of glucose from noncarbohydrate precursors such as amino acids and lactate.

glucose: hexose monosaccharide. Glucose is the commonest sugar in the blood and is the dominant cellular fuel in animals, being used in glycolysis to generate ATP and pyruvate, the latter fueling the Krebs cycle.

glucose carrier: plasma membrane protein that carries glucose into or out of cells. Some cells, such as skeletal muscle cells, will only translocate glucose carriers to their membranes when protein kinase B is active.

glyceride: compound formed by attaching units to a glycerol backbone. Triacylglycerols (previously called triglycerides) and phospholipids are glycerides.

glycerol: CH_2OH—$CHOH$—CH_2OH. The backbone to which acyl groups (fatty acid chains) are attached to make triacylglycerols and phospholipids.

glycogen: glucose polymer that can be quickly hydrolyzed to yield glucose; an $\alpha 1 \rightarrow 4$ polymer of glucose with $\alpha 1 \rightarrow 6$ branches.

glycogen phosphorylase: enzyme that releases glucose-1-phosphate monomers from glycogen. The glucose-1-phosphate is then converted to glucose-6-phosphate, which can be used in respiration or dephosphorylated to glucose for release into the blood.

glycogen phosphorylase kinase (phosphorylase kinase): kinase that is activated by the calcium–calmodulin complex and phosphorylates glycogen phosphorylase, activating the latter enzyme.

glycolysis: breakdown of glucose to pyruvate.

glycosidic bond: bond linking monosaccharide residues in which the carbon backbones are linked through oxygen and a water molecule is lost.

glycosylation: addition of sugar residues to a molecule. Both proteins and lipids can be glycosylated.

glyoxisome: type of peroxisome found in plant cells that is concerned with the conversion of fatty acids to carbohydrate.

glyoxylate: CHO—COO^-; an intermediate in various metabolic pathways including the glyoxylate shunt.

glyoxylate shunt: pathway found in plants and bacteria that allows acetyl-CoA to be converted into glucose.

GM (genetically modified): organism with a genome that has been modified by modern molecular techniques, usually by the addition of novel gene(s) or by swopping in new DNA to replace existing gene(s).

Golgi apparatus: system of flattened cisternae concerned with glycosylation and other modifications of proteins.

G protein (trimeric G protein): protein that links a class of metabotropic cell surface receptors with downstream targets. Trimeric G proteins comprise an α subunit that binds and hydrolyzes GTP and a $\beta\gamma$ subunit that dissociates from the α subunit while the latter is in the GTP-bound state. Important G proteins are G_q, which activates phospholipase Cβ, and G_s, which activates adenylate cyclase.

G_q: isoform of trimeric G protein that activates phospholipase Cβ and therefore generates a calcium signal.

grana: distinctive structures within the chloroplast formed by the stacking of the thylakoid membranes.

gratuitous inducer: inducer of transcription that is not itself metabolized by the resulting enzymes.

Grb2 (growth factor receptor binding protein number 2): linker protein that has an SH2 domain and is therefore recruited to phosphotyrosine, e.g., on receptor tyrosine kinases. Grb2 in turn recruits SOS, bringing SOS to the plasma membrane where it can act as a guanine nucleotide exchange protein (GEF) for Ras.

green fluorescent protein: fluorescent protein made by the jellyfish *Victoria victoria*. Unlike other colored or fluorescent proteins, it contains no prosthetic groups and therefore will fluoresce when expressed by any cell in which the gene is successfully inserted and expressed.

growth factor: paracrine transmitter that modifies the developmental pathway of the target cell, often by causing cell division.

growth factor receptor binding protein number 2 (Grb2): linker protein that has an SH2 domain and is therefore recruited to phosphotyrosine, e.g., on receptor tyrosine kinases. Grb2 in turn recruits SOS, bringing SOS to the plasma membrane where it can act as a guanine nucleotide exchange protein (GEF) for Ras.

G_s: isoform of trimeric G protein that activates adenylate cyclase and therefore causes an increase of cAMP concentration.

GTP: guanosine triphosphate; guanosine with three phosphates attached to the $5'$ carbon of ribose.

GTPase: enzyme that hydrolyzes GTP. The name is usually restricted to that family of proteins that bind GTP and adopt a new shape that can then activate target proteins. Once they hydrolyze the GTP they have bound, they switch back to the original form. Examples are EF-tu, Arf, Ran, dynamin, G_q, and G_s.

GTPase activating protein (GAP): protein that speeds up the rate at which GTPases hydrolyze GTP and therefore switch from the active to the inactive state.

guanine: one of the bases found in DNA and RNA; guanine is a purine.

guanine nucleotide exchange factor (GEF): protein that accelerates the rate at which GDP leaves a GTPase to be replaced by GTP, thus switching the GTPase from its inactive state to its active state.

guanosine: guanine linked to the sugar ribose; guanosine is a nucleoside.

guanosine diphosphate (GDP): guanosine with two phosphates attached to the $5'$ carbon of ribose.

guanosine triphosphate (GTP): guanosine with three phosphates attached to the $5'$ carbon of ribose.

guanylate cyclase: An enzyme that converts GTP to the intracellular messenger cyclic GMP (cGMP). One isoform of guanylate cyclase is activated by NO.

Guthrie test: test for phenylketonuria. Newborn babies' blood is tested for the presence of phenylalanine at unusually high concentration.

hairpin loop: loop in which a linear object folds back on itself. Used to describe the loop formed in a RNA molecule due to complementary base pairing.

haploid: containing a single copy of each chromosome, in humans, this means 23 chromosomes. Sperm and eggs are haploid while somatic cells contain one set of 23 from each parent and are referred to as being diploid.

head group: hydrophilic group found in phospholipids. The head group is attached to the glycerol backbone by a phosphodiester link. Examples are choline and inositol.

heel prick test (Guthrie test): test for phenylketonuria. Newborn babies' blood is tested for the presence of phenylalanine at unusually high concentration.

helicase: enzyme that helps unwind the DNA double helix during replication.

α helix: a common secondary structure in proteins, in which the polypeptide chain is coiled, each turn of the helix taking 3.6 amino acid residues. The nitrogen atom in each peptide bond forms a hydrogen bond with the oxygen four residues ahead of it in the polypeptide chain.

hemicellulose: polysaccharide component of the plant cell wall that links cellulose fibrils together.

hemoglobin: oxygen-carrying, iron-containing protein of the blood.

heterochromatin: that portion of the nuclear chromatin that is tightly packed. Much of the heterochromatin is repetitive DNA with no coding function.

heterozygote: individual whose two copies of a gene differ, e.g., one may be mutant. We have not used the term in this book but have used the simpler word *carrier* for the more restricted case where an individual is a heterozygote for a recessive gene.

hexokinase: enzyme that phosphorylates glucose on the number 6 carbon.

histone: positively charged protein that binds to negatively charged DNA and helps to fold DNA into chromatin.

histone octamer: two molecules each of histones H2A, H2B, H3, and H4, the whole forming a nucleosome.

homologous recombination: process in which a length of DNA with ends that are homologous to a section of chromosome swops in, replacing the existing length of DNA in the chromosome. Homologous recombination occurs naturally at chiasmata during crossing over during meiosis. It can also occur in some cells when they transfected with the appropriate exogenous DNA.

homozygote: individual whose two copies of a gene are identical. Most of us are homozygous for most of our genes.

hormone: long-lived transmitter that is released into the blood and travels around the body before being broken down.

hormone response element (HRE): specific DNA sequence to which a steroid hormone receptor binds.

housekeeping gene: gene that is transcribed into mRNA in nearly all the cells of a eukaryotic organism; in bacterial cells a housekeeping gene is one that is always being transcribed.

HRE (hormone response element): specific DNA sequence to which a steroid hormone receptor binds.

hybridization: association of unlike things. In molecular genetics, the association of two nucleic acid strands (either RNA or DNA) by complementary base pairing.

hydration shell: cloud of water molecules that surrounds an ion in solution.

hydrocarbon tail: long chain of carbon atoms with attached hydrogens found in phospholipids and triacylglycerols. The tail represents all of a molecule of fatty acid except the carboxyl group.

hydrogen bond: relatively weak bond formed between a hydrogen atom and two electronegative atoms (such as nitrogen or oxygen) where the hydrogen is shared between the other atoms.

hydrogen ion gradient: energy currency. Hydrogen ions are more concentrated outside bacteria and mitochondria than inside, and this chemical gradient is supplemented by a voltage gradient pulling hydrogen ions in. If hydrogen ions are allowed to rush in down their electrochemical gradient, they release 17,000 J/mol.

hydrolysis: breakage of a covalent bond by the addition of water. —H is added to one side, —OH to the other.

hydropathy plot: running average of side chain hydrophobicity along a polypeptide chain. From the hydropathy plot one can predict, e.g., membrane-spanning domains in integral membrane proteins.

hydrophilic: molecule or part of a molecule that can interact with water.

hydrophobic: a molecule or part of a molecule that will associate with other hydrophobic molecules in preference to water.

hydrophobic effect: tendency of hydrophobic molecules or parts of molecules to cluster together away from water, such as hydrophobic amino acid residues in the center of a protein, or the fatty acid chains in lipid bilayers.

hydroxyl group: —OH group. The term is specifically *not* used for an —OH that forms part of a carboxyl group —COOH.

hypoxanthine: purine that is used to make the nucleotide inosine. Inosine can pair with any of uracil, cytosine, or adenine.

imino acid: organic molecule containing both carboxyl (—COOH) and imino (—NH—) groups. Proline is an imino acid although it is usually called an amino acid.

immunofluorescence: use of fluorescently labeled antibodies to reveal the location of specific chemicals, e.g., in fluorescence microscopy or in western blotting.

inactivation (of voltage gated channels): long term closure, often because of blockage of the open channel with a plug that is attached to the cytosolic face of the protein.

indirect immunofluorescence microscopy: form of light microscopy in which a fluorescent secondary antibody is used to label a preapplied primary antibody specific for a particular protein or subcellular structure. Also called secondary immunofluorescence microscopy.

inducible operon: operon that is transcribed only when a specific substance is present.

initial velocity (of a reaction): rate at which an enzyme converts substrate to product in the absence of product, that is, at the onset of a reaction.

inositol: cyclic polyalcohol, $(CHOH)_6$, that forms the headgroup of the phospholipid phosphatidylinositol. Phosphorylation of inositol yields inositol trisphosphate.

inositol trisphosphate (IP_3, $InsP_3$): small phosphorylated cyclic polyalcohol that is released into the cytosol by the action of phospholipase C on the membrane lipid phosphatidylinositol bisphosphate and that acts to cause release of calcium ions from the endoplasmic reticulum.

inositol trisphosphate-gated calcium channel: channel found in the endoplasmic reticulum of many cells. The channel opens when inositol trisphosphate binds to its cytoplasmic aspect. It allows only calcium ions to pass.

insertional mutagenesis: insertion of additional nucleotides into a stretch of DNA.

in situ hybridization: binding of a particular labeled sequence of RNA (or DNA) to its matching sequence in the genome as a way of searching for the location of that sequence in a tissue sample.

$InsP_3$(inositol trisphosphate, IP_3): small phosphorylated cyclic polyalcohol that is released into the cytosol by the action of phospholipase C on the membrane lipid phosphatidylinositol bisphosphate and that acts to cause release of calcium ions from the endoplasmic reticulum.

insulin: hormone produced by endocrine cells in the pancreas. It activates its own receptor tyrosine kinase, which in turn acts mainly through activation of PI-3-kinase and hence protein kinase B.

insulin receptor: receptor tyrosine kinase specific for insulin, which acts mainly through activation of PI-3-kinase and hence protein kinase B.

insulin receptor substrate number 1 (IRS-1): protein phosphorylated on tyrosine by the insulin receptor. Once phosphorylated, it recruits PI-3-kinase, which can then be phosphorylated and hence activated.

integral protein (of a membrane): class of protein that is tightly associated with a membrane, usually because its polypeptide chain crosses the membrane at least once. Integral membrane proteins can only be isolated by destroying the membrane, e.g., with detergent. In contrast, peripheral membrane proteins are more loosely associated.

integrin: dimeric proteins with an extracellular domain that binds to extracellular matrix proteins and an intracellular domain that attaches to actin microfilaments.

intermediate filament: one of the filaments that makes up the cytoskeleton; composed of various subunit proteins.

intermembrane space: in organelles such as mitochondria, chloroplasts, and nuclei that are bound by two membranes, the intermembrane space is the aqueous space between the inner and outer membranes. The intermembrane space of nuclei is contiguous with the lumen of the ER. The intermembrane space of mitochondria has the ionic composition of cytosol because porin in the outer mitochondrial membrane allows solutes of $M_r < 10,000$ to pass.

interphase: period of synthesis and growth that separates one cell division from the next; consists of the G1, S, and G2 phases of the cell division cycle.

intracellular messenger: cytosolic solute that changes in concentration in response to external stimuli or internal events, and which acts on intracellular targets to change their behavior. Calcium ions, cyclic AMP and cyclic GMP are the three common intracellular messengers.

intracellular receptors: receptors that are not on the plasma membrane but that lie within the cell and bind transmitters that diffuse through the plasma membrane.

intron: in a eukaryotic gene, introns are those parts that are spliced out before the RNA leaves the nucleus. In contrast exons are those parts that after RNA processing leave the nucleus.

ion: charged chemical species. A single atom that has more or less electrons than are required to exactly neutralize the charge on the nucleus is an ion (e.g., Na^+, Cl^-). A molecule with one or more charged regions is also an ion (e.g., phosphate, HPO_4^{2-}; leucine, ^-OOC—$CH(CH_2.CH(CH_3)_2)$—NH_3^+.

ionotropic cell surface receptors: channels that open when a specific chemical binds to the extracellular face of the channel protein.

IP$_3$ (inositol trisphosphate, InsP$_3$): small phosphorylated cyclic polyalcohol that is released into the cytosol by the action of phospholipase C on the membrane lipid phosphatidylinositol bisphosphate and that acts to cause release of calcium ions from the endoplasmic reticulum.

IP$_3$ receptor: calcium channel in the membrane of the endoplasmic reticulum that opens when inositol trisphosphate binds to its cytosolic aspect.

IRS-1: protein phosphorylated on tyrosine by the insulin receptor. Once phosphorylated, it recruits PI-3-kinase, which can then be phosphorylated and hence activated.

isoelectric point: pH at which a protein or other molecule has no net charge.

isoforms: related proteins that are the products of different genes or of differential splicing of mRNAs from one gene.

isomers: different compounds with the same molecular formula. For example, glucose and mannose are both $C_6H_{12}O_6$ and are therefore isomers.

JNK: stress-activated protein kinase that stimulates cell repair but can also trigger apoptosis.

K$^+$/Na$^+$ ATPase (sodium/potassium ATPase): plasma membrane carrier. For every ATP hydrolyzed, three Na^+ ions are moved out of the cytosol and two K^+ ions are moved in.

Kartegener's syndrome: human condition in which the heart is on the wrong side. Male subjects may also be sterile. Caused by mutations in components of cilia and flagella.

keratin: protein that makes up the intermediate filaments in epithelial cells.

ketone: any chemical containing a carbon atom with single bonds to two other carbons and a double bond to an oxygen. Acetone (CH_3—CO—CH_3) and acetoacetate (CH_3—CO—CH_2—COO^-) are ketones.

ketone bodies: acetoacetate (CH_3—CO—CH_2—COO^-) and 3-hydroxybutyrate (CH_3—CHOH—CH_2—COO^-); circulating fuels formed from fats and used to fuel body tissues in times of starvation. Acetoacetate slowly loses carbon dioxide to form acetone.

ketosis: overproduction of ketone bodies that is seen in extreme starvation and in diabetes.

kinase: enzyme that phosphorylates a molecule by transferring a phosphate group from ATP to the molecule.

kinesin: molecular motor protein responsible for movement along microtubules in the anterograde direction.

kinetochore: point of attachment of the chromosome to the spindle. The kinetochore forms around the centromere.

K_M **(Michaelis constant):** substrate concentration at which one measures an initial velocity that is half as fast as the maximal velocity (V_m) of an enzyme reaction.

Krebs cycle (tricarboxylic acid cycle): series of reactions in the mitochondrial matrix in which acetate is completely oxidized to CO_2 with the attendant reduction of NAD^+ to NADH and FAD to $FADH_2$.

lac **(lactose) operon:** cluster of three bacterial genes that encode enzymes involved in metabolism of lactose.

lactose: disaccharide comprising galactose linked to glucose by a $\beta 1 \rightarrow 4$ glycosidic bond.

lactose intolerant: unable to hydrolyze dietary lactose.

lactose (*lac*) operon: cluster of three bacterial genes that encode enzymes involved in metabolism of lactose.

lagging strand: strand of DNA that grows discontinuously during replication; in contrast to the leading strand, which is synthesized continuously.

lamins: proteins that make up the nuclear lamina. Lamins are chemically related to intermediate filaments.

leading strand: strand of DNA that grows continuously in the $5'$ to $3'$ direction by the addition of deoxyribonucleotides.

ligand: when two molecules bind together, one (often the smaller one) is called the ligand, and the other (often the bigger one) is called the receptor.

light reactions: reactions in photosynthesis which use the energy of light to split water to release oxygen, to produce NADPH, and to move hydrogen ions across the thylakoid membrane setting up a gradient which is then used to drive the formation of ATP from ADP. The NADPH and ATP are used in the dark reactions.

linkage, linked (of genes): physical association of genes on the same chromosome. Linked genes tend to be inherited together.

linker DNA: stretch of DNA that separates two nucleosomes.

lipid bilayer: two layers of lipid molecules that form a membrane.

localization sequence (targeting sequence): stretch of polypeptide that determines the cellular compartment to which a synthesized protein is sent.

lumen: inside of a closed structure or tube.

lysosome: membrane-bound organelle containing digestive enzymes.

M-phase: the period of the cell division cycle during which the cell divides; M phase consists of mitosis and cytokinesis.

M-phase promoting factor (MPF): the complex of CDK1 and cyclin B that regulates the G2/M phase transition of the cell division cycle.

macromolecule: large molecule.

macrophage: phagocytic housekeeping cell that engulfs and digests bacteria and dead cells.

major groove (of DNA): larger of the two grooves along the surface of the DNA double helix.

mannose: hexose monosaccharide.

mannose-6-phosphate: mannose that is phosphorylated on carbon number 6. Mannose-6-phosphate is the sorting signal that identifies lysosomal proteins.

MAPK, MAP kinase (mitogen-associated protein kinase): enzyme that phosphorylates numerous targets, including transcription factors that trigger transcription of the cyclin-D gene.

MAPKK (MAP kinase kinase): enzyme that phosphorylates and activates MAP kinase, it itself being phosphorylated and hence activated by MAPKK kinase.

MAPKKK (MAP kinase kinase kinase): enzyme that phosphorylates and activates MAPK kinase. The most important isoform of MAPKKK is also called Raf. Raf is activated by the G protein Ras.

matrix: very vague term meaning a more or less closed location, often but not exclusively one that is a solid basis on which things can grow or attach. The term is used of the extracellular matrix in animals, formed of collagen and other fibers, and of the aqueous volume inside various organelles.

maximal velocity (of an enzyme-catalyzed reaction): limiting value of the initial velocity of a reaction as the substrate concentration is increased at constant enzyme concentration. Occurs when the enzyme is saturated with substrate.

meiosis: form of cell division that produces gametes, each with half the genetic material of the cells that produce them.

meiosis I and II: first and second meiotic divisions.

meiotic spindle: bipolar, microtubule-based structure on which chromosome segregation occurs during meiosis I and II.

membrane: bilayer formed from lipid plus protein.

messenger RNA (mRNA): RNA molecule that carries the genetic code. The order of bases on mRNA specifies the amino acid sequence of a polypeptide chain. In eukaryotes, the mRNA leaves the nucleus and is translated into protein in the cytoplasm.

metabolism: all of the reactions going on inside a cell.

metabotropic cell surface receptors: receptors that are linked to, and activate, enzymes.

metaphase: period of mitosis or meiosis at which the chromosomes align prior to separation at anaphase.

metaphase plate: equator of the mitotic or meiotic spindle; the point at which the chromosomes congregate at metaphase of mitosis or meiosis.

7-methyl guanosine cap: modified guanosine found at the $5'$ terminus of eukaryotic mRNA. A guanosine is attached to the mRNA by a $5'$-$5'$-phosphodiester link and is subsequently methylated on atom number 7 of the guanine.

Michaelis constant (K_M): substrate concentration at which one measures an initial velocity that is half as fast as the maximal velocity (V_m) of an enzyme reaction.

Michaelis–Menten equation: equation that defines the effect of substrate concentration on the initial velocity of an enzyme-catalyzed reaction:

$$v_0 = \frac{V_m\,[\text{S}]}{K_M + [\text{S}]}$$

where V_m is the limiting initial velocity obtained as the substrate concentration approaches infinity, [S] is the substrate concentration, and K_M is a constant called the Michaelis constant, equal to the substrate concentration at which one measures an initial velocity that is half as fast as the maximal velocity.

microarray: tiny glass wafer to which cloned DNAs are attached. Also known as gene chips or DNA chips.

microfilament: one of the major filaments of the cytoskeleton; also known as actin filament or F-actin; synonymous with the thin filament of striated muscle.

microsatellite DNA: repetitive DNA of unknown function comprising many repeats of a unit of four or fewer base pairs.

microtubular molecular motor: protein that moves vesicles along microtubules; examples are cytoplasmic dynein and kinesin.

microtubule: tubular cytoplasmic filament composed of tubulin. Tubulin is the major component of 9 + 2 cilia and flagella and of the mitotic and meiotic spindle.

microtubule organizing center (MTOC): structure from which cytoplasmic microtubules arise; synonymous with the centrosome.

microvilli (sing. **microvillus):** projections from the surface of epithelial cells that increase the absorptive surface; contain actin filaments.

minisatellite DNA: repetitious DNA of unknown function comprising up to 20,000 repeats of a unit of about 25 base pairs.

minor groove (of DNA) : smaller of the two grooves along the surface of the DNA double helix.

missense mutation: base change in a DNA molecule that changes a codon so that it now specifies a different amino acid.

mitochondrial inner membrane: inner membrane of mitochondria that is elaborated into cristae. The electron transport chain, and ATP synthase, are integral membrane proteins of the mitochondrial inner membrane.

mitochondrial matrix: aqueous space inside the mitochondrial inner membrane, where the enzymes of the Krebs cycle are located.

mitochondrial outer membrane: outer membrane of mitochondria, permeable to solutes of $M_r < 10,000$ because of the presence of the channel porin.

mitochondrion: cell organelle concerned with aerobic respiration.

mitogen: anything that promotes cell division. FGF and PDGF (fibroblast growth factor and platelet-derived growth factor) are potent mitogens for endothelial and smooth muscle cells.

mitogen-associated protein kinase (MAP kinase, MAPK): enzyme that phosphorylates numerous targets, including transcription factors that trigger transcription of the cyclin-D gene.

mitosis: type of cell division found in somatic cells, in which each daughter cell receives the full complement of genetic material present in the original cell.

mitotic spindle: microtubule-based structure upon which chromosomes are arranged and translocated during mitosis.

mole: amount of substance comprising 6.023×10^{23} (Avogadro's number) molecules. One mole has a mass equal to the value of the relative molecular mass expressed in grams.

monomer: single unit, usually used to refer to a single building block of a larger molecule. Thus DNA is formed of nucleotide monomers. By analogy, the word is sometimes used to describe proteins that act as a single unit to distinguish them from related proteins that act as larger units, so myoglobin is said to be monomeric by comparison with hemoglobin, which has four subunits. GTPases such as Ran, Arf, and Ras are sometimes called "monomeric G proteins" to distinguish them from trimeric G proteins such as G_q and G_s.

monosaccharide: sweet-tasting chemical with many hydroxyl groups that can adopt a form in which an oxygen atom completes a ring of carbons. All the monosaccharides in this book have the general formula $C_n(H_2O)_n$ where $n = 5$ (pentoses) or 6 (hexoses).

motif: recognizable conserved sequence of bases (in DNA) or amino acids (in a polypeptide). A motif in DNA may bind one transcription factor, e.g., 5'AGAACA3' binds the glucocorticoid receptor. A motif in a polypeptide may, e.g., fold in a particular way (e.g., a helix-turn-helix motif), or bind to a target (e.g., KDEL).

motoneuron (motoneurone): nerve cell that carries action potentials from the spine to the muscles. It releases the transmitter acetylcholine onto the muscle cells, causing them to depolarize and hence contract.

M phase: period of the cell division cycle during which the cell divides; consists of mitosis and cytokinesis.

MPF (M-phase promoting factor): complex of CDK1 and cyclin B that regulates the G2/M phase transition of the cell division cycle.

M-phase promoting factor (MPF): complex of CDK1 and cyclin B that regulates the G2/M phase transition of the cell division cycle.

M_r: Relative Molecular Mass (sometimes called "molecular weight" and abbreviated M_r or RMM) is the mass of one mole of a substance compared to the mass of one mole of hydrogen. It is dimensionless but is sometimes expressed in Daltons.

mRNA (messenger RNA): RNA molecule that carries the genetic code. The order of bases on mRNA specifies the amino acid sequence of a polypeptide chain.

MTOC (microtubule organizing center): structure from which cytoplasmic microtubules arise; synonymous with the centrosome.

muscarinic acetylcholine receptor: integral membrane protein that binds acetylcholine and then becomes a GEF for the trimeric G protein G_q, hence leading to the activation of phospholipase Cβ which in turn leads to release of calcium ions from the endoplasmic reticulum.

muscle: tissue specialized for generating contractile force.

mutation: inherited change in the structure of a gene or chromosome.

myelin: fatty substance that is wrapped around nerve cell axons by glial cells.

myosin: type of motor protein that moves along, or pulls on, actin filaments. The thick filaments in skeletal muscle are formed of myosin II molecules.

myosin II: class of myosin that forms the large filaments seen in striated muscle but that is also found in other cell types.

myosin V: class of myosin that carries cargo along actin filaments.

N-terminal (amino terminal): the end of a peptide or polypeptide that has a free α-amino group. This end is made first on the ribosome.

NADH (nicotinamide adenine dinucleotide, reduced form): combination of two nucleotides that is a strong reducing agent and one of the cell's energy currencies.

NADPH (nicotinamide adenine dinucleotide phosphate, reduced form): phosphorylated NADH; like NADH it is a strong reducing agent, but NADPH is not an energy currency, rather it is used in synthetic reactions in the cytoplasm.

Na^+/K^+ ATPase: plasma membrane carrier. For every ATP hydrolyzed, three Na^+ ions are moved out of the cytosol and two K^+ ions are moved in.

476 of its energy currencies

necrosis: cell death that is due to damage so severe that the cell cannot maintain the level of its energy currencies and therefore falls apart—distinct from apoptosis.

negative feedback: control system in which a change in some parameter activates a mechanism that reverses the change in that parameter; an example is the effect of tryptophan on expression of the *trp* operon.

negative regulation (of transcription): inhibition of transcription due to the presence of a particular substance, which is often the end product of a metabolic pathway.

Nernst equation: equation that allows the equilibrium voltage of an ion across a membrane to be calculated. The general form is

$$V_{eq} = \frac{RT}{zF} \log_e \left(\frac{[I_{outside}]}{[I_{inside}]} \right) \quad \text{volts}$$

where R is the gas constant (8.3 joules per degree per mole), T is the absolute temperature, z is the charge on the ion in elementary units, and F is 96,500 coulombs in a mole of monovalent ions. At 37°C the equation approximates to

$$V_{eq} = \frac{62}{z} \log_{10} \left(\frac{[I_{outside}]}{[I_{inside}]} \right) \quad \text{millivolts}$$

nerve cell: electrically excitable cell with a long axon specialized for transmission of (usually sodium) action potentials.

nerve growth factor (NGF): older name for neurotrophin 1.

nervous tissue: tissue formed of nerve and glial cells that carries out electrical data processing.

neurofilament: type of intermediate filament found in nerve cells.

neuron: another name for a nerve cell.

neurone: yet another name for a nerve cell.

neurotrophin: one of a family of paracrine transmitters that act upon nerve cells to prevent apoptosis and trigger differentiation. Neurotrophin 1 used to be called nerve growth factor or NGF.

NFAT: transcription factor that only translocates to the nucleus when dephosphorylated by the calcium–calmodulin activated phosphatase calcineurin.

NGF (nerve growth factor): older name for neurotrophin 1.

nicotinamide: base used in the double nucleotides NADH and NADPH.

nicotinamide adenine dinucleotide (reduced form) (NADH): combination of two nucleotides that is a strong reducing agent and one of the cell's energy currencies.

nicotinamide adenine dinucleotide phosphate (reduced form) (NADPH): phosphorylated NADH; like NADH it is a strong reducing agent, but NADPH is not an energy currency, rather it is used in synthetic reactions in the cytoplasm.

nicotinic acetylcholine receptor: integral membrane protein of the plasma membrane that forms an ion channel that allows both sodium and potassium ions to pass. When it binds acetylcholine it opens, which in turn leads to electrical depolarization.

nine plus two (9 + 2) axoneme: structure of cilium or flagellum; describes the arrangement of nine peripheral microtubules surrounding two central microtubules.

nitric oxide (NO): paracrine transmitter that acts on intracellular receptors, the most important of which is guanylate cyclase.

node (of a myelinated axon): gap between adjacent lengths of myelin sheath, where the nerve cell plasma membrane directly contacts the extracellular medium. Also called node of Ranvier, after its discoverer, Louis Ranvier.

nonpolar: covalent bonds in nonpolar molecules have electrons shared equally so that the constituent atoms do not carry a charge.

nonsense mutation: base change that causes a codon that coded for an amino acid to become a stop codon.

northern blotting: blotting technique in which RNAs, separated by size, are probed using a single-stranded cDNA probe or an antisense RNA probe.

N-terminal (amino terminal): end of a peptide or polypeptide that has a free α-amino group. This end is made first on the ribosome.

NTR: neurotrophin receptor. Although the Trk family are receptor tyrosine kinases for neurotrophins, the name NTR is usually reserved for the death domain receptor called p75 NTR.

nuclear envelope: double-membrane system enclosing the nucleus; it contains nuclear pores and is continuous with the endoplasmic reticulum.

nuclear lamina: meshwork of proteins lining the inner face of the nuclear envelope.

nuclear pores: holes running through the nuclear envelope that regulate traffic of proteins and nucleic acids between the nucleus and the cytoplasm.

nuclear pore complex: multiprotein complex that lies within and around the nuclear pore and that regulates import into and export from the nucleus in a process called gated transport.

nuclease: enzyme that degrades nucleic acids.

nucleic acid: polymer of nucleotides joined together by phosphodiester bonds; DNA and RNA are nucleic acids.

nucleoid: region of a bacterial cell that contains the chromosome.

nucleolar organizer regions: region of one or more chromosomes at which the nucleolus is formed.

nucleolus: region(s) of the nucleus concerned with the production of ribosomes.

nucleoside: purine, pyrimidine, or nicotinamide attached to either ribose or deoxyribose.

nucleosome: beadlike structure formed by a stretch of DNA wrapped around a histone octamer.

nucleotide: building block of nucleic acids; a nucleoside that is phosphorylated on its 5′ carbon atom.

nucleotide excision repair: process that removes a thymine dimer, together with some 30 surrounding nucleotides, from DNA. The gap is then repaired by the actions of DNA polymerase I and DNA ligase.

nucleus: cell organelle housing the chromosomes; enclosed within a nuclear envelope.

objective lens: lens of a light or electron microscope that forms a magnified image of a specimen.

obligate anaerobe: organism that is poisoned by oxygen so that it can only function anaerobically.

oil: triacylglycerol (triglyceride) that is liquid at room temperature. In contrast, fats are solid at room temperature.

Okasaki fragment: series of short fragments that are joined together to form the lagging strand during DNA replication.

oleic acid: $C_{17}H_{33}$—COOH; a fatty acid with a single double bond. The dominant acyl group in olive oil is derived from oleic acid.

olfactory neuron: nerve cell found in the nose, whose dendrites are sensitive to smell chemicals.

oligonucleotide: short fragment of DNA or RNA.

oligosaccharide: chain of up to 100 or so monosaccharides linked by glycosidic bonds.

oligosaccharide transferase: enzyme that adds an oligosaccharide group onto a protein.

oocyte: cell that undergoes meiosis to give rise to an egg.

open complex: complex between DNA and protein that causes the two strands of DNA to separate, e.g., during DNA replication and transcription.

open promoter complex: structure formed when the two strands of the double helix separate so that transcription can commence.

operator: DNA sequence to which a repressor protein binds to prevent transcription from an adjacent promoter.

operon: cluster of genes that encode proteins involved in the same metabolic pathway and that are transcribed as one length of mRNA under the control of one promoter. As far as we know, operons are only found in prokaryotes.

optical isomers: two molecules that differ only in that one is the mirror image of the other.

organelle: membrane-bound, intracellular structure such as a mitochondrion, chloroplast, lysosome, etc.

organic: in chemistry, an organic compound is one that contains carbon atoms. When applied to farming and food, organic means the avoidance of high-intensity agricultural techniques including genetic engineering.

organism: cell or clone of cells that functions as a discrete and integrated whole to maintain and replicate itself.

origin of replication: site on a chromosome at which DNA replication can commence.

osmolarity: overall strength of a solution; the more solute that is dissolved in a solution, the higher the osmolarity.

osmosis: movement of water down its concentration gradient.

outer doublet microtubules: paired microtubules that make the 9 of the $9 + 2$ axoneme in cilia and flagella.

oxidation: removal of electrons from a molecule, e.g., by adding oxygen atoms, which tend to take more than their fair share of electrons in any bonds they make. Removal of hydrogen atoms from a molecule oxidizes it. (Don't confuse with deprotonation, which is the loss of H^+).

β **oxidation:** process by which fatty acids are broken down into individual two-carbon units coupled to CoA to form acetyl-CoA. The process, which takes place in the mitochondrial matrix, generates both NADH and $FADH_2$.

oxidizing agent: agent that will act to lessen the share of electrons that the atoms of a molecule have. Oxidizing agents often work by adding oxygen atoms or by removing hydrogen atoms (Don't confuse with a base, which will accept H^+, not H). Examples are oxygen itself, NAD^+, and FAD.

P site (peptidyl site): the site on a ribosome that is occupied by the growing polypeptide chain.

p21^{CIP1}: inhibitor of cyclin-dependent kinases. p21^{CIP1}, which is upregulated by cell–cell contact, prevents cells entering the S phase of the cell cycle.

p38: a stress-activated protein kinase related to mitogen associated protein kinase (MAP kinase) but with a very different role: p38 is activated by cell stress and stimulates both cell repair and apoptosis.

p53: transcription factor that stimulates cell repair but also apoptosis. Cancer cells are frequently found to have mutated, nonfunctional p53 genes.

p75 neurotrophin receptor: death domain receptor for neurotrophins. Upon binding neurotrophins p75 activates caspase 8 and hence initiates apoptosis unless a countermanding signal to survive is present.

PAC (P1 artificial chromosome): cloning vector, derived from the bacteriophage P1, that is used to propagate, in *E. coli,* DNAs of about 150,000 bp.

pain receptor: nerve cell whose distal axon terminal is depolarized by potentially damaging events such as heat or stretching.

pain relay cell: this nerve cell, upon which a pain receptor synapses, carries the message on toward the brain.

paracrine transmitters: agonists released by cells into the extracellular fluid that can last many minutes and can therefore diffuse widely within the tissue before they are destroyed.

parallel β sheet: β sheet in which all the parallel polypeptide chains run in the same direction.

parenchyma cells: unspecialized plant cells whose cell walls are usually thin and bendable; the major site of metabolic activity and photosynthesis in leaves and green shoots.

patch clamp: technique in which a glass micropipette is sealed to the surface of a cell to allow electrical recording of cell properties.

PCR (polymerase chain reaction): method for making many copies of a DNA sequence where the sequence, at least at each end, is known.

PDGF (platelet-derived growth factor): paracrine transmitter that opposes apoptosis and promotes cell division in target cells such as endothelial cells and smooth muscle.

pectin: polysaccharide component of the plant cell wall; the amount of pectin in the cell wall determines its thickness.

pentose: monosaccharide with five carbon atoms.

pentose phosphate pathway: pathway that makes pentoses and generates reducing power in the form of NADPH by oxidizing glucose-6-phosphate.

peptide: short linear polymer of amino acids.

peptide bond: bond between amino acids. The bond is formed between the carboxyl group of one amino acid and the amino group of the next.

peptidyl site (P site): site on a ribosome that is occupied by the growing polypeptide chain.

peptidyl transferase: enzyme that catalyzes the formation of a peptide bond between two amino acids.

peripheral membrane protein: class of protein that is easily detached from a cell membrane, unlike integral membrane proteins, which can only be isolated by destroying the membrane, e.g., with detergent.

peroxisome: class of cell organelles of diverse function. Peroxisomes frequently contain the enzyme catalase, which breaks down hydrogen peroxide into oxygen and water.

pH: measure of the acidity of a solution, equal to minus the logarithm to base 10 of the hydrogen ion concentration in moles liter^{-1}. The smaller the pH value, the more acid the solution. A neutral solution has a pH of 7, that is, the H$^+$ concentration is 10^{-7} mol liter^{-1} or 100 nmol liter^{-1}.

PH domain: protein domain that binds phosphorylated inositols. The PH group on protein kinase B is recruited to the membrane by the intensely charged lipid phosphatidylinositol trisphosphate.

phage: short for bacteriophage, i.e., a virus that infects bacterial cells.

phase-contrast microscopy: type of light microscopy in which differences in the refractive index of a specimen are converted into differences in contrast.

PH domain: protein domain that binds phosphorylated inositols. The PH group on protein kinase B is recruited to the membrane by the intensely charged lipid phosphatidylinositol trisphosphate.

phenylketonuria: inherited disease in which the enzyme phenylalanine hydroxylase is missing or defective. Unless intake of phenylalanine is drastically curtailed, phenylalanine and its transamination product phenylpyruvate build up in the body and cause brain damage. The heel prick or Guthrie test, performed on all newborn babies, tests for this condition.

phloem: part of the plant vascular system; conveys the products of photosynthesis to their sites of use or storage.

phosphatase: enzyme that removes phosphate groups from substrates. Glucose-6-phosphate phosphatase and calcineurin are phosphatases operating on a sugar and on proteins, respectively.

phosphate: properly, a name for the ions $H_2PO_4^-$, HPO_4^{2-}, and PO_4^{3-}. The word is also very commonly used to mean the group

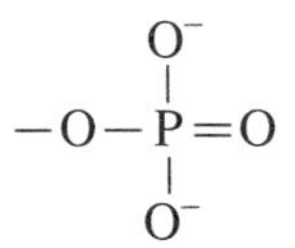

$$-O-\overset{\displaystyle O^-}{\underset{\displaystyle O^-}{P}}=O$$

and we have used this convention in this book. If one wants to specificically describe the group, *not* the ions, one can refer to a phosphoryl group (a term that does not include the leftmost oxygen in the diagram above).

phosphatidylinositol bisphosphate (PIP$_2$): membrane lipid. PIP_2 releases inositol trisphosphate into the cytosol upon hydrolysis by phospholipase C. Alternatively, PIP_2 can be further phosphorylated to yield phosphatidylinositol trisphosphate.

phosphatidylinositol trisphosphate (PIP$_3$): a membrane lipid. Its intensely charged headgroup can recruit proteins that contain PH domains to the plasma membrane.

phosphodiester link: link between two parts of a molecule in which a phosphate is attached through an oxygen atom to each of the two parts. In phospholipids, the headgroup is attached to glycerol by a phosphodiester link. In DNA and RNA, successive nucleotides are joined together by phosphodiester links.

phosphofructokinase: enzyme that phosphorylates fructose-6-phosphate to generate fructose-1,6-bisphosphate.

phosphoinositide 3-kinase (PI-3-kinase): enzyme that phosphorylates phosphatidylinositol bisphosphate on the 3 position of the inositol ring to generate the intensely charged lipid phosphatidylinositol trisphosphate (PIP_3). PIP_3 can then recruit proteins containing PH domains to the plasma membrane.

phospholipase C: enzyme that hydrolyzes the membrane lipid phosphatidylinositol bisphosphate (PIP_2) to release inositol trisphosphate into the cytosol. The β isoform is activated by the trimeric G protein G_q, while the γ isoform is activated by tyrosine phosphorylation.

phosphoryl group:

$$
\begin{array}{c}
O^- \\
| \\
-P=O \\
| \\
O^-
\end{array}
$$

phosphorylase: enzyme that cleaves a glycosidic bond by adding phosphate.

phosphorylase kinase (glycogen phosphorylase kinase): kinase that is activated by the calcium–calmodulin complex and which phosphorylates glycogen phosphorylase, activating the latter enzyme.

phosphorylated: having had a phosphate group added. The phosphate groups are usually substituted into a hydroxyl group to form a phosphoester bond but are sometimes substituted into an acid to form a phosphoanhydride or attached to a nitrogen atom to form a phosphoimide.

phosphorylation: addition of a phosphate group to a molecule. Usually substituted into a hydroxyl group to form a phosphoester bond. Sometimes substituted into an acid to form a phosphoanhydride, or attached to a nitrogen atom to form a phosphoimide.

photosynthesis: synthesis of complex organic molecules, and oxidation of water to release oxygen, that is driven by the energy of light.

phragmoplast: structure associated with the formation of a new cell wall during cytokinesis in plant cells.

PI 3-kinase (phosphoinositide 3-kinase): enzyme that phosphorylates phosphatidylinositol bisphosphate on the 3 position of the inositol ring to generate the intensely charged lipid phosphatidylinositol trisphosphate (PIP_3). PIP_3 can then recruit proteins containing PH domains to the plasma membrane.

PIP_2 (phosphatidylinositol bisphosphate): membrane lipid that releases inositol trisphosphate into the cytosol upon hydrolysis by phospholipase C.

pK_a: parameter equal to $-\log_{10} K_a$ and representing the pH at which the concentration of the dissociated acid is equal to the concentration of the undissociated acid.

PKA (protein kinase A; cAMP-dependent protein kinase): protein kinase that is activated by the intracellular messenger cyclic AMP. PKA phosphorylates proteins (e.g., glycogen phosphorylase kinase) on serine and threonine residues.

PKB (protein kinase B): protein kinase that is activated when it is itself phosphorylated; this in turn only occurs when PKB is recruited to the plasma membrane by phosphatidylinositol trisphosphate (PIP_3). PKB phosphorylates proteins on serine and threonine residues (e.g., the bcl-2 family protein BAD). An older name for PKB is Akt.

PKC (protein kinase C): protein kinase that is activated by a rise of cytosolic calcium concentration; it is also activated by diacylglycerol. PKC phosphorylates proteins on serine and threonine residues.

PKG (protein kinase G; cGMP-dependent protein kinase): protein kinase that is activated by the intracellular messenger cyclic GMP. PKG phosphorylates proteins (e.g., calcium ATPase) on serine and threonine residues.

plaque: area of dead bacteria in a lawn of live bacteria that is caused by infection of bacterial cells by bacteriophages.

plasmalemma: membrane that surrounds the cell. Also called the plasma membrane or the cell membrane.

plasma membrane: membrane that surrounds the cell. Also called the plasmalemma or the cell membrane.

plasmid: circular DNA molecule that is replicated independently of the host chromosome in bacterial cells.

plasmodesmata (singular, **plasmodesma**): type of cell junction unique to plant cells, which provides a much bigger hole for passage of substances between the cytoplasm of the two cells than gap junctions do.

plastoquinone: quinone electron carrier found in chloroplasts.

platelet: small fragment of cell that contains no nucleus but that has a plasma membrane and some endoplasmic reticulum. Platelets are critical in the process of blood clotting, and they also release platelet-derived growth factor.

platelet-derived growth factor (PDGF): paracrine transmitter that opposes apoptosis and promotes cell division in target cells such as endothelial cells and smooth muscle.

PLC (phospholipase C): enzyme that hydrolyzes the membrane lipid phosphatidylinositol bisphosphate (PIP_2) to release inositol trisphosphate into the cytosol. The β isoform is activated by the trimeric G protein G_q, while the γ isoform is activated by tyrosine phosphorylation.

polar: having covalent bonds in which the electrons are unequally shared, so that atoms have partial charges. Polar molecules can interact with water by electrostatic interactions and by hydrogen bonding.

poly adenosine tail: string of adenine residues added to the $3'$ end of a eukaryotic mRNA.

polyadenylation: process whereby a poly-A tail is added to the $3'$ end of a eukaryotic mRNA.

poly-A tail: string of adenine residues added to the $3'$ end of a eukaryotic mRNA.

polycistronic mRNA: mRNA that, when translated, yields more than one polypeptide.

polymer: chemical composed of a long chain of identical or similar subunits.

polymerase: enzyme that makes polymers, i.e., long chains of identical or very similar subunits. DNA and RNA polymerase, respectively, are involved in making DNA and RNA.

polymerase chain reaction (PCR): method for making many copies of a DNA sequence when the sequence, at least at each end, is known.

polynucleotide kinase: enzyme that adds a phosphate group to the $5'$ end of DNA or RNA.

polypeptide: polymer of more than 50 amino acids joined by peptide bonds.

polyploid: having three or more sets of chromosomes.

polyribosome (or **polysome**)**:** chain of ribosomes attached to an mRNA molecule.

polysaccharide: chain of more than 100 or so monosaccharides linked by glycosidic bonds.

polysome (or **polyribosome**)**:** chain of ribosomes attached to an mRNA molecule.

polyunsaturated: having several double bonds. Usually applied to fatty acids; triacylglycerols with polyunsaturated acyl (fatty acid) groups are liquid even at low temperatures.

porin: channel found in the mitochondrial outer membrane. It spends most of its time open and allows all solutes of $M_r < 10,000$ to pass.

positive feedback: process in which the consequences of a change act to increase the magnitude of that change, so that a small initial change tends to get bigger and bigger.

positive regulation (of transcription): process whereby transcription is activated in the presence of a particular substance.

postsynaptic cell: cell upon which a nerve cell releases its transmitter at a synapse.

potassium channel: channel in the plasma membrane of many cells that allows potassium ions to pass.

potassium/sodium ATPase (Na$^+$/K$^+$ ATPase): plasma membrane carrier. For every ATP hydrolyzed, three Na$^+$ ions are moved out of the cytosol and two K$^+$ ions are moved in.

preinitiation complex: complex formed between transcription factors and RNA polymerase at the promoter of a eukaryotic gene.

presynaptic terminal: region of an axon terminal that is specialized for exocytosis of transmitter.

primary antibody: first antibody applied to the preparation. In many cases the primary antibody is not itself labeled but must be revealed by applying a labeled secondary antibody.

primary cell wall: first layer of the plant cell wall.

primary endosome: acidic cell compartment with which coated vesicles fuse.

primary immunofluorescence: technique in which the location or presence of a chemical is revealed by treating the sample with a dye-labeled antibody.

primary structure (of a protein): sequence of amino acids held together by peptide bonds making up a polypeptide.

primase: enzyme that synthesizes the RNA primers needed for the initiation of synthesis of the leading and lagging DNA strands.

primer: short sequence of nucleic acid (RNA or DNA) that acts as the start point at which a polymerase can initiate synthesis of a longer nucleic acid chain.

primosome: protein complex, including the enzyme primase, that is involved in the synthesis of RNA primers for DNA replication.

prion: protein that is itself an infective agent for a disease.

profilin: type of actin-binding protein that regulates the assembly of actin filaments.

progenitor: ancestor; an individual from which others are descended. In this book we use the term *progenitor cell* to mean the cell that divides to give rise to two other cells (in mitosis) or four others (in meiosis).

progeny: children, offspring. In this book we use the term *progeny cells* to mean the products of cell division.

programmed cell death (apoptosis): process in which a cell actively promotes its own destruction, as distinct from necrosis. Apoptosis is important in vertebrate development, where tissues and organs are shaped by the death of certain cell lineages.

projector lens: lens of light or electron microscope that carries the image to the eye; more commonly known as the "eyepiece."

prokaryotic: type of cellular organization found in bacteria in which the cells lack a distinct nucleus and other organelles.

prometaphase: period of mitosis or meiosis that sees the breakdown of the nuclear envelope and the attachment of the chromosomes to the mitotic spindle.

promoter: region of DNA to which RNA polymerase binds to initiate transcription.

pronuclei: nuclei of the egg and sperm prior to fusion.

prophase: period of mitosis or meiosis in which the chromosomes condense.

prophase I: first prophase of meiosis.

prophase II: second prophase of meiosis.

prosthetic group: nonprotein molecule necessary for the activity of a protein. The concept overlaps with the concept of cofactor. The difference is just how tightly they are bound: a prosthetic group is very tightly bound and cannot be removed without at least partial unfolding of the protein. Examples are the heme groups of myoglobin, hemoglobin, and the cytochromes.

protein: polypeptide (a polymer of α-amino acids) that has a preferred way of folding.

protein engineering: designing a novel protein and then causing it to be built by altering the sequence of DNA.

protein kinase: enzyme that phosphorylates a protein by transferring a phosphate group from ATP to the molecule.

protein kinase A: protein kinase that is activated by the intracellular messenger cAMP. Protein kinase A phosphorylates proteins (e.g., glycogen phosphorylase kinase) on serine and threonine residues.

protein kinase B (PKB): protein kinase that is activated when it is itself phosphorylated; this in turn only occurs when PKB is recruited to the plasma membrane by phosphatidylinositol trisphosphate (PIP$_3$). PKB phosphorylates proteins on serine and threonine residues (e.g., the bcl-2 family protein BAD). An older name for PKB is Akt.

protein kinase C (PKC): protein kinase that is activated by a rise of cytosolic calcium concentration; it is also activated by diacylglycerol. PKC phosphorylates proteins on serine and threonine residues.

protein kinase G (PKG; cGMP-dependent protein kinase): protein kinase that is activated by the intracellular messenger cyclic GMP. PKG phosphorylates proteins (e.g., calcium ATPase) on serine and threonine residues.

protein phosphatase: enzyme that removes phosphate groups from proteins.

protein phosphorylation: addition of a phosphate group to a protein. The addition of the charge on the phosphoryl group can markedly alter the tertiary structure and therefore function of a protein.

protein targeting: delivery of proteins to their correct cellular location.

protein translocator: channel-like protein of the endoplasmic reticulum that allows polypeptide chains to cross the membrane as they are synthesized.

proteoglycans: heavily glycosylated proteins that contribute to the extracellular matrix.

proteome: complete protein content of the cell.

proteomics: study of the proteome. For example, one might compare the protein profiles of cells in two tissues.

protofilaments: chains of subunits that make up the wall of a microtubule.

protonated: having accepted an H^+. For example, the lactate ion, $CH_3CH(OH)COO^-$, will become protonated in acid solutions to become lactic acid, $CH_3CH(OH)COOH$. The word *proton* is being used as shorthand for *hydrogen atom nucleus* since, for the commonest isotope of hydrogen, the nucleus is a single proton.

proximal: close to the center.

pseudogene: gene that has mutated such that it no longer codes for a protein.

pseudopodium: projection extended by an amoeba or other crawling cell in the direction of movement.

P site (peptidyl site): site on the ribosome occupied by the growing polypeptide chain.

purine: nitrogenous base found in nucleotides and nucleosides; adenine, guanine and hypoxanthine are purines.

pyrimidine: nitrogenous base found in nucleotides and nucleosides; cytosine, thymine, and uracil are pyrimidines.

pyrophosphate:

$$^-O-\overset{\displaystyle O}{\overset{\|}{P}}-O-\overset{\displaystyle O}{\overset{\|}{P}}-O^-$$
$$\quad\ \ \ |\qquad\quad\ |$$
$$\quad\ \ OH\qquad\ O^-$$

the pyrophosphate ion is stable if kept away from biological tissue but is rapidly hydrolyzed to two phosphate ions by intracellular and extracellular enzymes.

quaternary structure: structure in which subunits of a protein, each of which has a tertiary structure, associate to form a more complex molecule. The subunits associate tightly, but not covalently and may be the same or different.

quiescent: at rest. Nondividing cells are often said to be quiescent.

Rab family: family of GTPases that mediate fusion of vesicles with other membranes.

Raf: isoform of MAPKKK, the enzyme at the top of the mitogen-associated protein kinase cascade.

Ras: GTPase that activates the MAP kinase pathway and hence promotes cell division. Ras is activated by SOS, its GTP exchange factor (GEF), which is in turn recruited to the plasma membrane via receptor tyrosine kinases and the adaptor protein Grb2.

RB: protein that binds to the transcription factor E2F-1 and therefore prevents transcription of proteins required for DNA synthesis. Phosphorylation of RB by CDK4 causes it to release E2F-1, allowing entry to S phase. Mutations in RB lead to the formation of the eye cancer called retinoblastoma.

reading frame: reading of the genetic code in blocks of three bases—there are three possible reading frames for each mRNA only one of which will produce the correct protein.

receptor: protein that specifically binds a particular solute. Receptors can be transmembrane, cytosolic, or nuclear proteins. Particular receptor proteins perform additional functions (ion channel, enzyme, activator of endocytosis, transcription factor . . .) that are activated by the binding of the solute.

receptor-mediated endocytosis: process in which ligands bind to specific receptors in the plasma membrane and trigger clathrin-mediated vesicle budding.

receptor tyrosine kinase: integral membrane protein with a binding site for transmitter on the extracellular aspect and tyrosine kinase catalytic ability on the cytosolic aspect. Binding of transmitter causes self-phosphorylation on tyrosine and hence recruitment of proteins with SH2 domains such as Grb2, phospholipase Cγ, and phosphoinositide 3-kinase, which may then themselves be phosphorylated. The PDGF receptor, the FGF receptor, the Trk family of receptors, and the insulin receptor are all receptor tyrosine kinases.

recessive: gene is recessive if its effects are hidden when the organism posesses a second, dominant, version. Recessive genes usually code for nonfunctional proteins, so that if the individual can make the protein using the other, working gene, no effects are seen.

recombinant DNA: artificial DNA molecule formed of DNA from two or more different sources.

recombinant plasmid: plasmid into which a foreign DNA sequence has been inserted.

recombinant protein: protein expressed from the foreign DNA inserted into a recombinant plasmid or other cloning vector. Recombinant proteins are often expressed in bacteria, yeast, insect, or mammalian cells.

recombination: "cut-and-paste" of DNA. Recombination occurs at chiasmata during meiosis and also in embryonic stem cells, allowing insertional mutagenesis.

recovery stroke: part of the beat cycle of a cilium in which the cilium is moved back into a position where it can push again; used in contrast to the effective stroke that generates the force.

reduction: addition of electrons to a compound, e.g., by the addition of hydrogen atoms or the removal of oxygen atoms.

refractive index: measure of the capacity of any material to slow down the passage of light.

regulated secretion: secretion that only occurs in response to a signal, such as a rise in the cytosolic concentration of calcium ions.

Relative Molecular Mass: mass of one mole of a substance compared to the mass of one mole of hydrogen. It is abbreviated M_r or RMM and is sometimes called "molecular weight." It is dimensionless but is sometimes expressed in Daltons.

repetitious DNA: DNA sequence that is repeated many times within the genome.

replication (of DNA): The process whereby two DNA molecules are made from one.

replication fork: Y-shaped structure formed when the two strands of the double helix separate during replication.

repressible operon: operon whose transcription is repressed in the presence of a particular substance; often the final product of the metabolic pathway.

RER (rough endoplasmic reticulum): portion of the endoplasmic reticulum associated with ribosomes and concerned with the synthesis of secreted proteins. Proteins destined to remain within the majority of single-membrane-bound organelles (Golgi, lysosomes, ...) are also made on the rough ER, as are integral proteins of these organelles and of the plasma membrane.

residue: that which is left over. In chemistry, when a molecule is built into a larger molecule with the loss of some part, the part that remains and forms part of the larger molecule is called a residue. For example, an amino acid that has been built into a polypeptide, with the loss of the elements of water, is said to be an amino acid residue.

resolving power: measure that defines the smallest object that can be distinguished using a microscope.

resting voltage: voltage across the plasma membrane of an unstimulated cell, typically -70 to -90 mV.

restriction enzyme endonuclease: enzyme that cleaves phosphodiester bonds within a specific sequence in a DNA molecule.

retinoblastoma: cancer of the eye, usually caused by a mutation in the *RB* gene.

retrograde: backward movement; when applied to axonal transport it means toward the cell body.

retrovirus: virus whose genetic information is stored in RNA.

reverse genetics: research that begins with a gene of known sequence but unknown effect and works toward deducing its effects and therefore function.

reverse transcriptase: enzyme of some viruses that copies RNA into DNA.

reverse transcription: process whereby RNA is copied into DNA.

ribonuclease: enzyme that cleaves phosphodiester links in RNA.

ribonuclease H: enzyme that cleaves phosphodiester links in an RNA molecule that is joined to a DNA molecule by complementary base pairing.

ribonucleic acid (RNA): polymer of ribonucleoside monophosphates; *see* **mRNA, tRNA,** and **rRNA**.

ribonucleic acid polymerase: enzyme that synthesizes RNA.

ribonucleic acid primer: short length of RNA, complementary in sequence to a DNA strand, that allows DNA polymerase III to attach and begin DNA synthesis.

ribonucleoside monophosphate: nitrogenous base attached to the sugar ribose that has one phosphate group on its $5'$ carbon atom; also known as a ribonucleotide.

ribose: pentose sugar used to make the nucleotides that form RNA.

ribosomal RNA (rRNA): RNA component of ribosomes. rRNA forms a major part of the ribosome and participates fully in the translation process.

ribozyme: RNA molecule with enzyme-like catalytic activity.

RNA (ribonucleic acid): polymer of ribonucleoside monophosphates. Cells contain three types of RNA: messenger RNA, ribosomal RNA, and transfer RNA.

RNA polymerase: enzyme that synthesizes RNA.

RNA primer: short length of RNA, complementary in sequence to a DNA strand that allows DNA polymerase III to attach and begin DNA synthesis.

RNA splicing: removal of introns from an RNA molecule and the joining together of exons to form the final RNA product.

rough endoplasmic reticulum (RER): portion of the endoplasmic reticulum associated with ribosomes and concerned with the synthesis of secreted proteins. Proteins destined to remain within the majority of single-membrane-bound organelles (Golgi, lysosomes, . . .) are also made on the rough ER, as are integral proteins of these organelles and of the plasma membrane.

rRNA (ribosomal RNA): RNA component of ribosomes. rRNA forms a major part of the ribosome and participates fully in the translation process.

ryanodine: plant toxin that binds to the ryanodine receptor, with complex effects on gating of the channel.

ryanodine receptor: calcium channel found in the membrane of the endoplasmic reticulum. In most cells it opens in response to a rise of calcium concentration in the cytosol. In skeletal muscle ryanodine receptors are directly linked to voltage-gated calcium channels in the plasma membrane and open when the latter open.

σ factor (sigma factor): subunit of bacterial RNA polymerase that recognises the promoter sequence.

S phase: period of the cell division cycle during which DNA replication occurs.

S value (sedimentation coefficient): a value that describes how fast macromolecules and organelles sediment in a centrifuge.

saltatory conduction: jumping of an action potential from node to node down a myelinated axon.

salt bridge (in protein structure): interaction between a positively charged amino acid residue (such as arginine) and a negatively charged residue (such as aspartate).

sarcomere: contractile unit of striated muscle.

sarcoplasmic reticulum: type of smooth endoplasmic reticulum found in striated muscle; concerned with the regulation of the concentration of calcium ions.

satellite DNA: DNA sequence that is tandemly repeated many times.

saturated (of fatty acids): containing no carbon–carbon double bonds.

saturation kinetics: are said to occur when the rate of a reaction approaches a maximum, limiting value as the concentration of reactant increases. The rate becomes limited by the availabilty of binding sites on the catalyst for the reactant.

scanning electron microscope (SEM): type of electron microscope in which the image is formed from electrons that are reflected back from the surface of a specimen as the electron beam scans rapidly back and forth over it. The scanning electron microscope is particularly useful for providing topographical information about the surfaces of cells or tissues.

Schwann cell: glial cell of the peripheral (outside the brain and spinal cord) nervous system.

SDS–PAGE (sodium dodecyl sulfate–polyacrylamide gel electrophoresis): technique for separating proteins by their relative molecular mass.

secondary antibody: in immunofluorescence, a labeled antibody that is used to reveal the location of an unlabeled primary antibody. Secondary antibodies are produced by one species of animal in response to the injection of another animal's antibodies, thus a "goat antirabbit" secondary will bind to any primary antibody that has been generated by a rabbit. Labeled secondary antibodies can be used to study many different questions because the specificity in the final image or western blot is provided by the primary antibody, not by the secondary.

secondary cell wall: layers of a plant cell wall formed external to the primary cell wall.

secondary immunofluorescence microscopy: form of light microscopy in which a fluorescent secondary antibody is used to label a preapplied primary antibody specific for a particular protein or subcellular structure. Also called indirect immunofluorescence microscopy.

secondary structure: regular, repeated folding of the backbone of a polypeptide. The side chains of the amino acids have an influence but are not directly involved. There are two common types of secondary structue: the α helix and the β sheet.

second messenger: another name for intracellular messenger (the "first messenger" being the extracellular transmitter chemical). A cytosolic solute that changes in concentration in response to external stimuli or internal events and that acts on intracellular targets to change their behavior. Calcium ions, cyclic AMP, and cyclic GMP are the three common intracellular messengers.

secretion: synthesis and release from the cell of a chemical.

secretory vesicles: vesicles derived from the Golgi apparatus that transport secreted proteins to the cell membrane with which they fuse.

sedimentation coefficient (S value): value that describes how fast macromolecules and organelles sediment in a centrifuge.

semiconservative replication: mode of replication used for DNA; both strands of the double helix serve as templates for the synthesis of new daughter strands.

serine–threonine kinase: enzyme that phosphorylates proteins on serine or threonine residues, by transferring the terminal phosphate of ATP to the amino acid side chain. With a few exceptions, protein kinases are either serine–threonine kinases (which can phosphorylate on serine or on threonine) or tyrosine kinases (which only phosphorylate on tyrosine).

seven-methyl guanosine cap: modified guanosine found at the 5′ terminus of eukaryotic mRNA. A guanosine is attached to the mRNA by a 5′-5′-phosphodiester link and is subsequently methylated on atom number 7 of the guanine.

SH2 domain: domain found in a number of proteins that binds to phosphorylated tyrosine. Many proteins with SH2 domains are recruited to receptor tyrosine kinases when the latter phosphorylate themselves in response to ligand binding. Important proteins with SH2 domains are Grb2, PI-3-kinase, and phospholipase Cγ.

β sheet: common secondary structure in proteins, in which lengths of fully extended polypeptide run alongside each other, hydrogen bonds forming between the peptide bonds of the adjoining strands.

Shine–Dalgarno sequence: sequence on a bacterial mRNA molecule to which the ribosome binds.

side chain (of amino acids): group attached to the α carbon of an amino acid.

sigma factor (σ factor): subunit of bacterial RNA polymerase that recognizes the promoter sequence.

signal peptidase: enzyme that cleaves the signal sequence from a polypeptide as it enters the lumen of the endoplasmic reticulum.

signal recognition particle: ribonucleoprotein particle that recognizes and binds to the signal sequence at the N terminus of a polypeptide.

signal recognition particle receptor: receptor on the endoplasmic reticulum to which the signal recognition particle binds during the process of polypeptide chain synthesis and import into the endoplasmic reticulum. Also called the "docking protein."

signal sequence: short stretch of amino acids found at the N terminus of polypeptides that targets them to the endoplasmic reticulum.

simple diffusion: not a specific term: we often say "simple diffusion" to emphasize that a solute movement is passive, down a concentration gradient, and does not require a carrier or channel.

single-strand binding protein: protein that binds to the separated DNA strands to keep them in an extended form during replication, thus preventing the double helix from reforming.

site-directed mutagenesis: technique used to change the base sequence of a DNA molecule at a specific site.

skeletal muscle cells: large multinucleate muscle cells that are attached to bone. Most cuts of meat are mainly skeletal muscle.

small nuclear RNAs (snRNAs): small RNA molecules found in the nucleus that play a role in RNA splicing.

smooth endoplasmic reticulum (SER): portion of the endoplasmic reticulum without attached ribosomes, among its functions are the synthesis of lipids and the storage and stimulated release of calcium ions.

smooth muscle: nonstriated muscle, found in many places in the body including blood vessels and the intestine.

smooth muscle cells: small muscle cells that lack the characteristic striations seen in skeletal and cardiac muscle.

SNARES: proteins that mediate fusion of vesicles with other membranes.

snRNAs (small nuclear RNAs): small RNA molecules found in the nucleus that play a role in RNA splicing.

sodium action potential: action potential driven by the opening of sodium channels and the resulting sodium influx.

sodium/calcium exchanger: carrier in the plasma membrane. Three sodium ions move into the cell down their electrochemical gradient and one calcium is moved out up its concentration gradient.

sodium gradient: energy currency. Sodium ions are more concentrated outside the cell than inside, and this chemical gradient is usually supplemented by a voltage gradient pulling sodium ions in. If sodium ions are allowed to enter down their electrochemical gradient, they release 15,000 J/mol.

sodium/potassium ATPase: plasma membrane carrier. For every ATP hydrolyzed, three Na^+ ions are moved out of the cytosol and two K^+ ions are moved in.

sodium pump (sodium/potassium ATPase): plasma membrane carrier. For every ATP hydrolyzed, three Na^+ ions are moved out of the cytosol and two K^+ ions are moved in.

solute: substance that is dissolved in a liquid.

somatic cells: cells that make up all the normal tissues of the human body; distinct from the germ cells that form the gametes (eggs and sperm).

sorting signal: section of a protein that causes the cell to direct the protein to a specific cell compartment such as the nucleus or mitochondrion. Sorting signals can be lengths of peptide (targeting sequences), such as the signal sequence that targets a protein to the endoplasmic reticulum, or can be the result of posttranslational modification, e.g., mannose-6-phosphate, which targets a protein to the lysosome.

sorting vesicle: vesicle that carries proteins from one membrane compartment to another.

SOS: guanine nucleotide exchange factor for a GTPase called Ras. SOS is recruited to the plasma membrane by binding to growth factor receptor binding protein number 2, which in turn binds activated receptor tyrosine kinases.

Southern blotting: blotting technique in which DNAs, separated by size, are probed using a single-stranded cDNA probe.

spatial summation: the process in which simultaneous release of synaptic transmitter by a number of presynaptic nerve cells depolarizes the postsynaptic neuron to threshold.

specific activity: ratio between activity and amount of material. For an enzyme it is the ratio between measured activity and mass of protein. It can be used as an indication of the purity of an enzyme preparation.

specificity constant: ratio between the catalytic rate constant (k_{cat}) and the Michaelis constant (K_M) for an enzyme. It is a rate constant with dimensions of liters mol^{-1} s^{-1} and is used to compare different substrates for the same enzyme or to compare the effectiveness of one enzyme with another.

spermatid: cell formed by meiosis that differentiates to form a spermatozoon.

spermatozoon: motile male gamete.

S phase: period of the cell division cycle during which DNA replication occurs.

splicesome: complex of proteins and small RNA molecules involved in RNA splicing.

squamous: flat. A term used of epithelial cells.

standard state: defined set of conditions under which to state the thermodynamics of a reaction. For example, $\Delta G^{o\prime}$ is determined for aqueous solutions at pH 7.0

start signal: start signal for protein synthesis is the codon AUG, specifying the incorporation of methionine.

stearic acid: $C_{17}H_{35}$—COOH; a fatty acid with no double bonds; i.e., it is fully saturated. The acyl group derived from stearic acid is common in animal fats and phospholipids.

stereo isomers: isomers (i.e., different compounds with the same molecular formula) in which the atoms have the same connectivity but that differ in the way the atoms are arranged in space.

steroid hormones: these transmitters act on intracellular receptors to activate transcription of particular genes. Glucocorticoids are one type of steroid hormone; the sex hormones (oestradiol, testosterone, progesterone) are another.

steroid hormone receptor: transcription factors that, upon binding their appropriate steroid hormone, activate transcription of particular genes.

sticky ends (of DNA): short single-stranded ends produced by cleavage of the two strands of a DNA molecule at sites that are not opposite to one another.

stop codon: codons UAA, UAG, and UGA are codons that signal protein synthesis to stop. Also known as termination codons.

stop signal: signal, to stop protein synthesis, given to the ribosome by a stop codon on mRNA.

stratified epithelium: type of epithelium consisting of several layers, such as the skin.

stress-activated protein kinase: kinase that stimulates cell repair but can also trigger apoptosis. p38 and JNK are stress-activated protein kinases.

stress fiber: bundle of actin filaments commonly seen in cultured, nonmotile animal cells.

striated muscle: striped muscle; includes skeletal and cardiac muscle.

stroma: volume within the chloroplast inner membrane but outside the thylakoids.

substrate: in normal English, a substrate is a solid base. In biology, the term is used to mean (1) a reactant in a reaction catalyzed by an enzyme, e.g., lactose is a substrate for β-galactosidase; (2) a base that cells grow and move on, e.g., collagen is a good substrate for cell attachment.

sucrose: disaccharide comprising glucose linked to fructose by an $\alpha 1 \leftrightarrow \beta 2$ glycosidic bond.

summation (at synapses): additive effects of more than one presynaptic action potential upon the postsynaptic voltage.

supercoiling: organization of a linear structure into coils at more than one spatial scale.

S value (sedimentation coefficient): value that describes how fast macromolecules and organelles sediment in a centrifuge.

Svedberg unit (S value): value that describes how fast macromolecules and organelles sediment in a centrifuge.

synapse: structure formed from the axon terminal of a nerve cell and the adjacent region of the postsynaptic cell. Transmitter released by the axon terminal diffuses across the synapse gap and acts upon the postsynaptic cell.

T (thymine): one of the four bases found in DNA; thymine is a pyrimidine.

tandem repeats: many copies of the same DNA sequence that lie adjacent to each other on the chromosome.

targeting sequence: stretch of polypeptide that determines the cellular compartment to which a synthesized protein is sent.

***TATA* box:** sequence found about 20 bases upstream of the beginning of many eukaryotic genes that forms part of the promoter sequence and is involved in positioning RNA polymerase for correct initiation of transcription.

taxol: compound obtained from the bark of the Pacific yew, *Taxus brevifolia;* binds to tubulin. Taxol is a powerful anticancer drug.

telomeres: specialized regions at the ends of eukaryotic chromosomes. Telomeres are rich in minisatellite DNA.

telophase: final period of mitosis or meiosis in which the chromosomes decondense and the nuclear envelope reforms.

telophase I: telophase of the first meiotic division (meiosis I).

telophase II: telophase of the second meiotic division (meiosis II).

temporal summation: the process in which action potentials in a presynaptic nerve cell occur at a high enough frequency that the depolarizations that they evoke in the postsynaptic cell do not have time to decay back to the resting voltage but rather ride on each other and depolarize the postsynaptic cell to threshold.

terminally differentiated: term that describes a cell that cannot return to the cell division cycle. Nerve cells are terminally differentiated; glial cells are not.

termination codon: codons UAA, UAG, and UGA are codons that signal protein synthesis to stop. Also known as stop codons.

terminator (of transcription): DNA sequence that, when transcribed into mRNA, causes transcription to terminate.

tertiary structure: three-dimensional folding of a polypeptide chain into a biologically active protein molecule. It usually includes regions of secondary structure. Interactions of the amino acid side chains are central in its formation.

thermodynamics: study of how energy affects matter and chemical reactions.

thermogenin: H^+ channel found in the inner mitochondrial membrane of brown fat cells. Its presence uncouples NADH oxidation from ATP synthesis, that is, the electron transport chain can operate, pumping H^+ out of the mitochondrial matrix, even when ATP synthase is not allowing H^+ back in.

thick filament: one of the two filaments that form the cytoskeleton of striated muscle; composed of the motor protein myosin II.

thin filament: one of the two filaments that form the cytoskeleton of striated muscle; composed of actin.

thiol group: —SH group.

threshold (voltage): plasma membrane transmembrane voltage at which enough calcium or sodium channels open to initiate an action potential.

thylakoid: folded inner membrane of the chloroplast; site of the light reaction of photosynthesis. Thylakoids are arranged in stacks called grana.

thymine: one of the four bases found in DNA; thymine is a pyrimidine.

tight junctions: type of cell junction in which a tight seal is formed between adjacent cells occluding the extracellular space.

tissue: group of cells having a common function.

topoisomerases: enzymes that cut and rejoin DNA strands. Topoisomerase I relieves simple torsional stress by cutting one strand, allowing rotation about the phosphodiester link of the other strand, then rejoining the cut strand. Topoisomerase II cuts both strands of a double helix and passes another complete double helix through the gap, keeping hold of the ends and rejoining them when the other strand has passed through. Topoisomerases are essential during DNA replication.

toxin: poison.

transcription: synthesis of an RNA molecule from a DNA template.

transcription bubble: structure formed when two strands of DNA separate and one acts as the template for synthesis of an RNA molecule.

transcription complex: A complex of RNA polymerase and various transcription factors.

transcription factor: protein (other than RNA polymerase) that is required for gene transcription.

trans: side from which material is removed. Of the Golgi apparatus, the surface from which vesicles bud to pass to the plasma membrane and to lysosomes.

transfected, transfection: cell, prokaryotic or eukaryotic, that has been infected by a foreign DNA molecule(s) is said to be transfected. The process of DNA infection is called transfection.

transfer RNA (tRNA): RNA molecule that carries an amino acid to an mRNA template.

transform, transformation: in addition to its common English meaning, this is used in molecular genetics to mean introduction of foreign DNA into a cell.

transgenic animal: animal carrying a gene from another organism; the foreign gene is usually injected into the nucleus of a fertilized egg.

trans-Golgi network: complex network of tubes and sheets that comprise the trans face of the Golgi apparatus. It is in the trans-Golgi network that proteins made on the rough endoplasmic reticulum are sorted as to their final destination.

translation: synthesis of a protein molecule from an mRNA template.

translocation: movement. When used of the ribosome, it means the movement, three nucleotides at a time, of the ribosome on the mRNA molecule.

transmembrane proteins: class of proteins that span the plasma membrane.

transmembrane translocation: form of protein transport in which unfolded polypeptide chains are threaded across one or more membranes as a simple polypeptide chain and then (re)folded at their final destination.

transmembrane voltage: voltage difference between one side of a membrane and the other. It is usually stated as the voltage inside with respect to outside.

transmission electron microscope: type of microscope in which the image is formed by electrons that are transmitted through the specimen.

transmitter: chemical that is released by one cell and that changes the behavior of another cell.

transpiration: loss of water from plant leaves.

transport vesicle: membrane vesicle that transports proteins from one membrane compartment to another.

triacylglycerol (triglyceride): three acyl (i.e., fatty acid) groups attached to a glycerol backbone by ester bonds. If the compound is liquid at room temperature, it is called an oil; if it is solid, it is called a fat.

tricarboxylic acid cycle (Krebs cycle): series of reactions in the mitochondrial matrix in which acetate is completely oxidized to CO_2 with the attendant reduction of NAD^+ to NADH and FAD to $FADH_2$.

triglyceride (triacylglycerol): three acyl (i.e., fatty acid) groups attached to a glycerol backbone by ester bonds. If the compound is liquid at room temperature, it is called an oil; if it is solid, it is called a fat.

trimeric: formed of three subunits.

trimeric G protein: protein with three subunits, α, β and γ, where the α subunit is a GTPase that dissociates from the $\beta\gamma$ units when it is in its GTP bound state. Both the α subunit, in its GTP bound state, and the now independent $\beta\gamma$ subunit, can activate target proteins. Examples are G_s, which activates adenylate cyclase, and G_q, which activates phospholipase $C\beta$.

trimethylamine: substance that gives rotting fish its characteristic smell. A base; when dissolved in water, trimethylamine will accept an H^+. Produced by the action of intestinal bacteria on trimethylamine N-oxide (from dietary fish), choline (from dietary phospholipid), or other chemicals containing a trimethylamine group. Trimethylamine is oxidized in mammalian liver by the action of flavin-containing monooxygenase. A mutation in the *FMO3* gene causes the disorder trimethylaminuria. Affected individuals excrete large amounts of trimethylamine in their urine, sweat, and breath and consequently have a fishy body odor. The structures of trimethylamine and protonated trimethylamine are

$$
\begin{array}{cc}
CH_3 & CH_3 \\
| & | \\
CH_3-N-CH_3 & CH_3-N^+-CH_3 \\
& | \\
& H
\end{array}
$$

trimethylamine N-oxide: The compound

$$
\begin{array}{c}
CH_3 \\
| \\
CH_3-N-CH_3 \\
\downarrow \\
O
\end{array}
$$

The arrow indicates a so-called dipolar bond between nitrogen and oxygen. This bond can also be represented as N^+—O^-. Trimethylamine N-oxide is generated from trimethylamine by the action of flavin-containing monooxygenase. A mutation in the *FMO3* gene causes the disorder trimethylaminuria. Affected individuals excrete large amounts of trimethylamine in their urine, sweat, and breath and consequently have a fishy body odor.

trisphosphate: having three phosphate groups attached at three different points on the molecule. Inositol trisphosphate and phosphatidylinositol trisphosphate are examples.

Trk: a family of receptor tyrosine kinases that bind growth factors of the neurotrophin class.

tRNA (transfer RNA): RNA molecule that carries an amino acid to an mRNA template.

troponin: calcium-binding protein found in muscle cells.

tryptophan (*trp*) operon: cluster of five bacterial genes involved in the synthesis of the amino acid tryptophan.

tubulin: protein that forms microtubules; exists as α, β, and γ subunits.

tumour: proliferative cell mass associated with many cancers.

turnover number: number of moles of substrate converted to product per mole of enzyme per unit time. Another term for catalytic rate constant, k_{cat}.

tyrosine kinase: enzyme that phosphorylates on tyrosine residues in proteins by transferring a phosphate group from ATP. In this book all the tyrosine kinases described are receptor tyrosine kinases.

U (uracil): one of the four bases found in RNA; uracil is a pyrimidine.

ultramicrotome: machine for cutting thin sections (<100 nm) for electron microscopy.

ultrastructure: fine structure of the cell and its organelles revealed by electron microscopy.

uncouples, uncoupler, uncoupled (of mitochondria): mitochondria are uncoupled when the tight link between the respiratory chain and ATP synthesis is broken. Since the link is the gradient of H^+ across the inner mitochondrial membrane, any chemical that allows H^+ ions to cross the inner mitochondrial membrane is an uncoupler. If the chemical is present in more than a small fraction of the cells of the body, it will kill, but expression of the protein uncoupler thermogenin in brown fat allows that tissue to generate heat.

unsaturated (of fatty acids): containing carbon–carbon double bonds.

untranslated sequence: sequence of bases in an mRNA molecule that does not code for protein. Untranslated regions are found at the $5'$ and $3'$ ends of an mRNA.

upstream: general term meaning the direction from which things have come. When applied to the DNA within and adjacent to a gene, it means lying on the side of the transcription start site that is not transcribed into RNA. When applied to signaling pathways, it means opposite to the direction in which the signal travels; thus the insulin receptor is upstream of protein kinase B.

uracil: one of the four bases found in RNA; uracil is a pyrimidine.

uracil–DNA glycosidase: DNA repair enzyme that recognizes and removes uracil from DNA molecules.

urea: H_2N—(CO)—NH_2, compound made in the liver that contains two nitrogen atoms but is less toxic than ammonia. Urea is chaotropic: at high concentration it reversibly denatures proteins.

urea cycle: metabolic cycle in the liver that makes urea that is then excreted in the urine.

vacuole: large membrane-bound compartment. Plant cells often contain a large vacuole filled with sugars and pigments.

Valium: antianxiety drug that increases the chance that the GABA receptor channel will open, allowing chloride ions to pass.

van der Waals force: weak close-range attraction between atoms.

vascular tissue: blood vessels. The term is also used to describe the water transporting and support tissue of plants that is composed of xylem and phloem.

vasoconstrictor: substance that constricts blood vessels.

vasodilator: substance that dilates (widens) blood vessels.

vector: Something that carries something else. The term is often used to describe a plasmid or bacteriophage that carries a foreign DNA molecule and is capable of independent replication within a bacterial cell.

vesicle: small closed bags made of membrane.

vesicular trafficking: precisely controlled movement of vesicles between different organelles and/or the plasma membrane.

villin: type of actin-binding protein that crosslinks actin filaments.

villus (plural **villi**): fingerlike extension of an epithelial surface that increases the surface area.

vimentin: protein that makes up the intermediate filaments in cells of mesenchymal origin such as fibroblasts.

virus: packaged fragment of DNA or RNA that uses the synthetic machinery of a host cell to replicate its component parts.

V_m: maximal velocity of an enzyme-catalyzed reaction: the limiting initial velocity obtained as the substrate concentration approaches infinity. V_m is also often used to mean transmembrane voltage.

VNTRs (variable number tandem repeats): DNA sequences that occur many times within the human genome. Each person carries a different number of these repeats.

voltage clamp: technique in which the experimenter passes current to one side of a membrane to artificially set the value of the transmembrane voltage to a desired level.

voltage-gated calcium channel: channel that is selective for calcium ions and that opens upon depolarization. Found in the plasma membrane of many cells.

voltage-gated sodium channel: channel that is selective for sodium ions and that opens upon depolarization. Found in the plasma membrane of nerve and muscle cells.

Wee1: protein kinase that phosphorylates and hence inactivates CDK1.

wobble (in tRNA binding): flexibility in the base pairing between the 5′ position of the anticodon and the 3′ position of the codon.

xeroderma pigmentosum: inherited human disease caused by defective DNA repair enzymes; affected individuals are sensitive to ultraviolet light and contract skin cancer when exposed to sunlight.

xylem: part of the plant vascular system; transports water from the roots to the leaves.

YAC (yeast artificial chromosome): cloning vector used to propagate DNAs, of about 500,000 bp, in yeast cells.

Z disc: disc that is set within and at right angles to the actin microfilaments in striated muscle, holding them in a regularly spaced array.

Z-DNA: left-handed helical form of DNA.

Zellweger's syndrome: inherited human disease resulting from aberrant targeting of proteins to the peroxisome.

zinc fingers: structural motif in some families of DNA binding proteins in which a zinc ion coordinated by cysteines and/or histidines stabilizes protruding regions that touch the edges of the base pairs exposed in the major groove of DNA.

INDEX

Underlined page numbers indicate a glossary entry, often a good place to begin reading about a topic.

Cell Biology: A Short Course, Second Edition, by Stephen R. Bolsover, Jeremy S. Hyams,
Elizabeth A. Shephard, Hugh A. White, Claudia G. Wiedemann
ISBN 0-471-26393-1 Copyright © 2004 by John Wiley & Sons, Inc.